Microbial Inhabitants of Humans

The indigenous microbiota (or "normal microflora") comprises those microbes that inhabit the healthy human body. Many of these organisms (e.g., *Streptococcus pyogenes, Staphylococcus aureus, Neisseria meningitidis, Haemophilus influenzae,* and *Escherichia coli*) are responsible for common, and sometimes life-threatening, infections of healthy individuals, whereas others cause disease only when the body's defenses are impaired.

This advanced textbook provides a unique overview of the microbial communities inhabiting those regions of the human body that are exposed to the external environment; these include the skin, eyes, oral cavity, and the respiratory, urinary, reproductive, and gastrointestinal tracts. To understand why particular organisms are able to colonize an anatomical region and why the resulting microbial community has a particular composition, an ecological approach is essential. Consequently, the key anatomical and physiological characteristics of each body site are described to show how these influence the nature of the environment at that site. The crucial roles of the indigenous microbiota in protecting against exogenous pathogens, regulating the development of our immune system and mucosae, and providing nutrients are also discussed. The involvement of these organisms in infections of healthy and debilitated individuals are described throughout, and methods of manipulating the composition of the indigenous microbiota for the benefit of human health are discussed. *Microbial Inhabitants of Humans* is a core textbook for advanced students taking courses in microbiology, medical microbiology, microbial ecology, and human biology.

Michael Wilson is Professor of Microbiology in the Faculty of Clinical Sciences and Head of the Department of Microbiology at the Eastman Dental Institute at University College London. He is the Editor-in-Chief of the journal *Biofilms* and has coauthored or edited several books, including *Bacterial Disease Mechanisms* (2002), *Bacterial Adhesion to Host Tissues* (2002), and *Medical Implications of Biofilms* (2003).

Microbial Inhabitants of Humans

Their ecology and role in health and disease

Michael Wilson
University College London

 CAMBRIDGE
UNIVERSITY PRESS

PUBLISHED BY THE PRESS SYNDICATE OF THE UNIVERSITY OF CAMBRIDGE
The Pitt Building, Trumpington Street, Cambridge, United Kingdom

CAMBRIDGE UNIVERSITY PRESS
The Edinburgh Building, Cambridge CB2 2RU, UK
40 West 20th Street, New York, NY 10011-4211, USA
477 Williamstown Road, Port Melbourne, VIC 3207, Australia
Ruiz de Alarcón 13, 28014 Madrid, Spain
Dock House, The Waterfront, Cape Town 8001, South Africa

http://www.cambridge.org

First published 2005

Printed in the United States of America

Typefaces Swift Light 9.5/13 pt. and Gill Sans   *System* LATEX 2$_\varepsilon$ [TB]

A catalogue record for this book is available from the British Library

Library of Congress Cataloguing-in-Publication Data available
Wilson, Michael, 1947-
    Microbial inhabitants of humans : their ecology and role in health and disease / Michael Wilson.
        p.   cm.
    Includes bibliographical references and index.
    ISBN 0-521-84158-5
    1. Body, Human – Microbiology.   2. Medical microbiology.   3. Microbial ecology.
    I. Title.
    QR46.W7493   2004
    612 – dc21   2004045927

ISBN 0 521 84158 5 hardback

For Andrew, Bernie, Caroline, Fionn, Margaret, Mike, Pippa, Richard, Sarah, and Wil

Long may your symbionts protect you

Contents

Chapter 3 | The eye and its indigenous microbiota 107

Chapter 5 | The urinary system and its indigenous microbiota

Preface

The *Guiness Book of Records* and the *Hitch Hikers Guide to the Galaxy* are both books that document many astounding facts concerning life, the universe, and everything else. However, as far as I am aware, in neither of these eminent publications is there any mention of two truly amazing observations: an adult human being consists of ten times as many microbial cells as mammalian cells, and he or she carries around approximately 1.25 kg of microbes. Knowing this, who could fail to be intrigued by the microbial component of that mammal–microbe symbiosis known to us as a "human being." The immediate questions prompted by this knowledge are usually along the lines of: (1) Which microbes are present? (2) How do they manage to survive? (3) What are they doing there? (4) Are they dangerous? Each of these questions about our "indigenous microbiota" is addressed in this book, and some of the answers will be surprising.

The complexity of the microbial communities found at many body sites is truly astounding; it has been estimated that we provide a home for at least 1,500 different microbial taxa that collectively contain more than 200 times as many genes as the human genome. Most of these organisms have not yet been grown in the laboratory, and so we know very little about them. However, modern molecular approaches are not only enabling us to detect their presence but, thanks to gene sequencing, also to gain some idea of their physiology and virulence potential. In order to try to understand how members of the indigenous microbiota manage to survive on our body and what they are up to while living there, an ecological approach has been adopted in this book – motivated by Pasteur's statement that, "The germ is nothing. It is the terrain in which it is found that is everything."

As for the question of the ability of such microbes to cause disease – the answer is a resounding "yes." Some of the deadliest diseases of humankind are caused by microbes that are normal inhabitants of various body sites of healthy individuals (e.g., *Neisseria meningitidis*, *Streptococcus pyogenes*, *Haemophilus influenzae*, *Staphylococcus aureus*, and *Streptococcus pneumoniae*). Amazingly, for most of our lives, the majority of individuals suffer no ill effect from harboring such organisms. Less virulent members of the indigenous microbiota are also a frequent cause of less life-threatening' but nevertheless debilitating diseases. In fact, the most common infectious diseases of mankind – caries, periodontal diseases, and urinary tract infections – are all caused by microbes indigenous to humans. Furthermore, advances in medicine and surgery have resulted in the increasing use of various devices and immunosuppressive therapies which have provided opportunities for many indigenous microbes to cause a variety of serious infections.

However, all is not doom and gloom, and it is important also to consider a question that is not often asked about our indigenous microbiota: Are they of any benefit? The answer, undoubtedly, is "yes." Not only do our indigenous microbiota protect us against exogenous pathogenic microbes, but they also provide us with as much as 10 percent of our energy requirements, supply a range of vitamins, and play a key role in the development of our immune system and mucosal surfaces.

Disappointingly, but not surprisingly, the attention that has been given to most members of the indigenous microbiota (other than disease-inducing species) by microbiologists and immunologists is insignificant compared with that directed toward exogenous pathogens. This has left us remarkably ignorant of the most intimate relationship that any of us will ever experience during our lifetime. The exhortation to "know thyself" must surely extend to that part of the human–microbe symbiosis (i.e., our symbionts) that can justifiably be termed "microbial self." It is the author's hope that this book will inspire others to take a greater interest in, as well as to cherish, our indigenous microbiota.

Abbreviations used for microbial genera

A.	*Actinomyces*		*L.*	*Lactobacillus*
Ab.	*Abiotrophia*		*Lac.*	*Lactococcus*
Acin.	*Acinetobacter*		*Leg.*	*Legionella*
Act.	*Actinobacillus*		*Lep.*	*Leptotrichia*
Aer.	*Aeromonas*		*Lis.*	*Listeria*
All.	*Alloiococcus*		*M.*	*Micrococcus*
B.	*Bacteroides*		*Mal.*	*Malassezia*
Bac.	*Bacillus*		*Mor.*	*Moraxella*
Bif.	*Bifidobacterium*		*Myc.*	*Mycoplasma*
Brev.	*Brevibacterium*		*Mycob.*	*Mycobacterium*
C.	*Corynebacterium*		*N.*	*Neisseria*
Camp.	*Campylobacter*		*P.*	*Propionibacterium*
Can.	*Candida*		*Pep.*	*Peptostreptococcus*
Cap.	*Capnocytophaga*		*Por.*	*Porphyromonas*
Chlam.	*Chlamydia*		*Pr.*	*Proteus*
Cit.	*Citrobacter*		*Prev.*	*Prevotella*
Cl.	*Clostridium*		*Ps.*	*Pseudomonas*
Col.	*Collinsella*		*Roth.*	*Rothia*
Des.	*Desulphovibrio*		*Rum.*	*Ruminococcus*
E.	*Escherichia*		*Sac.*	*Saccharomyces*
Eg.	*Eggerthella*		*Sal.*	*Salmonella*
Eik.	*Eikenella*		*Sel.*	*Selenomonas*
Ent.	*Enterococcus*		*Ser.*	*Serratia*
Enter.	*Enterobacter*		*Sh.*	*Shigella*
Eub.	*Eubacterium*		*Staph.*	*Staphylococcus*
F.	*Fusobacterium*		*Strep.*	*Streptococcus*
G.	*Gardnerella*		*T.*	*Treponema*
Gem.	*Gemella*		*Tr.*	*Trichomonas*
H.	*Haemophilus*		*Tur.*	*Turicella*
Hel.	*Helicobacter*		*U.*	*Ureaplasma*
K.	*Klebsiella*		*V.*	*Veillonella*
Kyt.	*Kytococcus*		*Y.*	*Yersinia*

An introduction to the human–microbe symbiosis

The first 9 months of our existence – the time we spend in our mother's womb – is the only period of our life during which we are free of microbes. Our delivery from this parasitic existence into the outside world exposes us to an enormous range of microbes from a variety of environments – our first encounter with life forms which have an anatomy, physiology, and metabolism very different from those of our own. Hence, our immediate companions on life's long journey include organisms from (1) the vagina, gastrointestinal tract (GIT), skin, oral cavity, and respiratory tract of our mother; (2) the skin, respiratory tract, and oral cavity of other individuals present at the delivery; (3) the instruments and equipment used during delivery; and (4) the immediate environment. These will include, therefore, not only microbes from other human beings, but also organisms from soil, water, and vegetation that may be present. All of the studies that have been carried out on neonates have shown that, within a very short time following delivery, microbes are detectable on most of those surfaces of the baby that are exposed to the external environment (i.e., the skin, respiratory tract, GIT, and oral cavity). Despite the fact that we are exposed to a wide variety of microbes at birth, only a limited number of species are able to permanently colonise the various body sites available, and each site is colonised predominantly by only certain microbial species (i.e., the microbes display "tissue tropism"). The organisms found at a particular site constitute what is known as the indigenous (or "normal") microbiota of that site. It is important to note that the term "indigenous microbiota" will include all of the bacteria, viruses, fungi, and protoctists that are able to colonise any of the body surfaces. However, the vast majority of studies undertaken so far have been concerned with identifying only the bacteria present at a particular site, and so we know very little about the distribution or frequency of occurrence of Archaea, viruses, fungi, or protoctists on healthy individuals. This book, therefore, is concerned only with the bacterial members of the indigenous microbiota and with those fungi (e.g., *Candida albicans* and *Malassezia* spp.) which the available data suggest are also indigenous to humans.

It is appropriate at this point to define what is meant by "symbiosis". Strictly speaking, the term means "living together" and so can be applied to any association between two (or more) organisms. However, it is possible to recognise at least three types of symbiosis: (1) mutualism – when both members of the association benefit, (2) commensalism – when one member benefits while the other is unaffected, and (3) parasitism – when one member suffers at the expense of the other. Confusingly, however, many scientists now use the term "symbiosis" to mean only the first of these three

Organ/system	Associated microbiota (grams wet weight)
eyes	1
nose	10
mouth	20
lungs	20
vagina	20
skin	200
intestines	1000

Table 1.1. Mass of the microbial communities associated with various body sites

possibilities (i.e., a mutually beneficial interaction between two (or more) organisms). In this book, symbiosis will be used in this sense while the term "mutualism" will be reserved for those mutually-beneficial relationships in which the association is obligatory (see Section 1.2.1). When the species comprising a symbiosis differ in size, the larger member is known as the host whereas the smaller is termed a "symbiont".

One of the many remarkable features of the microbiota of a particular anatomical location is the similarity of its composition among human beings worldwide despite the huge variations in the climate they are exposed to, the diet they consume, the clothes they wear, the hygiene measures they practice, and the lifestyle they have adopted. It would appear, therefore, that over many millennia humankind has co-evolved with some of the microbial life forms present on earth to form a symbiosis that is usually of mutual benefit to all of the organisms involved. However, this relationship between the indigenous microbiota and its human host is delicately balanced and can break down, resulting in an "endogenous" or "opportunistic" infection. This book is concerned with the indigenous microbiota of humans and describes (1) its development and composition at various body sites, (2) how its composition can be affected by various human activities, (3) the benefits it confers on its human host, (4) the diseases which it is able to cause, and (5) how its composition may be manipulated for the host's benefit.

1.1 Overview of the distribution and nature of the indigenous microbiota of humans

The indigenous microbiota of humans consists of a number of microbial communities, each with a composition characteristic of a particular body site. With few exceptions (the stomach and duodenum being two examples), the communities consist of large numbers of microbes and have a complex composition. As can be seen from Table 1.1, the microbial component of the average human being weighs approximately 1.25 kg. In terms of cell numbers, the figures are even more astonishing, with microbes outnumbering mammalian cells by a factor of 10 – the average human consists of 10^{13} mammalian cells and 10^{14} microbial cells. Some appreciation of the complexity of the indigenous microbiota can be gained by considering the number of different taxa (or phylotypes) that have been detected at various sites. Hence, the number of microbial taxa that are able to colonise the oral cavity has been estimated to be between

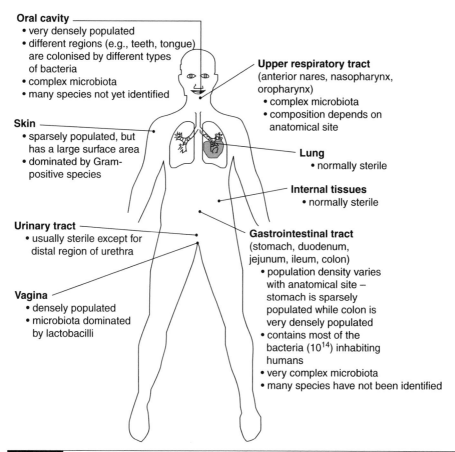

Oral cavity
- very densely populated
- different regions (e.g., teeth, tongue) are colonised by different types of bacteria
- complex microbiota
- many species not yet identified

Skin
- sparsely populated, but has a large surface area
- dominated by Gram-positive species

Urinary tract
- usually sterile except for distal region of urethra

Vagina
- densely populated
- microbiota dominated by lactobacilli

Upper respiratory tract (anterior nares, nasopharynx, oropharynx)
- complex microbiota
- composition depends on anatomical site

Lung
- normally sterile

Internal tissues
- normally sterile

Gastrointestinal tract (stomach, duodenum, jejunum, ileum, colon)
- population density varies with anatomical site – stomach is sparsely populated while colon is very densely populated
- contains most of the bacteria (10^{14}) inhabiting humans
- very complex microbiota
- many species have not been identified

Figure 1.1 The nature of the microbial communities found inhabiting various sites on the human body. Reproduced with permission from *Bacterial disease mechanisms; an introduction to cellular microbiology.* Wilson, M., McNab, R., and Henderson, B. 2002. Cambridge: Cambridge University Press.

500 and 700, whereas, for the colon, the number lies between 500 and 1,000 – these figures, however, are continually being revised upwards as detection methods improve. Fortunately, the numbers of different organisms detected in an individual at any one time are usually considerably lower – no more than approximately 100 in the more complex communities such as those found in the colon, dental plaque, and vagina.

Although many of those body surfaces that are exposed to the external environment are colonised by microbes, some are not (e.g., the lungs), and the population density of those sites that are colonised varies markedly from site to site (Figure 1.1). Hence, the oral cavity, the colon, and the vagina are densely colonised, whereas the eyes, stomach, and urethra have much sparser microbial communities. The density of colonisation and the community composition can vary enormously at different sites within an organ system. For example, the upper regions of the respiratory tract are more densely populated than the lower regions – in fact, the bronchi and alveoli are usually sterile. The skin is generally rather sparsely populated, but regions such as the axillae and the perineum support more substantial microbial communities. In the GI tract, the stomach, duodenum, and ileum have low population densities, whereas the jejunum, caecum, and colon are densely populated.

Table 1.2.	Problems with defining the indigenous microbiota of a body site

technical problems due to complexity of the microbial community

generally only small numbers of samples can be processed – limits the statistical reliability of the data obtained

difficulty in obtaining appropriate, uncontaminated samples from many body sites

variations between individuals related to genotype, age, sex, diet, hygiene practices, health status, type of clothing, occupation, prevailing climate, etc.

difficulties in comparing results obtained using different methodologies

changes in microbial nomenclature – renders comparisons with previous studies difficult

1.1.1 Difficulties associated with determining microbial community composition

Communities with a large diversity pose considerable technical problems when it comes to identifying all of the species present, and herein lies one of the problems associated with trying to define the indigenous microbiota of a body site. Until relatively recently, analysis of such communities relied on the cultivation of the species present. Such an approach is fraught with problems, and these are described in greater detail in Section 1.4.2. The application of modern molecular means of identifying microbes has added greatly to our knowledge (but not necessarily to our understanding) of the composition of the microbial communities inhabiting humans (Section 1.4.3). Unfortunately, however, few such studies have been carried out to date, and most of these have been restricted to samples taken from the oral cavity and the colon. It is important to emphasise at this point that, in addition to the technical difficulties associated with analysing such complex communities, there are a number of other problems inherent in attempting to determine the indigenous microbiota of a body site (Table 1.2). Firstly, regardless of whether culture-based or culture-independent methodologies are being used in a study, the work involved in processing a single sample is considerable, and this limits the number of samples that can be handled which, in turn, reduces the statistical reliability of the results obtained. Secondly, comparisons between studies are often difficult because of the different methodologies involved – not only between culture-based and culture-independent studies, but also among studies using similar approaches. Hence, culture-based studies often use different media with differing abilities to grow or select different species, whereas culture-independent studies often use primers or probes with different specificities. Changes in microbial nomenclature and taxonomy (particularly among the anaerobic Gram-positive cocci and rods and the anaerobic Gram-negative rods) have exacerbated the problem by making comparisons with previous studies difficult. While obtaining samples from some sites (e.g., the skin) is relatively easy, it can be extremely difficult to obtain samples from other sites. Hence, obtaining samples from the stomach and duodenum that are uncontaminated by microbes inhabiting adjacent sites is very difficult. This can be exacerbated by problems arising from the attitude of the individuals being sampled who are, naturally, reluctant to undergo any procedure that is uncomfortable, painful, or embarrassing. Studies have shown that the numbers and types of microbes present at a site may be affected by the age, gender, sexual maturity, diet, hygiene practices, type of clothing worn, occupation, prevailing climate, and so forth. This means that a properly designed study should minimise such variations between the participants in the study – this is seldom done because of the difficulty in recruiting sufficient numbers. Even if all of the previously described problems can be overcome, the scientific community is then faced with the problem of deciding whether or not a particular organism detected should be regarded as being a member of the indigenous microbiota of the site under investigation. This can be a very difficult and – because there are no rigid rules – controversial issue. If an organism A is isolated in large proportions from a particular body site in every participant in a large group of age- and gender-matched individuals and similar results are obtained on a number of different sampling occasions, then it would be reasonable to regard it as being a member of the indigenous microbiota of that site. However, what should be the status of organisms B and C if they are isolated from 50% and 5% of these individuals, respectively? Or what if B and C are isolated from all individuals on one occasion but not on another occasion? Attempts have been made to distinguish between microbes that are "residents" of a site and those that are "transients". Residents of a site should be able to grow and reproduce under the conditions operating at the site, whereas organisms that cannot do so, but are found at the site, are regarded as transients. However, the complexity of the microbial communities at many sites, the paucity of longitudinal studies of most sites, and the difficulties associated with trying to establish whether an organism is actively growing or reproducing at a site often make such distinctions difficult to make.

Once an organism has been designated as being a member of the indigenous microbiota of a body site, it is important to try and understand why it is present at that site. It is reasonable to assume that the organism must be adhering to some substratum within the site – this may be a host cell, the extracellular matrix, some molecule secreted by the host, some structure produced by the host (e.g., a tooth or hair), or another microbe. The predilection of many organisms for a particular host site has been known for many years, and this phenomenon is termed "tissue tropism". The presence of a receptor on a host tissue able to recognise the complementary adhesin on the bacterium is considered to be the mechanism underlying tissue tropism. However, this alone cannot explain the presence of an organism at a specific body site because it does not take into account the fact that, as well as acting as a substratum for adhesion, the site must also be able to satisfy all of the nutritional and other needs of the organism. Furthermore, the organism must also be able to withstand any antimicrobial defences being mounted by the host at that site. An understanding of such host–microbe interactions can be gained only by considering the anatomy and physiology of the site which are largely responsible for creating the unique environment existing there. As Pasteur remarked more than 120 years ago, "The germ is nothing. It is the terrain in which it is found that is everything". The author has tried, therefore, to provide information on the environmental factors operating at each of the body sites colonised by microbes. Unfortunately, in many cases, such data do not appear to be available – this being due to the difficulties in accessing the site or in analysing the small quantities of fluid and/or tissue that can be obtained from the site. Although the environment provided by the host is the dominant factor dictating whether or not an organism can colonise a particular site, once colonisation has occurred, the environment is altered by microbial activity. This results in the phenomenon of microbial succession in which organisms previously unable to colonise the original site are now provided with an

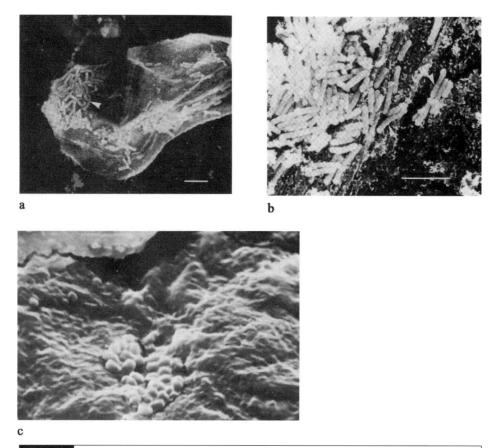

Figure 1.2 (a) Scanning electron micrograph showing a microcolony (arrowhead) of a *Lactobacillus* sp. on a uroepithelial cell. (b) Higher magnification image. The extracellular matrix enclosing the microcolony has collapsed during the dehydration stages essential for sample preparation and can be seen as accretions on the surface of some bacteria (arrowheads). Bar = 5 μm. (c) Microcolony on the surface of human skin. (a,b) Reproduced with permission of Lippincott Williams & Wilkins from: Adherence of cervical, vaginal and distal urethral normal microbial flora to human uroepithelial cells and the inhibition of adherence of Gram-negative uropathogens by competitive exclusion. Chan, R.C.Y., Bruce, A.W., and Reid, G. *Journal of Urology* 1983;131: 596–601. (c) Reprinted from: *Microbiology of human skin.* Noble, W.C. Copyright © 1974, with permission from Elsevier.

environment suitable for their growth and reproduction. This process is fundamental to understanding the development of microbial communities at the various body sites and will be referred to repeatedly throughout this book.

1.1.2 Structural aspects of residential microbial communities

As well as determining the numbers and types of microbes found at a particular anatomical site, it is also important to consider the structural organisation of the communities inhabiting these sites. Adhesion of an organism to a substratum is followed by its growth and reproduction – if the habitat has a suitable environment and can satisfy the nutritional requirements of the organism. This may result in the production of adherent microbial aggregates known as microcolonies, which are often enclosed within some microbial extracellular polymer (see Figures 1.2, 2.12, and 6.13). Microcolonies

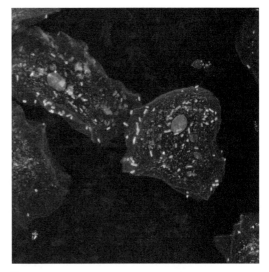

Figure 1.3 Epithelial cells from the cheek mucosa viewed by (a) confocal laser scanning microscopy and (b) scanning electron microscopy (bar = 10 μm). Pairs of bacteria and individual cells can be seen attached to the epithelial cells. Images kindly supplied by: (a) Dr. Chris Hope and (b) Mrs. Nicola Mordan, Eastman Dental Institute, University College London.

a

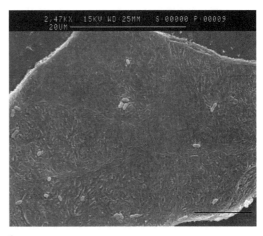

b

have been detected on the surface of the skin and on mucosal surfaces such as the respiratory, urogenital, and intestinal tracts. However, this does not happen in all cases, as the development of such aggregates is often limited by mechanical and hydrodynamic forces tending to disrupt or dislodge such structures (see Section 1.2.3). Furthermore, if the organism is motile, reproduction often leads to one or more of the daughter cells detaching and moving to another site within the habitat. Many epithelial cells, therefore, may only have small numbers of individual microbial cells on their surfaces (Figure 1.3). Another factor limiting the growth of microbial aggregates is that most of the surfaces exposed to the external environment (apart from the teeth) consist of epithelial cells which are continually being shed, taking the aggregates with them.

Sometimes however, the microcolony produced can grow further and develop into a larger structure known as a "biofilm" (Figure 1.4) – this occurs particularly on the non-shedding surfaces of the teeth and on mucosal surfaces with suitable anatomical features (e.g., the crypts of the tongue and tonsils and in the vagina). They are

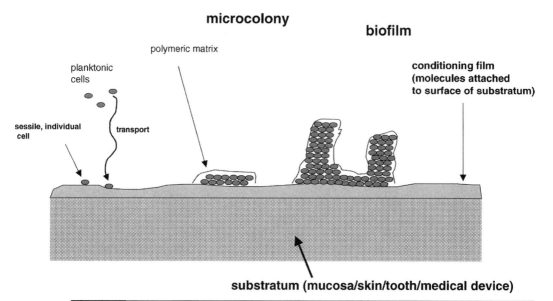

Figure 1.4 The various patterns of microbial colonisation that may be found on host tissues. Adhesion of single cells to the host tissue may lead to the production of microcolonies or biofilms.

also found on particulate matter in the colon and on medical devices and prostheses (e.g., catheters, artificial joints, limbs, and heart valves). A biofilm is defined as a matrix-enclosed microbial community attached to a surface. Because most surfaces in nature are coated with an adsorbed layer of macromolecules, the biofilm is usually attached to this layer (termed a "conditioning film") rather than directly to the surface itself. The matrix consists of polymers produced by the constituent microbes, as well as molecules derived from the host. An organism growing within a biofilm has a phenotype different from that which it displays when it grows planktonically (i.e., in an aqueous suspension) and the collective properties of a biofilm differ considerably from those of a simple aqueous suspension of the same organism(s) (Table 1.3). Furthermore, the utilisation of oxygen and nutrients from the environment by cells in the outermost layers of the biofilm, together with impeded diffusion of such molecules by the biofilm matrix, results in chemical and physicochemical gradients within the biofilm (Figure 1.5). Other gradients will be generated with respect to metabolites produced by the organism present inside the biofilm. Within the biofilm, therefore, an enormous variety of microhabitats exist, thereby providing conditions suitable for colonisation by a variety of physiological types of microbes.

Table 1.3.	General properties of biofilms

reduced susceptibility to antimicrobial agents
reduced susceptibility to host defence mechanisms
contain a range of microhabitats due to chemical and physico-chemical gradients
constituent organisms display novel phenotypes
facilitates nutritional interactions between constituent organisms
facilitates quorum sensing

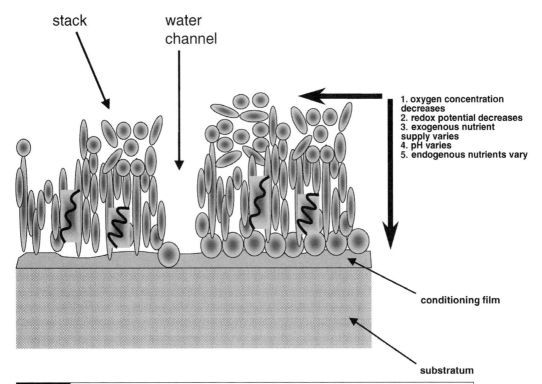

stack **water channel**

1. oxygen concentration decreases
2. redox potential decreases
3. exogenous nutrient supply varies
4. pH varies
5. endogenous nutrients vary

conditioning film

substratum

Figure 1.5 A wide range of microhabitats exist within a biofilm due to gradients in physicochemical factors (e.g., pH, redox potential), the partial pressure of gases such as oxygen and CO_2, the concentration of exogenous nutrients, and the concentration of metabolic end-products of the organism(s) within the biofilm.

Biofilms are highly hydrated structures, and the bacteria within them may occupy only between 10% and 50% of the total volume. This means that the staining and dehydration techniques used to prepare biofilms for examination by light and/or electron microscopy grossly distort their structure. Fortunately, the advent of confocal laser scanning microscopy (CLSM) – which enables the examination of biofilms in their native, hydrated state – has enabled a more accurate estimation of their structure and dimensions. Until CLSM began to be used for studying biofilm structure, there was little evidence that biofilms displayed any organised structure – bacteria were thought to be more or less randomly distributed throughout the matrix. However, CLSM (and other modern microscopic techniques such as differential interference contrast microscopy) has enabled us to view biofilms in their living, hydrated state, and this has revealed structures that are both complex and beautiful (Figure 1.6). Because a number of factors can affect biofilm structure, there is no single, unifying structure that can be said to characterise all biofilms. The key variables involved include the nature of the organism (or community), the concentration of nutrients present, the hydrodynamic properties of the environment, and the presence (and nature) of any mechanical forces operating at the site. Hence, the structure of a biofilm can range from the relatively featureless, flat type to one consisting of a more complex organisation involving tower-like "stacks" (consisting of microbes enclosed in an extracellular matrix) separated by water channels (Figure 1.6). The latter are characteristic of biofilms formed under the

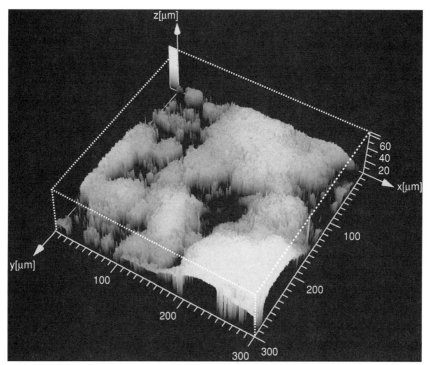

a

b

Figure 1.6 Confocal laser scanning micrograph (low magnification) of multi-species oral biofilms grown in the laboratory under conditions similar to those which exist *in vivo*. A fluorescent dye was used to stain the constituent organisms but not the extracellular biofilm matrix. (a) Three-dimensional image showing stacks of bacteria (up to 60 μm high) separated by water channels. (b) A section (160 μm $\times$ 160 μm) of a different part of the biofilm viewed from above. Grey-scale coding denotes the height of the constituent bacteria above the surface – the darker the colour, the nearer the bacteria are to the substratum. Images kindly supplied by: Dr. Chris Hope, Eastman Dental Institute, University College London.

following conditions: low nutrient concentration, high hydrodynamic shear stress, and the absence of mechanical abrasive and compressive forces. Depending on the microbial composition of the particular biofilm, the stacks may consist of a single species or of microcolonies of a number of different species. A microcolony forms at the particular location within a stack that has the appropriate combination of environmental factors (due to diffusion gradients, as mentioned previously) suitable for the survival and growth of that organism. The water channels may function as a primitive circulatory system, bringing fresh supplies of nutrients and oxygen while removing metabolic waste products. Further details of the structure and composition of biofilms are discussed in subsequent chapters.

1.1.3 Communication between members of microbial communities

Although it has been known for a long time that bacteria can sense, and respond to, their external environment, it has only recently been discovered that many species are also able to sense the presence of other bacteria. This phenomenon, known as "quorum sensing", involves the production of a low molecular mass auto-inducer, which diffuses out of the cell but which, on reaching a threshold concentration (due to the presence of a critical population density), can activate the transcription of certain genes. The nature of the auto-inducer depends on the particular species – in Gram-negative bacteria, it is usually an acyl-homoserine lactone. In Gram-positive organisms, the system is more complex and involves the active export of the auto-inducer (which is usually an oligopeptide) and a two-component signal transduction system. In general, the auto-inducer is species-specific, but recently a third type of system has been discovered which involves a furanosyl borate diester auto-inducer, and this enables inter-species quorum sensing. In effect, therefore, what this means is that bacteria have the ability to regulate the expression of certain genes in a population-dependent manner – a phenomenon of undoubted relevance to biofilms and microcolonies with their high bacterial density. Genes controlled by quorum sensing include those involved in biofilm formation, competence, and conjugation, as well as those encoding many virulence factors. The ability to limit gene expression until a large population has been reached is advantageous to the organism in a number of ways. Bacteria often derive their nutrients from complex polymers, and the degradation of such polymers requires the concerted secretion of enzymes from large numbers of cells. An individual cell, or a population in which only some of the members are secreting the appropriate enzymes, would not constitute an effective means of utilising the available nutrient resources. This applies to the quorum-dependent secretion of proteases by *Pseudomonas aeruginosa*. The advantage of competence and conjugation being regulated by a population-dependent process is obvious – DNA transfer is not possible in the absence of other cells. The ability to limit virulence factor secretion until a large number of bacteria are present could be a protective measure against host defence systems. Hence, if only a few bacteria were to secrete a particular virulence factor (small concentrations of which would be unlikely to cause serious damage), this could alert the host, which may then be able to dispose of this threat effectively – something it is less likely to be able to do if a large population is present. Some microbial activities and processes known to be regulated by quorum sensing are listed in Table 1.4.

Table 1.4. | Microbial activities and processes regulated by quorum sensing

mixed-species biofilm formation by *Porphyromonas gingivalis* and *Streptococcus gordonii*
bacteriocin (streptococcin) and haemolysin production by *Streptococcus pyogenes*
bacteriocin (salivaricin) production by *Streptococcus salivarius*
bacteriocin (plantaricin) production by *Lactobacillus plantarum*
iron acquisition and leukotoxin production by *Actinobacillus actinomycetemcomitans*
toxin production by *Clostridium perfringens*
cell division and morphology in *Escherichia coli*
competence in *Streptococcus pneumoniae*
protease and cytolysin production by *Enterococcus faecalis*
biofilm formation and competence in *Streptococcus mutans*
biofilm formation and bacteriocin production by *Staphylococcus epidermidis*
biofilm formation and exotoxin production by *Staphylococcus aureus*

1.2 | Environmental determinants affecting the distribution and composition of the indigenous microbiota

In order for an organism to become established as part of a community colonising a particular site on a human being, the environment of that site must be able to satisfy the organism's nutritional and physicochemical requirements, and the organism must be able to withstand any adverse features of the site, including the innate and acquired immune systems and the various mechanical and hydrodynamic microbe-removing systems present at certain sites (e.g., urination, coughing, and the mucociliary escalator). The presence and numbers of a particular microbe within a community are controlled by the type and quantity of nutrients present, its ability to tolerate the physicochemical factors operating there, and its ability to withstand any antimicrobial compounds or mechanical removal forces. The nutritional and physicochemical constraints operating to control microbial growth within a particular ecosystem are codified in two laws: Liebig's law of the minimum and Shelford's law of tolerance. Liebig's law of the minimum states that the total yield or biomass of any organism is determined by the nutrient present in the lowest concentration in relation to the requirements of the organism. In any given ecosystem, some nutrient will be present at a concentration that will limit the growth of a particular microbe. If the concentration of this nutrient is increased, then the population will grow until another nutrient becomes limiting, and so on. The nutrients available to an organism at a site on the human body are derived, in most cases, from two principal sources: the host and other microbes inhabiting the site. However, organisms inhabiting the GIT have an additional source of nutrients: the food ingested by the host. Shelford's law of tolerance relates to the non-nutritional factors that govern growth of an organism in an ecosystem (e.g., pH, temperature, redox potential) and states that, for survival and growth, each organism requires a complex set of conditions and that there are bounds for these factors outside of which an organism cannot survive or grow. An organism will only survive within an ecosystem, therefore, if each of the physicochemical conditions operating there remains within the tolerance range of that organism. Although the host "sets the scene" in terms of many of the physicochemical factors operating at a body site, the microbes present at that site often alter these factors dramatically.

Table 1.5. The main elements needed for microbial growth

Element	Function
carbon	major constituent of all cell components
oxygen	constituent of most organic cell components, constituent of water, used as an electron acceptor in aerobic electron transport
nitrogen	major constituent of proteins, nucleic acids, and some polysaccharides
hydrogen	major constituent of organic compounds and water
phosphorus	constituent of nucleic acids, phospholipids, ATP, and cell wall components such as teichoic acids
sulphur	constituent of certain amino acids (e.g., cysteine, methionine) and lipoic acid
potassium	maintenance of osmotic balance and an enzyme co-factor
sodium	involved in membrane transport
calcium	enzyme co-factor, constituent of bacterial spores
magnesium	enzyme co-factor
chlorine	important inorganic anion
iron	constituent of cytochromes and haem-containing enzymes

As well as being governed by nutritional and physicochemical factors, the presence of an organism in an ecosystem is also affected by biological factors such as the production of antimicrobial compounds by both the host and by other members of the resident microbial community of the site. Furthermore, the organism must, in some cases, be able to withstand mechanical removal forces generated by the flow of saliva, urine, and other fluids produced by the host.

All of the factors governing the ability of an organism to maintain itself at a body site and which determine the composition of the microbiota colonising that site are collectively termed "environmental determinants" or "environmental selecting factors" and will now be described.

1.2.1 Nutritional determinants

The main elements required for the growth of all microbes are well known and include – in decreasing order of their prevalence in a typical cell – carbon, oxygen, nitrogen, hydrogen, phosphorus, sulphur, potassium, sodium, calcium, magnesium, chlorine, and iron. The main cellular constituents in which these elements are found in a microbial cell are listed in Table 1.5. In addition, microbes need small quantities of a number of inorganic ions (known as trace elements or micronutrients), including cobalt, copper, manganese, zinc, and molybdenum. These are used as co-factors for various enzymes and are also important constituents of proteins and other cell components. Many organisms also need small quantities of organic growth factors such as amino acids, vitamins, fatty acids, or lipids.

Although there is a commonality among microbes in terms of the elements they need for growth and reproduction, there is enormous diversity with regard to the nature of the compounds (i.e., nutrients) that a particular organism can utilise as a source of each element. Microbes also differ with respect to the types of compounds that can serve as an energy source. Hence, a microbe will be able to colonise only those sites on a human being that can supply all of the elements needed for its growth in a

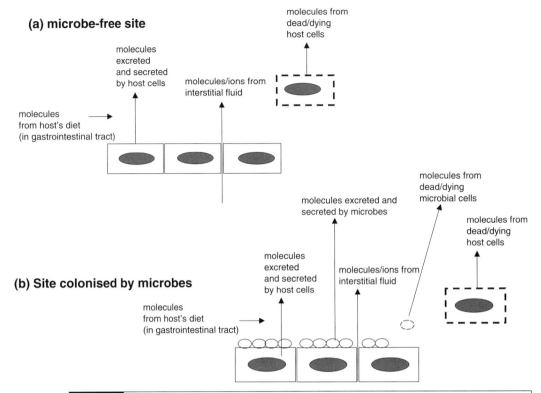

Figure 1.7 Nutrient sources available to microbes colonising (a) a sterile host tissue, and (b) a tissue with a resident microbiota.

form that it can utilise and that can also provide a suitable energy source. As previously mentioned, the main sources of nutrients at a particular body site are the host and other microbes. The host provides a range of potential nutrients in the form of the excretory and secretory products of live cells, transudates from the fluid which permeates tissues (interstitial fluid) and constituents of dead cells (Figure 1.7). In addition, food ingested by the host will also act as a source of nutrients for microbes inhabiting the GIT. The primary colonisers of any body site will be totally dependent on these host-derived nutrients. Once a site has been colonised, however, molecules excreted and secreted by the organisms present can act as additional nutrient sources. Inevitably, such nutrients will differ from those initially supplied by the host so that organisms with different nutritional requirements may be able to establish themselves at the site.

As will be seen in later chapters, the nutrients provided by the host vary enormously with the body site. Hence, lipids are important carbon sources on the skin, whereas on the respiratory mucosa, mucins and proteins are the main sources of carbon. In the caecum and ascending colon, the main carbon sources are carbohydrates derived from the host's diet, whereas in the descending colon, dietary proteins become increasingly important. This has a profound influence on the types of organisms that can colonise a particular site and become established as members of the site's microbiota. Whereas humans generously supply laboratory-grown bacteria with simple sugars such as glucose as a carbon and energy source, they rarely provide their own microbiota with such luxuries. Most often, the host-derived carbon sources available at a body site are more complex compounds such as polysaccharides, proteins, glycoproteins, fats, glycolipids, and lipoproteins. In order to utilise such substances, the microbe must first hydrolyse the molecule to smaller units that can be transported into the cell for subsequent metabolism. In some cases, the complexity of the carbon source is such that few single species are able to break down the molecule entirely, and the cooperation of several species is required. Nitrogen sources available at many sites include proteins, glycoproteins, lipoproteins, amino acids, urea, ammonium ions, and nitrate ions. Oxygen is, of course, present in water and most organic compounds and can be obtained from the atmosphere and/or host tissues at some body sites. Phosphorus is available in nucleic acids, phosphoproteins, phospholipids, phosphates, and so forth. Many of the other major elements required for microbial growth are present as inorganic salts or low molecular mass organic compounds. The dependence of microbes on small quantities of a variety of metal ions provides an opportunity for the host to control microbial growth by depriving them of such micronutrients. This is achieved by converting them to forms which many microbes are unable to use. The classic example of this is iron deprivation. Iron is an essential constituent of cytochromes and other enzymes, and most microbes need the element to be available in their environment at a concentration of approximately 10^{-6} M in order to grow and reproduce. Although there is an abundance of iron in the human body, most is bound to haemoglobin, myoglobin, and cytochromes. Any excess iron is complexed with the iron-binding proteins transferrin, lactoferrin, ovalbumin, and ferritin, leaving only minute quantities (approximately 10^{-15} M) of free iron available to microbes. In order to scavenge the limited amounts of the element available, microbes produce high-affinity, iron-binding compounds known as siderophores. Some organisms produce enzymes able to degrade host iron-binding compounds so liberating iron, others express receptors for these compounds and then remove the element from the bound complex. Furthermore, many organisms produce haemolysins which lyse red blood cells, thereby liberating iron-containing compounds such as haemoglobin from which the organism can obtain iron.

The host is the sole source of nutrients only for the pioneer organisms of a body site. Once that site has been colonised, substances excreted and secreted by microbes, as well as dead microbes, will serve as additional sources of nutrients and so the "menu" available is extended, thereby creating opportunities for colonisation by other organisms. Hence, microbes which would otherwise not have been able to inhabit the site become established there (such organisms are often termed "secondary colonisers"), and this results in "autogenic succession" (i.e., a change in the composition of a microbial community arising from microbial activities). A change in the composition of the community due to external, non-microbial factors is termed "allogenic succession". As the community increases in complexity, the variety of interactions between its members increases and these may have negative as well as positive outcomes (Table 1.6). Such nutritional interactions (along with other physicochemical interactions described later) contribute to the development of a microbial community at that site and ensure its stability, although gross changes imposed on the site by external forces can lead to its disruption.

Commensalism is a relationship between two organisms in which one benefits while the other is unaffected. Examples of this include (1) the production of amino acids or vitamins by one organism which are then used by another organism, (2) the degradation of polymers to release monomers which can be utilised by another organism, (3) the

Table 1.6.	Range of possible interactions that can occur between two microbial populations

Type of interaction	Examples
positive for one or both organisms	commensalism synergism (or protocooperation) mutualism
negative for one or both organisms	competition amensalism (or antagonism) predation parasitism
no interaction between organisms	neutralism

Note: The interactions involved may be nutritional, physicochemical, or biological.

production of substances by one organism that solubilise compounds which can then be utilised by another organism, and (4) the utilisation or neutralisation by an organism of a molecule that is toxic to another species. For example, the oral species *Prevotella intermedia* has a requirement for vitamin K which is produced by *Veillonella* spp. and by coryneforms. The growth of *Porphyromonas gingivalis* in the oral cavity is stimulated by a cytochrome b derivative produced by *Campylobacter rectus*. The oral species *Streptococcus mutans* requires *p*-aminobenzoate for growth, and this is produced by *Strep. sanguis*. An example of one organism utilising a substance that is toxic to another is provided by *Malassezia* spp., which metabolise lauric acid that is toxic to *Propionibacterium acnes*, thus enabling the survival of the latter organism on the skin where lauric acid is plentiful.

Synergism is a relationship in which both members benefit – when this involves a nutritional interaction, the term "syntrophism" is often used. An important example of syntrophism is cross-feeding. Cross-feeding occurs when one organism (A) utilises a compound which the other organism (B) cannot, but in doing so produces a new compound which organism B can utilise. The removal of the new compound by B is usually beneficial to A because it removes the negative feedback control exerted by the compound. For example, streptococci can utilise sugars as a carbon and energy source but *Veillonella* spp. cannot. One of the end-products of sugar metabolism by streptococci is lactate, the accumulation of which inhibits glycolysis. However, lactate can be used by *Veillonella* spp. as a carbon and energy source, and its utilisation removes its inhibitory effect on streptococcal glycolysis (Figure 1.8). Another example is the ability of *Enterococcus faecalis* to convert arginine to ornithine (something that *Escherichia coli* cannot do) which is then converted by *E. coli* to putrescine, a compound which both organisms can utilise. *Por. gingivalis* and *Treponema denticola* are unable to grow on a complex medium lacking haemin when inoculated individually, but can do so when inoculated together. Each organism produces an essential growth factor needed by the other – *Por. gingivalis* produces isobutyric acid, while *T. denticola* produces succinic acid. The presence of *Prev. intermedia* has been shown to stimulate the growth of *T. denticola* on human serum by cleaving the carbohydrate chains (which it metabolises further) from glycoproteins, thereby exposing the underlying peptide backbone, which can then be hydrolysed and used as a carbon and energy source by *T. denticola*. A number of polysaccharides present in food (starch, dextran, pullulan) can be degraded by *Strep. mutans*

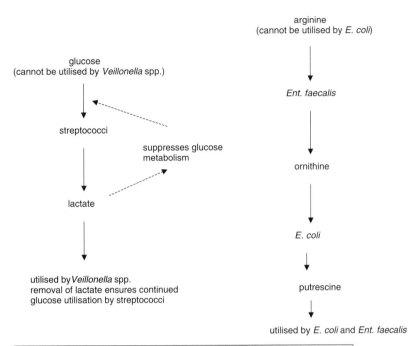

Figure 1.8 Examples of syntrophism (cross-feeding). See text for explanation.

in the oral cavity and the hydrolysis products utilised not only by this organism, but also by a range of oral organisms which do not produce polysaccharidases. Important sources of nutrients for microbes at most body sites are mucins. These are complex glycoproteins whose complete breakdown requires a variety of enzymes – sulphatases, sialidases, glycosidases, and proteases. Very few organisms produce this full complement of enzymes, so that the cooperation of a number of species is usually necessary to achieve complete degradation of this valuable nutrient source (see Section 1.5.3). The liberation of sulphate from mucins in the GIT (e.g., by *Bacteroides* spp.) enables the growth of *Desulphovibrio* spp., which use sulphate as an energy source. These are simple examples of what are often termed "food chains". However, these often involve more than two organisms and can result in complex "food webs" (see Figures 2.14 and 8.8).

Mutualism is a form of synergism in which the relationship between the two organisms is obligatory. However, the terms mutualism and synergism are often used interchangeably. Examples of true mutualism include the state of lysogeny established between a temperate phage and its bacterial host as well as the relationship between *Wolbachia* spp. and their filarial nematode hosts.

Competition is an interaction in which both microbes are adversely affected and occurs when each organism utilises the same nutrient – this may be a source of carbon, nitrogen, a trace element, or vitamin. Initially, competition for this nutrient results in both organisms growing at a slower rate or reaching lower population densities than they would have if they had been alone. Ultimately, however, one of the organisms will be excluded from the habitat, and this effect is an example of "competitive exclusion". For example, hydrogen is produced in the colon as a byproduct of microbial metabolism, and the two main users of the gas are the methanogens and sulphate-reducing bacteria. These two groups of organisms compete for hydrogen, and there is always an inverse relationship between their proportions in the colon. In continuous culture models of the colon, competition for a carbohydrate has been shown to result in suppression of the growth of both *E. coli* and *Clostridium difficile*.

Amensalism is not a nutritional interaction but will be described briefly here as an example of a negative interaction – it is a topic that will be dealt with in much greater detail in subsequent chapters. This occurs when one organism produces a substance that is toxic to another or creates an environment that is unsuitable for others. Examples include the production of bacteriocins by many organisms. Furthermore, the production of hydrogen peroxide by lactobacilli in the vagina excludes a number of organisms from this site. The acidic end-products of metabolism of some organisms can lead to an environment with a low pH, which is unsuitable for the growth of other species. Hence, the low pH on the skin surface is due, in part, to acid production by members of the skin microbiota, whereas the low pH of the vagina is partly attributable to the lactobacilli that reside there.

Parasitism involves one organism (the parasite) deriving its nutritional requirements from another (the host), which ultimately is harmed in some way (e.g., the bacteriophages [or phages] that parasitise bacteria). Although phages able to attack most members of the indigenous microbiota of humans have been detected, there is little information regarding their ability to affect the composition of the microbial community at a body site. There is interest in using phages to control infections caused by members of the indigenous microbiota and by exogenous pathogens. Predation involves the engulfing of one organism (the prey) by another (the predator) (e.g., bacteria are the major food source for protoctists, that is, eukaryotic microbes without a cell wall). Protoctists are found in the oral cavity (e.g. *Entamoeba gingivalis*) and the intestinal tract (e.g., *Dientamoeba fragilis*), but little is known of their effect on the bacterial communities of these sites.

Positive interactions tend to create new niches (note that the term "niche" means the functional role of an organism in a community [i.e., the sum of the processes it performs] and does not refer to its location in an ecosystem) to be filled by other potential colonisers of the site and thereby enable the utilisation of all of the nutritional resources available at the site. Negative interactions act as feedback mechanisms that ultimately limit the population of an organism in the community. The combination of these negative and positive nutritional interactions, together with interactions affecting the physicochemical features of the site (Figure 1.9), ultimately gives rise to a "climax community" in which the constituent organisms are in a state of dynamic equilibrium. Such a community is characterised by homeostatic mechanisms, which enable it to resist change and render it stable in terms of its composition – unless it is subjected to some catastrophic, externally generated change. An important aspect of such stability is "colonisation resistance", which prevents the establishment of a microbe that is not normally a member of the microbiota of that body site (i.e., an "allochthonous" species as opposed to organisms normally present at a site which are termed "autochthonous" species) – this is discussed in greater detail in Section 9.1. Each organism within a climax community occupies a niche (i.e., has a functional role) that contributes towards the stability of that community. This calls into question the validity of the search by some investigators for the organisms within a community that are "responsible" for colonisation resistance – such resistance is a consequence of the interactions between all members of the community.

primary colonisers

**oxygen consumption
production of CO$_2$
fall in Eh
provision of additional adhesion sites (i.e., bacteria and exopolymers)
pH alteration
production of additional nutrients (i.e., metabolic end-products, exopolymers)**

provides suitable conditions for secondary colonisers

further autogenic succession eventually results in a climax community

> **Figure 1.9** Autogenic succession within a habitat. The primary colonisers of a habitat alter the environmental conditions in a number of ways (e.g., by consuming nutrients and oxygen and by producing metabolic end-products and secreting macromolecules). Additional adhesion sites and nutrients and different environmental conditions (e.g., pH, Eh, oxygen concentration) are therefore created, thus enabling other organisms (secondary colonisers) to establish themselves at the site. The latter can themselves then alter the environment allowing different organisms to colonise the habitat. Eventually, a stable "climax community" is established at the site. Eh = redox potential.

1.2.2 Physicochemical determinants

Although a number of physicochemical factors are known to affect the growth of microbes, not all of these (e.g., atmospheric pressure, magnetic fields) have much relevance to the composition or growth of microbial communities inhabiting human beings. This is because we are limited with respect to the type and range of environments that can be tolerated, and homeostatic mechanisms ensure a fairly constant environment at most sites colonised by microbes. These homeostatic mechanisms are augmented by clothing, housing, air-conditioning, and so forth, which help to protect against the more extreme environments to which we may be exposed. However, despite the effectiveness of our homeostatic mechanisms, the conjunctiva and exposed regions of the skin will experience more dramatic fluctuations in temperature, humidity, and so forth, than sites such as the mucosal surfaces of the respiratory and intestinal systems. The most important physicochemical determinants affecting growth of the indigenous microbiota of humans are listed in Table 1.7.

Table 1.7.	Physicochemical determinants of growth of microbes inhabiting humans

temperature
pH
redox potential
atmospheric composition
water activity
salinity
light

Table 1.8. pH of different body sites in human beings

Site	pH	Predominant organisms colonising the site
stomach	1–5	*Helicobacter* spp., acidophilic streptococci
forehead	4.8–5.0	*Propionibacterium* spp., *Staphylococcus* spp.
caecum	5.7	anaerobes (e.g., *Bacteroides* spp., *Clostridium* spp.), facultative anaerobes (*E. coli*, lactobacilli, enterococci)
duodenum	5.7–6.4	acidophilic streptococci and lactobacilli
ileum	7.3–7.7	streptococci, coliforms, anaerobes (*Veillonella* spp., *Clostridium* spp., *Bacteroides* spp.)
descending colon	6.6	anaerobes – *Bacteroides* spp., *Eubacterium* spp., *Bifidobacterium* spp., *Clostridium* spp.
subgingival region	7.5–8.5	streptococci, *Actinomyces* spp., anaerobes (*Fusobacterium* spp., *Prevotella* spp., *Porphyromonas* spp., *Eubacterium* spp., *Veillonella* spp.)

The temperature of most body sites in human beings is approximately 37°C. This limits the type of microbe that can colonise humans to those species that are mesophiles (i.e., grow over the temperature range 25°–40°C and have an optimum growth temperature of approximately 37°C). The temperature of exposed areas of the skin (e.g., face and hands) tend to be a few degrees lower (approximately 33°C) than most other body sites and, although temperatures far lower than 37°C have been recorded at these sites in those living in polar and temperate climates, this does not appear to affect the types of organism colonising such sites.

In contrast to temperature, the pH of different body sites varies enormously and ranges from pH 1–2 in the stomach to alkaline values (pH 7.5–8.5) in the subgingival region (Table 1.8). This has a considerable effect on the types of organism that inhabit these sites. Interestingly, while very acidic pHs are found at a number of body sites (e.g., stomach, duodenum, caecum, skin), alkaline environments are infrequently encountered in the human body and are generally present only in the tear film, ileum, and in subgingival regions of the oral cavity. Microbes are broadly classified as being acidophiles, neutrophiles, or alkaliphiles, depending on the range of pHs over which they can grow and their optimum pH for growth. Acidophiles generally have an optimum pH for growth of less than 5.5, neutrophiles grow best over the pH range 5.4–8.0, and alkaliphiles have an optimum pH for growth greater than 8.0. Most members of the indigenous microbiota of humans are neutrophiles. Some organisms can tolerate extremely acidic and alkaline pHs, even though they may not be able to grow optimally at such pHs – such organisms are termed aciduric and alkaliduric, respectively. The term "acidogenic" refers to those organisms that produce acidic end-products of metabolism and include streptococci, lactobacilli, and propionibacteria. Although aciduric and acidophilic organisms, not surprisingly, are isolated from body sites with a low pH, not all organisms colonising acidic sites belong to either of these categories. Hence, *Helicobacter pylori*, which colonises the most acidic region of the body – the stomach – is a neutrophile, not an acidophile. This organism creates a microenvironment with a pH near neutral due to the production of ammonia from urea present in gastric

Table 1.9. Classification of microbes on the basis of their oxygen requirements

Microbial group	Description
obligate aerobes	require oxygen to grow
capnophiles	aerobes which grow best at CO_2 concentrations greater than those encountered in air – usually 5–10% CO_2
obligate anaerobes	do not grow in the presence of oxygen
facultative anaerobes	can grow in the presence or absence of oxygen
microaerophiles	grow best in the presence of low concentrations of oxygen (2–10%)

juices. While the pH of a body site is initially dictated by the host, the activities of the microbes colonising that site can also have a profound effect on its pH. Hence, the metabolic activities of microbes present on the skin and in the vagina play a major role in lowering the pH of these sites (see Sections 2.3 and 6.4.3.1.1). The generation of these low pHs affects the composition of the microbial communities inhabiting these sites by preventing the colonisation of those organisms unable to survive at such pHs and constitutes an example of amensalism. The pH of the colon gradually increases along its length to a more neutral value because of the production of ammonia due to amino-acid fermentation by resident bacteria. The presence of sucrose in the diet enables acid formation by oral streptococci and lactobacilli resulting in low pHs in the biofilms that form on the tooth surface. Although the growth of a particular organism is limited to a certain pH range, studies have shown that when the organism is part of a mixed-species community, the range of pHs over which it can grow is often extended considerably.

The oxygen content of the atmosphere present at a site is an important determinant of which organisms will be able to colonise that site and, in particular, of the type of microbes that can function as pioneer organisms. It is possible to distinguish a number of microbial groups on the basis of their relationship to molecular oxygen, and these are listed in Table 1.9. Interestingly, while humans inhabit an aerobic environment, the vast majority of microbes which inhabit them are either obligate or facultative anaerobes rather than obligate aerobes. Obligate aerobes, in fact, are not frequently encountered on any body surface, even at sites such as the skin and oral cavity, which are in constant contact with the atmosphere. In fact, humans can be considered to be very effective anaerobic incubators. At birth, all surfaces of an individual are aerobic so that only obligate aerobes or facultative anaerobes can colonise any body site. However, once a pioneer community has been established, the composition of the atmosphere at a site changes due to microbial consumption of oxygen and the production of CO_2. This creates a range of microhabitats at the site enabling colonisation by capnophiles, microaerophiles, and obligate anaerobes which, in turn, will bring about further alterations in the composition of the atmosphere. The ability of aerobes and facultative anaerobes to create environments suitable for the growth of obligate anaerobes and microaerophiles serves to illustrate the operation of both commensalism and autogenic succession within a habitat (Figure 1.9). Specific examples of this include the colonisation of the tooth surface by the facultatively anaerobic streptococci followed by the establishment of obligate anaerobes such as *Veillonella* spp. and *Fusobacterium* spp. Many other examples of autogenic succession induced by changes in the gaseous composition of a site are found in Chapters 2–8.

Related to, but not identical with, the oxygen content of a site is the redox potential (Eh). This is a measure of the reducing power of a system and has an important influence on the functioning of those enzyme reactions which involve the simultaneous oxidation and reduction of compounds. Some organisms can only accomplish such reactions in an oxidising environment, while for others a reduced environment is essential. The redox potential of a system is expressed in terms of its ability to donate electrons to a hydrogen electrode, which is considered to have an Eh of 0 mV. Any system/environment able to donate electrons to the electrode is designated as having a negative Eh (indicative of a reducing environment), while any system that accepts electrons has a positive Eh value and is oxidising. Obligate aerobes are metabolically active only in environments with a positive Eh, while obligate anaerobes require a negative Eh. Facultative anaerobes can function over a wide range of Eh values. Because of its powerful oxidising ability, the presence of oxygen in an environment exerts a dramatic effect on its Eh. Nevertheless, because of the rapid consumption of oxygen by respiring microbes, a site may be in contact with atmospheric oxygen, but may have a low oxygen content and an Eh which is sufficiently negative to permit the survival of obligate anaerobes. Examples of such sites include the biofilms on tooth surfaces (Section 8.4.3.1), many regions of the respiratory mucosa (Section 4.4.3.4), and hair follicles (Section 2.3). The production of a negative Eh by aerobic and facultative organisms, thereby creating an environment suitable for the growth of obligate anaerobes, is another example of commensalism resulting in autogenic succession.

CO_2 is an important component of a site's atmosphere and stimulates the growth of capnophilic organisms such as *Haemophilus* spp. and *Neisseria* spp. Such organisms grow poorly in atmospheres with a normal concentration of the gas and are dependent on CO_2 generated by host or microbial respiration.

The water activity (a_w) of a site is an indication of the proportion of water available for microbial activity and is invariably less than the total amount of water present. The water activity is affected by the concentration of solutes and also by the presence of surfaces. Pure water has an a_w of 1.0 and human cells require an a_w of 0.997 for growth. Many of the microbes colonising humans require an a_w of at least 0.96 for active metabolism (Table 1.10), and most of the sites available for colonisation can satisfy this requirement with the exception of many regions of the skin. Staphylococci, unlike many microbes, can grow when the a_w is as low as 0.85 and so are able to colonise dry regions of the skin such as the arm, leg, and palm of the hand (Section 2.3).

High salt concentrations are detrimental to many microbes because they result in dehydration and denaturation of proteins. The salt content of most human body sites is within the limits tolerated by most microbes and so does not exert a selective effect. The exception to this generalisation is the skin surface where high salt concentrations are found due to the evaporation of sweat. This selects for colonisation by halotolerant organisms such as staphylococci (Section 2.3).

Sunlight contains potentially damaging ultraviolet radiation and also can induce the generation of toxic free radicals and reactive oxygen species from certain compounds known as photosensitisers (Section 2.3). Both the skin and the eyes are exposed to

Table 1.10.	Water activity required to support the growth of various groups of microbes

Organism	a_w
Gram-negative rods	0.94–0.96
E. coli	0.95
Klebsiella aerogenes	0.94
Gram-positive bacilli	0.90–0.93
Gram-positive cocci	0.83–0.95
Micrococcus luteus	0.93
Staph. aureus	0.86
most CNS	0.85
halophiles	0.75–0.83

Note: CNS = coagulase-negative staphylococci.

light, but there is little evidence that light exerts any selective effects on the microbes inhabiting these sites.

1.2.3 Mechanical determinants

Certain regions of the body are subjected to mechanical forces that can affect the ability of microbes to colonise such sites. These regions include the GIT and the urinary tract. In the oral cavity, the stomach, and the upper regions of the small intestine, the flow of saliva or intestinal secretions create hydrodynamic shear forces that can sweep away microbes not attached to the mucosal surfaces. In the lower regions of the GIT (i.e., stomach, small and large intestines), peristalsis and other gut movements will also tend to remove unattached microbes. These mechanical forces will exert selection pressures favouring those organisms able to adhere to host surfaces in these regions. Similarly, in the urinary tract, the periodic flushing action of urine will remove microbes unable to adhere to mucosal surfaces. In the oral cavity, tongue movements and the chewing action of teeth will also generate considerable mechanical forces that are able to dislodge bacteria attached to mucosal and tooth surfaces. This encourages colonisation of sites protected from these forces such as those existing between the teeth and in the gingival crevice (Section 8.3.1).

While no strong hydrodynamic shear forces operate in the respiratory tract, the production of a mucus "blanket" that is continually propelled towards the oral cavity serves to trap and expel microbes arriving in this region – this system is known as the "mucociliary escalator" (Section 4.2). Only organisms able to adhere to the underlying epithelium or to the more static periciliary layer beneath the mucus layer are able to colonise the respiratory tract.

Finally, it is important to mention here that the microbial communities colonising any body site (other than the tooth surface) are attached to epithelial surfaces which are continually being shed, taking with them any adherent microbes. In the cases of the skin and urogenital tract, the microbial communities are shed almost directly into the environment – although they may be retained temporarily by clothing – while in the gut they become part of the lumenal microbiota for many hours before being shed into the environment. In the respiratory tract, the mucociliary escalator expels the communities into the oral cavity from where they can be swallowed or expelled into the external environment.

1.2.4 Biological determinants

The innate and acquired immune systems of humans are able to produce a variety of molecules and activated cells that kill microbes, inhibit their growth, prevent their adhesion to epithelial surfaces and neutralise the toxins they produce. While a detailed description of the innate and acquired immune responses is beyond the scope of this book, the effector molecules and effector cells of these systems which are present at surfaces colonised by indigenous microbes will be mentioned briefly here and are described in greater detail in appropriate sections of Chapters 2–8.

The innate immune system involves many different types of cells including epithelial cells, monocytes, macrophages, polymorphonuclear leukocytes (PMNs), natural killer (NK) cells, dendritic cells, and a number of different lymphocyte subpopulations which link the innate and acquired immune systems. Those responses that involve the cooperation of different cell types are coordinated by cytokines. This system results in the release of a large variety of effector molecules onto the external surfaces of epithelia, including a number of antimicrobial peptides, lysozyme, lactoferrin, lactoperoxidase, secretory phospholipase A_2, and collectins. The activities and functions of these molecules are described in greater detail in appropriate sections of Chapters 2–8. There is great interest in the use of many of these agents, particularly the antimicrobial peptides, for the prevention and treatment of infectious diseases, and their effectiveness in these respects has already been demonstrated in a number of clinical trials. The acquired immune system also involves a large number of different cell types and includes some of those involved in innate immunity, but the prime movers in this system are the B and T lymphocytes. Again, the response is coordinated mainly by cytokines. The most important effector molecule of the acquired immune response with respect to indigenous microbes is secretory IgA (sIgA) which, as its name implies, is able to reach the external surfaces of epithelia and so accumulate at sites of microbial colonisation. IgG is also present on mucosal surfaces but, except for the female reproductive tract, is present at a lower concentration than IgA. Of all the antibodies produced by humans, sIgA is produced in the greatest amount – between 5 and 15 g/day in adults. The molecule exists as a dimer linked by a 15 kDa protein known as the J chain and also has a proteinaceous "secretory component" attached, which is a fragment of the receptor involved in its transport across the epithelium. The secretory component may be important in protecting the molecule from degradation by bacterial proteases, which are produced by several pathogens. One of the principal functions of this antibody is to prevent bacteria adhering to host structures – this is achieved by the sIgA binding to bacterial adhesins, thereby preventing their interaction with receptors on host tissues. IgA is only a weak activator of complement and is a poor opsonin. The binding of IgA to microbes not only prevents them from adhering to the mucosal surface, but also can result in the formation of aggregates which are more easily removed by urine, saliva, tears, and so forth, than individual microbial cells.

Effector cells of the innate and acquired immune systems that may be present on epithelial surfaces in the absence of an infection include PMNs and macrophages, but little is known of their ability to affect the composition of the indigenous microbiota.

Macrophages are thought to be important in preventing colonisation of the lower regions of the respiratory tract by inhaled microbes.

While the innate and acquired immune systems are undoubtedly important in defence against pathogenic microbes, their role in the regulation of the indigenous microbiota is shrouded in speculation and controversy. This situation has arisen because most work on the immune responses to microbes has, not surprisingly, focused on their role in combatting pathogenic microbes rather than on their interactions with indigenous species. Little is known, therefore, of the mechanisms which enable the survival and long-term tolerance of indigenous microbial communities or why these organisms do not elicit a damaging chronic inflammatory response. That no inflammatory response is generated by such enormous numbers of microbes which possess constituents such as lipopolysaccharide, lipoteichoic acids, peptidoglycans, and proteins (known collectively as modulins) that are able to stimulate the release of pro-inflammatory cytokines from host cells is surprising. It has been proposed that the immune systems of the host do not respond to members of the indigenous microbiota (i.e., the host exhibits "tolerance"). This is supported by the results of a number of studies in humans and other animals, which have demonstrated that lymphocytes do not proliferate in response to their indigenous gut microbiota, whereas they do proliferate when challenged with microbes from the gut of other individuals of the same species. Furthermore, the immune response provoked in mice injected with murine strains of *E. coli* or *Bacteroides* spp. is considerably reduced, compared with that resulting from injection with the corresponding human strains of these organisms. It has been suggested that this tolerance is a consequence of the close similarities that exist between the surface antigens of many indigenous microbes and those of host tissues. This results in such organisms being recognized as "self" by the immune system which is beneficial for the microbes (as they are provided with a habitat suitable for their growth) and for the human host because it is provided with an indigenous microbiota conferring a number of benefits (described in Chapter 9). However, other studies have shown that some indigenous microbes can provoke humoral and other immunological responses and that some (but not all) organisms in the gut and on the skin surface are covered in sIgA. Furthermore, antibodies against a variety of indigenous microbes (e.g., *Veillonella* spp., *Bacteroides fragilis*, oral streptococci) have been detected in the sera of healthy individuals. While indigenous microbes may be able to induce the production of antibodies, these effector molecules do not appear to have a major role in regulating the composition of the indigenous microbiota. Hence, although deficiencies in sIgA production are fairly common, the composition of the communities colonising such individuals are not significantly different from those found in individuals without such deficiencies. Similarly, loss of T-cell function appears to have little effect on the composition of the intestinal microbiota.

The innate immune system, however, may play a dominant role in controlling colonisation and regulating the composition of the microbiota at a body site. Mucosal surfaces are known to secrete a number of antimicrobial compounds, and the mucosa of a particular site produces a characteristic range of such compounds, each of which has a distinct antimicrobial spectrum. Some of these compounds are constitutively expressed, while others are produced in response only to certain cytokines or to the presence of a particular organism (Section 1.5.4). It is likely, therefore, that the characteristic mixture of antimicrobial compounds produced at a particular body site would influence the type of microbes able to colonise that site. Other innate immune responses of epithelial cells to the indigenous microbiota are described in Section 1.5.4.

The production of hormones, and fluctuations in their concentrations, can exert a profound effect on the environment of certain body sites. For example, the increased production of sebum at puberty leads to dramatic changes in the skin environment (Sections 2.3, 2.4.1.5 and 2.5.1). The production of oestrogen and progesterone also alters the vaginal environment at different stages in the life of females (at menarch and menopause) as well as during the menstrual cycle of post-menarchal/pre-menopausal women (Sections 6.1 and 6.3).

Microbes arriving at a body site have to cope not only with host defence systems but also with antimicrobial compounds produced by those organisms already present. As mentioned in Section 1.2.1, an extensive range of antimicrobial substances (e.g., bacteriocins, fatty acids, hydrogen peroxide) is produced by microbes indigenous to humans, and these will be described in greater detail in appropriate sections of Chapters 2–8. There is great interest in capitalising on this phenomenon by employing microbes producing such compounds to prevent or treat infectious diseases – the use of these microbes in replacement therapy and as probiotics is discussed in detail in Chapter 10.

1.3 | Host characteristics affecting the indigenous microbiota

The influence of host factors on the composition of the community occupying a particular body site has been previously mentioned and will be emphasised throughout the rest of the book. However, it is, of course, important to remember that there is tremendous variation among individual human beings and, consequently, the environmental conditions provided by the host at a particular body site will certainly vary from person to person. Some of the factors that may affect the environmental conditions at a body site include age, gender, genotype, nutritional status, diet, health status, disability, hospitalisation, emotional state, stress, climate, geography, personal hygiene, living conditions, occupation, and lifestyle. Many of these factors are, of course, inter-related and some will affect all body sites, while others are more likely to influence only particular sites. Such person-to-person variations lead to differences in the composition of the microbiota of a site and create great difficulties in attempts to define the "indigenous microbiota" of a body site. Unfortunately, the effects on the indigenous microbiota of human beings of most of the factors listed previously have not been extensively investigated. Nevertheless, some information is available and some general points will now be made, while specific details are included in appropriate sections of Chapters 2–8.

1.3.1 Effect of age

Most of the data on the composition of the indigenous microbiota of a body site described in Chapters 2–8 relate to "healthy adults". However, the microbiota of many body sites appears to be different in very young and very old individuals. In the very young, such differences are a consequence of many factors, including an immature immune system, a milk-based diet, the absence of teeth, and behavioural factors. Hence, in children prior to weaning, the faecal microbiota is dominated by *Bifidobacterium* spp. in breast-fed infants, whereas in adults, *Bacteroides* spp. predominate (Section 7.4.2). In

Table 1.11. Factors that may contribute to alterations in the indigenous microbiota of elderly individuals

Factor	Consequences
immunosenescence	possibly influences microbiota of all sites
malnutrition	impaired immune response; affects composition of host secretions at most sites, thereby influencing microbiota
decreased mucociliary clearance	increased microbial colonisation of respiratory tract including species normally expelled
decreased gastric-acid production	increased colonisation of stomach including species normally killed
decreased urinary flow rate, post-void residual urine, increased bacterial adhesion to uroepithelium, prostatic hypertrophy in men	increased colonisation of urinary tract including species normally expelled
lack of oestrogen in females	decreased acid production in vagina, thereby enabling colonisation by species usually excluded
decreased intestinal motility, alterations in mucus composition	enables colonisation of small intestine

children without teeth, *Strep. sanguis* is absent from the oral cavity, whereas this is one of the dominant organisms present in the oral cavity of adults. The nasopharynx of infants is dominated by *Moraxella catarrhalis, Strep. pneumoniae*, and *Haemophilus influenzae*, whereas these organisms comprise much lower proportions of the nasopharyngeal microbiota of adults. The prevalence of *Neisseria meningitidis* in the nasopharynx is lower in infants than in young adults.

In the elderly, differences arise as a result of a decrease in the effectiveness of the immune system, dysfunctioning of many organ systems, poor nutrition, poor hygiene, and increasing use of medical devices and prostheses (e.g., catheters, dentures). There are several markers of the existence of a dysfunction of the immune response in the elderly (i.e., immunosenescence), including decreased antibody production, decreased numbers of circulating lymphocytes, impaired T-cell proliferation, and impaired phagocytosis and microbial killing by PMNs. However, whether this contributes to alterations in the composition of the microbiotas of body sites is not known. Malnutrition is a common problem in the elderly, and it has been estimated that between 10–25% of the elderly in industrialised countries have some nutritional defect. This will affect not only immune function, but also the composition of host secretions, which would be expected to have an impact on indigenous microbes. Other factors that are likely to affect the microbiotas of specific body sites in the elderly are listed in Table 1.11.

Changes in the composition of the microbiotas of a number of body sites in the elderly that have been reported include (1) a decrease in *Veillonella* spp. and bifidobacteria in the faecal microbiota; (2) an increase in the proportions of clostridia, lactobacilli, and enterobacteria in the faecal microbiota; (3) an increase in colonisation of the urinary tract – 35% and 20% of females and males aged 65 years or more in institutions have asymptomatic bacteriuria; (4) an increased colonisation of the oropharynx by

Gram-negative bacteria (e.g., *Klebsiella* spp, *E. coli*, and *Enterobacter* spp.), as well as by *Can. albicans*; (5) an increased prevalence of streptococci and enterobacteria on the skin; (6) an increase in the frequency of isolation of coryneforms and Gram-negative bacilli from the eye; and (7) an increase in the frequency of isolation of staphylococci and enterobacteria from the oral cavity.

1.3.2 Effect of host genotype

While a number of studies have investigated the effect of genotype on the susceptibility of individuals to infectious diseases, little is known about its effect on the indigenous microbiota. It would be expected that the environment of a particular body site would be more alike in individuals who have a high degree of genetic relatedness (because of anatomical and physiological similarities) than in those who were more distantly related. The faecal microbiotas of monozygotic twins (analysed by means of DGGE profiles of the amplicons from 16S rRNA genes – see Section 1.4.3) have been found to have a significantly higher degree of similarity than those of unrelated individuals. Furthermore, the degree of similarity of the microbiotas showed a significant correlation with the genetic relatedness of individuals. In a study of the cellular fatty-acid profiles of the faecal microbiota of mice, it was found that the profile obtained was related to MHC-encoded genes.

A considerable amount of information is now available showing that colonisation by *Helicobacter pylori*, as well as disease severity due to the organism, is related to host genotype (see Section 7.51 for details of infections due to this organism). Hence, *vacA* type s1a strains of the organism (which are isolated more frequently in adults with peptic ulcer disease and are associated with increased gastric epithelial damage) are found only in individuals of Asian descent. Other studies have shown positive associations between gastric colonisation by *Hel. pylori* and cytokine genotypes (polymorphisms in the promoter regions of the *IL-10* and *IL-4* genes) and blood group antigens (Lewis b antigen) and a negative association with an HLA allele (class II DR-DQ).

The nasal microbiota of identical twins has a much higher degree of similarity than that of non-identical twins. Furthermore, it has been shown that there is an association between the HLA phenotype and nasal carriage of *Staphylococcus aureus*. Hence, the HLA-DR3 phenotype is associated with a greater risk of carriage of the organism, while HLA-DR2, HLA-DR1, and HLA-Bw35 are associated with non-carriage.

ABH and related blood group antigens are glycosphingolipids that contain a variable carbohydrate chain protruding from the cell surface and a lipophilic region embedded in the cell membrane. These antigens may be present on the cell surfaces of epithelial cells of mucosal surfaces, as well as in the secretions covering such surfaces. Between 20% and 30% of individuals, however, are unable to express these antigens on cell surfaces or in secretions and are said to be "non-secretors". The susceptibility of females to recurrent urinary tract infections (UTIs) with *E. coli* is greater in non-secretors than in secretors. In non-secretors, the glycosphingolipids are sialylated instead of being fucosylated and so are not processed to produce the carbohydrates characteristic of the blood group antigens. Increased binding of *E. coli* to the uroepithelial cells of non-secretors has been reported, and this may account for their greater susceptibility to UTIs. This increased adhesion of the uropathogen may be because the sialylated glycosphingolipids (which are absent from the uroepithelial cells of secretors) are more

| Table 1.12. | Examples of the effect of gender on the indigenous microbiota |

males have a significantly higher risk than females of being colonised by *Hel. pylori*

males have a higher density of microbes on the skin than females – this may be linked to increased sebum and/or sweat production

the skin of males is more frequently colonised by *Acinetobacter* spp. than that of females

females have a greater risk of a urinary tract infection than males – this is largely related to anatomical factors

the proportion of *Prev. intermedia* in the gingival crevice is higher in female than male adolescents – this is probably related to the higher concentrations of oestradiol and progesterone

the carriage rate of *Staph. aureus* in the anterior nares is higher in males than in females

effective (or more numerous) receptors for the organism than the ABH antigens present on the cells of secretors.

1.3.3 Effect of gender

A number of differences between males and females have been observed regarding the composition of the microbiota of a number of body sites. Unfortunately, the reasons behind these have often not been determined but are likely to be multifactorial and involve anatomical, behavioural, hormonal, and other physiological factors. Some examples are shown in Table 1.12.

1.4 | Analytical methods used in characterising the indigenous microbiota

A variety of techniques have been used to study the indigenous microbiota of various anatomical sites in humans and each, of course, has its advantages and disadvantages.

1.4.1 Microscopy

Light microscopy is one of the simplest and most direct approaches used in the study of microbial communities. It has the great advantage of being able to reveal details of the physical structure of a community and the spatial arrangement of the constituent organisms. Furthermore, it provides a "gold standard" with respect to the total number of microbes present within a sample – this is often used as a yardstick for assessing the ability of other, less direct, techniques to detect all of the organisms present. In this way, it has been revealed that analysis of samples of faeces by culture may detect as few as 15% of the organisms present. Differential counts of the various morphotypes in a sample give some indication of the diversity of the microbiota, and this has proved useful for many years in the study of dental plaque communities – a high proportion of motile and spiral organisms in a subgingival plaque sample being indicative of pathology (i.e., periodontitis) at the site from which it was taken (Section 8.5.2.3). The analytical power of light microscopy can be enhanced in a number of ways. The use of vital stains, for example, can reveal the relative proportions of live and dead cells present in a

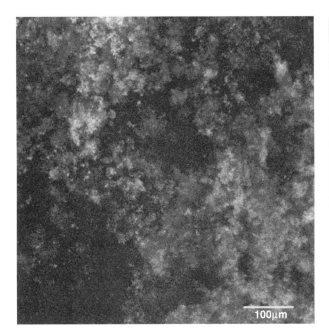

100μm

Figure 1.10 Low-resolution confocal micrograph of a multi-species biofilm consisting of oral bacteria grown in the laboratory under conditions similar to those which exist *in vivo*. A grey scale is used to denote the distance of the various regions of the biofilm from the microscope objective – the lighter the area, the greater the distance from the substratum. The biofilm can be seen to consist of numerous "stacks" of bacteria of different height separated by fluid-filled regions. Image kindly supplied by: Dr. Jon Pratten, Eastman Dental Institute, University College London.

sample. Furthermore, information regarding the identity of the organisms present (and their spatial relationships) can be obtained by using appropriate fluorescent-labelled antibodies and labelled oligonucleotide probes.

Confocal laser scanning microscopy is a technique that enables the examination of communities in their living, hydrated state and provides valuable information concerning the true spatial organisation of the constituent cells, as well as the overall shape and dimensions of the community (Figure 1.10). It is a technique that has revolutionised our understanding of the structures of those communities that exist as biofilms (e.g., on the surfaces of teeth, medical devices, and various mucosae). As in the case of light microscopy, additional information can be obtained by using vital stains, fluorescent-labelled antibodies and labelled oligonucleotide probes. Furthermore, information regarding the nature of the environment within the biofilm (e.g., pH, redox potential) can be obtained using appropriate probes. It is also possible to monitor gene expression within biofilms using reporter genes such as green fluorescent protein.

The use of new molecular techniques in conjunction with light microscopy is certain to prove extremely rewarding in studies of the indigenous microbiotas of humans. One such technique is fluorescent *in situ* hybridisation (FISH), which involves the use of fluorescent-labelled oligonucleotide probes targeting specific regions of bacterial DNA. The sample is permeabilised to allow access of the probes to the bacterial DNA, the probe is added and time allowed for hybridisation to take place, and then the sample is visualised by fluorescent microscopy. Currently, most of the probes used have been those designed to recognise genes encoding 16S ribosomal RNA (16S rRNA). The gene encoding 16S rRNA in a bacterium consists of both constant and variable regions. It is, therefore, possible to have regions that are highly specific for a particular bacterial species as well as regions that are found in all bacteria, regions that are found in only one bacterial genus, and regions found in closely related groups of bacteria. Probes can, therefore, be designed to enable the identification of individual

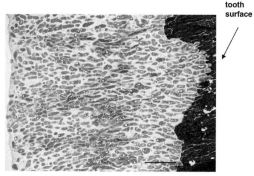

tooth surface

Figure 1.11 (a) Transmission electron micrograph of dental plaque. A variety of morphotypes can be seen, and the bacteria appear to be tightly packed (bar = 3 μm). (b) Scanning electron micrograph showing the surface layers of dental plaque. Various morphotypes can be seen, including long filamentous forms (bar = 50 μm). Images kindly supplied by Mrs. Nicola Mordan, Eastman Dental Institute, University College London.

a

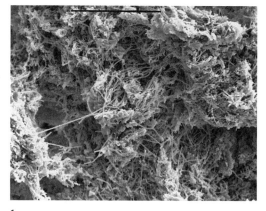

b

species, individual genera, certain related microbial groups, or even all bacteria. More recently, probes recognising mRNA have been used to identify which genes are being expressed in communities. An important advantage of this approach is that it can be automated and the resulting data processed using computerised image analysis software.

Transmission and scanning electron microscopy can provide information not obtainable by ordinary light microscopy (Figure 1.11). The high magnifications possible can be used to reveal details of microbial adhesins and adhesive structures. The organisms present can be identified using antibodies conjugated to electron-dense markers (e.g., gold, ferritin). However, an important disadvantage of electron microscopy is that specimen-processing and the accompanying dehydration alters the structure of the sample, thereby preventing elucidation of the exact spatial arrangement of cells within communities.

1.4.2 Culture

Most of our knowledge of the composition of indigenous microbial communities has come from using quantitative culture techniques. This involves some form of sample dispersion (e.g., by vortexing, shaking with beads), plating out the sample (and usually dilutions of it) on media (either selective or non-selective), incubation, subculture of isolated colonies, and then identification of each isolate. Given the complexity of the communities at most body sites, this is inevitably a very labour-intensive and painstaking task. There are also a number of inherent problems with the technique. First of all, if a non-selective medium is to be used, then one must be chosen that is capable of supporting the growth of all of the species likely to be present – this is virtually impossible given the disparate (and often very exacting) nutritional requirements of the members of such communities. Furthermore, it is difficult to provide the optimum environmental conditions (e.g., pH, oxygen content, CO_2 content) necessary to enable the growth of all of the different physiological types of microbes present. To do so would require the incubation of inoculated plates under a variety of conditions resulting in an enormous increase in the workload associated with the subsequent identification of isolates. Problems will, of course, arise as a result of some organisms growing faster than others resulting in overgrowth of plates and failure to isolate slow-growing organisms. In samples taken from sites with a very dense microbiota (e.g., the colon, vagina, dental plaque), it is essential to use dilutions of the sample to obtain isolated colonies for subsequent identification. This means that organisms present in low proportions are "diluted out" and so are rarely isolated. Many studies have used selective media instead of, or in addition to, non-selective media. These can be useful, but analysis of a complex microbiota requires the use of a number of selective media for the various groups of organisms present. However, no medium can be relied upon to be truly selective, and the inhibitory constituents may also have some adverse effect on the organisms for which the medium is supposedly selective. These problems all contribute to a greater workload which inevitably results in an increase in the number of errors, a decrease in the number of samples that can be processed and, hence, a decrease in the statistical reliability of the data obtained.

Comparison of samples analysed by culture and by microscopy have revealed that even the best culture methods seriously underestimate the number of organisms present in the microbiotas of certain body sites – particularly those from the GIT and oral cavity. The reasons for this are many and include: (1) the failure to satisfy the nutritional and/or environmental requirements for some of the organisms present, (2) the presence in the community of organisms in a "viable but not cultivable" state, (3) the failure to disrupt chains or clusters of organisms prior to plating out – this results in the production of only one "colony-forming unit" from a cluster or chain consisting of many viable bacteria, and (4) the death of viable cells during transportation and processing of the sample. Collectively, these difficulties have resulted in a serious underestimate of the number, and variety, of organisms in a sample taken from any environment, and it has been estimated that we are able to culture in the laboratory no more than 1–2% of the microbial species (which are thought to number between 10^5 and 10^6) present on planet Earth. Once individual isolates have been obtained, the next task is to identify each one. Traditionally, this has involved the use of a battery of morphological, physiological, and metabolic tests, which is very labour-intensive and often not very discriminatory. The use of commercially available kits for this purpose has made the process less technically demanding. Other phenotypic tests that have been used for identification purposes include cell-wall protein analysis, serology, and fatty-acid methyl-ester analysis. During the last few years, there has been a trend to increasing use of molecular techniques for identifying the organisms isolated, and one of these is based on the sequencing of genes encoding 16S rRNA. The gene is amplified by polymerase chain reaction (PCR) and the sequence of the resulting DNA determined and then compared with the sequences of the 16S rRNA genes of organisms that have been deposited in databases. If the sequence is ≥98% similar to that of one already in the database, then it is assumed that the gene is from the same species and, hence, the identity of an unknown organism can be established. The procedure is much simpler to perform than a battery of phenotypic identification tests and has the great advantage that it enables phylogenetic comparisons of the isolated organisms. However, some taxa are recalcitrant to PCR and some (e.g., many viridans streptococci) are so closely related that they cannot be differentiated using this approach.

Another approach involves colony hybridisation with nucleic-acid probes. Basically, the technique involves lysing an isolated colony and exposing it to a labelled oligonu-cleotide probe. Hybridisation is recognised by detection of the probe after washing – the label may be radioactive, enzymatic, or fluorescent. The probe can be designed to recognise a species, genus, or a group of organisms. In practise, the probing is carried out simultaneously on many colonies that have been transferred to a nitrocellulose membrane. A variety of other molecular techniques is used for the identification and further characterisation of isolated colonies, including pulsed-field gel electrophoresis (PFGE), ribotyping, multiplex PCR, and arbitrary-primed PCR.

1.4.3 Molecular approaches

Many of the problems inherent in culture-based approaches to analysing microbial communities can be circumvented by the use of molecular techniques. However, it must be pointed out that there are also a number of problems associated with the use of such approaches. The first stage in the analysis of a microbial community by a molecular technique is to isolate either DNA or RNA from the sample, and herein lies the first problem. Extraction of nucleic acids from microbes requires that the cells are lysed and the ease of lysis varies significantly among different organisms. Numerous protocols for the lysis of microbes present in samples have been devised and include the use of enzymes, chemicals, and mechanical methods. Care has to be taken that, once lysed, the nucleic acids do not undergo shearing and are not degraded by nucleases. The extracted nucleic acids can then be used in a variety of ways to reveal the identity of the microbes originally present in the sample and/or to produce a "profile" or "fingerprint" of the microbial community.

Universal primers can be used to amplify all of the 16S rRNA genes present in the DNA extracted from the sample and the amplified sequences are then cloned. The sequences of the clones are then determined and compared with sequences in databases. If a sequence is ≥98% similar to one already in the database, then it is regarded as being identical and so the corresponding organism can be assumed to have been present in the sample. In this way, the sequences of the 16S rRNA genes of all organisms present in the community (including those that cannot be cultivated in the laboratory) can be determined and, if these sequences match those of known organisms in databases, then the identities of all the organisms present will be revealed. However, studies of the faecal microbiota, for example, have revealed that not all of the sequences of the 16S rRNA genes obtained correspond to sequences in databases – in fact, as many as 75% of the sequences do not match those of known organisms.

Another useful approach is to separate the amplicons produced on a denaturing gradient gel. Although all of the amplicons have the same length, their different base compositions result in them having different melting points and so each will melt at a different point when run on a gel along which there is either a temperature

1 2 3 4 5

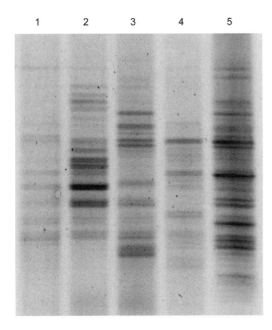

Figure 1.12 Polyacrylamide gel showing the DGGE profiles of five dental plaques sampled from the gingival crevice of different individuals. The figure shows bands corresponding to the 16S rDNA amplicons of the same molecular mass that have been separated by DGGE on a 10% polyacrylamide gel with a denaturant gradient ranging from 40–80% (100% denaturant corresponds to 7 M urea and 40% deionised formamide). The basis of this technique is that DNA amplicons that have different nucleotide sequences denature to differing extents in the presence of a given concentration of denaturing chemicals. The use of a G+C rich clamp on one of the PCR primers prevents the complete melting of the double-stranded DNA. The greater the A+T content of the DNA fragment, the lower its electrophoretic migration in a polyacrylamide gel. Image kindly supplied by: Mr. Gavin Gaffan, Eastman Dental Institute, University College London. DGGE = denaturing gradient gel electrophoresis.

gradient (temperature gradient gel electrophoresis – TGGE) or a gradient in the concentration of a denaturing agent such as urea or formamide (denaturing gradient gel electrophoresis – DGGE). The altered conformation of the DNA due to denaturation slows their migration, and this results in separation of the various amplicons. Staining of the DNA in the resulting gel reveals a banding pattern or "fingerprint" that is characteristic of that particular community (Figure 1.12). In order to prevent the two DNA strands from completely dissociating under denaturing conditions, a primer pair in which one of the primers has a G+C "clamp" attached to the 5′ end is used during the initial amplification stage. The individual bands can be cut out and each amplicon eluted, reamplified, sequenced, and identified using databases as previously described. Alternatively, the fingerprints produced from samples from the same individual obtained on different occasions can be compared and analysed for differences. Hence, bands appearing or disappearing with time can be sequenced to determine the gain or loss of an organism from the community. The method is also useful for comparing the microbiotas present at the same body site in different individuals. Comparisons are facilitated by computer and statistical analysis of the banding patterns obtained.

The extracted DNA can also be used in dot-blot hybridisation assays. In such assays, aliquots of the extracted DNA are spotted onto a nitrocellulose membrane to form a gridded array. This array can then be probed with labelled oligonucleotide probes designed to recognise a single species, a genus, or a group of related organisms. Alternatively, gridded arrays of DNA from a range of organisms can be prepared and these can be probed with the DNA extracted from the sample – once this has been labelled in some way.

1.4.4 Metabolic approaches

A very different approach to microbial community analysis is one which involves determining the metabolic capability of the community as a whole as well as its potential functional diversity. Both of these can be assessed by incubating samples of the community with a range of substrates (e.g., carbon sources) and determining which of these can be utilised. This results in a metabolic "fingerprint" indicating the range of substrates that can be utilised by the community – the greater the community diversity then the greater the number of substrates metabolised. The technique is known as community-level physiological profiling (CLPP). In practice, the assay is usually carried out in 96-well microlitre plates, each well containing a different substrate and some indicator dye (e.g., a tetrazolium salt), which changes colour when the substrate is metabolised. Inoculation of the wells with aliquots of the microbial community is followed by incubation and measurement of the pattern and intensity of the colour changes in the wells. In addition to providing information on the substrates actually used by a community, CLPP may be useful in monitoring the community's response to altered environmental conditions. There are, however, a number of problems inherent in the technique. Firstly, because the conditions of incubation provided in the laboratory are so different from those of the community's natural environment, it is very unlikely that the observed response would represent the metabolic capabilities of the community in its natural environment. Secondly, the observed substrate utilisation pattern will reflect the activities of only those organisms in the community that are metabolically active under the particular set of incubation conditions employed. Furthermore, the relative proportions of the constituent members of the community are likely to change during the period of incubation. Despite these drawbacks, this simple technique can provide some insight into the functional abilities of a community.

This type of approach has been used in a more limited way to establish some of the "microflora-associated characteristics" (MACs) of the host (Sections 9.2–9.4). Hence, a number of metabolic capabilities of the colonic microbiota have been assessed, including the protease, urease, glycosidase, bile salt hydrolase, and azoreductase activities, as well as the ability of these communities to produce short-chain fatty acids; to convert cholesterol to coprostanol; to breakdown mucin; to inactivate trypsin; and to produce gases such as methane, CO_2, and hydrogen. The determination of one, or a limited number, of such activities has been employed by investigators interested in a particular function of the colonic microbiota. However, a more extensive range of these MACs (i.e., "profiling") has been determined in some studies, particularly those involving the effects of antibiotics and diet on the colonic microbiota or those investigating the development of such communities.

1.5 | The epithelium – site of host–microbe interactions

The internal tissues of the human body are protected from the microbes present in the environment by a coating known as the epithelium. This consists of a layer of specialised cells, sometimes only one cell thick, which forms a continuous covering on all surfaces exposed to the external environment (i.e., the skin, eyes, and the respiratory, gastrointestinal, urinary, and genital tracts). One of its many functions is to exclude microbes from the underlying tissues but it is itself colonised by those microbes which are the subject of this book. Because the epithelium is the primary site of the interactions that occur between microbes and their human host (as well as the interactions that occur between the microbial colonisers of humans), it is appropriate to consider this tissue in greater detail. The only other surface on which host–microbe interactions occur is that of the tooth and its structure and is described in Chapter 8.

1.5.1 Structure of epithelia

Basically, two main types of epithelial surfaces can be distinguished – the dry epithelium (known as the epidermis) comprising the skin and the moist epithelia which cover the eyes and those internal body surfaces that are in communication with the external environment (i.e., the respiratory, gastrointestinal, urinary, and genital tracts). Moist epithelia are also referred to as mucosae (or mucous membranes) because they are invariably coated in a layer of mucus that consists primarily of glycoproteins known as mucins. The main cellular element of epithelial surfaces is the epithelial cell. In an epithelium, the constituent cells are joined together by a variety of junctions, including (1) tight junctions – these seal adjacent cells just below their apical surfaces (i.e., surfaces exposed to the external environment) preventing the passage of molecules between cells – they are impermeable to all but the smallest molecules; (2) gap junctions – these consist of channels between adjacent cells and allow the passage of molecules less than 1,000 Da between them; and (3) adherens junctions – these provide strong mechanical attachment between cells and are formed by linkage of adjacent transmembrane proteins known as cadherins.

The epithelium is attached to the underlying tissue via a layer of extracellular matrix (mainly collagen and laminin) known as the basal lamina. Attachment to this layer is mediated by hemidesmosomes. Epithelial cells vary in shape and may be flattened (squamous), cuboidal, or columnar, and their apical surface may be covered in fine, hair-like processes known as cilia. They may be present as a single layer of cells or they may form several layers, and some may undergo a process known as keratinisation in which the protein keratin is deposited in the cell which eventually dehydrates and dies. Epithelial cells that undergo keratinisation are known as keratinocytes and are particularly important constituents of skin on the surface of which they form a protective barrier, they are also found on certain mucosal surfaces (e.g., the tongue). The structure of skin differs in several respects from that of the mucosae and is described in greater detail in Chapter 2. The rest of this chapter will be concerned mainly with the mucosae found lining all of the other surfaces exposed to the external environment (i.e., the oral cavity and the gastrointestinal, respiratory, urinary, and genital tracts).

The main types of epithelia comprising mucosal surfaces are shown in Figure 1.13 and will now be described briefly. The epithelium has an apical surface facing the lumen of the particular body cavity and so is in contact with the external environment and a basal (or basolateral) surface which is in contact with the basal lamina (or basement membrane) and underlying connective tissue. Similarly, each cell is considered to have apical and basolateral surfaces. Simple squamous and columnar epithelia consist of a single layer of epithelial cells and are found lining many body cavities. In some cases, the cells have numerous microvilli to increase the surface area, and such epithelia are found in regions designed for absorption of substances (e.g., the small intestine). Epithelia consisting of several layers of cells with different shapes are known as transitional epithelia and are found lining cavities that are subject to expansion and contraction (e.g., the bladder). When the layers of cells progressively flatten, the epithelium is known as a "stratified squamous epithelium" and is characteristic of mucosae subjected

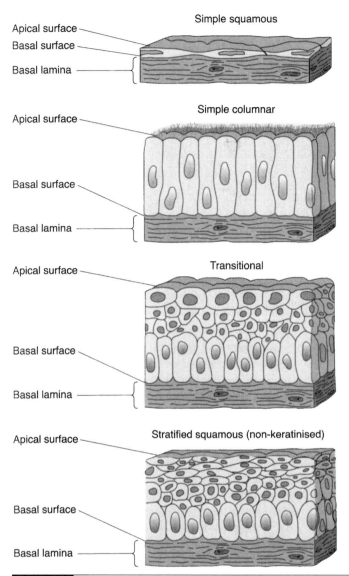

Apical surface
Basal surface
Basal lamina

Simple squamous

Apical surface

Simple columnar

Basal surface

Basal lamina

Apical surface

Transitional

Basal surface

Basal lamina

Apical surface

Stratified squamous (non-keratinised)

Basal surface

Basal lamina

Figure 1.13 The different types of epithelial surfaces present in the human body. A description of each type is given in the text. Reproduced with permission from: *Bacterial disease mechanisms; an introduction to cellular microbiology*. Wilson, M., McNab, R., and Henderson, B. 2002. Cambridge: Cambridge University Press.

to mechanical abrasion (e.g., the oral cavity and vagina). Keratinisation of the outer layers of cells may occur, and the epithelium is then said to be a "keratinised, stratified, squamous epithelium", examples of which include the skin and the hard palate of the oral cavity.

Simple squamous epithelia consist of a single layer of flattened cells and they line surfaces that are designed for absorption but do not experience wear and tear (e.g., the alveoli of the lungs). Simple columnar epithelia consist of a single layer of columnar cells and are found lining much of the digestive tract. They also line some regions of the respiratory tract where they may be covered in short, hair-like projections known as cilia. Ciliated epithelia constitute an important defence system because the cilia

Figure 1.14 Scanning electron micrograph of the respiratory mucosa showing the apices of goblet cells (G) and cilia (C). Secretory granules can be seen within the goblet cells beneath their cytoplasmic membranes. Bar = 30 μm. Reproduced with permission from: Airway mucosa: secretory cells, mucus, and mucin genes. Jeffery, P.K. and Li, D. *European Respiratory Journal* 1997;10:1655–62.

beat in a coordinated manner, thereby propelling a layer of mucus containing trapped bacteria along the epithelial surface and, ultimately, ejecting this from the body cavity (Section 4.2). When more than one layer of cells is present, the epithelium is described as being "stratified" and is found in regions subjected to wear and tear. When the cells are different shapes, it is known as a transitional epithelium and this lines organs, such as the bladder, that are subject to change in shape. Sometimes an epithelium consists of a single layer of different types of cells, some of which are small and do not reach the epithelial surface, whereas others are large and do reach the surface. The overall appearance often resembles a stratified epithelium but, because all of the constituent cells are in contact with the basement membrane, such an epithelium is termed a "pseudo-stratified epithelium". As well as providing a barrier to the ingress of microbes to underlying tissues, epithelial cells have a variety of other functions, including the secretion and absorption of molecules and ions and the production of antimicrobial peptides and other effector molecules of the host defence system.

As well as epithelial cells, the epithelium of most mucosal surfaces contains one or more additional cell types. Cells capable of secreting the constituents of mucus are invariably present and are known as goblet cells (Figure 1.14). A number of other cells involved in host defence may also be found, including intra-epithelial lymphocytes and dendritic cells, both of which are involved in the acquired immune response. Furthermore, the epithelial surface is invariably punctuated by the openings of a variety of glands. Other cells may also be present, depending on the particular mucosal surface; these are mentioned in subsequent chapters.

1.5.2 The epithelium as an excluder of microbes

There are a number of ways in which the epithelium excludes microbes from the rest of the body (Table 1.13). Firstly, it acts as a physical barrier preventing the penetration of

Table 1.13.	The epithelium as a microbe-excluder

provides a physical, impermeable barrier preventing access to underlying tissues

secretes mucus (on mucosal surfaces) which prevents access of microbes to underlying epithelial cells, traps microbes and facilitates their expulsion

undergoes keratinisation (on skin surfaces) providing a dry, chemically- and physically-resistant surface that is inhospitable to microbes

undergoes desquamation (exfoliation), thereby removing any microbes that have succeeded in colonising the epithelial surface

produces a range of antimicrobial peptides and proteins which are microbicidal or microbiostatic

microbes to the underlying tissues. Secondly, on the mucosal surfaces, the mucus layer secreted by the epithelium hinders access to the underlying epithelial cells, thereby helping to prevent permanent attachment of microbes to these cells. The mucus layer is continually expelled from the body along with any entrapped microbes. On the surface of the skin, the epithelial cells undergo keratinisation which results in a layer of dry, dead cells which microbes find difficult to colonise. Thirdly, the outermost cells of the skin and mucosa are continually being shed and replaced from below so that any microbes that do become attached are physically ejected from the body along with the shed epithelial cell. Finally, the epithelium secretes a range of antimicrobial peptides and proteins, which are able to either kill microbes or inhibit their growth. Further details of these antimicrobial mechanisms are provided in appropriate sections of succeeding chapters.

1.5.3 Mucus and mucins

The mucus layer is an important component of the host defence system but it is also a major source of nutrients for microbes at several body sites (e.g., the respiratory, genital, and urinary tracts). This illustrates an important aspect of the interaction between humans and microbes – both indigenous species and those that cause infections – that is, no matter what type of defence system the host has developed, some microbes will have evolved means of neutralising it or utilising it for their own benefit.

As mentioned previously, all mucosal surfaces of the body are covered by a gel known as mucus, the constituents of which are produced by the goblet and epithelial cells. Mucus has a number of important protective functions. Firstly, it contains receptors for the adhesins of a variety of microbes and so "traps" them when they arrive at a site, thereby preventing them from adhering to the underlying epithelium and colonising the site. Secondly, it contains a range of antimicrobial compounds (e.g., antibodies, enzymes, peptides) produced by the host epithelium and so is able to kill many of the entrapped organisms. The accumulation of these compounds in the mucus layer also increases their effectiveness because higher concentrations are achieved than would be the case if they were allowed to diffuse away from the mucosal surface. Thirdly, its continual production and expulsion from the body results in the ejection of microbes, their components, and their products. Fourthly, it protects the epithelium against chemical and physical damage by a number of materials (e.g., gastric acid, enzymes, particulate matter). Finally, it has lubricating and moisturising functions which facilitate movement of materials such as food, fluids, and chyme.

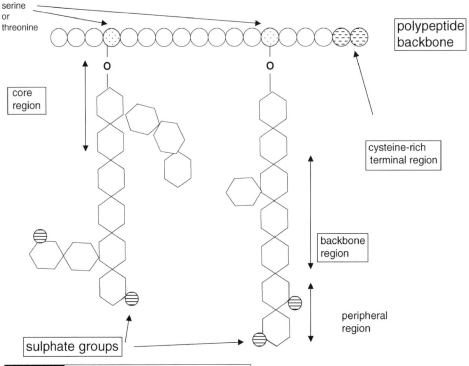

Figure 1.15 Generalised structure of a mucin molecule.

The main constituents of mucus are mucins which are usually present at a concentration of between 2% and 10% (w/v). These are glycoproteins containing high proportions of carbohydrates – usually between 70% and 85% (w/w). The mucins are unusual glycoproteins in that most of the carbohydrate side-chains are linked to the protein at serine and threonine residues via an oxygen atom (i.e., they are "O-glycosylated"), although N-glycosylation also occurs. The structure of a typical mucin molecule consists of a protein to which carbohydrate side-chains are linked by O-glycosylation and/or N-glycosylation. The protein backbone consists of several thousand amino-acid residues (in the case of MUC2, a major type of mucin found in the GIT, the number of amino acids is 5,179) and contains regions with many oligosaccharide side-chains and other regions without such side-chains. The oligosaccharide-rich regions are resistant to proteases, whereas the other regions are protease-sensitive. The oligosaccharide-containing regions of the protein are rich in serine, threonine, and proline. The side-chains usually consist of between two and twelve residues from a restricted range of sugars – usually galactose, fucose, N-acetylglucosamine, N-acetylgalactosamine, mannose, and sialic acids. The region of the side-chain involved in linkage to the protein is known as the "core" which is itself linked to the "backbone", which consists of a number of repeating disaccharide units containing galactose and N-acetylglucosamine (Figure 1.15). The terminal sugars are known as "peripheral" residues and often consist of sialic acid or sulphated sugars. The peripheral regions are often antigenic and contain the ABH or Lewis blood group determinants. The molecular mass of the individual glycoprotein molecules is usually in the region of 10^5 – 10^6 Da, but they are generally present as even larger molecular mass polymers due to the formation of disulphide bonds linking

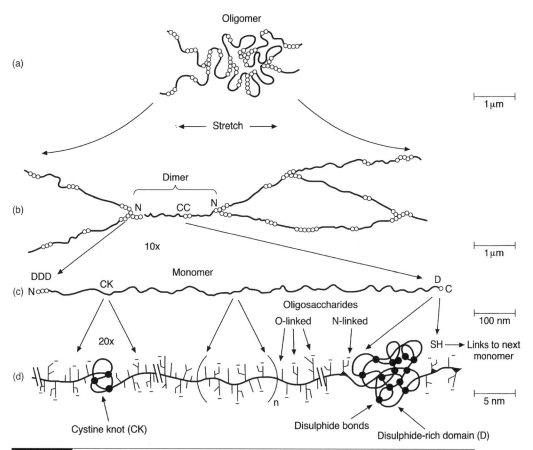

Figure 1.16 Monomeric and oligomeric structures of mucin. In (a), a number of mucin monomers (denoted by lines) are linked together (linkages denoted by circles) in an oligomeric gel. In (b), the gel has been stretched to show more clearly the linkages between the individual monomers. N and C denote the N- and C-termini of the individual mucin monomers. In (c), an individual monomer is schematically denoted showing the presence of the D domains which are involved in forming disulphide bonds between monomers. In (d), the structure of the monomer is shown in more detail. It should be noted that the monomer contains many O- and N-linked oligosaccharides. Reproduced with permission from: *Bacterial disease mechanisms; an introduction to cellular microbiology.* Wilson, M., McNab, R., and Henderson, B. 2002. Cambridge: Cambridge University Press.

the protein constituents of neighbouring molecules (Figure 1.16). These polymers may be several micrometres in length and form a viscoelastic gel in an aqueous medium. The whole polymer has a "bottle-brush" structure with the carbohydrate side-chains projecting as "bristles" from the central protein backbone. This structure is maintained by repulsive forces operating between the negatively charged carbohydrate side-chains and between the side-chains and the protein backbone. The side-chains also protect the protein from degradation by proteases and so help preserve the integrity of the mucus gel. Some of the mucins produced by the epithelium do not form polymers and remain covalently attached to the epithelial cells, and constitute what is known as a "glycocalyx". The glycocalyx interacts non-covalently with the mucus layer so helping it to remain associated with the mucosal surface (Figure 1.17). The mucus layer often has a monomolecular layer of lipids on its external surface which renders it hydrophobic. Apart from mucins and lipids, mucus also contains exfoliated cells; the contents of

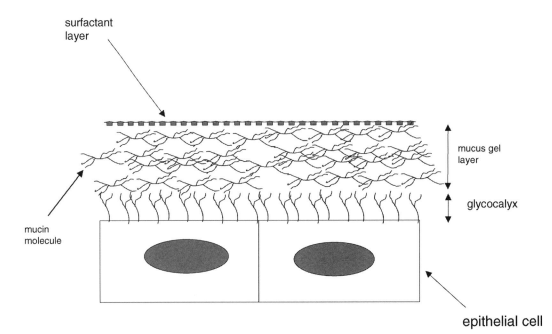

Figure 1.17 A typical mucosal surface is covered with a layer of mucus, which often has a thin lipid layer on its outer surface. The mucins of the mucus layer interact with the membrane-bound mucins (often referred to as the "glycocalyx") of the epithelial cells.

dead, lysed host cells; antibodies; and a range of antimicrobial compounds produced by the mucosa. The nature and function of these antimicrobial compounds are described in appropriate sections of Chapters 2–8. In some regions (e.g., the cervix and respiratory tract), the mucus gel is continually propelled along the mucosal surface by means of cilia – this is described in greater detail in Sections 4.2 and 6.2.1.

The protein backbones of the various mucins produced are encoded by a family of nine *MUC* genes; which of these genes are expressed depends on the particular body site. Hence, MUC2 is the main type of mucin protein produced in the intestinal tract whereas the expression of six *MUC* genes has been detected in the cervix – with different patterns of expression at different phases of the menstrual cycle. Differences also exist with regard to the glycosylation patterns of a particular type of mucin protein – hence, the composition of the carbohydrate side-chains of a mucin with a MUC1-protein backbone will be different in, for example, the respiratory and intestinal tracts. Certain mucin protein types are found only in mucus (e.g., MUC5B), while others are usually present only in mucins attached to epithelial cells (e.g., MUC1).

Mucus undoubtedly has an important role to play in host defence systems. However, the presence of such an abundant potential source of nutrients renders this complex polymer an attractive target for microbes, and a number of species have evolved the capability of utilising these molecules as a source of carbon, nitrogen, and energy. This is particularly important at those body sites where there may be few alternative nutrient sources (e.g., the urinary and respiratory tracts). However, the structural complexity of this family of polymers means that the complete degradation of any one of them by a single microbial species is unlikely. Such degradation entails the production of a whole range of enzymes in a certain order because regions of the molecule only

Table 1.14. | Enzymes required for the complete degradation of mucins

Type of enzyme	Role in mucin degradation
sulphatases	removal of terminal sulphate residues, thereby exposing underlying sugars rendering them more susceptible to the action of glycosidases
sialidases (neuraminidases)	removal of terminal sialic acid residues; this exposes underlying sugars to the action of glycosidases; the sialic acid itself can be further degraded by acetylneuraminate pyruvate lyase to N-acetylmannosamine which can be used as carbon and energy sources by some bacteria
exoglycosidases	cleave sugars from side-chains (e.g. β-D-galactosidase, N-acetyl-β-D-galactosaminidase, α-fucosidase, N-acetyl-β-D-glucosaminidase
endoglycosidases	cleave entire side-chain from the peptide backbone or attack the side-chain at sites other than the terminal residue – this may occur before or after the side-chain has been cleaved from the protein
peptidases/proteases	cleave at non-glycosylated regions; degrade protein backbone after side-chains have been removed

become accessible once others have been removed – this is more readily accomplished by microbial consortia rather than by individual species. Nevertheless, there are a limited number of microbes that can achieve complete degradation of a mucin (e.g., the intestinal organisms *Ruminococcus torques, Rum. gnavus*, and a *Bifidobacterium* sp.). Many other species possess a more limited repertoire of enzymes and accomplish partial degradation but, in doing so, not only obtain sufficient nutrients for their own needs, but also produce a mucin fragment that can be utilised by another organism and so on until, ultimately, the consortium has degraded the whole molecule. The range of enzymes needed to achieve complete degradation of a mucin is shown in Table 1.14. The ability to degrade mucin, entirely or partially, has been detected in microbes or microbial consortia inhabiting all mucosal sites of the body, and these are described further in appropriate sections of Chapters 2–8. The complete removal of mucus from a mucosal surface would obviously leave the host vulnerable to microbial colonisation and would have other harmful consequences. However, this does not appear to be a very common event which means that mucus utilisation by the indigenous microbiota must occur at the same rate as mucus production by the host – another example of the balanced relationship that exists between the host and its indigenous microbes.

1.5.4 Innate and acquired immune responses at the mucosal surface

Epithelial cells are in continuous contact with complex microbial communities which, for most of the time, consist of members of the indigenous microbiota of the particular site. As mentioned previously, all such microbes contain molecules (i.e., modulins) able to induce the release of pro-inflammatory cytokines from epithelial cells and instigate an inflammatory response. In order to avoid the detrimental consequences of a constant inflammatory state throughout the mucosae, the epithelium at a particular site must be able to recognise the indigenous microbiota of that site and suppress any inflammatory response to it (i.e., it must display "tolerance" to its indigenous microbiota).

However, the mucosa also needs to respond rapidly to the arrival of potentially harmful organisms and so must be able to recognise such organisms and distinguish them from members of its indigenous microbiota. The mechanisms responsible for such tolerance and discrimination are only now beginning to be elucidated and what little is known has generally come from studies involving the intestinal mucosa. To date, five main mechanisms have been identified. The first of these relates to the ability of human cells to recognise conserved microbial structural components known as "pathogen-associated molecular patterns" (PAMPs). PAMPs include molecules such as LPS, LTA, peptidoglycan, lipoproteins, and proteins (i.e., modulins). And these are recognized by a system involving proteins known as Toll-like receptors (TLRs). The term PAMP, it must be said, is something of a misnomer because the molecules included in this term are also present in the vast majority of non-pathogenic species – a more appropriate terminology would be MAMPs (i.e., microbe-associated molecular patterns). At present, ten different TLRs have been identified and each recognises one or more MAMPs. For example, TLR4, TLR2, and TLR5 recognise LPS, peptidoglycan, and flagellin, respectively. In the case of LPS, a co-receptor (CD14) is necessary to activate signalling via TLR4. Interaction of the TLR with its ligand activates a signalling pathway in the host cell, which induces some response such as the production of cytokines and macrophage activation. Combinations of TLRs are thought to be used by host cells to recognise a specific microbe, or type of microbe, and so may be able to distinguish between indigenous microbes and exogenous pathogens. Depending on the nature of the microbe, a specific set of TLR signalling pathways would be activated and the net signal generated may be recognised by the host as being characteristic of a "friendly" or "aggressive" microbe and an appropriate response generated. In intestinal epithelial cells, however, TLR and/or CD14 expression is strongly down-regulated; therefore, these cells do not respond to PAMPs such as LPS so that no inflammatory response is elicited. Although this may account for the non-inflammatory relationship that exists between members of the indigenous microbiota and intestinal epithelial cells, it does not explain how the mucosa can mount an inflammatory response when it is needed (i.e., when challenged by pathogenic microbes). One possibility is that the recognition/response system may be located intracellularly or on the basolateral surface of the epithelial cells. It would, therefore, be activated only when a pathogen invaded the cell or when it gained access to its basolateral surface. There is some evidence that such a system exists. Hence, invasion of an epithelial cell by *Shigella flexneri* (an organism responsible for dysentery) or injection of LPS into the cell results in the induction of the transcription factor NF-κB and the signalling molecule JNK, thereby generating an inflammatory response. The LPS is probably recognised by a protein known as caspase-activating and recruitment domain 4 (CARD4) which, therefore, functions like an internal TLR. Exposure of the basolateral (but not the apical) surface of an intestinal epithelial cell to flagellin, the structural component of the flagella of *Salmonella typhi* (the causative agent of typhoid fever), has also been shown to induce an inflammatory response. Although LPS is usually unable to stimulate an inflammatory response when it is in contact with the apical surface of an epithelial cell, it can do so in uroepithelial cells when it is complexed with another virulence factor – the type I fimbriae of uropathogenic *E. coli*. It would appear, therefore, that epithelia do not respond to inflammatory microbial constituents when they are present on the apical surface (i.e., when they are constituents of the indigenous microbiota) but will respond to such molecules if they are in the form of a complex

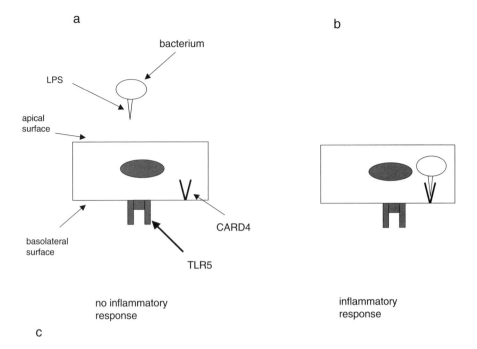

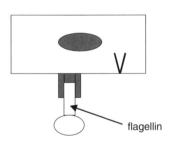

Figure 1.18 The role of Toll-like receptors in mediating tolerance to the indigenous microbiota of the intestinal tract. (a) The TLR for LPS (TLR4) and other PAMPs are down-regulated on the apical surface, but may be present on the basolateral surface (e.g., TLR5 which recognises flagellin). The co-receptor for LPS (CD14) is also not expressed on the apical surface of intestinal epithelial cells. LPS on a bacterium in the intestinal lumen is, therefore, not recognised, and there is no inflammatory response. (b) If an organism invades the cell, the LPS is recognised by CARD4 (the intracellular equivalent of a TLR) and an inflammatory response is induced. (c) If bacterial flagellin gains access to the basolateral surface, it is recognised by TLR5 and so induces an inflammatory response. LPS = lipopolysaccharide; PAMPs = pathogen-associated molecular patterns; TLR = Toll-like receptor.

Table 1.15. Antimicrobial peptides/proteins produced by epithelia

Antimicrobial peptide/protein	Site of production	Activities other than as an antimicrobial
HBD-1	E, S, OC, RT, St, I, UT, FGT	
HBD-2	E, S, OC, RT, I, FGT	
HBD-3	S, OC, RT, FGT	
HBD-4	St, RT, FGT, MGT	chemoattractant for monocytes
HE2β1	OC, RT, MGT	
HD5	I, FGT	
HD6	I	
histatins	OC	suppress bacteria-induced cytokine induction
LL-37	S, OC, RT, I, FGT, MGT	neutralises some biological activities of LPS and LTA
hepcidin	UT	
lysozyme	S, E, OC, RT, I, FGT, MGT	agglutinates bacteria
lactoferrin	E, OC, RT, I, FGT, MGT	iron-binding; prevents bacterial adhesion; enhances activity of natural killer cells
adrenomedullin	S, OC, St, I	
histone H1	I	
SLPI	E, OC, RT, FGT, MGT	neutralises some biological activities of LPS
SPLA$_2$	E, RT, I, MGT	
dermcidin	S	
elafin	RT	neutralises some biological activities of LPS; inhibits human neutrophil elastase

Notes: E = eye; FGT = female genital tract; HBD = human β-defensin; HD = human α-defensin; HE = human epididymis secretory protein; I = intestines; MGT = male genital tract; OC = oral cavity; RT = respiratory tract; S = skin; SLPI = secretory leukocyte protease inhibitor; SPLA$_2$ = secretory phospholipase A$_2$; St = stomach, UT = urinary tract.

with virulence factors of pathogenic species (e.g., the fimbriae of uropathogenic *E. coli*) or when they gain access to the cell interior or reach the basolateral surface – that is, when pathogenic species invade the mucosa or gain access to the basolateral surface via a breach in the epithelium (Figure 1.18).

A second mechanism involves the induction of an immunosuppressive effect. Intestinal pathogens can induce an inflammatory response by activating the transcription factor NF-κB, one of the consequences being the release of the neutrophil chemokine interleukin-8 (IL-8), which attracts PMNs to the site. A number of enteric organisms have been shown to inhibit activation of NF-κB by intestinal pathogens and inflammatory stimuli such as tumour necrosis factor-α and so prevent release of IL-8. Whether this is due to the production of an immunosuppressive factor by non-pathogenic microbes or is a host-cell product induced in response to stimulation by such organisms remains to be established. The net effect is that the epithelium is maintained in a hypo-responsive state by the presence of members of the indigenous microbiota (i.e., it exhibits tolerance to these microbes).

In a third mechanism, indigenous gut microbes were found to induce only a transient release of pro-inflammatory cytokines from intestinal epithelial cells. This was quickly suppressed by macrophages from the lamina propria. In contrast, the macrophages were unable to suppress the release of pro-inflammatory cytokines induced by exogenous pathogens.

Fourthly, certain indigenous intestinal species such as *Lactobacillus* spp. are able to induce the release of transforming growth factor (TGF)-β from intestinal epithelial cells. TGF-β is a key regulator of the immune response and has an overall anti-inflammatory effect.

Finally, antimicrobial peptides and proteins are only now being appreciated as an important means by which mucosal surfaces regulate their indigenous microbiotas. A wide range of antimicrobial peptides are known to be produced by epithelial cells, and these are listed in Table 1.15 and described in greater detail in subsequent chapters. Although these molecules are best known for their ability to inhibit or kill microbes, some are also able to exert other effects, such as neutralising the biological activities of modulins such as LPS. Each antimicrobial peptide has a particular antimicrobial spectrum, and each mucosal surface produces only some of these peptides. Furthermore, some peptides are produced constitutively, while others are induced (or up-regulated) in response to certain microbial species, and some peptides display synergy with others. It is tempting, therefore, to suggest that the particular mixture of peptides secreted by a particular mucosal surface plays some role in dictating the composition of the microbial community at that site.

1.6 | Further Reading

Books

Atlas, R.M. and Bartha, R. (1997). *Microbial ecology: fundamentals and applications.* Boston: Addison-Wesley.

Grubb, R., Midtvedt, T., and Norin, E. (eds). (1988). *The regulatory and protective role of the normal microflora.* New York: Stockton Press.

Hill, M.J. and Marsh, P.D. (1990). *Human microbial ecology.* Boca Raton: CRC Press.

Rosebury, T. (1962). *Microorganisms indigenous to man.* New York: McGraw-Hill Book Company.

Skinner, F.A. and Carr, J.G. (eds.). (1974). *The normal microbial flora of man.* London: Academic Press.

Tannock, G.W. (1995). *Normal microflora.* London: Chapman and Hall.

Tannock, G.W. (ed.). (1999). *Medical importance of the normal microflora.* Dordrecht: Kluwer Academic Publishers.

Reviews and Papers

Akira, S. and Hemmi, H. (2003). Recognition of pathogen-associated molecular patterns by TLR family. *Immunology Letters* **85**, 85–95.

Bauer, W.D. and Robinson, J.B. (2002). Disruption of bacterial quorum sensing by other organisms. *Current Opinion in Biotechnology* **13**, 234–237.

Belley, A., Keller, K., Gottke, M., Chadee, K., and Goettke, M. (1999). Intestinal mucins in colonization and host defense against pathogens. *American Journal of Tropical Medicine and Hygiene* **60** (4 Suppl), 10–15.

Beutler, B. (2003). Innate immune responses to microbial poisons: discovery and function of the Toll-like receptors. *Annual Review of Pharmacology and Toxicology* **43**, 609–628.

Biancone, L., Monteleone, I., Del Vecchio Blanco, G., Vavassori, P., and Pallone, F. (2002). Resident bacterial flora and immune system. *Digestive and Liver Disease* **34** (Suppl 2), S37–S43.

Blaut, M., Collins, M.D., Welling, G.W., Dore, J., van Loo, J., and de Vos, W. (2002). Molecular biological methods for studying the gut microbiota: the EU human gut flora project. *British Journal of Nutrition* **87** (Suppl 2), S203–S211.

Bohannan, B.J., Kerr, B., Jessup, C.M., Hughes, J.B., and Sandvik, G. (2002). Trade-offs and coexistence in microbial microcosms. *Antonie Van Leeuwenhoek* **81**, 107–115.

Boman, H.G. (2000). Innate immunity and the normal microflora. *Immunological Reviews* **173**, 5–16.

Corfield, A.P., Carroll, D., Myerscough, N., and Probert, C.S. (2001). Mucins in the gastrointestinal tract in health and disease. *Frontiers in Bioscience* **6**, D1321–D1357.

Davies, C.E., Wilson, M.J., Hill, K.E., Stephens, P., Hill, C.M., Harding, K.G., and Thomas, D.W. (2001). Use of molecular techniques to study microbial diversity in the skin: chronic wounds re-evaluated. *Wound Repair and Regeneration* **9**, 332–340.

Deplancke, B. and Gaskins, H.R. (2001). Microbial modulation of innate defense: goblet cells and the intestinal mucus layer. *American Journal of Clinical Nutrition* **73**, 1131S–1141S.

Devine, D.A. (2004). Cationic antimicrobial peptides in regulation of commensal and pathogenic microbial populations. In *Mammalian host defense peptides*, D.A. Devine and R.E.W. Hancock (eds.). Cambridge: Cambridge University Press.

Devine, D.A. and Hancock, R.E. (2002). Cationic peptides: distribution and mechanisms of resistance. *Current Pharmaceutical Design* **8**, 703–714.

Donabedian, H. (2003). Quorum sensing and its relevance to infectious diseases. *Journal of Infection* **46**, 207–214.

Donlan, R.M. (2002). Biofilms: microbial life on surfaces. *Emerging Infectious Diseases* **8**, 881–890.

Donlan, R.M. and Costerton, J.W. (2002). Biofilms: survival mechanisms of clinically relevant microorganisms. *Clinical Microbiology Reviews* **15**, 167–193.

Donskey, C.J., Hujer, A.M., Das, S.M., Pultz, N.J., Bonomo, R.A., and Rice, L.B. (2003). Use of denaturing gradient gel electrophoresis for analysis of the stool microbiota of hospitalized patients. *Journal of Microbiological Methods* **54**, 249–256.

Dunn, A.K. and Handelsman, J. (2002). Toward an understanding of microbial communities through analysis of communication networks. *Antonie Van Leeuwenhoek* **81**, 565–574.

Elson, C,O. and Cong, Y. (2002). Understanding immune-microbial homeostasis in intestine. *Immunologic Research* **26**, 87–94.

Emonts, M., Hazelzet, J.A., de Groot, R., and Hermans, P.W. (2003). Host genetic determinants of *Neisseria meningitidis* infections. *Lancet Infectious Diseases* **3**, 565–577.

Fujimoto, C., Maeda, H., Kokeguchi, S., Takashiba, S., Nishimura, F., Arai, H., Fukui, K., and Murayama, Y. (2003). Application of denaturing gradient gel electrophoresis (DGGE) to the analysis of microbial communities of subgingival plaque. *Journal of Periodontal Research* **38**, 440–445.

Ganz, T. (2003). Defensins: antimicrobial peptides of innate immunity. *Nature Reviews Immunology* **3**, 710–720.

Gavazzi, G. and Krause, K.-H. (2002). Aging and infection. *Lancet Infectious Diseases* **2**, 659–666.

Greene, E.A. and Voordouw, G. (2003). Analysis of environmental microbial communities by reverse sample genome probing. *Journal of Microbiological Methods* **53**, 211–219.

Hall-Stoodley, L., Costerton, J.W., and Stoodley, P. (2004). Bacterial biofilms: from the natural environment to infectious diseases. *Nature Reviews Microbiology* **2**, 95-108.

Hebuterne, X. (2003). Gut changes attributed to aging: effects on intestinal microflora. *Current Opinion in Clinical Nutrition and Metabolic Care* **6**, 49–54.

Hecht, G. (1999). Innate mechanisms of epithelial host defense: spotlight on intestine. *American Journal of Physiology* **277**, C351–C358.

Heine, H. and Lien, E. (2003). Toll-like receptors and their function in innate and adaptive immunity. *International Archives of Allergy and Immunology* **130**, 180–192.

Hertz, C.J. and Modlin, R.L. (2003). Role of Toll-like receptors in response to bacterial infection. *Contributions in Microbiology* **10**, 149–163.

Janssens, S. and Beyaert, R. (2003). Role of Toll-like receptors in pathogen recognition. *Clinical Microbiology Reviews* **16**, 637-646.

Knowles, M.R. and Boucher, R.C. (2002). Mucus clearance as a primary innate defense mechanism for mammalian airways. *Journal of Clinical Investigation* **109**, 571–577.

Koczulla, A.R. and Bals, R. (2003). Antimicrobial peptides: current status and therapeutic potential. *Drugs* **63**, 389–406.

Lawrence, J.R. and Neu, T.R. (1999). Confocal laser scanning microscopy for analysis of microbial biofilms. *Methods in Enzymology* **310**, 131–144.

Lerat, E. and Moran, N.A. (2004). The evolutionary history of quorum-sensing systems in bacteria. *Molecular Biology and Evolution* **21**, 903–913.

Lillehoj, E.R. and Kim, K.C. (2002). Airway mucus: its components and function. *Archives of Pharmaceutical Research* **25**, 770–780.

McCartney, A.L. (2002). Application of molecular biological methods for studying probiotics and the gut flora. *British Journal of Nutrition* **88** (Suppl 1), S29–S37.

McFarland, L.V. (2000). Normal flora: diversity and functions. *Microbial Ecology in Health and Disease* **12**, 193–207.

McGlauchlen, K.S. and Vogel, L.A. (2003). Ineffective humoral immunity in the elderly. *Microbes and Infection* **5**, 1279–1284.

Moter, A. and Gobel, U.B. (2000). Fluorescence in situ hybridization (FISH) for direct visualization of microorganisms. *Journal of Microbiological Methods* **41**, 85–112.

Muyzer, G. (1999). DGGE/TGGE – a method for identifying genes from natural ecosystems. *Current Opinion in Microbiology* **2**, 317–322.

Parsek, M.R. and Singh, P.K. (2003). Bacterial biofilms: an emerging link to disease pathogenesis. *Annual Review of Microbiology* **57**, 677–701.

Pasteur, L. (1880). De l'attenuation virus du cholera des poules. *Comptes Rendus de l'Academie des Sciences* **91**, 673–680.

Podbielski, A. and Kreikemeyer, B. (2004). Cell density-dependent regulation: basic principles and effects on the virulence of Gram-positive cocci. *International Journal of Infectious Diseases* **8**, 81–95.

Preston-Mafham, J., Boddy, L., and Randerson, P.F. (2002). Analysis of microbial community functional diversity using sole-carbon-source utilisation profiles – a critique. *FEMS Microbiology Ecology* **42**, 1–14.

Ranjard, L., Poly, F., and Nazaret, S. (2000). Monitoring complex bacterial communities using culture-independent molecular techniques: application to soil environment. *Research in Microbiology* **151**, 167–177.

Relman, D.A. (2002). New technologies, human-microbe interactions, and the search for previously unrecognized pathogens. *Journal of Infectious Diseases* **186** (Suppl 2), S254–S258.

Riley, M.A. and Wertz, J.E. (2002). Bacteriocins: evolution, ecology, and application. *Annual Review of Microbiology* **56**, 117–137.

Ryley, H.C. (2001). Human antimicrobial peptides. *Reviews in Medical Microbiology* **12**, 177–186.

Shapiro, H.M. (2000). Microbial analysis at the single-cell level: tasks and techniques. *Journal of Microbiological Methods* **42**, 3–16.

Smith, V. H. (2002). Effects of resource supplies on the structure and function of microbial communities. *Antonie Van Leeuwenhoek* **81**, 99–106.

Stoodley, P., Sauer, K., Davies, D.G., and Costerton, J.W. (2002). Biofilms as complex differentiated communities. *Annual Review of Microbiology* **56**, 187–209.

Takeda, K., Kaisho, T., and Akira, S. (2003). Toll-like receptors. *Annual Review of Immunology* **21**, 335–376.

Tannock, G.W. (2001). Molecular assessment of intestinal microflora. *American Journal of Clinical Nutrition* **73** (2 Suppl), 410S–414S.

Theron, J. and Cloete, T.E. (2000). Molecular techniques for determining microbial diversity and community structure in natural environments. *Critical Reviews in Microbiology* **26**, 37–57.

Toivanen, P., Vaahtovuo, J., and Eerola, E. (2001). Influence of major histocompatibility complex on bacterial composition of fecal flora. *Infection and Immunity* **69**, 2372–2377.

Torsvik, V., Daae, F.L., Sandaa, R.-A., and Ovreas, L. (1998). Novel techniques for analysing microbial diversity in natural and perturbed environments. *Journal of Biotechnology* **64**, 53–62.

Underhill, D.M. and Ozinsky, A. (2002). Toll-like receptors: key mediators of microbe detection. *Current Opinion in Immunology* **14**, 103–110.

Vaughan, E.E., Schut, F., Heilig, H.G., Zoetendal, E.G., de Vos, W.M., and Akkermans, A.D. (2000). A molecular view of the intestinal ecosystem. *Current Issues in Intestinal Microbiology* **1**, 1–12.

Vimr, E.R. (1994). Microbial sialidases: does bigger always mean better? *Trends in Microbiology* **2**, 271–277.

Vimir, E.R., Kalivoda, K.A., Deszo, E.L., and Steenbergen, S.M. (2004). Diversity of microbial sialic acid metabolism. *Microbiology and Molecular Biology Reviews* **68**, 132–153.

Wilson, M. (2001). Bacterial biofilms and human disease. *Science Progress* **84**, 235–254.

Xavier, K.B. and Bassler, B.L. (2003). LuxS quorum sensing: more than just a numbers game. *Current Opinion in Microbiology* **6**, 191–197.

Xu, J. and Gordon, J.I. (2003). Honor thy symbionts. *Proceedings of the National Academy of Sciences USA* **100**, 10452–10459.

Yang, D., Biragyn, A., Kwak, L.W., and Oppenheim, J.J. (2002). Mammalian defensins in immunity: more than just microbicidal. *Trends in Immunology* **23**, 291–296.

Zhou, J. (2003). Microarrays for bacterial detection and microbial community analysis. *Current Opinion in Microbiology* **6**, 288–294.

Zoetendal, E.G., Akkermans, A.D.L., Akkermans-van Vliet, W.M., de Visser, J.A.G.M., and de Vos, W.M. (2001). The host genotype affects the bacterial community in the human gastrointestinal tract. *Microbial Ecology in Health and Disease* **13**, 129–134.

The skin and its indigenous microbiota

The skin, together with its accessory structures (hair, nails, glands, sensory receptors, muscles, and nerves), constitute what is known as the integumentary system. The skin itself is an organ composed of several tissues (epidermal, connective, nervous, and muscular) and is one of the largest organs of the body in terms of its surface area (approximately 1.75 m^2) and weight (approximately 5 kg). It has a variety of functions, chief among which is protecting underlying tissues from microbes.

2.1 | Anatomy and physiology of human skin

It is important to realise that the structure of skin is not uniform over the whole body surface, and profound differences occur at different body sites. Obvious examples are the presence or absence of hair and sudoriferous glands. Nevertheless, certain features are common to skin regardless of its location. Hence, it is basically composed of two layers – an inner dermis and an outer epithelium known as the epidermis (Figure 2.1). The epidermis is a keratinised, stratified, squamous epithelium within which five layers can be distinguished (Table 2.1 and Figure 2.2). Its thickness varies from 0.5 to 3 mm, depending on its location and, being the outermost layer of the skin, it is obviously an important site for microbial colonisation. The most common cell of the epidermis (comprising approximately 90% of all the cells of the skin) is the keratinocyte. New keratinocytes are constantly being produced in the stratum basale and, as they are pushed towards the surface, they undergo a process known as keratinisation. This involves the production of a protein, keratin, and the eventual death of the cell. The dead, keratinised cells (known as squames) comprise the outer layers of the epidermis (with lipids filling the intercellular spaces) and are gradually sloughed off – a process termed desquamation. It has been estimated that the skin surface of the average adult is composed of approximately 2×10^9 squames. It takes between 2 and 4 weeks for the transfer of a cell from the basal layer to the outermost layer and, as a result of this process, it has been estimated that the stratum corneum is renewed every 15 days. The keratin present in the cells protects the underlying tissues from heat, chemicals, and microbes. Melanocytes are the next most common cells of the epidermis. These have long slender projections and produce the brown-black pigment melanin which is transferred to keratinocytes where it absorbs ultraviolet light, thus protecting the skin from its damaging effects. The only other cells present in the epidermis are Langerhans cells (which are involved in the immune response to microbial invaders) and Merkel

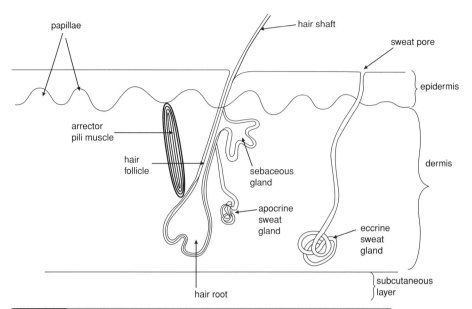

Figure 2.1 Diagrammatic representation of the main structures comprising human skin.

cells (which are associated with sensory neurons and are involved in the sensation of touch).

The dermis has a far more complex structure (Figure 2.1) and consists of (1) connective tissue containing collagen and elastin fibres giving the skin elasticity; (2) small, finger-like projections (papillae) which protrude into the epidermis and contain nerve endings sensitive to touch, heat, and pain; (3) hair follicles; (4) arrector pili muscles – for controlling hair movement; (5) sebaceous glands; (6) sudoriferous glands; (7) nerves; (8) adipose tissue; and (9) capillaries and veins. The presence or absence, as well as the concentrations of these various components and the thickness of the dermis depend very much on the particular anatomical site. For example, the dermis is very thick on the palms of the hands and soles of the feet, but is very thin on the eyelids and

Table 2.1. Main characteristics of the various layers of the epidermis

Layer	Description
stratum basale	single layer of keratinocytes capable of cell division; also contains melanocytes, Langerhans cells, and Merkel cells
stratum spinosum	between 8 and 10 layers of multi-sided keratinocytes with spiky projections; cells secrete lipids into intercellular spaces; Langerhans cells and projections from melanocytes are also present
stratum granulosum	approximately five layers of flattened keratinocytes; cells have granules containing lipids that are secreted to provide a water-repellent sealant
stratum lucidum	contains about five layers of flat dead cells; present only in skin of the palms and soles of the feet
stratum corneum	consists of about 15 layers of flat dead cells with a variety of lipids occupying the intercellular spaces

Figure 2.2 Cross-section through the epidermis of the human fingertip showing the different epidermal layers; stratum basale (B), stratum spinosum (S), stratum granulosum (G), stratum lucidum (L), and stratum corneum (C). Reproduced with permission from: *The skin microflora and microbial skin disease.* Noble, W.C. (ed). 1992. Cambridge: Cambridge University Press.

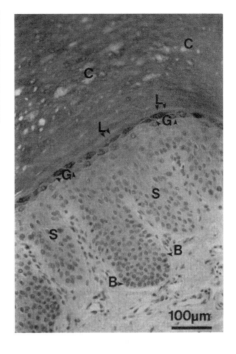

scrotum; sudoriferous glands are particularly plentiful on the forehead and axillae, but are absent from the margins of the lips; hair follicles are plentiful on the head, eyebrows, and external genitalia, but absent from the palms, soles, and eyelids. From the point of view of microbial colonisation, the most important structure in the dermis is the hair follicle with its associated glands because it is open to the external environment (Figures 2.1 and 2.3). Hair is present on most skin surfaces, and it has a variety

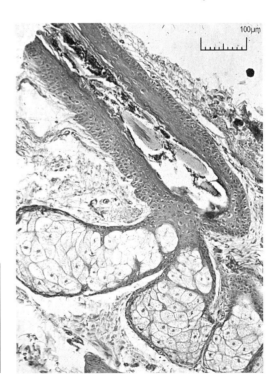

Figure 2.3 Stained longitudinal section through a single pilosebaceous follicle showing the multi-lobular sebaceous gland and the pilosebaceous duct which has been colonised by bacteria. Reproduced with kind permission of Kluwer Academic Publishers from: The human cutaneous microbiota and factors controlling colonisation. Bojar, R.A. and Holland, K.T. *World Journal of Microbiology & Biotechnology* 2002;18:889–903, Figure 8.

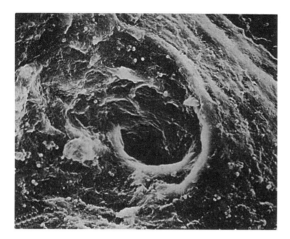

Figure 2.4 Scanning electron micrograph of a sweat duct in the skin from a human foot. Note the presence of bacteria – mainly as single cells, pairs, or small groups. Reprinted with permission of Blackwell Publishing from: The demonstration of bacteria on and within the stratum corneum using scanning electron microscopy. Malcolm, S.A. and Hughes, T.C. *British Journal of Dermatology* 1980;102:267–275.

of functions, including protection of the head from solar radiation and the eyes and respiratory system from particulate matter. A hair consists of dead keratinised cells – the portion protruding from the skin is known as the shaft, while that within the skin is called the root. Each hair follicle is surrounded by touch-sensitive hair root plexuses and has an "arrector pili" muscle. The hair usually extends from the skin at an angle, but the arrector pili muscle responds to stresses such as cold and fright by contracting so that the hair becomes perpendicular to the skin surface. The density of hair follicles depends on the anatomical location; the highest density is on the scalp which has between 300 and 500 follicles/cm^2. Many (but not all) hair follicles have one (or sometimes two) associated sebaceous gland, the duct of which opens into the follicular canal. This gland secretes a lipid-rich fluid known as sebum, which protects the hair from drying out, prevents excessive moisture loss from the skin surface, keeps the skin soft, and inhibits the growth of certain microbes. Some glands open directly onto the skin surface rather than into the hair follicle. Sebaceous glands are present on most skin surfaces, except for the palms of the hands and the soles and dorsum of the feet. Their concentration depends on the anatomical site, with the greatest density being found on the scalp, forehead, face, and upper chest where there are between 400 and 800 glands/cm^2. Elsewhere, there are approximately 100 glands/cm^2.

Like sebaceous glands, sudoriferous (sweat-producing) glands are of two types – one type (apocrine) opens into the follicular canal, whereas the other (eccrine) opens directly onto the skin surface via a sweat pore. Apocrine glands have a restricted distribution and are found only in the axillae and perineum and, in a modified form as ceruminous glands, in the external ear canal. They secrete a viscous, odouriferous material of uncertain composition. These glands do not function until puberty, and their secretions are thought to have some sexual function which is now vestigial (e.g., as a sexual attractant, territorial marker). Eccrine glands are found in most areas of the body, but are particularly numerous on the forehead, in the axillae, on the palms of the hands, and on the soles of the feet (Figure 2.4). In these regions, their density may be as high as 620 glands/cm^2. They are least abundant on the back, where their density is approximately 10-fold lower. These glands secrete a hypotonic solution (sweat) which helps to regulate body temperature, eliminates waste materials, and carries mediators of the innate and acquired immune response.

Table 2.2.	Antimicrobial defence mechanisms of human skin

airflow across body surfaces
low moisture content
intact stratum corneum
acidic pH
shedding of squames
high salt content of some regions
production of antibacterial fatty acids and lipids
acidified nitrite
nitric oxide
lysozyme
antimicrobial peptides
immunoglobulins

2.2 | Antimicrobial defence mechanisms of the skin

The first line of defence against microbial colonisation of the skin is the continuous flow of air across the body surfaces (Table 2.2). This helps to prevent airborne microbes and microbe-containing particles from settling on the skin surface, but would provide no defence against transfer by direct contact with microbe-laden individuals or objects. Microbes that reach the skin surface are faced with an intact layer of overlapping dead keratinised cells, with any gaps between them filled by lipid-rich material. This intact stratum corneum acts as a barrier to the penetration of organisms into the more nutrient-rich underlying tissues (Figure 2.5). Few organisms can degrade keratin which, therefore, is an effective barrier material. The material between the dead cells contains free fatty acids, triglycerides, wax esters, sterols, alkanes, ceramides, squalene, phospholipids, and glycosphingolipids produced by keratinocytes and the sebaceous glands. The presence of layers of dead keratinocytes and lipids renders the surface of the skin very dry and, therefore, limits microbial growth. Furthermore, the skin has a low pH at most body sites (see Figure 2.8) due to acidic substances secreted by glands, excreted by epidermal cells, and produced by microbes. This low pH inhibits the growth of a wide range of microbes. Apart from their pH-lowering effect, many of the substances present on the skin surface also have direct antimicrobial activity. Of these, the free fatty acids, particularly lauric and myristic acids, are the most effective antimicrobials and have

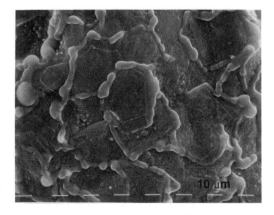

Figure 2.5 Scanning electron micrograph of the surface of bovine skin after freeze-drying to preserve lipids. Keratinised cells can be seen with gaps between them filled by lipid-rich material. Reproduced with permission from: *The skin microflora and microbial skin disease*. Noble, W.C. (ed). 1992. Cambridge: Cambridge University Press.

a wide spectrum of activity against members of the resident microbiota and against transients. Linoleic and linolenic acids also have antimicrobial activity, particularly against transients such as *Staph. aureus*. Furthermore, one of the main products of hydrolysis of the sphingolipid ceramide is sphingosine, which is very effective at killing *Staph. aureus*. Sphingosine is present at a concentration of 268 μM in healthy skin, which is approximately 20-fold higher than its minimum inhibitory concentration (MIC) for *Staph. aureus*. Some of the saturated free fatty acids, particularly lauric acid, suppress the growth of *Propionibacterium acnes*. Lauric acid – as well as caproic, butyric, and myristic acids – is inhibitory to (in decreasing order of susceptibility) *Streptococcus pneumoniae*, *Streptococcus pyogenes*, corynebacteria, micrococci, *Candida* spp., and staphylococci.

Although the relatively low moisture content of the stratum corneum and its acidity serve to restrict microbial growth, certain sites (particularly the inter-triginous regions) have a higher moisture content and a neutral or slightly alkaline pH because of the accumulation of skin secretions, and such regions have correspondingly denser microbial populations. Evaporation of water from sweat leaves behind a high concentration of solutes – particularly NaCl – and this can result in an aqueous environment with a high osmolarity which restricts the growth of many microbes, particularly Gram-negative bacteria. For those organisms that manage to overcome these obstacles, that is not the end of the story because the constant shedding of squames, taking with them any adherent microbes, is another important means of limiting microbial colonisation of the skin.

As well as these defence mechanisms which are specific to the skin, the usual innate and acquired host antimicrobial defence systems are operative. The skin-associated lymphoid tissue is involved in the generation of humoral and cell-mediated immune responses and consists of (1) Langerhans cells and dermal dendritic cells, both of which are antigen-presenting cells and circulate between the skin and the lymph nodes; (2) keratinocytes and endothelial cells which produce cytokines; and (3) lymphocytes. Interestingly, although Langerhans cells comprise less than 5% of epidermal cells, they cover almost 25% of the skin surface because of their long processes. In this way, they form a virtually continuous network that can capture almost any antigen entering the skin. Both IgA and IgG antibodies are secreted by the eccrine glands and are deposited on the skin surface where they can exert their antimicrobial effects, as well as preventing microbial adhesion. Immunohistochemical studies have shown that many organisms present on the skin surface are coated with IgA, IgG, or IgM antibodies.

With regard to innate immunity, keratinocytes can produce a range of antibiotic peptides, including human β-defensin (HBD)-1, HBD-2, HBD-3, LL-37, and adrenomedullin. HBDs are cysteine-rich, cationic peptides with a broad antimicrobial spectrum at micromolar concentrations, but their activity decreases with increasing ionic strength and are generally not active at high salt concentrations (>100 mM NaCl). They display activity against Gram-positive and Gram-negative bacteria, fungi, and viruses. HBD-1 is produced constitutively and is active against Gram-positive and Gram-negative species, including *Staph. aureus, E. coli, Listeria monocytogenes, K. pneumoniae, Ps. aeruginosa*, and *Can. albicans*. It is active at low pHs, but is inactivated by high salt concentrations (>100 mM). The production of HBD-2 has been investigated in some detail, and this has resulted in some interesting findings. Some organisms able to cause skin infections either stimulate HBD-2 production by keratinocytes, but are resistant to its effect (e.g., *Staph. aureus*) or are poor inducers of its production (e.g., *Strep. pyogenes*) while being susceptible to its effects. Either of these tactics would help the organism to survive on the skin surface. In contrast, organisms such as *E. coli* and *Ps. aeruginosa*, which are not often present on healthy skin, are consistent inducers of, but highly susceptible to, HBD-2. Although *Staph. epidermidis* (invariably present on healthy skin, but rarely a cause of skin infections) induces HBD-2 expression in keratinocytes, it is relatively resistant to its antibacterial effects. This could account, in part, for its ubiquity on the skin surface. HBD-2 is also active against *Ent. faecalis* and *Can. albicans* and displays synergic antimicrobial activity with lysozyme against a range of bacteria. HBD-2, like HBD-1, is not active at high salt concentrations. The production of HBD-3, like that of HBD-2, is up-regulated in response to bacteria. It is active against *Staph. aureus, Strep. pyogenes, Ps. aeruginosa, Ent. faecium, E. coli*, and *Can. albicans*. LL-37 is a small peptide containing thirty-seven amino-acid residues, which is released from its precursor – human cationic antimicrobial protein – by hydrolysis. It is a member of the cathelicidin family of antimicrobial peptides and is present at low levels in the epidermis of normal individuals, but its synthesis by keratinocytes is markedly up-regulated in inflammatory conditions. It exerts a cidal effect alone, and in synergy with lysozyme, against *Strep. pyogenes, E. coli, Staph. aureus, Strep. pneumoniae, Staph. epidermidis, Ent. faecalis, Ps. aeruginosa, Lis. monocytogenes, Sal. typhimurium*, and *Bacillus megaterium*. LL-37 has also been shown to bind and neutralise the biological activities of LPS and lipoteichoic acid and is chemotactic for human peripheral monocytes, neutrophils, and CD4 T lymphocytes. Expression of LL-37 has been shown to be able to protect mice against skin invasion by *Strep. pyogenes*. Adrenomedullin is a multi-functional peptide produced by a variety of epithelial cells, which can inhibit the growth of, or kill, a variety of bacteria, including *P. acnes, Staph. aureus*, and *M. luteus*. Although it is constitutively produced by skin keratinocytes, its production is up-regulated by *P. acnes, Staph. aureus*, and *M. luteus*.

It has recently been shown that sweat contains a protein known as dermcidin, which undergoes proteolytic cleavage to produce a 47-amino-acid peptide with activity against *E. coli, Ent. faecalis, Staph. aureus*, and *Can. albicans*. Importantly, the peptide retains its activity at the low pHs and high salt concentrations that exist in human sweat, and is also effective at concentrations (1–10 μg/ml) known to be present in sweat. Dermcidin is produced constitutively. The presence in sweat of this protein (together with lysozyme, discussed later) may account, in part, for the failure of microbes to colonise the sudoriferous glands.

Lysozyme is present in sweat and is also produced by keratinocytes and staphylococci. Estimates of its concentration in skin range from 14 to 120 μg/g wet weight of skin depending on the site. Lysozyme is an enzyme that cleaves the $\beta 1 \rightarrow 4$ glycosidic bond between N-acetylglucosamine and N-acetylmuramic acid in peptidoglycan and is effective mainly against Gram-positive bacteria – the peptidoglycan of Gram-negative species being protected by the outer membrane. At low pHs, the antibacterial activity of lysozyme is enhanced by monovalent ions, such as chloride, bicarbonate, and thiocyanate. Of the resident cutaneous microbiota, micrococci are the most susceptible to the enzyme, which may account, in part, for the generally low population densities of these organisms on the skin. Another possible antimicrobial defence system in skin involves the generation of reactive nitrogen intermediates (RNIs) by acidification of nitrite ions. The nitrate present in sweat can be converted to nitrite by a number of cutaneous microbes and, at the low pH of the epidermis, this is converted nonenzymatically to a variety of RNIs, including nitrous acid, dinitrogen tetroxide, and peroxynitrite. Acidified nitrite is toxic to a number of skin organisms, including *P. acnes, Staph. aureus, Can. albicans*, and several dermatophytes. Nitric oxide, an important

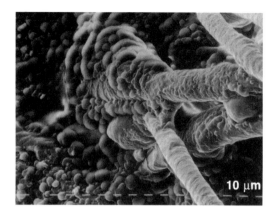

Figure 2.6 A scanning electron micrograph of the surface of sheep skin after fixation and freeze-drying showing a hair pore sealed by layers of hardened lipid. Reproduced with permission from: *The skin microflora and microbial skin disease.* Noble, W.C. (ed). 1992. Cambridge: Cambridge University Press.

regulator of numerous mammalian physiological processes, is continually released from the epidermis. As well as its regulatory activities, the gas is also a potent antimicrobial agent and is effective against *Ps. aeruginosa, Staph. aureus, Salmonella* spp., *Strep. agalactiae, K. pneumoniae, E. coli*, and *Mycob. tuberculosis*.

2.3 | Environmental determinants at different regions of the skin

More appears to be known about the environment of the skin than almost any other anatomical region. This is mainly because the skin is far more accessible than most other regions of the body – this means that measurements can be made and samples obtained not only very easily but, just as importantly, with a minimum of discomfort and embarrassment to the individual. As has been described in the previous section, the surface of the skin consists predominantly of dead, keratinised cells. The gaps between these cells are "sealed" by a water-proof, lipid-containing material secreted by the sebaceous glands (Figure 2.5). Depending on the particular anatomical location, hairs may also be present, the gap surrounding the hair in the follicular canal again being sealed by lipid-rich material (Figure 2.6). Periodically, again depending on the anatomical site, the surface will be pitted by the openings of sebaceous and eccrine glands. While this generalised description of the skin surface applies to many regions, the environmental conditions operating at this surface will depend on the particular anatomical site. The main factors affecting microbial survival and growth on the skin are shown in Table 2.3. Before each factor is discussed, however, it is worth pointing out two characteristics of the skin that exert a broad selecting effect on the type of organisms that can survive on its surface. Firstly, the skin surface is, in general, a

Table 2.3.	Main factors influencing microbial survival and growth on the skin

temperature
moisture content
pH
oxygen and carbon-dioxide concentration
light
nutrient availability
interactions (beneficial and antagonistic) with other microbes
host defence systems

Table 2.4. Survival times of Gram-negative bacteria on the surface of human skin

Organism	Survival time (minutes)
E. coli	<10
Salmonella typhi	<10
Salmonella enteritidis	<10
Serratia marcescens	10–20
Ps. aeruginosa	10–20

Note: The hands of volunteers were submerged in a broth culture of each test organism and then swabbed at various time intervals to determine the number of surviving organisms.

relatively dry environment and, secondly, any fluids on its surface have a relatively high osmotic pressure. Both of these factors favour the survival of Gram-positive bacteria and tend to exclude Gram-negative species. Many staphylococci, for example, can withstand high osmotic pressures, which accounts, in part, for their prevalence on the skin surface. In contrast, studies have shown that many Gram-negative species from other body sites, as well as exogenous organisms, are unable to survive for long on the skin surface (Table 2.4).

While the temperature of the internal organs is maintained at a constant $37°C$, that of the skin is invariably lower than this and also varies with the anatomical location (Figure 2.7). Hence, the axillae and groin tend to have the highest temperatures, whereas

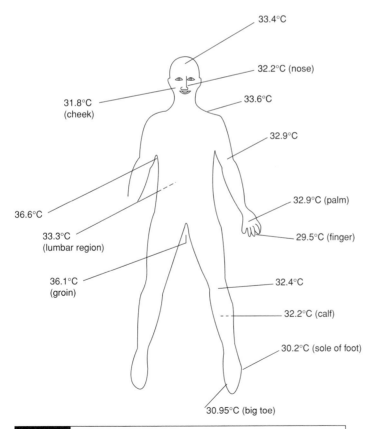

33.4°C
32.2°C (nose)
31.8°C (cheek)
33.6°C
32.9°C
32.9°C (palm)
36.6°C
33.3°C (lumbar region)
29.5°C (finger)
36.1°C (groin)
32.4°C
32.2°C (calf)
30.2°C (sole of foot)
30.95°C (big toe)

Figure 2.7 The temperature of various regions of the skin surface in adults.

the extremities (particularly the toes and fingers) can be up to 7°C lower than this. The temperature of the skin is also affected by that of the environment, by exercise, and by clothing. Variations in temperature of a few degrees centigrade from one site to another are, in general, unlikely to exert a dramatic selection pressure on the cutaneous microbiota because most microbes colonising humans can grow well over the range of temperatures encountered. However, such temperature variation may affect the microbiota of a site indirectly. Hence, those regions that have a high temperature will experience increased sweat production, which will create an environment with not only a higher moisture content, but also with a different range of nutrients and/or antimicrobial compounds, different osmolarity, and different pH. Such changes will certainly affect the composition of the microbial community able to exist at the site. Certain pathogenic bacteria and fungi have optimum growth temperatures considerably lower than those of most microbes colonising humans, and the existence and survival of such organisms on the skin will be affected by its temperature. Examples of such organisms include *Mycobacterium marinum, Mycob. haemophilum, Mycob. leprae*, and many dermatophytic fungi.

The availability of moisture at a given region of the skin has a dramatic effect on the numbers and types of microbes present and, in fact, is one of the main factors governing the distribution of the skin microbiota. The water content of the stratum corneum is generally low – approximately 15% by weight. However, sweat production can increase the moisture content at the surface, and which can be even greater in occluded regions from which sweat cannot easily evaporate. Hence, moist regions such as the axillae and toe webs have much greater population densities than exposed, dry areas like the palms of the hands. The moisture content of an environment is expressed in terms of the water activity (a_w) – this is related to the relative humidity, osmotic pressure, and temperature of the environment (Section 1.2.2). Most bacteria can grow only in an environment with an a_w above 0.950. However, staphylococci and some micrococci can survive and grow in environments with an a_w as low as 0.86 and 0.90, respectively. Corynebacteria, Gram-negative bacteria, and fungi all prefer moist environments and are less resistant to dessication than staphylococci and micrococci. The main sources of water at the surface of the skin are the atmosphere, the sudoriferous glands, and transepidermal water loss (TEWL) (i.e., the diffusion of water from the blood to the skin surface). However, the relative contribution of each of these to the water content of a particular site varies, depending on the atmospheric conditions, the density of sudoriferous glands, and the extent of any physical exertion being undertaken. Furthermore, TEWL displays a circadian rhythm, being at a maximum in the evening and at a minimum in the morning. Skin squames are very hygroscopic and can absorb 3–4 times their own weight of water from the atmosphere. The quantity of water absorbed and, consequently, the relative humidity of the skin microenvironment are, therefore, likely to be affected by the prevailing atmospheric conditions – particularly the humidity and temperature. The amount of water produced by the sudoriferous glands, as well as by TEWL, depends on the atmospheric conditions and the temperature and degree of physical exertion of the individual. Evaporation of the aqueous component of sweat will deposit on the skin surface those solutes originally present in the sweat. Hence, the concentration of NaCl, for example, will increase dramatically in regions with a high density of eccrine glands because the NaCl content of sweat can be as high as 100 mM.

The stratum corneum is exposed to the atmosphere and so microbes on its surface have ready access to oxygen and carbon dioxide. Supplies of these gases are also available by diffusion from capillaries in the dermis and dissolved in glandular secretions. While predominantly an aerobic environment, reduced levels of oxygen exist in certain micro-environments of the skin. Hence, within spatially restricted regions such as hair follicles and in the depths of the stratum corneum, host-cell respiration will deplete the local oxygen concentration and increase the levels of carbon dioxide. This, together with aerobic respiration by resident microbes, could provide microaerophilic and/or anaerobic regions suitable for the growth of microaerophiles and obligate anaerobes. Regions of low oxygen tension will also be present within the microcolonies found on the skin surface (Section 1.1.2). Propionibacteria, which are microaerophilic, are major colonisers of hair follicles, although they are also present on the skin surface. Obligate anaerobes such as *Clostridium* spp., *Bacteroides* spp., and *Peptostreptococcus* spp., although regularly isolated from certain skin regions (particularly the perineum), are transients rather than permanent skin residents.

The main environmental determinants operating at the various regions of the body are basically the same (e.g., pH, nutrient content, oxygen content, etc.) (Section 1.2). However, two exceptions to this generalisation are the skin and the eyes, because in these regions there is the possibility of an additional selecting factor – light. Sunlight can exert an antimicrobial effect in three main ways. First of all, light in the ultraviolet region (290–400 nm) can kill microbes directly by damaging DNA. In this respect, light with a wavelength of 290–320 nm (i.e., ultraviolet B) is more potent than ultraviolet A (320–400 nm). Certain compounds, particularly carotenoids, can protect against the harmful effects of ultraviolet light, and it may be that carotenoid-containing organisms will be at an advantage on those regions of the skin that are continually exposed to sunlight. The only evidence in support of this comes from a single study, which showed that frequent sunbathers had significantly higher proportions of carotenoid-containing bacteria on the forehead, arms, and back than individuals who were infrequent sunbathers. Light with a longer wavelength does not exert a direct antimicrobial effect, but can be absorbed by certain compounds (known as photosensitisers), which release singlet oxygen and free radicals that can kill microbes – a process known as lethal photosensitisation. Examples of photosensitisers include porphyrins and cyanins, which are present in a wide range of microbes. Propionibacteria, in particular, contain porphyrins which could, potentially, be activated by sunlight resulting in lethal photosensitisation, and this has been demonstrated in the laboratory. However, this would appear not to affect bacterial populations on the skin because regions which receive high light doses (e.g., the face) are heavily colonised by propionibacteria. Finally, sunlight stimulates the synthesis of previtamin D_3 in the skin, and this has been shown to exert an antibacterial effect against certain bacteria, including *Mycob. tuberculosis*. While it has been shown *in vitro* that many skin microbes are susceptible to killing by ultraviolet light and by lethal photosensitisation, there is little evidence (other than the study referred to herein) to suggest that these effects operate *in vivo*.

The acidic nature of the skin surface has been recognised since 1928 when it was picturesquely termed the "acid mantle" and recognised as being an important anti-microbial defence mechanism. As might be expected, the skin pH is affected by a number of factors, including the anatomical location, and Figure 2.8 shows the pH at

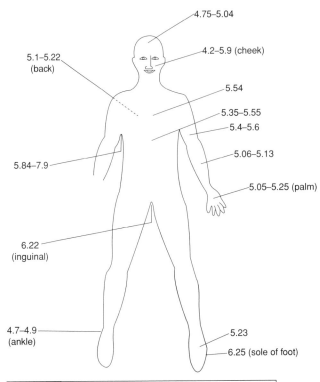

Figure 2.8 pH of various regions of the skin surface in adults.

different locations in adults. Although the surface of the skin is acidic, the pH of the lower layers of the epidermis containing living cells is actually much higher and can reach 7.4. The change in pH is not regular across the whole of the epidermis and, in fact, there is little change in pH until the stratum granulosum is reached, after which point the pH rises steadily. The exact causes of this pH gradient and the low pH of the skin surface are unknown. A number of acidic compounds have been detected on the skin surface, including (1) lactic acid and amino acids (from sweat and as a result of microbial activities), (2) fatty acids (from sebum), and (3) a group of compounds derived from the degradation of filaggrin (a protein abundantly present in keratinocytes) during the keratinisation process – urocanic acid (UCA), pyrrolidone carboxylic acid, and amino acids. UCA is also a major contributor to the acidity of the outer layers of the stratum corneum. The maintenance of a low pH within the epidermis is important to provide the optimal conditions for those enzymes involved in keratinisation and in the formation of the barrier lipids (ceramides) and their arrangement into lamellar structures. Many factors influence the pH of the skin, and these are listed in Table 2.5.

The main sources of nutrients for members of the cutaneous microbiota are usually considered to be (1) compounds present in sweat and sebum, (2) compounds released by dead or dying keratinocytes and microbes, and (3) compounds secreted by viable keratinocytes and microbes. However, because the epidermis is not permeated by a vascular network, resident keratinocytes obtain their nutrients by diffusion from interstitial fluid – the latter may, therefore, act as an additional source of nutrients for microbes on the skin surface. There is, therefore, a wide range of nutrients present on and in the stratum corneum and hair follicles to enable growth of resident organisms. Some

Table 2.5.	Factors that affect the surface pH of skin
Factor	**Effect on pH**
age	generally higher in the elderly; approximately neutral in neonates
circadian rhythm	maximum in early afternoon and minimum at night
anatomical site	higher values found in intertriginous areas (e.g., axillae, groin, toe webs)
sex	conflicting data – probably little difference.
race	African-Americans have lower pH than Caucasians

of the most abundant sources of nutrients are in the form of polymers, including proteins (e.g., keratin and collagen), lipids, DNA, and RNA. The stratum corneum, in fact, consists mainly of protein (70%) and lipids (15%), with the remainder being predominantly water, together with small quantities of other substances. Little information is available regarding the presence of polysaccharides in the epidermis, but the surface of keratinocytes has been reported to be coated with polysaccharide-containing material, and glycogen is present at low concentrations (<0.5 μg/mg wet weight), although higher concentrations are found in eccrine sweat glands and in hair follicles.

Although sweat consists primarily of water (99–99.5% by weight), it contains a very wide range of solutes. Table 2.6 lists the approximate concentrations of the various sweat components, but it is important to realise that these vary markedly with the rate of sweat production. For example, at the maximum sweat rate, the NaCl content can be 10-fold higher than that found at normal sweat rates. Furthermore, the concentration of the various solutes varies markedly with age, sex, and the anatomical location of the sweat glands. The main components of sweat are urea, proteins, lactic acid, and a wide range of amino acids. Vitamins reported to be present include thiamine, riboflavin, nicotinic acid, pantothemic acid, pyridoxine, inositol, p-aminobenzoic acid, and ascorbic acid. The pH of sweat varies with its rate of production and ranges from 5.0 at low sweat rates to 6.5–7.0 at the highest rates. The average adult has between 2 and 4 million eccrine glands, and the total mass of these is approximately 100 g, which is about the same as a kidney. The rate of sweat production varies enormously, depending on the temperature and humidity of the external environment and on the physical exertion being undertaken. It can be as high as 2–3 litres per hour for short periods.

Table 2.6.	Substances present in human sweat			
>99 μg/ml	10–99 μg/ml	1.0–9.9 μg/ml	0.1–0.99 μg/ml	<0.1 μg/ml
urea	glucose	uric acid	bromide	iodine
proteins	pyruvic acid	creatinine	fluoride	magnesium
lactic acid	urocanic acid	potassium	calcium	manganese
serine	ammonia	sodium	iron	cadmium
ornithine	sulphate	copper	zinc	lead
glycine	phosphate	several amino acids	cystine	nickel
	chloride			acetylcholine
	many amino acids			vitamins

Note: Reproduced with permission from *The skin microflora and microbial skin disease*. Noble, W.C. (ed). 1992. Cambridge: Cambridge University Press.

Table 2.7. | Composition of sebum from different skin locations

Compound	Back	Scalp	Face
glycerides	46.4	31.7	35.4
free fatty acids	16.0	29.6	27.2
wax esters	21.5	20.2	22.6
squalene	11.4	12.8	11.6
sterols and sterol esters	4.7	5.7	3.2

However, the maximum daily output is estimated to be approximately 15 litres, with most individuals in temperate climates producing approximately 1 litre per day.

Little information is available concerning the composition of apocrine secretions because the duct of the aprocrine gland opens into the follicular canal and so the secretions are difficult to distinguish from those of the sebaceous gland. Nevertheless, it is thought that apocrine secretions contain mucoproteins, proteins, carbohydrates, lipids, and ammonia. Because of the restricted distribution of apocrine glands, their secretions will be able to serve as nutrients only to microbes in the axillae and on the perineum, genitalia, and nipples.

Sebum is the lipid-rich product of the sebaceous glands. These glands are distributed over most of the body surface, but their highest density is on the face, forehead, and scalp. They are also numerous on the lower back and the chest. The composition of sebum varies with the anatomical location of the sebaceous glands (Table 2.7) and also with the age and sex of the individual. However, it consists mainly of glycerides, free fatty acids, wax esters, and squalene. It also contains vitamin E. Sebum is not the only source of lipids on the skin surface; some are produced by keratinocytes and are also released from them when they die. The composition of the lipid fraction present in the stratum corneum of forearm skin is shown in Table 2.8. Of these, the polar lipids and glycosphingolipids are produced almost exclusively by the keratinocytes. The skin also contains a range of trace elements, including chromium (1.2 $\mu g/g$), molybdenum (1.7 $\mu g/g$), and selenium (1.2 $\mu g/g$), which are used by many bacteria and fungi.

Table 2.8. | Composition of lipids in the stratum corneum of the forearm

Compound	Proportion (%)	Concentration in extracellular space (mg/ml)
free fatty acids	25.0	52.0
ceramides	21.0	44.0
free sterols	20.0	42.0
triglycerides	12.0	25.0
glycosphingolipids	6.3	13.0
unidentified	5.0	10.0
polar lipids	3.8	7.9
n-alkanes	3.1	6.5
squalene	1.9	4.0
sterol esters	1.3	2.7

Note: Reproduced with permission from: *In vitro* and *in vivo* antistaphylococcal activity of human stratum corneum lipids. Miller. S.J., Aly, R., Shinefeld, H.R., and Elias, P.M. *Archives of Dermatology* 1988;124:209–215.

| Table 2.9. | Extracellular enzymes produced by members of the cutaneous microbiota |

Microbes known to produce enzyme	Extracellular enzyme			
	Lipase	Protease	keratinase	DNAse
Staph. epidermidis	+	+	+	+
Staph. haemolyticus	+	NA	+	+
Staph. hominis	+	NA	+	NA
M. luteus	NA	+	+	NA
M. lylae	NA	+	+	NA
P. avidum	+	+	NA	+
P. granulosum	+	NA	NA	+
P. acnes	+	+	NA	+
Corynebacterium spp.	+	−	NA	+
Brevibacterium spp.	−	+	+	+
Dermabacter hominis	NA	+	NA	+
Malassezia spp.	+	+	−	NA
Acinetobacter spp.	+	+	NA	NA

Note: + = produced by some species or strains; − = not produced by most species or strains; NA = data not available.

Many members of the cutaneous microbiota produce extracellular enzymes able to hydrolyse host macromolecules to low molecular mass compounds that can be transported into the cell to serve as nutrients (Table 2.9). The abundance of such enzymes will, therefore, ensure a plentiful supply of amino acids (from keratin and other proteins), purines, pyrimidines, phosphate and sugars (from DNA and RNA), glycerol and fatty acids (from lipids), and phosphate (from phosphorylated compounds). The skin, therefore, can supply, either directly or indirectly, a wide range of nutrients for the growth of its resident microbiota; examples of the most important of these are given in Table 2.10.

An additional source of nutrients on the skin surface is, of course, the microbiota itself, and some examples of possible nutritional interactions between cutaneous microbes are given in Section 2.4.4.

Major environmental determinants also include the antimicrobial host defences (described in Section 2.2), as well as the antagonistic activities of members of the skin microbiota. Examples of the latter are given in Section 2.4.4.

2.4 | The indigenous microbiota of the skin

2.4.1 Main characteristics of key members of the cutaneous microbiota

The organisms most frequently present on the skin surface are those belonging to the genera *Corynebacterium, Staphylococcus, Propionibacterium, Micrococcus, Malassezia, Brevibacterium, Dermabacter,* and *Acinetobacter.*

2.4.1.1 *Corynebacterium* spp.

There is ample scope for confusion regarding this genus, especially when the older literature is consulted. Studies of the cutaneous microbiota often make reference to diphtheroids, large-colony diphtheroids, small-colony diphtheroids, coryneforms,

Table 2.10. Major sources of the nutrients required by members of the cutaneous microbiota

Class of nutrient	Compound	Source	Microbes known to utilise/require this nutrient
carbon/energy source	glucose	sweat; hydrolysis of glycoproteins	staphylococci, micrococci, propionibacteria, corynebacteria, brevibacteria
	ribose	hydrolysis of DNA and RNA	some staphylococci, some micrococci, some propionibacteria, *Acinetobacter* spp.
	glycerol	lipid hydrolysis	staphylococci, micrococci, propionibacteria
	amino acids	protein hydrolysis; sweat	staphylococci, micrococci, lipophilic coryneforms, some propionibacteria, *Acinetobacter* spp.
	fatty acids	lipid hydrolysis; sweat	some staphylococci, propionibacteria, *Acinetobacter* spp., brevibacteria, micrococci, lipophilic coryneforms, *Malassezia* spp.
	lactic acid	sweat	staphylococci, micrococci, *Acinetobacter* spp.
nitrogen source	NH_4^+	sweat	staphylococci, micrococci, aerobic coryneforms, *Malassezia* spp., *Acinetobacter* spp.
	amino acids	protein hydrolysis, sweat	staphylococci, some micrococci, propionibacteria, *Malassezia* spp., *Acinetobacter* spp.
	urea	sweat	*Corynebacterium* spp., staphylococci, *Brevibacterium* spp.
essential amino acids	amino acids	protein hydrolysis sweat, cerumen	staphylococci, propionibacteria, *Malassezia* spp.
phosphorus source	phosphate	hydrolysis of DNA and RNA, sweat	all organisms
vitamins	biotin	sweat	most staphylococci, micrococci, propionibacteria, some aerobic coryneforms
	thiamine	sweat	staphylococci, micrococci, propionibacteria, some aerobic coryneforms

Table 2.11. Cutaneous *Corynebacterium* spp. and some of their main characteristics

| | Main characteristics | | | | |
Species	Atmospheric requirements	Lipid requirement	Oxidative or fermentative	Acid from glucose	Growth in 6.5% NaCl
C. amycolatum	facultative	no	fermentative	yes	yes
C. jeikeium	obligate aerobe	yes	oxidative	yes	no
C. minutissimum	facultative	no	fermentative	yes	yes
C. striatum	facultative	no	fermentative	yes	yes
C. urealyticum	obligate aerobe	yes	oxidative	no	yes
C. xerosis	facultative	no	fermentative	yes	yes
CLC group	facultative	yes	oxidative	no	yes

Note: CLC = cutaneous lipophilic corynebacteria.

lipohilic coryneforms, coryneform bacteria, aerobic corynebacteria, anaerobic corynebacteria, corynebacteria, and *Corynebacterium* spp. without defining exactly what is meant by the term being used. The term "coryneform" as now used refers to any non-acid-fast, non-branching, non-sporing, pleomorphic, Gram-positive rod whether aerobic or anaerobic. The term is synonymous with "diphtheroid" because the previous description also applies to the important pathogen, *Corynebacterium diphtheriae*. In the past, any "coryneform" found on the skin was regarded as being a member of the genus *Corynebacterium*, but now four genera of cutaneous bacteria (as well as many non-cutaneous genera) with the aforementioned morphology are recognized: *Corynebacterium*, *Propionibacterium*, *Brevibacterium*, and *Dermabacter*. More than twenty other genera of "coryneforms" are recognized and, although these do not reside on the skin, some are colonisers of other anatomical regions.

The defining characteristics of members of the genus *Corynebacterium*, in addition to the morphological features listed previously, are (1) G+C content of 46–74 mol%; (2) presence of meso-diaminopimelic acid, short-chain (C_{22}–C_{36}) mycolic acids, and dehydrogenated menaquinones with 8 or 9 isoprene units; (3) the cell wall contains arabinogalactan; (4) catalase-positive; (5) non-motile; and (6) aerobes or facultative anaerobes. Although there are at least fifty-nine species within the genus, only six species and a group of organisms known as cutaneous lipophilic corynebacteria (CLC – for which the designation "*C. lipophilicus*" has been suggested) are regularly detected on skin (Table 2.11). The cutaneous *Corynebacterium* spp. vary with regard to their atmospheric requirements. *C. jeikeium* and *C. urealyticum* are obligate aerobes and their ability to colonise the hair follicles may be impaired by the low oxygen content of this region. The remaining cutaneous species are facultative anaerobes and, provided that other requirements are satisfied, would be able to colonise any of the skin regions. Apart from *C. jeikeium*, they are all halotolerant and can grow at NaCl concentrations as high as 10% and are able to colonise regions of the skin with a high salt content (e.g., occluded regions such as the axillae and toe webs). They can grow over a wide temperature range (15–40°C), but grow optimally at 37°C.

Corynebacterium spp. (apart from *C. urealyticum*) can use carbohydrates and amino acids as sources of carbon and energy. Some amino acids are also essential growth requirements and are incorporated into cellular protein. Glucose and a range of amino acids are present in sweat, and the many proteolytic enzymes produced by the cutaneous microbiota liberate amino acids from skin proteins. Unlike the other cutaneous species, the lipophilic corynebacteria require a number of vitamins, including riboflavin, nicotinamide, thiamine, pantothenate, and biotin. Many of these are present in sweat, and others are liberated by dying keratinocytes.

Corynebacteria adhere well to human epidermal keratinocytes *in vitro*, and bacterial densities range from 4 to 25 per epithelial cell, depending on the particular species. *C. minutissimum* adheres in greater numbers than other isolates. The presence of mannose, galactose, or *N*-acetylglucosamine reduces the number of adherent organisms considerably, implying that molecules containing these sugars are the main receptors for the bacteria. Fibronectin may also function as a receptor because the presence of this protein significantly reduces bacterial adhesion. Corynebacteria are also able to adhere to a wide range of lipids known to be secreted in sebum, so it is possible that adhesion occurs via these lipids. Little is known regarding the adhesins responsible for mediating attachment of *Corynebacterium* spp. to the skin.

C. minutissimum causes erythrasma (Section 2.5.8), while *C. jeikeium* causes a wide range of serious infections in immunocompromised individuals and those with medical devices. *C. urealyticum* is responsible for urinary tract infections in individuals who have been in hospital for a long period, are immunocompromised, have been catheterised, or are elderly. *C. xerosis* is associated with axillary odour. Little is known regarding the virulence factors of any of these organisms.

2.4.1.2 *Propionibacterium* spp.

Propionibacteria are non-motile, non-sporing, fermentative, Gram-positive bacilli. The G+C content of their DNA is 57–68 mol%. Their exact atmospheric requirements remain a matter of controversy, and they are often described as obligate anaerobes or as microaerophiles. This may explain, in part, their preference for hair follicles where the oxygen levels are likely to be reduced. However, they are also found on the skin surface where, presumably, oxygen utilisation by aerobes (e.g., micrococci, brevibacteria, and some *Corynebacterium* spp.) and facultative anaerobes (staphylococci and some *Corynebacterium* spp.) will help to provide an oxygen-depleted environment. Four species of the genus are found on the skin, and three of these (*P. acnes, P. avidum*, and *P. granulosum*) are catalase-positive, while the remaining species (*P. propionicum*) is catalase-negative. *P. acnes* is the most prevalent member of the genus on human skin and inhabits predominantly the hair follicles where it can be present at a concentration of 10^5 cfu/follicle. *P. granulosum* prefers a similar habitat to *P. acnes*, whereas *P. avidum* frequents regions that are rich in eccrine sweat glands and *P. propionicum* is found mainly on the eyelids.

Surprisingly little is known with regard to the means by which propionibacteria adhere to skin. A cutaneous isolate of *P. granulosum* is known to bind to lactosylceramide, but not to glycolipids that do not contain lactose. *P. acnes* can adhere to oleic acid, which is an important constituent of sebum, and this fatty acid also promotes co-aggregation of the organism. Such co-aggregation, and the association of the resulting microcolony with the oleic acid in the follicle, would help to physically maintain the organism within this habitat. Microcolony formation would also aid in establishing – at least within the centre of such aggregates – the microaerophilic environment needed by the organism. The organism can also bind to fibronectin, and this may be important in mediating attachment during medical-device-associated infections.

Adhesion of the organism to human epidermal keratinocytes has been studied *in vitro*, and the mean number of bacteria adhering per keratinocyte was found to range from 0.1 to 5, depending on the particular strain.

Propionibacteria can grow over a wide range of pHs (4.5–8.0), but their optimum pH for growth is 5.5–6.0. With respect to pH, therefore, they are ideally suited to grow in most skin regions. With regard to carbon and energy sources, propionibacteria can utilise free fatty acids and glycerol (but this is a poor carbon and energy source compared with glucose), as well as sugars for these purposes, and the main end-products of glucose metabolism are propionic and acetic acids. They produce a number of lipases that can liberate fatty acids from the lipids present in sebaceous glands and those secreted onto the skin surface in sebum and sweat. *P. acnes* and *P. avidum* also produce proteases that can liberate arginine from skin proteins, and this amino acid can be used as a carbon and energy source. RNA can also be used by propionibacteria as a carbon and energy source. Propionibacteria vary with regard to their amino-acid requirements for growth, *P. acnes* appears to be the least fastidious and can grow in media containing only eight amino acids. The proteases produced by most of the propionibacteria will generate the amino acids necessary for growth, and sweat, of course, also contains a range of these compounds (Table 2.6). As well as being incorporated into cellular proteins, they can also be used as a source of nitrogen, carbon, and energy. Propionibacteria require a number of vitamins for growth, including biotin, nicotinamide, pantothenate, and thiamine. Biotin and thiamine are known to be present in sweat (Table 2.6), as are several other vitamins. Propionibacteria can inhibit the growth of a number of organisms due to their production of propionic acid, bacteriolytic enzymes, and bacteriocins.

P. acnes is the causative agent of a number of infections, including acne, endocarditis, endophthalmitis, osteomyelitis, and infections of implanted devices and wounds. A number of putative virulence factors have been identified in the organism, including proteinases, hyaluronidase, lipase, phospholipase C, sialidase (neuraminidase), and acid phosphatase. The organism can also induce the release of a number of inflammatory cytokines, including interleukin (IL)-1β, IL-6, IL-8, tumour necrosis factor-α (TNFα) (from macrophages and monocytes), and IL-1β (from keratinocytes). The factor responsible for the induction of cytokine release from macrophages and monocytes is thought to be a peptidoglycan-polysaccharide complex. It is also able to subvert certain host-defence mechanisms (e.g., it can resist killing by phagocytes and can survive in macrophages).

P. granulosum is associated with a similar range of infections as *P. acnes*, while *P. propionicum* causes mainly canaliculitis and dacryocystitis.

2.4.1.3 *Staphylococcus* spp.

Staphylococci are non-motile, non-sporing, facultatively anaerobic Gram-positive cocci (0.5–1.5 μm in diameter) which occur singly, in pairs, or in clusters. They have a G+C content of 30–39 mol%. The genus consists of thirty-five species, of which nearly half can be found on human skin. They are catalase-positive, but most do not have an oxidase and they can ferment a number of sugars to produce acid – mainly lactic acid. A major division of the genus is made on the basis of the production of the enzyme coagulase – strains that produce coagulase are recognised as belonging to the species *Staph. aureus*, *Staph. intermedius*, or *Staph. delphni*, while those that do not

Table 2.12. | Distribution of *Staphylococcus* spp. on the skin surface

Region	Principal species	Other species found in high proportions
scalp	*Staph. capitis*	*Staph. epidermidis*
scalp of pre-adolescent children	*Staph. epidermidis* *Staph. hominis*	*Staph. haemolyticus*
forehead	*Staph. capitis*	*Staph. epidermidis* *Staph. hominis* *Staph. haemolyticus*
cheek	*Staph. epidermidis*	*Staph. capitis* *Staph. hominis*
external auditory canal	*Staph. auricularis* *Staph. epidermidis*	*Staph. capitis*
auricle	*Staph. capitis*	*Staph. epidermidis*
axillae	*Staph. epidermidis*	*Staph. saprophyticus* *Staph. haemolyticus* *Staph. hominis*
perineum	*Staph. epidermidis* *Staph. hominis*	*Staph. haemolyticus* *Staph. saprophyticus*
arms	*Staph. epidermidis* *Staph. hominis*	*Staph. haemolyticus*
legs	*Staph. epidermidis* *Staph. hominis*	*Staph. haemolyticus*
toe webs	*Staph. epidermidis*	*Staph. haemolyticus* *Staph. cohnii* *Staph. hominis*

are known collectively as coagulase-negative staphylococci (CNS). The main habitats of *Staph. aureus* are the anterior nares; this organism is described in greater detail in Chapter 4. In this chapter, only the CNS is described further. CNS produce a range of hydrolytic enzymes, including proteases, lipases, keratinases, and nucleases, which enables them to utilise a variety of host polymers as nutrients. They can grow over the temperature range 10–45°C, but growth is most rapid between 30°C and 37°C. They can also grow over a wide pH range (pH 4.0–9.0), although optimum growth occurs between pH 7.0–7.5. They are halotolerant organisms – all species can grow at 10% NaCl, and many can grow at concentrations as high as 15% – a property which enables them to grow on the skin surface. Furthermore, they can grow in environments with low water activities, which also contributes to their ability to grow on skin. Although most of the CNS can be isolated from almost any skin region, individual species generally show a predilection for a particular site. The distribution of the main CNS inhabiting the skin is summarised in Table 2.12. Strains of *Staph. epidermidis* exhibit marked differences in their ability to adhere to skin. In a study of twenty-eight strains of the organism, the number of bacteria adhering to each epidermal cell in intact human skin sections ranged from three to sixteen. Three patterns of attachment were apparent: (1) bacteria adhered to the surfaces of superficial skin cells, but not to deeper cells; (2) adhesion was uniform throughout the section; and (3) bacteria adhered only to the cell–cell junctions of superficial and deeper cells (Figure 2.9). This implies the existence of three different receptors for the organism, but the nature of these, and of the adhesins to which they bind, remains unknown. *Staph. epidermidis* has a number of

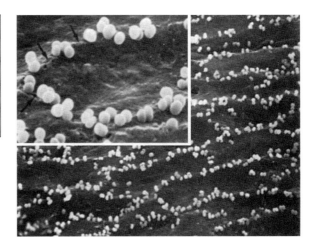

Figure 2.9 Scanning electron micrograph showing the preferential adhesion of some strains of *Staph. epidermidis* to cell–cell junctions. The arrows in the inset indicate the cell–cell junctions. Reprinted with permission of Blackwell Publishing from: The adhesion of coagulase negative staphylococci to human skin and its relevance to the bacterial flora of milk. Brooker, B.E. and Fuller, R. *Journal of Applied Bacteriology* 1984;57:325–332.

adhesins (Table 2.13) and their involvement in the pathogenesis of device-associated infections is described in Section 2.5.2. The role that any of these play in other infections due to *Staph. epidermidis* is not known. Potential virulence factors of *Staph. epidermidis* include a number which helps it to subvert the host defence systems (Table 2.14). Hence, it produces a cysteine protease which can inactivate secretory IgA and IgM. The enzyme can also degrade fibrinogen and fibronectin, thus enabling it to spread through tissues. The extracellular slime produced by the organism is able to decrease phagocytosis by polymorphonuclear leucocytes (PMNs) and macrophages, and also interferes with the chemotactic response of PMNs. It can also survive in peritoneal macrophages. One of the main defence mechanisms of skin is the presence of antimicrobial fatty acids on its surface. *Staph. epidermidis* produces a fatty acid-modifying enzyme (FAME) that esterifies fatty acids to cholesterol which renders them non-toxic. Eighty-eight percent of *Staph. epidermidis* strains have been found to produce FAME. A metalloprotease is also secreted which degrades elastin, thereby contributing to its ability to invade host tissue. Compared with *Staph. aureus*, the organism is not highly toxigenic and appears to produce only one toxin, δ-toxin. This toxin is able to lyse red blood cells, but is also a constituent of an inflammatory polypeptide complex (phenol-soluble modulin – PSM) secreted by the organism. PSM can induce the release of TNFα, IL-1β, and IL-6 from monocytes. The toxin is produced by most strains of *Staph. epidermidis* and also by *Staph. saprophyticus* and *Staph. haemolyticus*. The peptidoglycan and lipoteichoic acid of the organism can also induce the release of TNFα, IL-1β, and IL-6 from monocytes. The overproduction of these cytokines prolongs and exacerbates the inflammatory

Table 2.13. Adhesins of *Staphylococcus epidermidis* and their receptors

Adhesin	Receptor
autolysin	vitronectin
fibrinogen-binding protein	fibrinogen
accumulation-associated protein	other *Staph. epidermidis* cells
polysaccharide intercellular adhesion	other *Staph. epidermidis* cells, red blood cells
lipase	collagen

Table 2.14. | Virulence factors (other than adhesins) of *Staphylococcus epidermidis*

Putative virulence factor	Activity
cysteine protease	degrades human sIgA, IgM, serum albumin, fibrinogen, and fibronectin
metalloprotease	degrades elastin and casein
fatty-acid–modifying enzyme	esterifies fatty acids to cholesterol, thereby inactivating their antibacterial activity
δ-toxin	lyses red blood cells; forms part of an inflammatory polypeptide complex (phenol-soluble modulin)
extracellular slime	decreases phagocytosis by PMNs and macrophages; interferes with chemotaxis by PMNs

Note: PMNs = polymorphonuclear leukocytes.

response, which ultimately results in tissue damage. *Staph. epidermidis* is able to produce a number of lantibiotics, including epidermin, epilancin K7, and Pep5. These very stable antibacterial peptides with broad-spectrum activities may confer a competitive advantage over other organisms seeking to become established on the skin surface.

As well as being a member of the cutaneous microbiota (particularly on the feet), *Staph. saprophyticus* is regularly present in the rectum and in the vagina. It is an important uropathogen, and its role in urinary tract infections is described in Sections 2.5.5 and 7.5.7.1. It is able to bind to exfoliated uroepithelial cells, to the human urethra in organ culture, and to epithelial cell lines; a number of adhesins have been identified which may mediate such attachment. One of these, Aas, is a 160-kDa autolysin that binds to fibronectin on the surface of uroepithelial cells (Figure 2.10). Maximum expression of the protein (which also has haemagglutinating properties) occurs under anaerobic conditions. The amino-acid sequence of Aas is homologous to the autolysins Atl and AtlE from *Staph. aureus* and *Staph. epidermidis*, respectively. The surface-associated fibrillar protein, Ssp, is another adhesin that enables binding of the organism to epithelial cells. This protein is present as fibrils on the surface of the cell and tends to form clumps. There is also some evidence that the lipoteichoic acid functions as an adhesin. Once attached to the uroepithelium, the organism produces a number of exoenzymes, including an elastase, a urease, and a lipase. The products released by the actions of

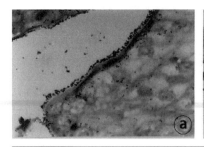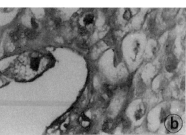

Figure 2.10 Electron micrographs showing: (a) binding of *Staph. saprophyticus* to uroepithelial cells and (b) absence of adherent bacteria induced by pre-treatment of the uroepithelial cells with an antibody to the haemagglutinin of the organism. Reproduced with permission from: Meyer, H.W., *et al. Infection and Immunity* 1996;64:3893–3896.

these enzymes cause severe damage to the epithelium. The urease is probably the most important of these virulence factors because mutants lacking the enzyme exhibit reduced virulence in animal models. The organism also produces substances that are mitogenic for T cells.

2.4.1.4 *Micrococcus* spp.

Although the older literature recognises at least ten species of micrococci, the genus has now been split into six genera: *Micrococcus* (*M. luteus*, *M. lylae*, and *M. antarcticus*), *Kocuria* (containing the former *M. roseus*, *M. varians*, and *M. kristinae*), *Kytococcus* (containing the former *M. sedentarius*), *Nesterenkonia* (containing the former *M. halobius*), *Dermacoccus* (containing the former *M. nishinomiyaensis*), and *Arthrobacter* (containing the former *M. agilis*). Of these, *M. luteus*, *M. lylae*, and *Kocuria varians* are the most frequently encountered on human skin.

Micrococci are morphologically similar to staphylococci – the main distinguishing features being that micrococci are obligate aerobes, they produce oxidase, and they have a G+C content of 66–75 mol%. They are halotolerant (up to 7.5% NaCl), which contributes to their ability to survive on skin. They can grow over the temperature range of 25–37°C. They can use carbohydrates and amino acids as carbon and energy sources, and most strains produce proteases and keratinases and so can utilise these skin polymers as nutrient sources. As well as being able to utilise glucose and glycerol as carbon sources, they can also use propionate, lactate, and acetate for this purpose, which means that they can make use of the metabolic end-products of other organisms as nutrients. Most species require some amino acids as growth factors – commonly arginine, cysteine, methionine, and tyrosine. Vitamin requirements vary with the species. Micrococci rarely cause infections in humans.

The genus *Kocuria* belongs to the family *Micrococcaceae* and consists of non-motile, non-sporing, aerobic Gram-positive cocci. The G+C content of their DNA is 66–75 mol%. Two of the five species belonging to this genus, *Kocuria varians* and *Kocuria kristinae*, have been isolated from the skin of healthy humans.

2.4.1.5 *Malassezia* spp.

This is a genus of lipophilic, dimorphic yeasts which normally inhabits the skin of humans and other warm-blooded animals. Until 1995, there was considerable confusion regarding the taxonomy and nomenclature of this genus, and the yeast and mycelial forms were regarded as different organisms and placed in different genera. The yeast forms were considered to be members of the genus *Pityrosporum*, while the mycelial forms were placed in the genus *Malassezia*. Since 1996, seven different species have been recognised within the genus, all of which can be found on the skin of healthy individuals, and these are: *Mal. furfur*, *Mal. pachydermatis*, *Mal. sympodialis*, *Mal. globosa*, *Mal. slooffiae*, *Mal. restricta*, and *Mal. obtusa*. In culture and on healthy skin, *Malassezia* spp. are usually present in the yeast form, which consists of round, oval, or cylindrical cells (Figure 2.11). They reproduce by budding, which leaves a distinctive scar on the mother cell following separation of the daughter cell. The pseudomycelial form is often present in diseased skin, but can also be seen in culture. The cell wall is much thicker than that found in other yeasts, is multi-layered, and consists mainly of mannoproteins, lipids, and chitin. The organisms are difficult to grow, and this has hindered studies of their physiology.

Figure 2.11 Scanning electron micrograph of a *Malassezia* sp. isolated from human skin. Reproduced with kind permission of Kluwer Academic Publishers from: The human cutaneous microbiota and factors controlling colonisation. Bojar, R.A. and Holland, K.T. *World Journal of Microbiology & Biotechnology* 2002;18:889–903, Figure 4.

Malassezia spp. can grow under aerobic, microaerophilic, and anaerobic atmospheres, although they are usually cultured aerobically. Provided that other conditions are suitable, they can therefore colonise any region of the skin. They are unable to ferment sugars, but can use lipids as the sole source of carbon and energy. All species, except *Mal. pachydermatis*, require preformed fatty acids for growth – these must have carbon-chain lengths greater than 10 and are incorporated directly into cellular lipids. These compounds may be obtained directly from sebum or as products of the hydrolysis of lipids (present in sebum as well as produced by, and derived from, keratinocytes), brought about by their own lipases or those secreted by most other cutaneous organisms (staphylococci, propionibacteria, and *Corynebacterium* spp). The fatty acid and lipid requirements of these yeasts would explain, in part, their preferential colonisation of skin regions that are rich in sebaceous glands. Ammonium ions and amino acids are used as nitrogen sources. Both of these are present in sweat, while ammonium ions are also produced from the urea present in sweat by the action of ureases from *Corynebacterium* spp., staphylococci, and *Brevibacterium* spp. Amino acids are also liberated from skin proteins by the action of the numerous proteases produced by most skin organisms, including *Malassezia* spp. Methionine is the preferred source of sulphur for these organisms, but they can also utilise cysteine – both of these amino acids are found in sweat and would be present in protein hydrolysates. The organisms do not appear to require any vitamins for growth. Nothing is known about their salt tolerance. *Malassezia* spp. are found on the skin of 75–98% of the population and in >90% of adults. Although the organism is present on infants, colonisation increases after puberty when the sebaceous glands become active and the concentration of skin lipids increases. Although they may be isolated from most parts of the body, they are present in higher numbers on the face, scalp, chest, and upper back due to the presence of numerous sebaceous glands in these regions. They inhabit the stratum corneum and hair follicles, where they obtain the lipids essential for their growth. Healthy individuals are usually colonised by between one and three *Malassezia* spp. – the most common being *Mal. restricta, Mal. globosa*, and *Mal. sympodialis. Mal. furfur* and *Mal. sloffiae* can also be detected on the skin of healthy individuals, but less frequently.

A number of lipases, a lipoxygenase, and a phospholipase are secreted by these organisms, and great interest has been shown in the latter as a possible virulence factor. The phospholipase releases arachidonic acid from epithelial cells, and the metabolites of this compound can induce inflammation of the skin. This could be the means by which the organism plays a role in a number of inflammatory skin conditions, including atopic dermatitis, seborrhoeic dermatitis, and folliculitis (Section 2.5.7). One of the metabolic products, azelaic acid, has antibacterial and antifungal activities, and can also decrease the production of reactive oxygen species in neutrophils. The yeast is also able to induce the release of the pro-inflammatory cytokines IL-6, IL-1β, and TNFα from human peripheral blood mononuclear cells. Interestingly, at higher concentrations, the yeast is able to inhibit cytokine release from these cells.

2.4.1.6 *Acinetobacter* spp.

These are aerobic Gram-negative coccobacilli which are widely distributed in the environment and are frequently found on human skin. Until the late 1980s, only one species was recognised in the genus, and this was further classified into two subspecies: *Acin. calcoaceticus* var. *anitratus* and *Acin. calcoaceticus* var. *lwoffii*. Now, however, at least nineteen species have been distinguished on the basis of DNA–DNA hybridisation tests. The most frequently isolated species from skin are *Acin. lwoffii* and *Acin. johnsonii*. Other species, especially those involved in nosocomial infections such as *Acin. baumannii*, are rarely found on the skin. *Acinetobacter* spp. have a G+C content of between 39 and 47 mol%, are obligate aerobes, non-motile, catalase-positive, and oxidase-negative. Because of their strict requirement for oxygen, colonisation of oxygen-depleted regions, such as the deeper regions of hair follicles, is restricted. They are, therefore, found primarily on the skin surface. The organisms grow over the temperature range of 20–42°C, with optimum growth occurring at 33–35°C so that their growth rate will generally be optimal at many skin regions. *Acinetobacter* spp. can utilise a wide range of organic compounds as carbon and energy sources. However, glucose generally cannot be used for this purpose, and it is thought that the main carbon and energy sources utilised are organic acids (e.g., acetate, lactate, and pyruvate), amino acids, and alcohols. Lactate and pyruvate are present in sweat, while acetate and lactate are the metabolic end-products of many cutaneous microbes. Amino acids are present in sweat and are liberated from proteins by microbial proteases. The organisms can grow on ammonium ions as the sole nitrogen source, and these are plentiful in the cutaneous environment. They do not require any vitamins for growth. Their salt tolerance varies with many strains able to grow in 6% NaCl.

The organisms are found on the skin of approximately 25% of the population, and they are particularly prevalent in moist regions such as the axillae, perineum, and toe webs. *Acin. lwoffii* has occasionally been isolated from the blood of patients suffering from vascular catheter-associated infections, but, in general, the cutaneous species of the genus rarely cause infections.

2.4.1.7 *Brevibacterium* spp.

This genus consists of obligately aerobic, non-motile, catalase-positive, Gram-positive bacilli. They are oxidative, tolerate high salt concentrations (up to 15%), and have an optimum growth temperature of between 30°C and 37°C. Their G+C content is 64 mol%, and they contain mesodiaminopimelic acid. At least seven species are recognised, of which the following are regularly present on the skin: *Brev. epidermidis*, *Brev. otitidis*, *Brev. mcbrellneri*, and *Brev. casei*. The fact that they are obligate aerobes would limit their ability to colonise hair follicles where oxygen concentrations are depleted. A range of organic compounds can be used as carbon and energy sources, including glucose, acetate, and lactate. Glucose is present in sweat, while acetate and lactate are the end-products of metabolism of many skin inhabitants, including propionibacteria and staphylococci. While amino acids can also be used as carbon and energy sources, some are essential growth requirements and are incorporated into cellular protein. Sweat contains a range of amino acids and the numerous proteolytic enzymes produced by *Brevibacterium* spp. and other members of the cutaneous microbiota liberate amino acids from skin proteins.

All four cutaneous species produce a range of potent proteolytic enzymes which liberate volatile sulphur compounds from proteins, and these are thought to be responsible for the characteristic cheese-like odour of their cultures. The production of such compounds is also thought to contribute to foot odour. Interestingly, it is known that some malarial mosquitos preferentially bite around the feet and ankles and are also known to be attracted to Limburger cheese. Their attraction to human feet and ankles may be due to colonisation of these regions by odour-producing brevibacteria. *Brevibacterium* spp. also produce keratinases.

Brevibacteria are occasionally responsible for a number of infections of humans, including osteomyelitis, peritonitis, and septicemia. Their most significant virulence factor is considered to be protease production.

2.4.1.8 *Dermabacter hominis*

This species is the sole member of the genus *Dermabacter*. It is a non-motile, catalase-positive, oxidase-negative, Gram-positive bacillus which ferments glucose, lactose, sucrose, and maltose; hydrolyses aesculin; and decarboxylates lysine and ornithine. Acetate and lactate are the main end-products of glucose metabolism. It has a G+C content of 62 mol%, contains mesodiaminopimelic acid, but does not contain mycolic acids. To date, there have been no reports of its involvement in any infectious disease. The organism produces proteases, DNases, and an amylase, and the action of these enzymes on skin macromolecules would provide a range of potential nutrients for itself and other organisms. Furthermore, the acetate and lactate it produces from glucose could also act as nutrient sources for other skin microbes, such as *Acinetobacter* spp.

2.4.2 Acquisition of the cutaneous microbiota

Whereas the skin of babies delivered by caesarean section is sterile, that of babies born by vaginal delivery is colonised by organisms from the birth canal. At the time of birth, the skin of neonates is very different from that of adults because it is coated with a lipid-rich material known as the vernix caseosa and is thinner, less hairy, and has fewer sudoriferous and sebaceous glands than that of adults. It also has a much higher pH (7.4) than the skin of adults. Following delivery, the neonate is exposed to potential microbial colonisers from a variety of sources, including its mother, other individuals present at the birth, and the environment. Even when only the mother is considered, the range of potential colonising microbes is very wide as a result of maternal handling, caressing, and kissing of the baby and so will include organisms present on various regions of the skin (hands, breasts, arms, face), as well as those in the oral cavity. In a study of ninety-four neonates sampled within 2.5 hours of birth, no microbes were cultivated from between 6% and 13% of individuals, depending on the sampling

Table 2.15. Frequency of colonisation of various skin sites in 94 neonates

Organism	Sampling site				
	Gabella	Umbilicus	Axilla	Groin	Scapula
Staph. epidermidis	68	51	73	65	63
Bac. subtilis	63	53	49	47	67
E. coli	1	4	11	10	0
Enterobacter spp.	2	4	4	5	0
Proteus spp.	2	5	5	7	2
Staph. aureus	0	0	0	0	0
No growth	9	11	6	7	13

Note: The sites were sampled within 2.5 hours of birth. Figures indicate the % of neonates colonised by the organism.

site (Table 2.15). The most frequently isolated organisms were *Staph. epidermidis* and *Bacillus subtilis* (Table 2.15). Many other studies have also found that CNS predominate at a number of skin sites in neonates immediately after birth. What is striking is that, given the huge numbers of lactobacilli in the birth canal, very few studies have reported the presence of lactobacilli on neonatal skin. Interestingly, while diphtheroids may be present in small numbers during the first 2–3 hours following birth, their numbers then increase dramatically over the next 12 hours at all skin sites, so they are frequently detected in studies involving neonates more than a few hours old.

In a study of the development of the skin microbiota of neonates during their first 6 weeks, it was found that those sites that were colonised shortly after birth were only sparsely populated (Table 2.16). Hence, 2 hours after birth, the total number of organisms present ranged from 36 to 51 cfu/cm^2. Within 24 hours, the axillae and groin were heavily colonised, while the scalp had a lower microbial density. The degree of colonisation increased steadily until after 6 weeks when the microbiotas were similar to those found in comparable sites of adults. The most frequently isolated organisms on all occasions were the CNS. Full identification of the CNS found on the skin of neonates has revealed that the predominant species are *Staph. epidermidis*, *Staph. haemolyticus*, and *Staph. hominis*, regardless of the skin site. Furthermore, these species continue to predominate during at least the first 8 months of life.

Although many of the organisms colonising the skin of neonates are derived from the mother's birth canal, skin, and saliva, the frequent isolation of organisms such as *Bac. subtilis* (see Table 2.15) implies that colonisation of the skin by environmental organisms is a common event. Furthermore, studies have shown that not all of the

Table 2.16. Density of microbial colonisation of various skin sites in neonates

Time after birth	Microbial density (cfu/cm^2) on:		
	Axilla	Groin	Scalp
24 hours	10^3	10^3	540
2 days	10^4	10^4	525
5 days	10^4	10^4	2.7×10^3
6 weeks	9.8×10^4	3.2×10^5	1.8×10^5

CNS present on the neonate are derived from the mother, suggesting that colonisation by organisms from other individuals also takes place. Immediately after birth, when the cutaneous microbiota is very sparse, neonates are, therefore, in danger of becoming colonised by pathogenic organisms – especially if birth has taken place in a hospital. Premature neonates who are hospitalised in a neonatal intensive care unit represent a population who, in comparison with healthy full-term neonates, have minimal contact with their mother and other family members, but maximum contact with hospital staff and medical equipment – both of which are potential sources of pathogenic and/or antibiotic-resistant organisms. In a study of the skin microbiota (eleven different sites) of ten such neonates (aged 4–5 days), the mean density of colonisation ranged from 10^3 to 10^6 cfu/cm^2, with the umbilicus, skinfolds, and armpits being the most densely colonised. As in healthy full-term neonates, CNS were found to be the predominant organism isolated from each of the sites. Other organisms found included *Staph. aureus*, anaerobes (*Bacteroides* spp., *Propionibacterium* spp., *Peptostreptococcus* spp.), streptococci, corynebacteria, and Gram-negative bacilli. Antibiograms of the 256 staphylococci isolated revealed that many were resistant to a wide range of antibiotics, including penicillin G (96% resistant), erythromycin (52%), methicillin (31%), gentamicin (28%), lincomycin (27%), and cotrimoxazole (21%). The most resistant species were *Staph. epidermidis* and *Staph. haemolyticus*, with many of the former being multi-resistant. Such high levels of antibiotic resistance, together with the frequent occurrence of multi-resistant strains, reflect the exposure of these neonates to the hospital environment. Data obtained in this study, however, represent only a "snapshot" of the skin microbiota on that particular sampling occasion. A longitudinal study involving repeated sampling of hospitalised premature neonates over more than a month revealed some very interesting findings. Firstly, the density of colonisation of the various sites was approximately 100-fold lower than that found in healthy full-term neonates. Secondly, although CNS dominated the skin microbiota, a variety of species (a total of 13) was present, as well as other organisms, including *Micrococcus* spp., aerobic coryneforms, *Propionibacterium* spp., streptococci, *Bacillus* spp., *Malassezia* spp., *Klebsiella* spp., and *E. coli*. Unlike the situation in infants and adults, the numbers and relative proportions varied markedly from day to day. Thirdly, many of the isolated organisms were antibiotic-resistant, but the proportion of such organisms varied from day to day and did not correlate with antibiotic usage. These findings suggest that the skin microbiota of hospitalised neonates consists of an unstable, complex, transient microbiota derived mainly from hospital staff and is liable to change each time the neonate is handled.

2.4.3 Community composition at different sites

In general, the skin provides a relatively inhospitable environment for microbial growth in that it is quite dry with little free water being available (in contrast to the moist mucosa of other body sites); has a low pH and a high osmolality; is exposed to the external environment and consequently is subject to fluctuations in temperature, ion concentrations, osmolality, and water content; and is subjected to radiation and mechanical stress. For these reasons, the cutaneous microbiota is dominated by Gram-positive species which, mainly because of their cell-wall structure, are more able to withstand these conditions and the fluctuations that occur in them.

Before describing the microbiotas of specific skin regions, some general comments are worth making with regard to the types of organisms found on the skin surface and within the hair follicles (i.e., the two main structures that are available for microbial colonisation). Among the most frequently isolated bacteria from both of these regions are CNS. Although at least eighteen different species of CNS have been detected, the most common isolate is *Staph. epidermidis*, which usually constitutes approximately 50% of the staphylococci present on the skin. Other species frequently isolated are *Staph. hominis, Staph. haemolyticus, Staph. capitis,* and *Staph. warneri. Staph. aureus* is not regularly isolated from most regions of the skin and cannot grow and replicate in the skin environment; therefore, it is regarded as a transient. Aerobic coryneforms (i.e., *Corynebacterium* spp. and *Brevibacterium* spp.) are major inhabitants of the skin surface rather than the hair follicles and are particularly prevalent in moist intertriginous regions. Propionibacteria are found mainly in hair follicles, but are also present on the skin surface; they dominate the microbiota of sebum-rich areas of skin on the head, chest, and back – the predominant species being *P. acnes*. Micrococci, particularly *M. luteus*, are regarded as members of the indigenous cutaneous microbiota and are found mainly on the skin surface. Gram-negative bacteria are not usually present on the skin, with the exception of *Acinetobacter* spp., which may be found on the skin surface, particularly in moist intertriginous areas. The cutaneous microbiota includes fungi belonging to the genus *Malassezia*, which inhabit the hair follicles, as well as the skin surface. Because the skin is exposed to the external environment and is breached by openings (e.g., mouth, nose, rectum, urethra) into anatomical regions with their own microbiotas, a variety of organisms from such sites may be found on the skin surface. If such organisms are unable to grow and reproduce under the environmental conditions provided by the skin and cannot permanently colonise this organ (or a region of it), they are known as "transients". Well-known transients on skin include *Staph. aureus*, faecal organisms (particularly prevalent on the perineum), and environmental organisms such as *Bacillus* spp. and *Ps. aeruginosa*.

A number of techniques have been used to establish the exact location of bacteria on and within the epidermis. Hence, successive stripping of the skin with adhesive tape, scanning electron microscopy, and microscopy of sections through the epidermis have shown that most organisms are present in the outermost layers of the stratum corneum and in the hair follicles, with the latter being the major site of colonisation. Bacteria have been found within the stratum corneum up to a depth of six cell layers, but this may not be the limit of penetration. Most organisms within or on the stratum corneum appear to be present as microcolonies, which are usually no more than two cells thick. The number of cells in such colonies varies with the site and the individual, and range from less than 10 to 10^5 (Figure 2.12). Not all skin squames are colonised by microbes and, although the follicles constitute a major site of microbial colonisation, less than 50% of normal follicles contain microbes. The sudoriferous glands appear not to be colonised by microbes. In a study of follicles isolated from skin biopsies, three main groups of organisms were detected – propionibacteria, staphylococci, and *Malassezia* spp. – and the proportions of isolated follicles colonised by these organisms were 12%, 4%, and 13%, respectively. The dominant organisms in the follicles were propionibacteria – the mean viable count of bacteria in the colonised follicles was 2.6×10^5, 5.5×10^3, and 10^2 cfu for propionibacteria, staphylococci, and *Malassezia* spp., respectively. The distribution of two

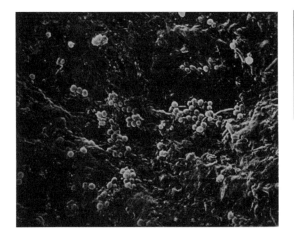

Figure 2.12 Scanning electron micrograph showing microcolonies of bacteria on the surface of the skin of a human foot. Reprinted with the permission of Blackwell Publishing from: The demonstration of bacteria on and within the stratum corneum using scanning electron microscopy. Malcolm, S.A. and Hughes, T.C. *British Journal of Dermatology* 1980;102:267–275.

of the predominant groups of skin organisms – staphylococci and propionibacteria – within pilosebaceous follicles has been studied in samples of thoracic skin taken from cadavers. Although this showed that the distribution varies substantially, propionibacteria were, in general, present in a narrow band within the follicle, whereas staphylococci were more broadly distributed. The propionibacterial "band" occurred at different depths within different follicles.

The composition and population density of the cutaneous microbiota vary markedly between different anatomical sites and, in the case of a particular site, between individuals. Both the diversity and population density at a particular site are, of course, governed by the environmental determinants operating at the site. Despite the large number of environmental variables to be considered, the population density and composition of the microbiota at a site are affected mainly by the number and density of sebaceous and sudoriferous glands at that site (Table 2.17). This is because (1) these glands are important sources of microbial nutrients; (2) the sudoriferous glands are a major source of free water; (3) the secretions of both glands contain antimicrobial substances; (4) temperature regulation can result in the production of large quantities of sweat; and (5) sebum produced by the sebaceous glands contributes to the "cement" between squames and is involved in controlling transepidermal water loss and, therefore, water availability on the surface as well as microbial penetration into the epidermis. Variations in the number and density of these glands, therefore, will affect many of the key environmental determinants at the skin surface, including temperature, water content, range and concentration of nutrients, osmolality, pH, and the range and concentration of antimicrobial substances. Occlusion (i.e., covering) of sites, whether this is anatomical (e.g., the axillae, submammary regions) or due to clothing, will also affect the local environment because it will hinder evaporation of water, encourage the accumulation of secretions, alter pH, and so forth. Some broad patterns emerge related to these factors, shown in Table 2.17. The actual population density of the cutaneous microbiota varies markedly with the anatomical location and among individuals when the same site is considered. Hence, from Figure 2.13, it can be seen that the number of propionibacteria present on the cheek in a group of 761 individuals ranged from <10 per cm^2 to >10,000,000 per cm^2. Such variations are not confined to this particular anatomical site nor to these particular organisms, but are typical of any skin site and any organism studied.

	Important environmental	
Region	determinants	Effect on microbiota
head	many sebaceous and sudoriferous glands	high population density; dominated by propionibacteria; few corynebacteria
axillae	many sebaceous and sudoriferous glands; partially occluded, therefore increased moisture, temperature, and pH	high microbial density; higher numbers of moisture-requiring corynebacteria, fungi, and *Acinetobacter* spp.
perineum	partially occluded, therefore increased moisture and temperature	high microbial density; higher numbers of moisture-requiring corynebacteria, fungi, and *Acinetobacter* spp.
toe webs	partially occluded, therefore increased moisture and temperature	high microbial density; higher numbers of moisture-requiring corynebacteria, fungi, brevibacteria, and *Acinetobacter* spp.
arms and legs	few sebaceous glands; no sudoriferous glands; relatively dry regions	low microbial density; mainly staphylococci and micrococci; very few fungi
hands	no sebaceous glands; exposed area, therefore low water content	mainly staphylococci; few fungi, corynebacteria, or propionibacteria

Table 2.17. General patterns of colonisation by cutaneous microbes related to the distribution of sebaceous and sudoriferous glands, moisture content, and temperature of the region

Because of space restrictions, it is not possible to describe the microbiota of each of the many different regions of the skin. Nevertheless, those described include regions from each of the main types of cutaneous environments present in humans: oily, moist, and dry. Oily regions include the head, neck, upper back, and trunk. Moist regions include the axilla, groin, perineum, and toe interspace. Dry regions include

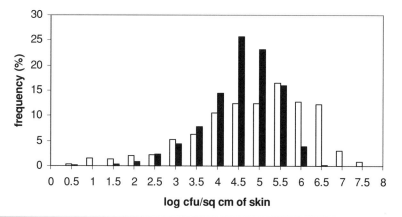

Figure 2.13 Population distribution of *Propionibacterium* spp. (open bars) and *Micrococcus* spp. (black bars) on the cheek of 761 individuals. Reproduced with kind permission of Kluwer Academic Publishers from: The human cutaneous microbiota and factors controlling colonisation. Bojar, R.A. and Holland, K.T. *World Journal of Microbiology & Biotechnology* 2002;18:889–903, Figure 7.

Table 2.18. Prevalence and population density of microbes on the scalp

Organism	Frequency of isolation (%)	Population density (cfu/cm^2)
P. acnes	100	2.2. $\times$ 10^5
staphylococci	100	2.0 $\times$ 10^5
CLC group	27	4.2 $\times$ 10^4
P. granulosum	29	1.8 $\times$ 10^4
C. minutissimum	7	4.2 $\times$ 10^3
Malassezia spp.	100	1.4 $\times$ 10^3
P. avidum	0	0
C. xerosis	0	0
Brev. epidermidis	0	0

Notes: Data are based on studies carried out by different investigators using different numbers of subjects and so must be regarded as only an approximate guide. All studies were carried out on adults and, generally, involved both males and females. The absence of an organism from the table does not necessarily mean that it is not a member of the scalp microbiota. CLC = cutaneous lipophilic corynebacteria.
Frequency of isolation = percentage of the study population from which the organism was isolated.

the arms and legs. It must be cautioned that few studies have involved a qualitative and quantitative analyses of all of the microbes present at a particular skin site – this would represent an enormous undertaking. Many studies, however, have determined the prevalence of a particular organism (or group of organisms) in a population and this has, on occasion, been supplemented by data concerning the population density of the organism under investigation. In many of the tables that follow, the data presented have been obtained from several studies, and these will often have been carried out using different sampling procedures, population groups, and identification schemes. The data must, therefore, be regarded as indicative of trends rather than representing precise values suitable for detailed comparison. Furthermore, the absence of a particular organism from a table does not necessarily mean that the organism was not present at that site; it is more likely that it was not looked for in the investigations that have been used to compile the table.

2.4.3.1 Scalp

The scalp has a high density of sebaceous glands and, in those who are not bald, the abundant hair covering traps moisture and a layer of air resulting in a relatively warm and moist environment, compared with more exposed, hairless regions of the skin such as the forehead. The microbiota of this region is dominated by propionibacteria, particularly P. acnes, and staphylococci (Table 2.18). The predominant staphylococcus is Staph. capitis, which usually constitutes at least 80% of the staphylococci on the scalp. Malassezia spp. are invariably present, but generally at lower densities.

2.4.3.2 Forehead

The important features of the environment of the skin of the forehead are that it has high densities of both sebaceous and eccrine glands, has a very acidic pH (one of the most acidic regions of the skin), and is an exposed area and thus has a more variable temperature than many other regions. The microbiota of the forehead is dominated by

| Table 2.19. | Prevalence and population density of microbes on the forehead |

Organism	Frequency of isolation (%)	Population density (cfu/cm^2)
P. acnes	100	5.0×10^5
staphylococci	100	4.5×10^4
P. granulosum	29	1.2×10^4
P. avidum	0	0
CLC group	27	1.6×10^4
C. minutissimum	7	1.6×10^2
C. xerosis	0	0
Brev. epidermidis	0	0
Malassezia spp.	88	4.8×10^3
Acin. lwoffii	53	ND

Notes: Data are based on studies carried out by different investigators using different numbers of subjects and so must be regarded as only an approximate guide. All studies were carried out on adults and, generally, involved both males and females. The absence of an organism from the table does not necessarily mean that it is not a member of the microbiota of the forehead. CLC = cutaneous lipophilic corynebacteria; ND = not determined. Frequency of isolation = proportion of the study population from which the organism was isolated.

propionibacteria, particularly *P. acnes* (Table 2.19). CNS are invariably present, the dominant species of which are *Staph. capitis, Staph. epidermidis*, and *Staph. hominis. Malassezia* spp. are also usually present. Other organisms frequently present are *M. luteus* and *M. lylae*. The most frequently isolated coryneforms are members of the CLC group.

2.4.3.3 Toe interspace
This is an occluded region and is characterised by a relatively high moisture content, temperature, and pH. Sebaceous glands, hair follicles, and apocrine glands are absent, although eccrine glands are present. The high moisture content of the region favours colonisation by coryneforms and Gram-negative bacteria. In a study involving 60 healthy adults, aerobic coryneforms and CNS were isolated from all subjects, and these two microbial groups constituted the numerically dominant members of the microbiota of the site (Table 2.20). The main aerobic coryneforms found in this region are cutaneous lipophilic corynebacteria, *C. minutissimum*, and *Brev. epiderrmidis*. Other organisms frequently isolated are anaerobic coryneforms, micrococci, Gram-negative bacteria, *Malassezia* spp., and anaerobic Gram-positive cocci. However, all of these organisms are present in much lower numbers than the aerobic coryneforms and CNS.

2.4.3.4 Perineum
The perineum is an occluded region and has a plentiful supply of apocrine, eccrine, and sebaceous glands. Furthermore, contamination by faeces and/or urine (particularly in infants) may supplement the usual sources of nutrients available to microbes inhabiting the region. It is, therefore, a warm, moist, nutrient-rich environment and supports a dense microbiota. Because of its proximity to the rectum, members of the intestinal microbiota can frequently be recovered. In females, the proximity of the vagina also means that lactobacilli are also often present. The perineal microbiota, therefore, tends to have a greater species diversity than other skin regions, although

Table 2.20. | Prevalence and population density of microbes present in the fourth toe cleft of 60 adults

Organism	Frequency of isolation (%)	Population density (cfu/cm^2)
aerobic coryneforms	100	6.0×10^6
CNS	100	1.4×10^6
Staph. epidermidis	71	5.1×10^5
Staph. haemolyticus	42	2.1×10^5
Staph. cohnii	25	3.7×10^5
Staph. hominis	25	2.8×10^5
Staph. warneri	25	1.7×10^5
anaerobic coryneforms	77	90
Micrococcus spp.	36	1.3×10^4
Malassezia spp.	32	20
anaerobic Gram-positive cocci	27	1.9×10^3
Gram-negative organisms	25	500

Notes: Reprinted with the permission of Blackwell Publishing from: The cutaneous microbiology of normal human feet. Marshall, J., Leeming, J.P., and Holland, K.T. *Journal of Applied Bacteriology* 1987; 62;139–146. CNS = coagulase-negative staphylococci.
Frequency of isolation = proportion of the study population from which the organism was isolated.

many of the organisms isolated are likely to be transients rather than residents. Aerobic coryneforms dominate the perineal microbiota (Table 2.21). The predominant CNS are *Staph. epidermidis* and *Staph. hominis*. *Staph. aureus* is also frequently isolated from this region.

2.4.3.5 Axillae

Axillae are occluded regions with abundant hair follicles, as well as eccrine and apocrine glands – they are, therefore, warm and moist environments. Furthermore, there is a plentiful supply of sebum from the numerous sebaceous glands. An axilla, therefore, is a densely colonised area and is, in fact, one of the most densely populated regions of the skin. In a study of healthy adult males and females, members of the *Micrococcaceae* were present in all subjects, and coryneforms were also frequently present. These two groups of organisms also comprised the greatest proportions of the axillary microbiota (Table 2.22). The total counts were monitored on a daily basis for 5 days and were found to be stable. The number of organisms on individual hairs was also determined and found to be very low, with an average of 327 cfu of aerobes and 4 cfu of anaerobes per hair. Identification of members of the *Micrococcaceae* revealed that most (51%) were *Staph. epidermidis* or *Staph. saprophyticus* (29%), while the remainder consisted of *Staph. aureus* (10%) and micrococci (10%). Large-colony diphtheroids are those coryneform species that are facultatively anaerobic and do not require lipids for growth (e.g., *C. minutissimum*, *C. xerosis*, and *Brevibacterium* spp.), and this group was often detected and comprised a substantial proportion of the microbiota. The Gram-negative rods isolated consisted of *E. coli* (36%), *Klebsiella* spp. (25%), *Proteus* spp. (18%), *Enterobacter* spp. (13%), and *Acinetobacter* spp. (8%). The axillae are, therefore, important sites of carriage of *Staph. aureus* and are among the few cutaneous sites frequently colonised by Gram-negative rods.

Table 2.21. Prevalence and population density of microbes on the perineum

Organism	Frequency of isolation (%)	Population density (cfu/cm^2)
P. acnes	62	7.8×10^3
P. granulosum	9	4.5×10^3
P. avidum	44	2.6×10^3
CLC group	100	4.2×10^3
C. minutissimum	63	2.3×10^7
C. jeikeium	17	4.5×10^6
C. xerosis	47	3.9×10^6
Brev. epidermidis	40	1.9×10^6
Malassezia spp.	69	ND
Acin. lwoffii	3	ND
Staph. aureus	62	ND
coagulase-negative staphylococci	58	ND
streptococci	30	ND
E. coli	33	ND

Notes: Data are based on studies carried out by different investigators using different numbers of subjects and so must be regarded as only an approximate guide. All studies were carried out on adults and generally involved both males and females. The absence of an organism from the table does not necessarily mean that it is not a member of the perineal microbiota. CLC = cutaneous lipophilic corynebacteria; ND = not determined.
Frequency of isolation = proportion of the study population from which the organism was isolated.

Interestingly, as well as finding that coryneforms and Gram-positive cocci are the dominant axillary organisms in healthy adult populations, many studies have also reported that the majority of individuals fall into one of two groups – one in which the axillary microbiota is dominated by coryneforms and one in which Gram-positive cocci

Table 2.22. Axillary microbiota in adult males (128) and females (77)

Organism	Males Frequency of isolation Population density		Females Frequency of isolation Population density	
	%	cfu/cm^2	%	cfu/cm^2
total aerobes	100	6.9×10^5	100	8.9×10^5
Micrococcaceae	100	1.2×10^5	100	3.6×10^5
lipophilic diphtheroids	85	2.5×10^5	66	2.3×10^5
large-colony diphtheroids	26	2.7×10^4	25	3.7×10^4
GNR	20	2.3×10^3	19	2.1×10^3
propionibacteria	70	5.1×10^3	47	1.7×10^4
P. acnes	47	7.2×10^3	30	1.8×10^4
P. avidum	34	4.2×10^3	21	1.5×10^4
P. granulosum	8	4.1×10^3	5	4.5×10^3

Notes: GNR = Gram-negative rods.
Frequency of isolation = proportion of the study population from which the organism was isolated.

Table 2.23. Prevalence and population density of microbes present on the soles of the feet of 60 adults

Organism	Frequency of isolation (%)	Population density (cfu/cm^2)
CNS	100	3.0×10^5
Staph. epidermidis	59	9.3×10^4
Staph. hominis	54	3.9×10^4
Staph. haemolyticus	48	6.9×10^4
Staph. cohnii	36	8.3×10^4
Staph. warneri	20	4.2×10^4
aerobic coryneforms	97	1.6×10^4
Micrococcus spp.	78	1.3×10^3
anaerobic coryneforms	47	20
Aerococcus spp.	19	7.8×10^2
Gram-negative organisms	14	1.1×10^2
Malassezia spp.	7	5

Notes: Reprinted with the permission of Blackwell Publishing from: The cutaneous microbiology of normal human feet. Marshall, J., Leeming, J.P., and Holland, K.T. *Journal of Applied Bacteriology* 1987;62:139–146. CNS = coagulase-negative staphylococci.
Frequency of isolation = proportion of the study population from which the organism was isolated.

predominate. Hence, in a study of the axillary microbiota of 285 adults, coryneforms were dominant in 47%, whereas Gram-positive cocci were dominant in 43% of individuals. The remaining 10% had approximately equal numbers of each group of organisms. Furthermore, those dominated by coryneforms had a much greater population density (1.26×10^6 cfu/cm^2) than those dominated by cocci (0.24×10^6/cm^2). Eighty-three percent of the coryneforms were *Corynebacterium* spp., 5% were *Brevibacterium* spp., and 12% were other coryneforms. It has been suggested that different rates of eccrine sweating may account for these differences, with coryneforms preferring an environment with a high moisture content. Interestingly, individuals with a coryneform-dominated axillary microbiota are more likely to have underarm odour (Section 2.5.9).

2.4.3.6 Sole of the foot

The sole of the foot differs from many other skin regions in having no hair follicles, no sebaceous glands, and no apocrine glands. It does, however, have a high density of eccrine glands. The environment of this region is markedly altered in a large proportion of the world's population by the wearing of socks and shoes, which confer on this body site the characteristics of an occluded region (i.e., an increased moisture content, temperature, pH, and carbon-dioxide concentration). This results in significant differences between the microbiota of this body site and that of anatomically similar regions, such as the palm of the hand. The microbiota is dominated by CNS and aerobic coryneforms – these being the most frequently isolated microbial groups and comprising the greatest proportions of the cultivable microbiota (Table 2.23). In contrast, organisms characteristic of sebum-rich regions (propionibacteria and *Malassezia* spp.) are less frequently isolated and, when present, comprise a much lower proportion of the microbiota.

Table 2.24.	Composition of the subungual microbiota in 26 adults		
Organism or group	Frequency of isolation (%)	Number of viable bacteria present (cfu)	Most frequently isolated species
coagulase-negative staphylococci	100	$2.5 \times 10^3 - 1.2 \times 10^6$	*Staph. haemolyticus* *Staph. epidermidis*
yeasts and other fungi	69	$1 - 1.9 \times 10^3$	*Candida parapsilosis* *Rhodotorula rubra*
coryneforms	46	$59 - 4.4 \times 10^5$	lipophilic diphtheroids *C. minutissimum*
Gram-negative bacilli	42	$3 - 2.9 \times 10^4$	*Pseudomonas* spp. *Enterobacter* spp.
Staph. aureus	8	$6.2 \times 10^3 - 5.5 \times 10^4$	not applicable
other organisms	42	$1 - 891$	not identified

Notes: Data are the mean values for five fingers per subject.
Frequency of isolation = proportion of the study population from which the organism was isolated.

2.4.3.7 Forearm and leg

These are relatively dry regions with very few eccrine or apocrine glands and only a low density of sebaceous glands and, consequently, have a lower density of colonisation than most other skin regions. The population density usually lies between 10^2 and 10^3 cfu/cm^2, and approximately 90% of the cultivable organisms are staphylococci – the rest being aerobic coryneforms and propionibacteria. The predominant staphylococcal species are *Staph. epidermidis* and *Staph. hominis*.

2.4.3.8 Hands

A hand provides a number of different microbial habitats, and these vary because of differences in the distribution of eccrine and sebaceous glands and the presence of occluded regions (e.g., beneath the fingernails). Because of their exposed nature, they are subject to large temperature variations. In a study of the microbes recovered from the whole of the hand, CNS were isolated from all of a group of thirty adult volunteers. Cutaneous lipophilic diphtheroids were the next most frequently isolated (73%), followed by large-colony diphtheroids (i.e., *C. xerosis*, *C. minutissimum*, and *Brev. epidermidis*; 33%), yeasts (27%), facultative Gram-negative bacilli (23%), and *Staph. aureus* (17%). *Staph. epidermidis* was the most frequently isolated CNS, followed by *Staph. hominis*.

The region beneath the fingernails (subungual space) is densely colonised, with one study reporting that, in a group of twenty-six adults, the mean number of viable organisms present was 4.6×10^4 per subungual space. CNS were isolated from all individuals and always constituted the highest proportion of organisms present in each sample (Table 2.24). Because this is an occluded region, the presence of facultatively anaerobic, Gram-negative bacilli and fungi would be expected, and both of these groups of organisms are found in many individuals.

The presence of large numbers of microbes in this region, together with the fact that it is relatively inaccessible to standard hand-washing procedures, suggests that the subungual space could act as a reservoir of organisms that could be transmitted

Table 2.25. Cultivable microbiota of the auditory canal and cerumen from 164 healthy individuals

Organism	Proportion (%) of total cultivable isolates from	
	Canal	Cerumen
Staph. auricularis	21	23
Staph. epidermidis	17	13
Staph. capitis	13	17
Turicella otitidis	12	12
Alloiococcus otitis	7	9
Corynebacterium spp.	5	7
Gram-negative bacteria	5	1
Staph. warneri	4	3
M. luteus	3	1
fungi	3	7
Staph. caprae	2	3
Staph. aureus	2	0.3
Bacillus spp.	2	2
Brevibacterium spp.	1	2

to other individuals and to surfaces, objects, and so forth – this is of particular importance for health-care personnel.

2.4.3.9 Outer ear

The auricle of the ear has a microbiota similar to that of other relatively dry, exposed
skin regions. However, the external auditory canal offers a very different environment
to potential microbial colonisers. The most obvious difference from other skin regions
is the presence of ceruminous glands, which open into the follicular canal. The combined secretions of the sebaceous and ceruminous glands are known as cerumen (ear
wax), which forms a protective, acidic layer lining the surface of the auditory canal.
The cerumen gradually moves towards the external opening of the ear canal and is
sloughed off, carrying with it any attached microbes. As well as containing lipids with
antimicrobial activity, cerumen has a low pH and contains lysozyme and immunoglobulins and is, therefore, capable of killing or inhibiting a number of microbes. However,
it also contains many other compounds which can serve as microbial nutrients – these
include a wide range of amino acids, lipids, and trace elements.

In a study involving 164 individuals, the microbiotas of the cerumen and the auditory canal were found to be very similar with, in each case, staphylococci comprising
60% and 62% of the isolates, respectively (Table 2.25). *Staph. auricularis* was the most
frequently isolated species from both types of specimens, followed by *Staph. epidermidis*
and *Staph. capitis*. The next most frequently isolated microbes were, in both cases, *Turicella otitidis* and *Alloiococcus otitis*. *Tur. otitidis* is the only species of the genus *Turicella*.
The organism is a non-fermentative coryneform that has also been implicated in the
pathogenesis of otitis media. *All. otitis* is an obligately aerobic, Gram-positive coccus
that has also been shown to be one of the causative agents of otitis media.

In a recent investigation involving twenty healthy individuals, a culture-independent
analysis (involving polymerase chain reaction amplification and sequencing of 16S

Table 2.26.	Culture-independent analysis of samples from the auditory canal of 20 adults

Organism	Proportion (%) of total clones
All. otitis	56.71
Tur. otitidis	20.44
Staph. auricularis	9.85
C. auris	3.11
Tilletiaria anomala	3.02
P. acnes	1.63
Corynebacterium spp.	0.79
M. obscurus	0.46
Stephanoascus ciferrii	0.37
Thrypochthonius tectorum	0.37
Staph. epidermidis	0.28
Fusobacterium periodonticum	0.19

rRNA genes) was carried out on samples taken from the auditory canal. A total of 2,150 clones were obtained from the group, and the proportions of each type are shown in Table 2.26. In most individuals, the microbiota was found to be relatively simple, and the sequences present in the greatest proportions were those corresponding to *All. otitis*, *Tur. otitidis*, and *Staph. auricularis*. These sequences were also the most frequently detected in the samples and were found in 85%, 65%, and 60% of individuals, respectively. *All. otitis* and *Tur. otitidis* are both fastidious organisms, which may explain why they appeared to have such low prevalences in the previously described culture-based study (Table 2.25). While the microbiota of the ear canal was found to include organisms such as CNS and *Corynebacterium* spp. that are regularly found in other skin regions, both the molecular- and culture-based studies revealed the presence of two organisms (*All. otitis* and *Tur. otitidis*) that are unique to this region. Unfortunately, little is known of the nutritional or general growth requirements of these organisms, and it is difficult, therefore, to speculate on what elements of the environment of the ear canal are responsible for enabling them to become established as significant members of the microbiota of this region.

2.4.4 Interactions among members of the cutaneous microbiota

Given the diversity of the cutaneous microbiota, the existence of interactions between some of its members would not be surprising. However, although many instances of antagonistic behaviour of one organism towards another have been reported, there are surprisingly few examples of beneficial interactions. This is probably a reflection of our greater interest in finding, and exploiting, antimicrobial agents. Nevertheless, some important beneficial interactions between cutaneous microbes have been reported. Although the skin is an aerobic environment, microaerophilic propionibacteria constitute one of the dominant groups of cutaneous microbes. Their ability to dominate the cutaneous microbiota is dependent, in part, on the presence of aerobic (micrococci, *C. jeikeium* and *C. urealyticum*, *Brevibacterium* spp., and *Acinetobacter* spp.) and facultative (staphylococci, *Dermabacter hominis*, *Corynebacterium* spp., and *Malassezia* spp.) organisms

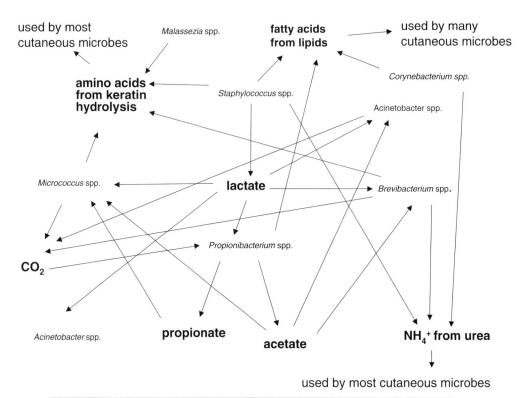

Figure 2.14 Possible nutritional interactions between members of the cutaneous microbiota.

which consume oxygen and create the optimal atmospheric conditions necessary for the growth of propionibacteria. This phenomenon is regularly encountered *in vitro* when samples taken from the skin are incubated aerobically – propionibacteria readily grow on such plates. However, once the propionibacteria have been subcultured to obtain pure cultures, they fail to grow aerobically in the absence of other organisms. When it comes to the provision of nutrients, a number of beneficial interactions can be envisaged, although whether these are examples of commensalism or synergism remain to be established (Figure 2.14). For example, propionibacteria excrete propionic and acetic acids, which can be used as carbon and energy sources by micrococci, *Acinetobacter* spp., and brevibacteria, while lactate produced by staphylococci can serve as a carbon and energy source for *Acinetobacter* spp. and micrococci. Many organisms can benefit from the amino acids produced by the hydrolysis of keratins by staphylococci, micrococci, and *Brevibacterium* spp., and by the hydrolysis of other proteins by numerous cutaneous organisms.

Some of the fatty acids present in skin are inhibitory, while others stimulate growth, and the effect of a particular fatty acid is very much species-dependent. It is possible that one organism can utilise a fatty acid that is inhibiting another and enable the latter to survive and proliferate in what would normally be an inhibitory environment. For example, *P. acnes* is inhibited by lauric acid and is unable to colonise regions containing high concentrations of this fatty acid. However, lauric acid can be utilised by *Malassezia* spp. whose presence, therefore, can enable the survival of *P. acnes*.

In the competition for space and nutrients that occurs between microbes occupying the same habitat, a number of strategies have been developed which enable one type of

Table 2.27.	Antagonistic substances produced by members of the cutaneous microbiota

Antimicrobial compound	Examples
carbon dioxide	produced by many bacteria – can inhibit growth of dermatophytes
lysozyme	produced by staphylococci – kills micrococci, *Brevibacterium* spp., and *Corynebacterium* spp.
proteases	produced by *P. acnes* – kills other *Propionibacterium* spp. and some staphylococci
propionic acid	produced by propionibacteria – inhibits many other species, particularly at the low pHs found on skin surface
acetic acid	produced by propionibacteria – inhibits many other species, particularly at the low pHs found on skin surface
fatty acids liberated from lipids	produced by many skin organisms – kill streptococci and Gram-negative species
hydrogen peroxide	produced by streptococci – kills many other species
bacteriocins	produced by staphylococci, *Corynebacterium* spp., and *Brevibacterium* spp. – inhibit or kill many cutaneous organisms

microbe to out-compete others. These include the production of antimicrobial agents, interference with adhesion mechanisms, alteration of the environment, and depletion of essential nutrients. Most attention has focussed on the production of antimicrobial agents because of the possible use of such compounds for the treatment of infections in humans and other animals. The range of compounds produced by cutaneous microbes that can inhibit or kill others is very wide and includes hydrogen peroxide; end-products of metabolism such as acetic, lactic, and propionic acids; carbon dioxide; enzymes; antibiotics; and bacteriocins (Table 2.27). The ability of cutaneous bacteria to inhibit the growth of other cutaneous species is widespread. For example, in one study of isolates from twenty patients, 21% of *Micrococcaceae*, 5% of aerobic diphtheroids, and 7% of propionibacteria were found to be capable of inhibiting the growth of other cutaneous bacteria. The bacteriocins produced by skin bacteria have received particular attention, and a large number have been isolated and characterised. While these compounds were initially regarded as having quite a narrow antimicrobial spectrum, being active mainly against members of the same species as the producer organism, it is obvious that this concept is no longer tenable, and some have been shown to have very broad-spectrum activity.

While an extraordinary range of antagonistic substances are produced by skin bacteria, there is little evidence that such compounds have any effect on the composition of the cutaneous microbiota *in vivo*. Furthermore, studies designed to ascertain whether the expected domination of the cutaneous microbiota by antagonist-producing strains did occur *in vivo* have produced conflicting results. Nevertheless, some investigations have shown that "inhibitory" strains of bacteria can be used to control skin infections, which are described in Section 10.4.1. There is evidence that corynebacteria exert

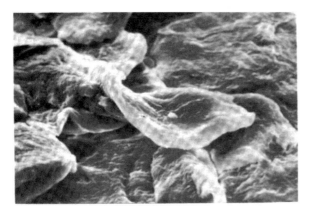

Figure 2.15 Scanning electron micrograph showing squames on the skin surface. Reprinted from: *Microbiology of human skin*. Noble W.C. Copyright © 1974, with permission from Elsevier.

some control of *Staph. aureus* colonisation of skin. For example, *Staph. aureus* has been applied to the skin of human volunteers and its persistence monitored 24 hours later. The cutaneous microbiota of those individuals (approximately one third) on whom the organism did not persist had a significantly higher proportion of coryneforms (34%) than did that (12%) of the individuals on whom the organism did survive. The means by which coryneforms achieve this effect is not known, but could involve the production of antagonistic substances, the utilisation of some essential nutrient, the blocking of adhesion sites, and so forth.

2.4.5 Dissemination of organisms from the skin

Cutaneous microbes can be transferred directly to other parts of the body, to other individuals, to objects, etc., by direct contact. Because the hands are the body parts that most frequently come into contact with other regions of the body, other individuals, and objects, most studies have investigated the transfer of microbes from these sites. For example, it has been shown that Gram-positive and Gram-negative bacteria, as well as viruses, are readily transferred from the hands to the lips of an individual. In an experiment in which the fingertips of volunteers were inoculated with 10^6 cfu of *M. luteus* or *Serratia rubidea* and then placed against the lips, 41% of *M. luteus* and 34% of *Ser. rubidea* were transferred to the lips during a 10-second contact period. Most studies of the transmission of bacteria from the skin have been carried out in hospitals, and molecular typing of isolates has demonstrated that a variety of organisms can readily be transferred from the hands of health-care workers to patients – often with disastrous consequences when the organisms transferred are pathogens such as *Staph. aureus*, *Cl. difficile*, and multi-resistant strains of Gram-negative bacilli.

Another major means of dissemination involves the shedding of squames with their attached organisms (Figure 2.15). While this happens spontaneously due to drying of the skin, most squames are generated as a result of friction between clothing and the skin surface. The number of squames shed by an individual is greatly increased during exercise and while undressing (e.g., a single act of undressing releases as many as 500,000 squames). On average, an adult sheds approximately 10^8 squamous particles per day, of which approximately 4% contain viable microbes. The median diameter of such microbe-bearing particles is approximately 13 μm, each of which carries approximately five viable microbes. Males appear to disperse more organisms than females,

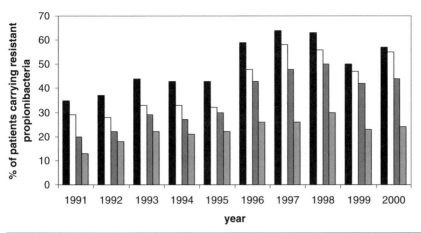

Figure 2.16 Prevalence of antibiotic-resistant propionibacteria on the skin of acne patients during the years 1991–2000. Bars represent the proportion of patients carrying propionibacteria resistant to any antibiotic (■), erythromycin (□), clindamycin (■), tetracycline (▨). Reproduced with permission of Blackwell Publishing from: Prevalence of antibiotic-resistant propionibacteria on the skin of acne patients: 10-year surveillance data and snapshot distribution study. Coates, P., Vyakrnam, S., Eady, E.A., Jones, C.E., Cove, J.H., and Cunliffe, W.J. *British Journal of Dermatology* 2002; 146:840–848.

and the number of organisms dispersed is greater from above than from below the waist. Certain individuals are more prolific dispersers of cutaneous microbes than others and are known as "dispersers" or "shedders". The dispersal of skin microbes in this way is an important means of disease dissemination and has received particular attention from the point of view of the spread of organisms such as *Staph. aureus* in hospitals. In this respect, young males are significantly more often found to be dispersers of the organism than females. In a study involving seventy-two adults, it was found that up to 11 cfu were shed per m^3 of air during undressing and that males dispersed between 2.5 and 5 times as many particles carrying viable bacteria than did females. Although hair does not provide a conducive environment for microbial growth, it is often colonised by microbes which are, of course, dispersed when the hairs are shed.

2.4.6 Effect of antibiotics and other interventions on the indigenous microbiota of the skin

2.4.6.1 Antibiotics

For more than 30 years, antibiotics have been the mainstay of treatment for acne. The antibiotics used include tetracyclines, erythromycin, and clindamycin, and these are frequently prescribed for relatively long periods of time (i.e., several months or even years). In the United Kingdom in 2000, general practitioners wrote more than 2.6 million prescriptions for antibiotics for acne patients – a quarter of these being for topically applied antibiotics. Such widespread, persistent use of antibiotics has, inevitably, resulted in the emergence of antibiotic-resistant skin bacteria. The development of such resistance in propionibacteria has been monitored over the past 10 years in a study of more than 4,000 acne patients (approximately equal numbers of males and females) attending outpatient clinics in Leeds, UK (Figure 2.16). The figure shows that over the 10-year period from 1991 to 2000, the general trend has been an increase in the proportion

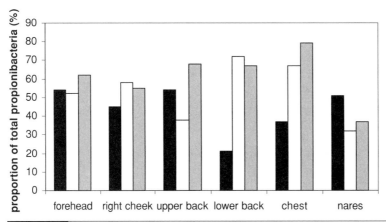

Figure 2.17 Prevalence of propionibacteria resistant to tetracycline (black bars), erythromycin (open bars), or clindamycin (grey bars) at different body sites in 4,000 acne patients. The proportion of propionibacteria resistant to each antibiotic is expressed as a percentage of the total cultivable propionibacteria at the site.

of patients harbouring propionibacteria resistant to erythromycin, clindamycin, and tetracycline. The proportion of patients colonised by strains resistant to each of the antibiotics approximately doubled over the 10-year period, with resistance to erythromycin being the most prevalent and resistance to tetracycline the least prevalent. The fall in the proportion of patients harbouring antibiotic-resistant strains from 1998 to 1999 reflects an attempt to curb the worrying increase in antibiotic-resistant strains by reducing the use of topical erythromycin and clindamycin. On one of the sampling occasions, the distribution of antibiotic-resistant propionibacteria was studied in seventy-two of the patients, and the results are shown in Figure 2.17. Although the proportion of resistant strains varied with the site and the antibiotic, in two-thirds of the sites more than 50% of the propionibacteria present at the site were resistant to at least one of the antibiotics. The population density of antibiotic-resistant strains was between 10^3 and 10^5 cfu/cm^2 at all sites for each of the antibiotics. Such high densities favour transmission of the antibiotic-resistant strains to new hosts.

With regard to cutaneous staphylococcal communities, in general, antibiotics do not change the relative proportions of the various species at a particular site. However, there are marked changes in the strain composition of the species and their antibiotic susceptibilities. Hence, in general, the number of different strains of a particular species at a site decreases and strains susceptible to the antibiotic administered are gradually replaced by those which are resistant. These resistant strains are usually acquired from other individuals or from the environment, and such strains can often transfer their resistance genes to other members of the cutaneous microbiota. Administration of penicillin, erythromycin, or tetracycline leads to resistance to the administered antibiotic, as well as to other antibiotics in this group. Clindamycin administration leads to a high proportion of cutaneous staphylococci becoming multiply resistant to macrolide, lincosamide, and streptogramin type B antibiotics. *Staph. haemolyticus*, in particular, is a species that often develops multiple resistance, whereas *Staph. capitis* and *Staph. auricularis* tend not to do so.

The effect of the administration of other antibiotics on the cutaneous microbiota are summarised in Table 2.28.

Table 2.28. Effect of antibiotic administration on the cutaneous microbiota

Antibiotic	Study population	Skin region investigated	Effects observed
amoxycillin (systemic)	infants	perineum	decrease in propionibacteria, streptococci, and staphylococci; increase in *Candida albicans*
pivmecillinam (systemic)	adults	beneath wing of nose	decrease in propionibacteria
		armpit	no effect
silver sulphadiazine (topical)	adults	various	skin surface sterilised in 40% of sites
rifampin/naficillin	adults	perianal	increase in rifampin-resistant CNS; 80% of these were also resistant to methicillin and gentamicin

Note: CNS = coagulase-negative staphylococci.

2.4.6.2 Occlusion

Occlusion of the skin (e.g., by dressings) has a profound effect on the micro-environment of the area of skin beneath the occluding material. As evaporative water loss is reduced, the water content of the occluded region increases, and this affects its temperature, pH, osmolality, etc. The effects of covering the forearm with plastic film on the pH and microbiota of skin are shown in Table 2.29. After only 1 day, the total viable count had increased 10,000-fold and continued to increase until day 4. As well as this quantitative change, the composition of the microbiota also showed marked changes. Although the proportion of CNS showed little difference over the 5 days, the proportions of micrococci and lipophilic coryneforms tended to increase, while the proportion of non-lipophilic coryneforms decreased. Although Gram-negative rods comprised only a very small proportion of the microbiota (never exceeding 0.03%), the frequency of detection increased steadily over the 5-day period from 0% prior to application of the plastic film to 60% on day 5. Other similar studies have found that Gram-negative bacteria comprise an increasing proportion of the microbiota following occlusion. These changes in the microbiota are a consequence of profound changes in the cutaneous environment under the occlusion. Hence, both the pH and moisture content increase dramatically. The transepithelial water loss was found to increase over the whole

Table 2.29. Effects of the presence of a plastic film on the forearm in ten healthy adults

Time (days)	pH	Total count (cfu/cm^2)	CNS	Micrococci	Lipophilic coryneforms	Non-lipophilic coryneforms
					(% of microbiota)	
0	4.4	1.8×10^2	63	6	0	17
1	–	1.4×10^6	73	9	0	19
2	5.2	1.4×10^7	64	23	10	0
3	6.6	1.5×10^7	60	23	14	0
4	7.1	9.8×10^7	67	10	19	4
5	6.4	7.5×10^6	63	11	26	0

Note: CNS = coagulase-negative staphylococci.

period, which reflects an increase in the water content of the skin – after 2 days, the skin was essentially saturated. Within a few hours of application of the plastic film, the carbon-dioxide concentration increased from 0.1% to between 5% and 7% (i.e., it equilibrated with the concentration of the gas present in the underlying tissues). The results of this, and other similar studies, demonstrate the preference of micrococci, lipohilic coryneforms, and Gram-negative rods for regions with an increased moisture content and/or pH (e.g., the axillae and perineum).

2.4.6.3 Skin cleansing

The importance of the transmission of an infectious disease by the direct transfer of the causative organism from the hands of those infected with, or carrying, the organism to non-infected individuals has been recognised for almost 150 years. There followed tremendous interest in optimising hand cleansing as a means of preventing disease transmission, and the literature on this subject is now enormous. Regular hand washing is an effective means of removing transient organisms from the skin and is undoubtedly important in the prevention of gastrointestinal and respiratory infections in the home and cross-infection in hospitals. However, this section focuses not on skin hygiene as a means of preventing disease transmission, but on the effect of skin hygiene measures on the cutaneous microbiota.

Washing with soap and water removes dirt and particulate matter and their associated microbes. It may also reduce the total number of microbes colonising the skin (by mechanically removing keratinocytes with attached microbes), although there is uncertainty with regard to exactly how effective the procedure is. Many studies have been carried out by applying cultures of known organisms to the skin surface and then studying the effectiveness of their removal by washing. These studies have revealed that washing is a very effective means of removing such organisms, although some of the observed reductions in the recovery of these organisms may be due to the powerful antimicrobial action of healthy skin which may have killed many of the organisms applied to its surface. The observed "poor recovery" of the organism from the skin (i.e., indicative of the effectiveness of washing) may simply be attributable to the fact that there were very few microbes left alive! In contrast, there is evidence that washing may be less effective in removing those microbes that are naturally present on the skin surface. In a recent study, washing with soap and water was found to remove 90% of micrococci that had been inoculated onto the hands, but removed less than 50% of the microbes already present on the skin surface. Although hand washing reduces the number of transient organisms (including potentially pathogenic species) contaminating the skin and and so is of benefit to the individual, the procedure may be of less benefit to the community. Hence, it has been shown that, after hand washing, there is a 17-fold increase in the dispersal of microbes from the hands, thereby increasing the potential for the transmission of microbes (and potential pathogens) to other individuals. The increase in the numbers of microbes dispersed following washing may be a consequence of the de-fatting action of the procedure (so increasing the number of squames that can be shed) and/or the disruption of microcolonies on the skin surface (so broadening the distribution of microbes on the surface and increasing the likelihood of a squame carrying a viable organism). Bathing and showering similarly increase the dispersal of cutaneous organisms into the air.

Table 2.30.	Effects of washing on the skin

removes dirt and particulate matter
removes outer layers of stratum corneum
removes skin lipids
increases pH
increases transepidermal water loss
increases moisture content
induces skin irritation

As has been described in Section 2.4.3, a number of factors are important in determining the nature of the cutaneous microbiota, and these include the water content and pH of the skin, the presence of lipids on its surface, and desquamation. Because washing affects all of these in some way, it has the potential to affect not only the composition of the cutaneous microbiota, but also the dispersal of skin microbes. The effects of washing on the skin are many and varied, and will depend on factors such as the manner in which it is carried out, the time taken for the procedure, how frequently the procedure is performed, and the nature of the cleansing product involved. Soap is the most commonly used product for washing and, because of its alkaline nature and its ability to remove lipids and fatty acids, the skin pH may be increased to neutral or alkaline values which can persist for several hours. Furthermore, the water used during washing increases the degree of hydration of the skin. Unfortunately, for many individuals, soap can act as a skin irritant – the response varying markedly among the population. The main effects of washing on the skin are summarised in Table 2.30. While most of the factors listed in this table could have a quantitative and/or qualitative effect on the cutaneous microbiota, any changes that do occur are generally transient, and the microbiota usually returns to normal values within a few hours. However, the skin of the hands, which may be washed very often by some individuals, may remain at an alkaline pH for long periods of time, and this would be expected to affect the composition of the cutaneous microbiota. The results of at least one study of the effects of repeated skin washing with soap have shown that the raised pH is accompanied by an increase in the proportions of propionibacteria on the skin regions studied – the forehead and the forearm. Similar effects are found when neutral or alkaline detergents are used rather than soap. Because of the irritant effect of soap, individuals working in occupations where repeated hand washing is carried out (e.g., health-care workers) often develop contact dermatitis – the reported prevalence of this condition ranges from 10 to 45%. As well as being unpleasant for the individual, damaged skin offers more attachment sites for bacteria, particularly pathogenic species such as *Staph. aureus*, and this may contribute to the high rates of carriage of such organisms by some health-care workers. It is more difficult to remove bacteria from damaged than healthy skin by washing and, furthermore, damaged skin sheds more bacteria than healthy skin. Both of these factors would increase the risk of cross-infection in hospitals.

Increasingly, manufacturers of skin hygiene products are incorporating antimicrobial agents into their products. Typical agents include chlorhexidine and triclosan, which have broad-spectrum antimicrobial properties and would be expected to affect the cutaneous microbiota. The use of hygiene products containing these antimicrobial agents certainly results in a substantial reduction in the number of organisms on the skin. However, there is concern about the development of resistance to such agents, and this has already been reported in *Staph. aureus*. Alcohol-based hand rinses are an effective alternative to such formulations because they have a broad-spectrum antimicrobial activity, and they have the advantage that there is no risk of resistance development. Because they do not involve washing or drying, their use is less damaging to the skin than washing with soap and water. Furthermore, unlike washing with soap and water, alcohol rinses do not result in increased skin shedding after use.

2.5 | Diseases caused by members of the cutaneous microbiota

2.5.1 Acne

Acne vulgaris is a chronic inflammatory condition of the pilosebaceous units which affects nearly all adolescents and adults at some time in their lives. Although the overall health of the affected individual is not impaired, acne is not a trivial disease because it can result in permanent cutaneous scars and psychological problems. The disease has a complex aetiology and involves abnormal keratinisation, hormonal function, bacterial growth, and immune hypersensitivity.

A possible role for bacteria in the pathogenesis of acne appeared to be firmly established with the finding that antibiotics such as tetracycline and erythromycin were beneficial in treating the disease. It was then observed that treatment with tetracycline resulted in a decrease in *P. acnes* and in the concentration of free fatty acids on the skin surface. It was suggested, therefore, that the fatty acids resulting from the lipolytic activity of *P. acnes* were responsible for the characteristic inflammation accompanying the disease. This also fitted in with the observation that both sebum production and the density of skin colonisation by *P. acnes* increase at puberty, which is also when acne generally first makes its appearance. However, the organism is also abundant on the skin of acne-free individuals, and this is difficult to reconcile with its being the causative agent of the disease. *P. acnes* (and/or *P. granulosum*) is now not regarded as the cause of acne, but as being a contributing factor to the inflammation associated with the disease. The first stage in the disease is comedogenesis (i.e., the formation of follicles that are distended with sebum and keratinocytes). These comedones are formed as a result of abnormal desquamation at the neck of the follicle, which causes blockage and the subsequent accumulation of sebum. *P. acnes* can exacerbate this abnormal desquamation, possibly by inducing overexpression of IL-1α by keratinocytes and/or macrophages. At this stage, the lesion is not inflamed and two types of comedones can be distinguished – closed and open. Closed comedones are those in which the follicular orifice is closed – these are often black in colour due to the accumulation of melanin and are known as blackheads. Open comedones (or whiteheads) are those in which the follicular orifice is narrowed rather than closed entirely. Within the comedone, *P. acnes* activates complement via the classical and alternative pathways, thereby releasing the potent chemoattractant C5a. It also produces a number of substances with chemoattractant properties and induces the release of IL-1, IL-8, and TNFα from macrophages. Consequently, PMNs and lymphocytes migrate into and accumulate within the swollen follicle. Their entry may be aided by bacterial proteases and hyaluronate lyase, which increase the permeability of the follicular epithelium. Complement

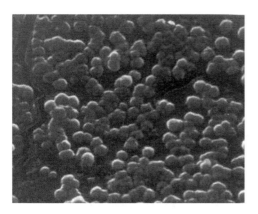

Figure 2.18 Scanning electron micrograph showing colonisation of a central venous catheter by a coagulase-negative staphylococcus. Magnification = ×500. Reproduced with permission from the Hospital Infection Society from: Implant infections: a haven for opportunistic bacteria. Schierholz, J.M. and Beuth, J. *Journal of Hospital Infection* 2001;49:87–93.

activation, together with the release of pro-inflammatory cytokines, results in inflammation and gives rise to the characteristic inflamed pustule. These processes are aided by the closed nature of the system; hence, secreted products and end-products of metabolism which would usually be flushed out of the follicle in its normal state accumulate within the closed follicle and so reach much higher concentrations.

The disease responds to antibiotics such as tetracycline, erythromycin, and clindamycin – all of which are effective against *P. acnes*. The beneficial effects of antibiotic therapy are thought to stem from a reduction in the numbers and, hence, the pro-inflammatory potency of *P. acnes*. Even at sublethal concentrations, the antibiotics can exert a beneficial effect because they reduce the production of pro-inflammatory substances by the organism. However, the widespread use of antibiotics for the treatment of this common disease has resulted in increased resistance of *P. acnes* to tetracycline, erythromycin, and clindamycin (Section 2.4.6.1). Topical retinoids are also effective against the disease because of their ability to prevent the blockage of follicles.

2.5.2 Intravascular catheter-associated infections

Intravascular catheters are widely used devices for the administration of medications, blood products, nutritional solutions, and fluids. It has been estimated that approximately 50% of hospitalised patients are given such a device during their stay. Because they are inserted through the skin and into a blood vessel, they provide a potential pathway from the skin to the blood. Their use, therefore, is associated with a variety of infections, including local infection, thrombophlebitis, bloodstream infections, endocarditis, osteomyelitis, and abscesses in distant organs. The magnitude of the problem can be gauged from the fact that, of the annual 200,000 nosocomial bloodstream infections in the United States (with a mortality of 10–25%), most are attributable to the use of these devices. The causative organism in 50–70% of these cases is *Staph. epidermidis*, which colonises the catheter and, in many cases, forms a biofilm on its surface (Figure 2.18). A biofilm consists of a surface-associated microbial community enclosed within a polymeric matrix which, as a consequence of this mode of growth, has a number of unique attributes (Section 1.1.2). Most importantly, from the infectious diseases point of view, a biofilm is very difficult to remove, is very resistant to host defence mechanisms, and is remarkably refractory to antimicrobial agents. Once a biofilm has formed on a catheter, often the only recourse is to remove the catheter and insert a new one.

Table 2.31.	Incidence of infection of implantable devices	
Implant	Site	Infection rate
prosthetic hip/knee joint	hip or knee	1–3%
vascular grafts	thoracic aorta	1%
	abdominal aorta/groin	3–6%
prosthetic heart valves	heart	1.5%
hydrocephalus shunts	brain or abdomen	3–20%

This is inconvenient for the patient and clinical staff, is unpleasant for the patient, and is costly. Following insertion of a catheter, cutaneous microbes at the site of insertion adhere to its surface. In the case of *Staph. epidermidis*, a number of adhesins are involved, including teichoic acid, an autolysin (AtlE), two surface-associated fimbria-like proteins (staphylococcal surface protein-1 and protein-2), and capsular polysaccharide/adhesin. The adherent microbes may then colonise the surface of the catheter, which is embedded in the tissue. Because this will be coated with platelets, plasma proteins, and tissue proteins (e.g., fibrin, fibronectin, collagen, thrombospondin, and laminin), adhesion is usually to this "conditioning film" rather than to the catheter material itself. *Staph. epidermidis* is able to adhere to vitronectin via its AtlE; to fibronectin via its teichoic acid; to fibrinogen via a fibrinogen-binding protein; and to fibrinogen, fibronectin, and vitronectin via another autolysin, AaE. Once attached, the bacteria grow and reproduce to form a biofilm, cell–cell adhesion within the biofilm being mediated by a polysaccharide intercellular adhesin, which is anchored to the bacterial cell surface by an accumulation-associated protein. Recent work has shown that biofilm formation is regulated by quorum-sensing mechanisms (Section 1.1.3). During and after biofilm formation, inflammation of and damage to adjacent tissues occurs as a result of the production of a number of virulence factors (Table 2.14).

2.5.3 Infections associated with implanted prosthetic devices

A large variety of implanted devices are now in common use and include hip, knee, and other joint prostheses; hydrocephalus shunts; prosthetic heart valves; pacemakers; vascular grafts; artificial urinary sphincters; intraocular lenses; and intraspinal drug delivery devices. Unfortunately, infection can occur during the implantation of these devices, and the infection rates range from 1% in operations involving hip and knee joints to 20% in the case of hydrocephalus shunts (Table 2.31). The main source of infection is the patient's skin, although other sources include the environment and the clinical personnel performing the operation. In order to reduce the risk of infection, the patient's skin in the area of the incision is subjected to rigorous disinfection, which virtually eliminates microbes from the skin surface. However, large numbers of viable organisms persist in the hair follicles, and these recolonise the skin surface within 15 minutes. These organisms then gain access to the incision and from there to the implanted device. The organism most frequently associated with implant infections is *Staph. epidermidis*, although other cutaneous organisms, particularly corynebacteria and propionibacteria, are also frequently involved (Table 2.32). In fact, a recent study has shown *P. acnes* to be responsible for more than half of infected hip prostheses that had to be removed – in 70% of these cases, it was the sole infecting organism.

Table 2.32.	Microbes frequently responsible for infection of implants

Type of implant	Infecting organism
prosthetic hip/knee joint	*Staph. epidermidis, Staph. aureus*; *Peptostreptococcus* spp., streptococci, AGNB, *P. acnes*
vascular grafts	*Staph. epidermidis, Staph. aureus*, AGNB
hydrocephalus shunts	*Staph. epidermidis, Staph. aureus*, coryneforms, AGNB
prosthetic heart valves	*Staph. epidermidis, Staph. aureus*, oral streptococci, enterococci

Note: AGNB = anaerobic Gram-negative bacilli.

2.5.4 Wound infections

A wound is a breach in the skin which allows access of microbes to the warm, moist, and nutritious subcutaneous environment. Two broad types of wound are recognised – acute and chronic. Acute wounds are those arising from external damage to intact skin and include burns, cuts, abrasions, bites, surgical wounds, knife and gun-shot injuries, etc. Chronic wounds are those arising from some predisposing condition (e.g., peripheral vascular disease, metabolic diseases such as diabetes mellitus) that eventually damage the integrity of the skin, resulting in leg ulcers, foot ulcers, pressure sores, etc. Once the skin has been breached, cutaneous microbes can take advantage of the situation and become opportunistic pathogens (Table 2.33). Both types of wounds may, of course, also become infected by environmental organisms and/or microbes from other anatomical regions. The characteristic features of a wound infection are the production of a purulent discharge or a spreading erythema indicative of cellulitis. Most infections are polymicrobial and frequently involve both aerobic and anaerobic organisms, with a combined population density of at least 4.6×10^5 cfu/cm^2.

2.5.5 Urinary tract infections

Although most urinary tract infections (UTIs) are caused by the intestinal organism *E. coli*, a significant proportion (10–15% in females) is attributable to *Staph. saprophyticus*. This organism, as well as being a resident of the intestinal tract, can be cultured from the skin of between 4% and 9% of adults, and generally comprises approximately 5% of the total staphylococci present. Interestingly, in females, the proportion of *Staph. saprophyticus* increases around puberty (when it accounts for 13% of staphylococci) and then decreases steadily until the age of 50, after which age it is rarely found. This shows some correlation with the frequency of UTIs in females that are due to *Staph.*

Table 2.33.	Cutaneous microbes associated with wound infections

Microbe	Type of wound infected
coagulase-negative staphylococci	acute, chronic
Micrococcus spp.	chronic
C. xerosis	chronic
Corynebacterium spp.	acute, chronic
P. acnes	acute, chronic

Table 2.34. Cutaneous infections due to *Malassezia* spp.

Disease	Causative organism(s)
atopic dermatitis	*Mal. furfur* > *Mal. globosa* > *Mal. sympodialis* > *Mal. slooffiae*
seborrhoeic dermatitis	*Mal. restricta, Mal. globosa*
pityriasis versicolor (temperate climates)	*Mal. globosa*
pityriasis versicolor (tropical climates)	*Mal. furfur, Mal. globosa*
folliculitis	*Mal. furfur, Mal. globosa, Mal. pachydermatis*

saprophyticus. Further details about UTIs due to this organism may be found in Section 7.5.7.1.

2.5.6 Infective endocarditis

Infective (or bacterial) endocarditis is a life-threatening infection of the endocardial surface of the heart which is caused mainly by viridans streptococci (Section 8.5.3), CNS, and *Staph. aureus*. The characteristic lesion (known as a "vegetation") consists of a collection of platelets, fibrin, microorganisms, and inflammatory cells. Although it can affect a number of sites on the heart, it most commonly involves the heart valves. Endocarditis due to CNS in individuals without prosthetic heart valves (known as "native-valve endocarditis") is quite rare and accounts for approximately 5% of cases of all native-valve infections. Such infections are usually caused by *Staph. epidermidis*, although *Staph. warneri* and *Staph. lugdunensis* may also be involved. The mortality rate is high (up to 36%), and valve replacement is necessary in approximately one-quarter of cases. In contrast to the low incidence of native-valve endocarditis due to CNS, as many as 50% of infections of prosthetic heart valves are caused by these organisms – the species responsible is usually *Staph. epidermidis*. Antimicrobial chemotherapy is, in most cases, inadequate and usually surgical intervention is necessary.

2.5.7 Diseases caused by *Malassezia* spp.

Malassezia spp. are associated mainly with diseases of the skin, although they have been shown to be responsible for systemic infections in neonates who were administered lipids through intravenous catheters. The association of the various species with skin diseases in humans is shown in Table 2.34. Pityriasis versicolor is characterised by scaly patches of various colours usually on the upper trunk (both chest and back), but can also spread to the neck and abdomen. It affects between 30% and 40% of the population of tropical climates but only 1–4% of those in temperate climates. *Mal. globosa* appears to be the main causative agent in both temperate and tropical climates, although *Mal. furfur* may also be involved in cases in tropical climates. The mycelial form of the yeast appears to be associated with the disease, but which factors induce its conversion from the yeast form remain to be established. The disease can be treated with either systemic or topical antifungal agents such as ketoconazole and azoles. However, recurrence of the infection is very common, occurring in 60% of cases in the first year and 80% after 2 years.

Seborrhoeic dermatitis affects between 1% and 3% of the population and involves scaling and inflammation of sebum-rich areas such as the scalp, eyebrows, and the paranasal and mid-thoracic regions. Dandruff, which affects between 5% and 10% of adults, is regarded as a mild form of the disease, and an association between *Malassezia* spp. and dandruff was first proposed as long ago as 1874. The involvement of *Malassezia* spp. in seborrhoeic dermatitis is still controversial but many studies have demonstrated the successful treatment of the condition with antifungal agents and a corresponding decrease in the numbers of these yeasts. *Mal. restricta* and *Mal. globosa* are the yeasts most frequently associated with the condition, although some studies have implicated *Mal. furfur* and *Mal. sympodialis*. The condition occurs more frequently in immunosuppressed individuals, implying the involvement of immune mechanisms in disease pathogenesis. However, results from studies of humoral and cell-mediated immune responses in patients with the disease have proved contradictory.

Overgrowth of a number of *Malassezia* spp., mainly *Mal. furfur, Mal. globosa*, and *Mal. pachydermatis*, in hair follicles can result in folliculitis. The inflammation may be a response to yeast metabolic products or could be triggered by free fatty acids liberated by the action of yeast lipases. The condition occurs most frequently on the back and chest and is particularly common in tropical climates. The disease responds quickly to antifungal agents.

Several studies have demonstrated that *Malassezia* spp., especially *Mal. furfur*, can act as allergens in patients with atopic dermatitis and can aggravate the condition. Patients with the condition, unlike healthy controls, have specific serum IgE antibodies against *Malassezia* spp., and a number of yeast antigens (including proteins and carbohydrates) have been implicated in the pathogenesis of the disease.

2.5.8 Erythrasma

This is a superficial infection due to *C. minutissimum* which occurs mainly in intertriginous sites, such as between the toes, the axillae, submammary regions, and crural areas. This organism is found in the intertriginous regions of approximately 20% of healthy individuals, where it is present at a density of approximately 10^4 per cm^2. The characteristic scaly lesions of erythrasma are produced when the population density of the organism has increased to a level approximately 20-fold greater than normal. In up to 30% of patients, the infection is accompanied by an infection with *Can. albicans* or a dermatophyte. Most commonly, it affects the toe webs and, in fact, it constitutes the most common bacterial infection of the foot. The reported incidence of the disease ranges from 5% to 58%, depending on the population studied – in healthy young adults it is approximately 19%. Pre-disposing factors include advanced age, poor hygiene, warm climate, institutionalisation, obesity, and diabetes mellitus. The condition can be successfully treated with a number of systemic and topical antimicrobial agents – the most effective being oral erythromycin. Little is known about the pathogenesis of the disease. The organism appears to invade the stratum corneum and, in doing so, it changes to a long, filamentous form. The tissue destruction accompanying the infection may be due to the secretion of keratinolytic enzymes by the organism.

2.5.9 Odour

Brev. epidermidis can be isolated from the lesions of athlete's foot and is responsible for the odour accompanying the infection. It produces several proteolytic enzymes which can release L-methionine from proteins, which is then converted into methane thiol and other volatile sulphur compounds that are responsible for the characteristic pungent odour. Axillary odour is attributable mainly to the microbial conversion of androstadienol and androstadienone (which are secreted by the apocrine glands) to 5α-androstenone, 5α-androst-16-en-3α-ol, and 5α-androst-16-en-3α-ol. These conversions appear to be carried out by aerobic corynebacteria and, in particular, *C. xerosis*.

2.5.10 Pitted keratolysis

This condition is characterised by pitted erosion of the stratum corneum and usually involves the soles of the feet or, occasionally, the palms of the hands. The pitting is thought to be the result of the action of proteolytic enzymes. The disease is often accompanied by a pungent odour and occurs mainly in individuals who wear shoes or boots for prolonged periods of time (e.g., soldiers and miners). The wearing of shoes for long periods of time results in increased hydration of the skin, increased pH, and dramatic increases in the number of microbes inhabiting the region (Section 2.4.6.2). A number of organisms, particularly coryneforms, have been associated with the condition but, more recently, *Kytococcus sedentarius* has been implicated as the aetiological agent. This is a proteolytic organism which, when inoculated onto feet and occluded, can induce pitted keratolysis. The organism is a member of the indigenous microbiota of the foot, where it is generally present in low numbers. However, the organism is alkaliphilic, and its growth rate increases markedly to a maximum at a pH of between 8 and 9. It also produces a number of proteases, including two keratin-degrading enzymes which have optimum activities at slightly alkaline pHs. The rise in pH and increased moisture content of occluded feet could, therefore, provide an environment in which the numbers of *Kyt. sedentarius* increase rapidly, resulting in the release of keratinases which bring about the degradation of the stratum corneum characteristic of pitted keratolysis.

2.5.11 Trichomycosis

This is an infection of the hair shaft that occurs mainly in the axillae and pubic region and is associated with poor hygiene. The organisms involved have not been definitively identified, but are thought to include several *Corynebacterium* spp. and possibly staphylococci. These form a biofilm on the outside of the hair shaft and eventually invade the hair itself (presumably by secreting keratinases), rendering it brittle. The disease is accompanied by an offensive odour attributable to compounds produced by microbes from sweat components.

2.6 | Further Reading

Books

Lesher, J.L., Aly, R., Babel, D.E., Cohen, P.R., Elston, D.M., and Tomecki, K.J. (eds). (2000). *An atlas of microbiology of the skin*. Boca Raton: CRC Press.

Maibach, H. and Aly, R. (ed). (1981). *Skin microbiology: relevance to clinical infection*. New York: Springer-Verlag.

Noble, W.C. (ed.). (1992). *The skin microflora and microbial skin disease*. Cambridge: Cambridge University Press.

Reviews and Papers

Ashbee, H.R. and Evans, E.G. (2002). Immunology of diseases associated with *Malassezia* species. *Clinical Microbiology Reviews* **15**, 21–57.

Ashbee, H.R., Leck, A.K., Puntis, J.W., Parsons, W.J., and Evans, E.G. (2002). Skin colonisation by *Malassezia* in neonates and infants. *Infection Control and Hospital Epidemiology* **23**, 212–216.

Bergogne-berezin, E. and Towner, K.J. (1996). *Acinetobacter* spp. as nosocomial pathogens: microbiological, clinical, and epidemiological features. *Clinical Microbiology Reviews* **9**, 148–165.

Berlau, J., Aucken, H., Malnick, H., and Pitt, T. (1999). Distribution of *Acinetobacter* species on skin of healthy humans. *European Journal of Clinical Microbiology and Infectious Diseases* **18**, 179–183.

Bojar, R.A. and Holland, K.T. (2002). The human cutaneous microbiota and factors controlling colonisation. *World Journal of Microbiology & Biotechnology* **18**, 889–903.

Bowler, P.G., Duerden, B.I., and Armstrong, D.G. (2001). Wound microbiology and associated approaches to wound management. *Clinical Microbiology Reviews* **14**, 244–269.

Boyce, J.M. and Pittet, D. (2002). Guideline for hand hygiene in health-care settings. *American Journal of Infection Control* **30**, S1–S46.

Brook, I. (2000). The effects of amoxicillin therapy on skin flora in infants. *Pediatric Dermatology* **17**, 360–363.

Brook, I. (2002). Secondary bacterial infections complicating skin lesions. *Journal of Medical Microbiology* **51**, 808–812.

Chiller, K., Selkin, B.A., and Murakawa, G.J. (2001). Skin microflora and bacterial infections of the skin. *Journal of Investigative Dermatology Symposium Proceedings* **6**, 170–174.

Coates, P., Vyakrnam, S., Eady, E.A., Jones, C.E., Cove, J.H., and Cunliffe, W.J. (2002). Prevalence of antibiotic-resistant propionibacteria on the skin of acne patients: 10-year surveillance data and snapshot distribution study. *British Journal of Dermatology* **146**, 840–848.

Crespo, E.V. and Delgado, F.V. (2002). *Malassezia* species in skin diseases. *Current Opinion in Infectious Diseases* **15**, 133–142.

Davies, C.E., Wilson, M.J., Hill, K.E., Stephens, P., Hill, C.M., Harding, K.G., and Thomas, D.W. (2001). Use of molecular techniques to study microbial diversity in the skin: chronic wounds reevaluated. *Wound Repair and Regeneration* **9**, 332–340.

Dinulos, J.G.H., Mentele, L., Fredericks, L.P., Dale, B.A., and Darmstadt, G.L. (2003). Keratinocyte expression of human β defensin 2 following bacterial infection: role in cutaneous host defence. *Clinical and Diagnostic Laboratory Immunology* **10**, 161–166.

Eady, E.A., Gloor, M., and Leyden, J.J. (2003). *Propionibacterium acnes* resistance: a worldwide problem. *Dermatology* **206**, 54–56.

Frank, D.N., Spiegelman, G.B., Davis, W., Wagner, E., Lyons, E., and Pace, N.R. (2003). Culture-independent molecular analysis of microbial constituents of the healthy human outer ear. *Journal of Clinical Microbiology* **41**, 295–303.

Fredricks, D.N. (2001). Microbial ecology of human skin in health and disease. *Journal of Investigative Dermatology Symposium Proceedings* **6**, 167–169.

Funke, G., von Graevenitz, A., Clarridge, J.E., III, and Bernard, K.A. (1997). Clinical microbiology of coryneform bacteria. *Clinical Microbiology Reviews* **10**, 125–159.

Gupta, A.K. and Kohli, Y. (2004). Prevalence of *Malassezia* species on various body sites in clinically healthy subjects representing different age groups. *Medical Mycology* **42**, 35–42.

Gupta, A.K., Madzia, S.E., and Batra, R. (2004). Etiology and management of seborrheic dermatitis. *Dermatology* **208**, 89–93.

Hadaway, L.C. (2003). Skin flora and infection. *Journal of Infusional Nursing* **26**, 44–48.

Hendolin, P.H., Karkkainen, U., Himi, T., Markkanen, A., Ylikoski, J., Stroman, D.W., Roland, P.S., Dohar, J., and Burt, W. (2001). Microbiology of normal external auditory canal. *Laryngoscope* **111**, 2054–2059.

Higaki, S., Kitagawa, T., Morohashi, M., and Yamagishi, T. (1999). Distribution and antimicrobial susceptibility of coagulase-negative staphylococci from skin lesions. *Journal of International Medical Research* **27**, 191–195.

Holland, K.T. and Bojar, R.A. (2002). Cosmetics: what is their influence on the skin microflora? *American Journal of Clinical Dermatology* **3**, 445–449.

Huebner, J. and Goldmann, D.A. (1999). Coagulase-negative staphylococci: role as pathogens. *Annual Review of Medicine* **50**, 223–236.

Jung, K., Brauner, A., Kuhn, I., Flock, J.I., and Mollby, R. (1998). Variation of coagulase-negative staphylococci in the skin flora of healthy individuals during one year. *Microbial Ecology in Health and Disease* **10**, 85–90.

Larson, E. (2001). Hygiene of the skin: when is clean too clean? *Emerging Infectious Diseases* **7**, 225–230.

Laube, S. and Farrell, A.M. (2002). Bacterial skin infections in the elderly: diagnosis and treatment. *Drugs and Aging* **19**, 331–342.

Leyden, J.J. (2001). The evolving role of *Propionibacterium acnes* in acne. *Seminars in Cutaneous Medicine and Surgery* **20**, 139–143.

Ljubojevic, S., Skerlev, M., Lipozencic, J., and Basta-Juzbasic, A. (2002). The role of *Malassezia furfur* in dermatology. *Clinics in Dermatology* **20**, 179–182.

Lubbe, J. (2003). Secondary infections in patients with atopic dermatitis. *American Journal of Clinical Dermatology* **4**, 641–654.

Moreillon, P., and Que, Y.A. (2004). Infective endocarditis. *Lancet* **363**, 139–149.

Nagase, N., Sasaki, A., Yamashita, K., Shimizu, A., Wakita, Y., Kitai, S., and Kawano, J. (2002). Isolation and species distribution of staphylococci from animal and human skin. *Journal of Veterinary Medical Science* **64**, 245–250.

Nizet, V., Ohtake, T., Lauth, X., Trowbridge, J., Rudisill, J., Dorschner, R.A., Pestonjamasp, V., Piraino, J., Huttner, K., and Gallo, R.L. (2001). Innate antimicrobial peptide protects the skin from invasive bacterial infection. *Nature* **414**, 454–457.

O'Gara, J.P. and Humphreys, H. (2001). *Staphylococcus epidermidis* biofilms: importance and implications. *Journal of Medical Microbiologoy* **50**, 582–587.

Rippke, F., Schreiner, V., and Schwanitz, H.J. (2002). The acidic milieu of the horny layer: new findings on the physiology and pathophysiology of skin pH. *American Journal of Clinical Dermatology* **3**, 261–272.

Schittek, B., Hipfel, R., Sauer, B., Bauer, J., Kalbacher, H., Stevanovic, S., Schirle, M., Schroeder, K., Blin, N., Meier, F., Rassner, G., and Garbe, C. (2001). Dermcidin: a novel human antibiotic peptide secreted by sweat glands. *Nature Immunology* **2**, 1133–1137.

Solberg, C.O. (2000). Spread of *Staphylococcus aureus* in hospitals: causes and prevention. *Scandinavian Journal of Infectious Diseases* **32**, 587–595.

Sugita, T., Suto, H., Unno, T., Tsuboi, R., Ogawa, H., Shinoda, T., and Nishikawa, A. (2001). Molecular analysis of *Malassezia* microflora on the skin of atopic dermatitis patients and healthy subjects. *Journal of Clinical Microbiology* **39**, 3486–3490.

Terpstra, S., Noordhoek, G.T., Voesten, H.G.J., Hendriks, B., and Degener, J.E. (1999). Rapid emergence of resistant coagulase-negative staphylococci on the skin after antibiotic prophylaxis. *Journal of Hospital Infection* **43**, 195–202.

Thestrup-Pedersen, K. (1998). Bacteria and the skin: clinical practice and therapy update. *British Journal of Dermatology* **139** (Suppl 53), 1–3.

Till, A.E., Goulden, V., Cunliffe, W.J., and Holland, K.T. (2000). The cutaneous microflora of adolescent, persistent and late-onset acne patients does not differ. *British Journal of Dermatology* **142**, 885–892.

von Eiff, C., Peters, G., and Heilmann, C. (2002). Pathogenesis of infections due to coagulase-negative staphylococci. *Lancet Infectious Diseases* **2**, 677–685.

Vowels, B.R., Feingold, D.S., Sloughfy, C., Foglia, A.N., Konnikov, N., Ordoukhanian, E., Starkey, P. and Leyden, J.J. (1996). Effects of topical erythromycin on ecology of aerobic cutaneous bacterial flora. *Antimicrobial Agents and Chemotherapy* **40**, 2598–2604.

Vuong, C. and Otto, M. (2002). *Staphylococcus epidermidis* infections. *Microbes and Infection* **4**, 481–489.

3

The eye and its indigenous microbiota

The eye is a sense organ designed to focus light onto receptor cells and to control the amount of light reaching these receptors. Its optimal functioning requires the presence of certain accessory structures which include the eyebrows, eyelids, eyelashes, and the lacrimal apparatus.

3.1 | Anatomy and physiology of the eye

The basic anatomy of the eye and its accessory structures are shown in Figure 3.1. The eyeball is approximately 2.5 cm in diameter and is bounded by a white, fibrous, protective layer of connective tissue (the sclera) at the back and by a transparent fibrous layer (the cornea) at the front. Apart from the cornea, the front of the eye is covered by a transparent layer of modified skin (the conjunctiva), which also lines the eyelid. Inside the sclera is a layer of vascular tissue (the choroid), which supplies nutrients to it. The region of the choroid beneath the cornea is modified to form the iris and ciliary body, which are muscular structures involved in controlling, respectively, the amount of light entering the eye and the shape of the lens. The innermost layer of the eye is the retina, which extends back from the ciliary body to cover the rear three-quarters of the surface of the eyeball. The retina contains the light-sensitive cells and has its own blood and nerve supply. The interior of the eye is filled with fluid. In front of the lens (which focuses incoming light onto the retina) is the anterior cavity, which is filled with aqueous humour. This fluid is similar in composition to cerebrospinal fluid, and supplies nutrients and oxygen to the cornea, iris, and lens. Behind the lens lies the vitreous chamber, which is filled with a transparent gel known as vitreous humour – its function is to maintain the shape of the eyeball and to keep the retina flush against the choroid.

The function of the eyebrows and eyelashes is to protect the eyeballs from the direct rays of the sun, foreign objects, and perspiration. The eyelids also protect the eyes from excessive light and physical injury, but they have an additional function in that they spread lubricating fluid (i.e., tears) over the eyeballs during blinking. The lacrimal apparatus consists of a collection of glands, ducts, and canals whose function is to produce and supply tears to the eyeballs and then to collect any excess fluid which drains into the nasal cavity (Figure 3.2). Tears have lubricating, moistening, cleaning, and protective functions.

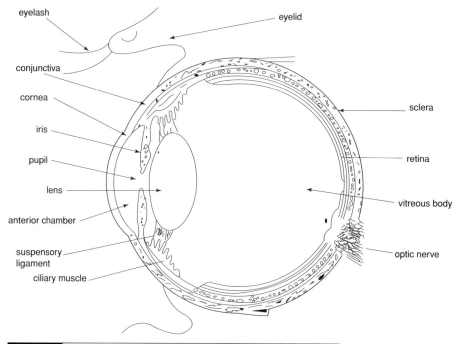

Figure 3.1 Main anatomical features of the eye and its associated structures.

3.2 | Antimicrobial defence mechanisms of the eye

The eyelids constitute an important component of the ocular antimicrobial defence system for a number of reasons. Firstly, the blink reflex (usually 12 blinks/minute) provides protection against foreign objects which would invariably be contaminated with microbes. Secondly, while moving over the cornea, it removes foreign debris (together with any associated microbes) and desquamated cells. Finally, it distributes tears (total volume approximately 7 μl/eye) with their broad-spectrum antimicrobial properties

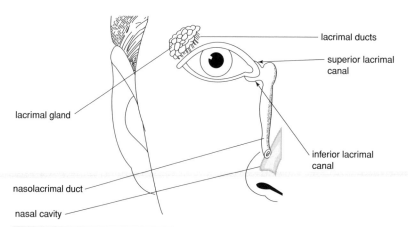

Figure 3.2 The lacrimal apparatus. Tears are produced by the lacrimal gland and flow through the lacrimal duct onto the surface of the eyeball. Excess tears drain into the nasal cavity via the lacrimal canals and nasolacrimal duct.

as a thin film (known as the "tear film") over the surface of the eye. The importance of this component of the defence system can be appreciated by considering the effects that suppression of the reflex has on ocular health. Hence, bacterial conjunctivitis and keratitis are frequent problems in comatose patients or those with Bell's palsy.

The conjunctival epithelium is intermediate in type between the keratinised squamous epithelium of the skin and typical mucosal epithelia, although it most resembles the latter. It contains numerous mucin-secreting goblet cells, as well as intraepithelial lymphocytes and dendritic cells. It consists of between five and seven layers of cells and provides a very effective barrier against microbial invasion, and very few microbes are able to infect the deeper ocular tissues unless this barrier has been damaged in some way. It is covered by the tear film which protects it from dehydration, microbes, and particulate matter (Figure 3.3). There is some uncertainty with regard to the exact thickness of the tear film and estimates range from 7 to 40 μm. It comprises three layers (Figure 3.4). The outermost layer (0.6–2.0 μm thick) consists of a mixture of phospholipids (in contact with the aqueous layer) and neutral lipids (at the air interface) and has three main functions. Firstly, it prevents evaporation of the underlying aqueous layers and, due to its surfactant properties, facilitates spreading of the aqueous layers. Finally, it traps small dust particles, thereby protecting the conjunctival epithelium from potentially damaging abrasive particles. The lipids are produced by the Meibomian glands of the upper eyelid and comprise mainly sterol esters, wax esters, and phospholipids. Underlying the lipid layer is the aqueous layer (4–7 μm thick), which is produced by the lacrimal glands and contains effector molecules of the innate and adaptive immune defence systems (e.g., antibodies, enzymes, antimicrobial peptides). It also contains nutrients and end-products of epithelial cell metabolism. Its main functions are to hydrate and to provide nutrients and oxygen to the conjunctival epithelium and also to protect it from microbes and from physical damage. The innermost layer is known as the mucous layer and is produced by the goblet cells of the conjunctiva. It is the thickest of the three layers and contains mucins (mainly MUC5AC), which can bind microbes, thus preventing them from adhering to the underlying ocular structures. Furthermore, the mucins form a complex network covering the conjunctiva, and this traps exfoliated epithelial cells, foreign particles, microbes, and debris. Blinking causes the network to collapse into a single strand which is pushed towards the inner corner of the eye. It then becomes compacted into a small clump and is pushed onto the skin where it dries and eventually either falls off or is removed by rubbing. Hence, foreign particles, microbes, and debris are regularly removed from the conjunctival surface.

Because of evaporation of the tear film, the temperature of the conjunctiva and underlying cornea are usually several degrees below body temperature, and this could reduce the rate of microbial growth. For example, when the ambient temperature is 20°C, the temperature of the corneal surface is approximately 32°C in still air. Evaporation is increased with increasing air movement so that even lower temperatures will be reached in a windy environment.

As can be seen from Table 3.1, the tear film contains a number of antimicrobial compounds, and studies have shown that tears are rapidly bactericidal for a range of organisms, including *Staph. aureus* and *Strep. agalactiae*. Lactoferrin is present in tears at a concentration of approximately 2.2 mg/ml, which is much higher than that found in serum and it is one of the most abundant proteins of the tear film. The protein

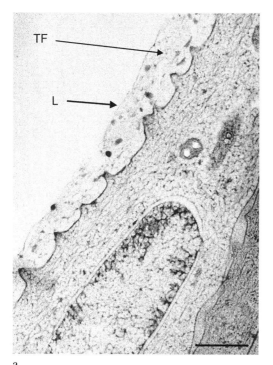

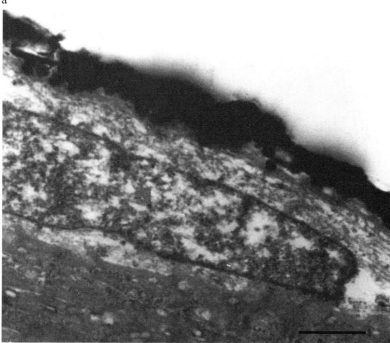

Figure 3.3 (a) Transmission electron micrograph of the mouse corneal epithelium showing the tear film (TF), including the lipid layer (L) on its outer surface. (b) Another sample after fixation in glutaraldehyde and tannic acid showing the tear film as a dark layer on the surface of the corneal epithelium. Bar = 1.0 μm. Reprinted from: *Investigative ophthalmology and visual sciences.* Vol. 44, pp. 3520–3525. Tran, C.H., Routledge, C., Miller, J., Miller, F., and Hodson, S.A. Copyright © 2003, with permission from The Association for Research in Vision and Ophthalmology.

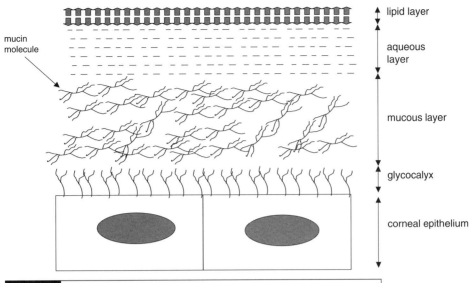

mucin
molecule

lipid layer

aqueous
layer

mucous layer

glycocalyx

corneal epithelium

Figure 3.4 Diagram showing the three layers of the tear film. See text for explanation.

has a very high affinity for iron and exists in two forms: iron-free (apo-lactoferrin) and iron-saturated (holo-lactoferrin). Its high affinity for iron ensures that very little free iron is available to any microbes present in the tear film and thus restricts their growth. In addition, lactoferrin exerts a number of additional antimicrobial effects, including enhancing the activity of natural killer cells and preventing bacterial adhesion to host cells. Apo-lactoferrin can also kill certain organisms, including *Strep. mutans, Strep. bovis,*

Table 3.1. Antimicrobial compounds present in the tear film

Component	Function
lysozyme	antibacterial
lactoferrin	iron-binding; antibacterial; prevents bacterial adhesion
transferrin	iron-binding
secretory phospholipase A_2	bactericidal against many Gram-positive species
secretory leukocyte protease inhibitor	microbicidal; inhibits pro-inflammatory activities of bacterial components
β-defensins	microbicidal
lactoperoxidase	microbicidal
caeruloplasmin	chelates Cu ions; acts as a superoxide dismutase
prealbumin	enhances lysozyme activity
fibronectin	facilitates phagocytosis
sialin	binds microbes
plasminogen activator	chemoattractant for leukocytes
β-lysin	antibacterial
fatty acids	microbicidal to streptococci, fungi
antibodies (mainly IgA)	prevent bacterial adhesion; involved in opsonisation

Vibrio cholerae, *E. coli*, *Bac. subtilis*, and *Lis. monocytogenes*. Furthermore, proteolytic cleavage of the molecule produces an antimicrobial peptide, lactoferricin, which can kill a range of organisms, including *E. coli*, *Can. albicans*, *Sal. enteritidis*, *Klebsiella pneumoniae*, *Proteus vulgaris*, *Yersinia enterocolitica*, *Pseudomonas aeruginosa*, *Campylobacter jejuni*, *Staph. aureus*, *Strep. mutans*, *Strep. bovis*, *Bac. subtilis*, *C. diphtheriae*, *Lis. monocytogenes*, and *Clostridium perfringens*. Lactoferrin also acts synergically with other antimicrobial proteins and peptides, including lysozyme, IgA, LL-37, and human β-defensins resulting not only in a greater potency than that displayed by the sum of the individual activities of the two compounds, but also a broader antimicrobial spectrum. Tears contain high concentrations (approximately 1.7 mg/ml) of lysozyme, the antibacterial properties of which have already been described in Section 2.2. Lactoperoxidase catalyses the reaction between hydrogen peroxide and thiocyanate (both of which are present in tears), resulting in the production of hypothiocyanite which is active against *Staph. aureus*, *Staph. epidermidis*, various streptococci, *H. influenzae*, *E. coli*, *Prevotella* spp., *Por. gingivalis*, and *Pseudomonas* spp. Secretory leukocyte proteinase inhibitor (SLPI), is an inhibitor of serine proteases and protects tissues against these enzymes which are released by neutrophils during inflammation. It is also able to bind to lipopolysaccharide and neutralise some of its pro-inflammatory activities. As well as these anti-inflammatory activities, SLPI has antimicrobial activity and has been shown to kill *E. coli*, *Ps. aeruginosa*, and *Staph. aureus* at micromolar concentrations. β-Lysin can disrupt the membrane integrity of a number of bacterial species and is particularly effective against micrococci. It can act synergistically with lysozyme resulting in degradation of both the cell wall and membrane of susceptible species. Another important antimicrobial constituent of tears is secretory phospholipase A_2, which is effective against many Gram-positive organisms, including staphylococci, streptococci, micrococci, enterococci, and *Lis. monocytogenes*. The enzyme is present in tear fluid at a concentration of 55 μg/ml, which is more than 10-fold greater than that found in serum – concentrations of between 15 and 80 ng/ml are sufficient to kill *Staph. aureus*. A number of antimicrobial peptides are produced by the conjunctival epithelium and by the nasolacrimal glands. These include human β-defensin 1 and β-defensin 2, which have activity against a number of Gram-positive and Gram-negative bacteria, fungi, and viruses (Section 2.2).

A major component of the ocular defence system is the secretory IgA (sIgA) present in tears – this constitutes approximately 17% of the total protein content of tears. IgG and IgE are also present, but at lower concentrations, while IgM and IgD are rarely detected. Secretory IgA is generated principally by the common mucosal immune system, which is independent of systemic immunity. In this system, induction of the immune response is thought to occur mainly in the gut-associated lymphoid tissue (GALT) and the bronchus-associated lymphoid tissue (BALT). The antigen-stimulated B cells produced in these tissues are then transported to the lacrimal glands, where they undergo clonal expansion and maturation to produce mainly IgA plasma cells and smaller numbers of IgM and IgG plasma cells. Whether or not the lymphoid tissue associated with the eye can itself generate sIgA in response to antigens via the conjunctival surface has not been established. This would require transportation of the antigen across the conjunctival epithelium, and the latter does not appear to have specialised antigen-sampling cells such as M cells. Furthermore, the ability of the lymphoid tissue of the lacrimal glands to process and present antigens appears to be limited. Nevertheless, antigenic material applied to the eye is known to stimulate an ocular secretory immune response. Evidence

Table 3.2. | Effects of overnight eye closure on the composition of the tear film

Property	Change after overnight sleep
pH	decreases
osmolarity	decreases
total protein concentration	2- to 3-fold increase
sIgA concentration	approximately 10-fold increase
lysozme concentration	decreases
lactoferrin concentration	decreases
albumin concentration	increases
plasmin activity	approximately 10-fold increase
polymorphonuclear neutrophil concentration	increases
complement C3c concentration	increases
glucose concentration	no change

suggests that the antigenic material enters via the nasolacrimal duct and is then transported to the GALT and BALT, resulting in the production of antigen-stimulated B cells. The sIgA produced has a number of protective functions – it can inhibit the adhesion of microbes to ocular surfaces, cause microbial aggregation, thereby facilitating their removal in tears and can neutralise toxins and viruses.

Periods of prolonged eye closure, such as sleep, result in a dramatic decrease in the rate of tear production. It might be expected, therefore, that the composition of the tear film would change after overnight eye closure; this is, indeed, the case. After sleep, the properties of the tear film change in a number of ways, and these are summarised in Table 3.2. The composition of the tear film after overnight closure in fact is characteristic of that of a subclinical inflammatory state with high levels of sIgA, polymorphonuclear leukocytes, and complement C3c. During the prolonged eyelid closure that accompanies sleep, the loss of many protective functions provided by blinking and consequent renewal of the tear film renders the conjunctiva vulnerable to infection. However, the changes listed in Table 3.2 that occur during sleep puts the ocular environment into a state of readiness to combat any microbial assault and provides an alternative antimicrobial protection system to that present during periods when the eye is open.

As well as these elaborate antimicrobial defence systems present on the conjunctival surfaces, the cornea contains immunoglobulins, predominantly IgG and IgM, together with complement components, thereby providing protection against any microbes that may have overcome the outer ocular defences.

Because tear fluid contains only low concentrations of microbial nutrients (see Table 3.3), has a low temperature (3–4°C below body temperature), and has a wide range of antimicrobial components (Table 3.1), it is an inhospitable environment for many microbes. Consequently, the conjunctivae of many individuals are either sterile or harbour only sparse microbial populations.

3.3 | Environmental determinants at different regions of the eye

Most of the accessory structures of the eye (i.e., eyebrows, eyelids, and eyelashes) are colonised by microbes, but the environmental factors operating at these sites are similar to those that operate either at the skin (in the case of the eyelids) or hair (in the case

Table 3.3. Composition of the tear film	
Component	Concentration
total protein	0.3–2.0 g/100 ml
albumin	0.1–0.2 g/100 ml
immunoglobulins	0.07–0.4 g/100 ml
lysozyme	0.17 g/100 ml
lactoferrin	0.22 g/100 ml
Na^+	120–160 mmol/L
Cl^-	118–135 mmol/L
HCO_3^-	20–25 mmol/L
K^+	20–42 mmol/L
Mg^{2+}	0.7–0.9 mmol/L
Ca^{2+}	0.5–1.1 mmol/L
glucose	0.2–0.5 mmol/L
lactate	1–5 mmol/L
urea	54 mg/100 ml
cholesterol	0.02–0.2 g/100 ml
other lipids	?
fatty acids	?
enzymes (lactate dehdrogenase, amylase)	?
amino acids	8 mg/100 ml
anti-proteinases	6 mg/100 ml
antimicrobial peptides and proteins	see Table 3.1

Note: ? = values not known.

of eyelashes and eyebrows) and will not be discussed further. The microbiotas of these sites are also what would be expected of skin and its associated structures and will not be discussed further. The main structure of the eye itself that is colonised by microbes is the conjunctiva. Tear fluid is the main source of host-derived nutrients for microbes colonising this region, and its composition is shown in Table 3.3. Basically, tears have a high water content, are isotonic (their osmolality is approximately 320 mOsm/kg), and have a pH of between 7.14 and 7.82. Although a wide range of substances is present, they are usually at a low concentration rendering the fluid isotonic. For example, although more than sixty proteins have been detected, the total protein content is within the range of 0.3–2.0% (w/v). Under normal circumstances, tears are produced at a rate of approximately 1.2 μl/minute, but this rate can increase dramatically in response to physical and emotional factors. For example, one teardrop running down the cheek of someone who is crying contains 50 μl, and many of these drops are produced per minute. Although both the glucose and lactate present can be utilised by a wide range of organisms as carbon and energy sources, their low concentrations mean that the main sources of nutrients for microbes living on the conjunctival surface are likely to be the proteins present in tear fluid, the mucins in the mucus layer of the tear film, and the lipids of the outer layer of the tear film. Members of the ocular microbiota able to degrade these macromolecules (sometimes only partially, in the case of mucins) and thereby provide amino acids, sugars, and fatty acids to serve as carbon, nitrogen, and energy sources for themselves and other members of the ocular microbiota are listed in Table 3.4. A number of bacteria isolated from the conjunctivae of healthy adults, mainly viridans streptococci, are able to grow on ocular mucins. The sialidases

| Table 3.4. | Hydrolysis of macromolecules by members of the ocular microbiota |

Macromolecule	Organisms able to hydrolyse the macromolecule
mucin	viridans streptococci, *Strep. pneumoniae*, *P. acnes*
protein	*Staph. epidermidis*, *Staph. aureus*, *P. acnes*, *M. luteus*, *Strep. pneumoniae*, viridans streptococci, lactobacilli, *H. influenzae*
lipid	*Staph. epidermidis*, *Staph. aureus*, *P. acnes*, *Corynebacterium* spp.

Note: Often, a single species can only partially hydrolyse a macromolecule – this is usually the case for mucins.

they secrete cleave sugar residues from the mucin molecule, and these are used as carbon and energy sources. The rest of the mucin molecule may then be degraded further by these organisms, or by others with the appropriate hydrolytic enzymes (i.e., glycosidases and proteases). Urea can be hydrolysed by *Staph. epidermidis, Staph. aureus, M. luteus*, some *Corynebacterium* spp., and some viridans streptococci – resulting in the liberation of ammonia and the formation of ammonium ions which can be used as nitrogen sources by a variety of organisms.

The tear film is slightly alkaline, with a mean pH of approximately 7.5, although this value increases slightly with increasing tear flow rate. During prolonged eyelid closure, the pH decreases to approximately 7.25 due to the dissolution of trapped carbon dioxide produced by the cornea. This pH is suitable for the growth of many organisms present on the skin, which is the main source of ocular microbes. While the temperature of most body sites remains at a fairly constant value of 37°C, the temperature of the conjunctiva is usually several degrees lower – except in very hot and humid conditions. The main causes of heat loss from the eye are evaporation of the tear film, convection, and radiation. The ocular surface, more than any other region of the body, is subject to marked temperature changes due to variations in the temperature, humidity, and velocity of the surrounding air. While the temperature of the skin is also affected by these environmental factors, heat loss in this case is usually minimised by clothing. An ambient temperature of 20°C in still air results in a conjunctival temperature of 32°C, but this is unlikely to have a dramatic effect on the growth of most mesophilic organisms. However, when the air temperature falls to 6°C, the temperature of the conjunctivae drops to approximately 27°C in still air and to 17°C when the air velocity is 4 m/second. Given that the growth rate of many microbes halves for each 10°C fall in temperature, such temperature decreases would have a profound effect on the microbial growth. Furthermore, certain organisms, such as *Mycobacterium tuberculosis*, have a narrow temperature range for growth (30–39°C), while some streptococci (e.g., *Strep. mutans* and *Strep. sobrinus*) and Gram-negative species (e.g., *Pseudomonas cepacia, Alcaligenes faecalis*) do not grow well below 30°C. Thus, the lower conjunctival temperature could select against the survival of such organisms on the ocular surface.

Given the very large surface-area-to-volume ratio of the tear film, it is not surprising that the film is highly aerobic – its dissolved oxygen content is generally at least 75% of that of oxygen-saturated water. The survival of anaerobes and microaerophiles (e.g., *Propionibacterium* spp. and *Peptostreptococcus* spp.) will be limited to microhabitats with a low oxygen content created by oxygen-consuming aerobes and facultative anaerobes.

As described in Section 3.2, the conjunctival surface is equipped with a wide range of antimicrobial mechanisms that will have to be overcome before microbial colonisation can occur.

3.4 | The indigenous microbiota of the eye

There is continuing debate as to whether the organisms that can be detected on the outer surfaces of the eye constitute a resident microbiota or are merely transients. Two pieces of evidence in favour of the latter hypothesis are (1) samples taken from the conjunctivae of as many as 40% of healthy individuals do not yield any cultivable microbes, and (2) the organisms that can be cultivated are usually typical of those found on the skin. However, remarkably few studies have employed modern molecular detection methods in the analysis of such samples so that the apparently "sterile" ocular samples obtained may be an artefact arising from the inadequate culture media used and/or the presence of uncultivable or difficult-to-grow species in the microbiota. For example, in a recent study of the conjunctival microbiota, no bacteria were cultivated from 55% of samples in which bacteria were detected by polymerase chain-reaction amplification of 16S rRNA genes. Furthermore, while many (but not all) of the organisms that are cultured from the conjunctivae can also be found on the skin, not all of the organisms found on the skin can be detected on the conjunctivae. Consequently, while the skin may be the source of many of the organisms detected on the surface of the eye, the ocular environment selects for the survival of only some of these species, and these are regularly and consistently detectable on the conjunctivae.

3.4.1 Main characteristics of key members of the ocular microbiota

Most of the organisms comprising the ocular microbiota are also present on the skin and have already been described in Chapter 2. Although *Strep. pneumoniae* and viridans streptococci are also found on the conjunctivae, they are best known as members of the microbiota of the respiratory tract and are described in detail in Chapter 4.

The major adhesins and their receptors (where known) of members of the ocular microbiota that are relevant to colonisation of the eye are shown in Table 3.5. Other virulence factors of these organisms are described in appropriate sections of Chapters 2 and 4.

3.4.2 Acquisition of the ocular microbiota

The nature of the ocular microbiota of neonates is profoundly affected by their mode of delivery. Hence, in neonates delivered vaginally, the first colonisers of the conjunctiva are predominantly members of the genital microbiota of the mother. In contrast, those delivered by caesarean section have a much sparser conjunctival microbiota, and the organisms present are cutaneous species. In a study of 100 neonates, the mean numbers of bacterial species per conjunctiva in vaginally delivered individuals and those delivered by caesarean section were 1.84 and 0.50, respectively, while the numbers of

Table 3.5. Adhesins of members of the ocular microbiota and their receptors

Organism	Possible adhesin	Receptor
Staph. aureus	fibronectin-binding protein	fibronectin on conjunctival epithelial cells
	vitronectin-binding protein	vitronectin on conjunctival epithelial cells
	collagen-binding protein	collagen exposed when conjunctiva is damaged
coagulase-negative staphylococci	fibronectin-binding protein	fibronectin on conjunctival epithelial cells
	vitronectin-binding protein	vitronectin on conjunctival epithelial cells
streptococci	fibronectin-binding protein	fibronectin on conjunctival epithelial cells
	vitronectin-binding protein	vitronectin on conjunctival epithelial cells
H. influenzae	fimbrial adhesins	sialic acid-containing molecules and sulphated glycosaminoglycans

viable bacteria present on each conjunctiva were 1,790 cfu and 272 cfu, respectively. The conjunctival microbiota of the vaginally delivered neonates was dominated by lactobacilli and bifidobacteria, whereas in neonates delivered by caesarean section, propionibacteria and *Corynebacterium* spp. predominated (Table 3.6). Furthermore, in 80% of the neonates delivered by caesarean section, the conjunctivae were sterile, whereas in vaginally delivered individuals only 20% of the conjunctivae were sterile. These findings have two major implications. Firstly, in those neonates delivered by caesarean section, the absence of a protective ocular microbiota in 80% of individuals would increase the risk of colonisation of the conjunctiva by potentially pathogenic, multiple drug-resistant organisms often present in hospital environments (e.g., methicillin-resistant *Staph. aureus*). It is of concern that, in those infants delivered by caesarean section who did have bacteria on their conjunctivae, the organisms present appeared to have been

Table 3.6. Conjunctival microbiota of neonates delivered vaginally (56 individuals) and by caesarean section (30 individuals)

Organism	Mean number of viable bacteria per conjunctiva (cfu)	
	Vaginal delivery	Caesarean delivery
Lactobacillus spp.	718	0
Bifidobacterium spp.	499	0
E. coli	217	0
coryneforms	176	0.03
Propionibacterium spp.	83	2
Staph. epidermidis	82	0.10
Bacteroides spp.	21	0
Corynebacterium spp.	0.8	0.33
Staph. aureus	0.8	0
Aerococcus spp.	0.29	0

derived from the skin of their mother or clinical personnel. These individuals may be carriers of MRSA or other pathogenic organisms. In contrast, those neonates delivered vaginally had a highly varied microbiota characteristic of that found in the genital tract of the mother. Such a varied microbiota would provide protection against coloni-sation by exogenous pathogens. However, had the lower genital tract of the mother been colonised by organisms such as *Chlamydia* spp., *Neisseria gonorrhoeae*, or other pathogens, then vaginal delivery would increase the chances of conjunctival colonisation by these well-known ocular pathogens. Conjunctivitis occurs in 12% of neonates and is most frequently caused by *N. gonorrhoeae* or *Chlamydia trachomatis*. Administration of silver ni-trate to the conjunctivae of neonates has been an effective means of preventing these infections since 1881.

Within 5 days of birth, the conjunctival microbiota changes to one that is gener-ally dominated by coagulase-negative staphylococci (CNS), coryneforms, streptococci (including *Strep. pneumoniae*), and propionibacteria. These organisms are acquired pri-marily by transfer from the skin and respiratory tract of the infant, although the skin and respiratory tract of the parents and other individuals are other likely sources. During the next two decades, the composition of the conjunctival microbiota remains fairly constant, except that the frequency of isolation of *Strep. pneumoniae* decreases.

3.4.3 Composition of the indigenous microbiota of the eye

Microbes can be cultivated from most conjunctivae, but in different studies, the pro-portion of samples from which no organisms could be grown ranged from 0% to 47%. In individuals free of any eyes disease, the indigenous microbiotas of the left and right eyes are remarkably similar. Furthermore, the microbiotas of the lid margins and conjunctivae are also very similar, although the actual numbers of bacteria iso-lated tend to be highest at the lid margin. The eyelids support a microbial community identical to that found on adjacent skin regions. This section, therefore, describes only the microbiota of the conjunctival surface. The number of organisms cultivated from the conjunctival surface varies enormously, ranging from several hundred to more than 5×10^4 cfu in different studies. In general, only a limited number of species are present – usually fewer than three. The species most frequently isolated are CNS, coryne-forms, and *P. acnes*, and these generally account for 80–90% of the total cultivable microbiota (Table 3.7).

Unfortunately, few studies have undertaken speciation of the CNS isolated. However, in one study where this was undertaken, it was found that the proportions of the various species were as follows: *Staph. epidermidis*, 57%; *Staph. capitis*, 13%; *Staph. simulans*, 9%; *Staph. hominis*, 2%; *Staph. warneri*, 2%; *Staph. saprophyticus*, 2%; and unclassifiable staphylococci, 15%. Of interest is the finding that all of the corynebacterial members of the normal ocular microbiota appear to be lipophilic and have been identified as CDC Group G: *C. macginleyi*, *C. afermentans* subsp. *lipophilum*, *C. accolens*, and *C. jeikeium*. These lipophilic species are not frequently isolated from any other site on the human body, supporting the view that the conjunctiva does indeed have its own distinctive, indigenous microbiota.

As previously described in Section 3.2, the composition of tears alters during pe-riods of eye closure, and this might be expected to change the composition of the conjunctival microbiota. Although few studies have addressed this possibility,

Table 3.7. Organisms present on the conjunctivae of healthy adults

Organism	Frequency of isolation (%)	Proportion of cultivable microbiota (%)
coagulase-negative staphylococci	30–83	39–40
coryneforms	6–52	30–37
P. acnes	44	21
streptococci	1–26	0–1
Micrococcus spp.	14–22	0–1
Staph. aureus	2–36	3–17
Gram-negative rods	4–10	0–1
Gram-positive anaerobic cocci	6	0–1
Neisseria spp.	1–2	0–1
H. influenzae	<1	2–7

Note: Frequency of isolation = % of the study population from which the organism was isolated.

those that have been carried out point to an increase in the total microbial load rather than to any change in the composition of the microbiota. The conjunctival microbiota alters with age, with a trend towards a greater frequency of isolation of coryneforms and decreased frequency of *Strep. pneumoniae* with increasing age (Figure 3.5). Also, Gram-negative bacteria tend to be isolated more frequently from older individuals.

3.4.4 Interactions among members of the ocular microbiota

The dominant genera of the ocular microbiota, the staphylococci and coryneforms, as well as the propionibacteria, are all known producers of bacteriocins. However, the role of these antagonistic compounds in maintaining the composition of the ocular microbial community has not been investigated. Similarly, metabolic end-products

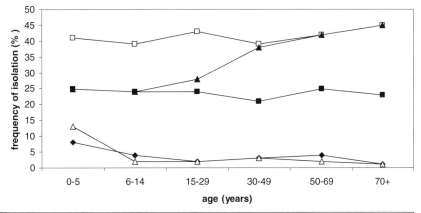

Figure 3.5 Changes in the frequency of isolation of various organisms from the conjunctivae with age. □ = *Staph. epidermidis*, ■ = *Staph. aureus*, ▲ = coryneforms, △ = *Strep. pneumoniae*, ◆ = viridans streptococci.

(e.g., acetic and lactic acids) of ocular microbes, as well as acting as nutrients for other species, may be responsible for excluding certain organisms from the conjunctival surface. That such antagonistic effects operate *in vivo* has been shown in a number of investigations. For example, the results of a study of organisms isolated from the conjunctivae of 410 healthy adults showed that when *Staph. aureus* is present on the conjunctiva, the chances of finding either CNS or coryneforms are significantly reduced. Some strains of *Staph. aureus* have been shown to produce a bacteriocin (Bac 1829) that is inhibitory to a number of *Corynebacterium* spp. Another bacteriocin of *Staph. aureus*, BacR1, is inhibitory to several *Staphylococcus* spp., as well as a variety of *Corynebacterium* spp. The study also found a synergistic relationship between CNS and corynebacteria – when either of these groups of organisms was present, the chances of finding members of the other group were significantly enhanced. This has also been observed in a number of other studies. Although the basis of this synergy has not been established, it is known that some *Corynebacterium* spp. are dependent on certain fatty acids for growth and that some staphylococci have lipases which can hydrolyse lipids, thereby releasing these fatty acids. It may be, therefore, that the staphylococci can enhance the growth of such corynebacteria. However, what benefit the staphylococci derive from the corynebacteria is not known. Ocular strains of micrococci have also been shown to secrete a number of antimicrobial peptides. These are active mainly against Gram-positive species and show very high activity against micrococci, corynebacteria, *Sarcina* spp., and *Bacillus* spp. and moderate activity against streptococci, clostridia, and mycobacteria. Their effect, if any, on the composition of the ocular microbiota has not been established.

3.4.5 Dissemination of organisms from the eye

The continuous flow of tears transports organisms from the conjunctivae through the inferior lacrimal canal and into the nasal cavity. From here, they may be propelled into the nasopharynx by mucociliary action. They may then be swallowed or expelled into the environment during sneezing and coughing. Excessive tear production will also transport ocular microbes onto the skin surface. Rubbing the eyes, of course, will transfer microbes onto the hands from where they can be disseminated further.

3.4.6 Effect of antibiotics and other interventions on the ocular microbiota

3.4.6.1 Antibiotic administration

Little is known with regard to the effect of antibiotic administration on the ocular microbiota. Because of the low population density and the predominance of Gram-positive species, administration of antibiotics such as penicillins, erythromycin, clindamycin, and cephalosporins can often eliminate the indigenous conjunctival microbiota. However, there is no evidence that the conjunctiva becomes colonised by allochthonous species as a consequence of this. Trachoma is the leading cause of blindness due to infection worldwide. As part of a programme to eliminate the disease, the World Health Organisation has started the mass distribution of systemic antibiotics. One of the most effective antibiotics against the causative agent, *Chlam. trachomatis*, is azithromycin, which is able to eliminate the organism after only a single dose. In a study of the effects of a single dose of azithromycin on the conjunctival microbiota of children in a trachoma-endemic area of Nepal, there was a significant increase in the isolation rate of azithromycin-resistant *Strep. pneumoniae* after treatment. Tetracycline is also used to treat trachoma and other eye infections, and approximately 75% of ocular isolates of *Staphylococcus* spp. are now reported to be resistant to this antibiotic.

3.4.6.2 Contact lens wear

Contact lenses (CLs) are a popular means of vision correction and are worn by approximately 85 million individuals worldwide. The majority of CL wearers use either rigid gas-permeable or soft hydrogel lenses. There are a number of ways in which these lenses can be worn: (1) daily wear with disinfection at night and re-use the next day, (2) daily disposable wear – a new lens is used each day, (3) extended wear – the lens is used continuously for 1 week and then replaced, and (4) continuous wear – the lens is used continuously for 1 month and then replaced. The wearing of a CL might be expected to alter the ocular microbiota for a number of reasons. Firstly, it would offer an additional substratum for microbial adhesion and thereby possibly encourage colonisation by microbes other than members of the indigenous microbiota, or else it could disturb the composition of the resident microbial community. Secondly, it may alter the ocular environment in some way, hence inducing a shift in the ocular microbiota. Studies have shown that bacteria do adhere to the CL and that the extent of colonisation depends on the nature of the material used. With regard to the CL disturbing the ocular environment, it has been shown that the oxygen content of the tear film underlying the lens may be markedly reduced – again, this is related to the nature of the material used in the manufacture of the CL. Studies have shown that the numbers of organisms present on the eyelids and conjunctivae increase over a 12-month period in those individuals using a daily-wear schedule, but not in those using an extended-wear schedule. The increase in numbers, however, occurs only in species that are members of the indigenous microbiota. Of concern is the finding that while the composition of the ocular microbiota of the daily-wear users is not altered, that of the extended-wear users is altered – the main change being an increase in the frequency of isolation of potential pathogens (i.e., Gram-negative rods) from both the conjunctivae and eyelids. The potentially pathogenic organisms include species belonging to the following genera: *Pseudomonas, Achromobacter, Acinetobacter, Klebsiella, Serratia*, and *Xanthomonas*. It is relevant to note that the environmental organism *Ps. aeruginosa* is an efficient coloniser of many types of CL material and that its viability is unaffected by many CL disinfecting solutions. It can, therefore, form a biofilm on the CL. CL wearers have an increased risk of developing microbial keratitis, and in 70% of such cases, the causative agent is *Ps. aeruginosa* (Section 3.5.3). Another important cause of keratitis associated with CL wear is the protozoan *Acanthamoeba*.

3.5 | Diseases caused by members of the ocular microbiota

The most important diseases caused by members of the ocular microbiota are summarised in Table 3.8.

Table 3.8. Eye infections in which members of the ocular microbiota are implicated

Disease	Causative organism
conjunctivitis	*Staph. aureus*, coagulase-negative staphylococci, *Strep. pneumoniae*, viridans streptococci, coryneforms, *Propionibacterium* spp.
blepharitis	*Staph. aureus*, coagulase-negative staphylococci
keratitis	*Staph. aureus*, *Strep. pneumoniae*
endophthalmitis	Coagulase-negative staphylococci, *Staph. aureus*, viridans streptococci, enterococci
orbital cellulitis	*Staph. aureus*, *Strep. pneumoniae*

3.5.1 Conjunctivitis

This is one of the most frequently occurring ocular infections and is often caused by viruses. However, bacteria are the causative agents in as many as 80% of cases. Exogenous pathogens are frequently responsible for the disease, and in neonates the causative agent is often *N. gonorrhoeae* or *Chlamydia trachomatis*, while in older individuals only the latter organism is usually involved. Nevertheless, members of the indigenous microbiota are often associated with the disease, with *Staph. aureus* and CNS being the most frequent causative agents. In children, *Strep. pneumoniae, H. influenzae*, and *Moraxella* spp. are often responsible for the infection. While these are principally members of the respiratory microbiota, *Strep. pneumoniae* is also a member of the ocular microbiota. The strains of *H. influenzae* responsible for conjunctivitis are frequently identical to those present in the nasopharynx of the infected child. In a recent study of the aetiology of purulent conjunctivitis, 16S rRNA gene sequencing and DGGE fingerprinting revealed that in 40% of the twenty-nine cases studied, the infections were polymicrobial. The most frequently detected genera were *Staphylococcus* and *Corynebacterium*, followed by *Propionibacterium* and *Streptococcus* – all of these being members of the ocular microbiota. Other occasional isolates belonged to the genera *Bacillus, Acinetobacter, Brevundimonas, Pseudomonas*, and *Proteus*.

In neonates, the most frequent causes of conjunctivitis are the exogenous pathogens *N. gonorrhoeae* and *Chlamydia trachomatis*, as well as endogenous organisms, such as *E. coli, Staph. aureus*, and *H. influenzae*.

3.5.2 Blepharitis

Infection of the eyelids, blepharitis, is one of the most common ocular infections. It is usually caused by *Staph. aureus* or CNS and can often lead to the formation of a stye which is a localised painful infection of the follicles. Sometimes the meibomian glands are also involved, resulting in the formation of a cyst.

3.5.3 Keratitis

This is an infection of the cornea and is usually due to the herpes simplex or herpes zoster viruses. However, bacteria can be the causative agent in immunosuppressed individuals or in those who have suffered damage to the eye. In such cases, *Staph. aureus* or *Strep. pneumoniae* are usually involved. Individuals using CLs are also at risk of developing keratitis. In those wearing rigid gas-permeable lenses, the incidence of keratitis can be up to 0.04%, while in those using soft hydrogel lenses, it may be as high as 0.6%. A wide range of organisms have been isolated from the disease lesion, but in approximately two-thirds of cases, the causative agent is the environmental organism *Ps. aeruginosa*. Members of the indigenous ocular microbiota implicated include CNS, *Corynebacterium* spp., *Propionibacterium acnes, Staph. aureus*, viridans streptococci, and *H. influenzae*. However, these organisms are much less frequently involved.

The first step in the sequence of events leading to keratitis in CL wearers is contamination of the CL. Within 24 hours of its insertion, more than 90% of the CL is covered by a conditioning film consisting of glycoproteins, proteins, and lipids derived from the tear film. Bacteria may, therefore, adhere either to the lens material itself or to this conditioning film. The sources of adherent organisms include the eye, the skin of the fingers and thumbs (during handling of the lens), the environment, and the case used for storage and/or cleaning of the CL. Once the lens has been colonised, biofilm formation may take place, resulting in very high microbial densities on the surface. The biofilm formed is often resistant to the disinfection and cleaning procedures routinely used for CL maintenance, and further growth may take place in the CL case during such procedures. Furthermore, it has been estimated that as many as 70% of CL wearers do not comply with the cleaning procedures recommended by manufacturers, and this exacerbates the problem. The CL case itself may be a source of infecting organisms, as it has been shown that as many as 81% of cases are themselves contaminated. Although contamination of, and biofilm formation on, the CL is a prerequisite for the initiation of keratitis, it has been shown that the disease will not occur unless the cornea is already damaged in some way. Such damage may occur during the insertion or removal of the CL, as a consequence of the reduced oxygen content beneath the CL during wear, or because of the toxicity of residual disinfectant/cleanser from the disinfection process. Bacteria present on the CL then adhere to the damaged epithelium. The presence of the CL alters the mucin layer of the tear film, which is responsible for preventing bacterial adhesion to the epithelium. Furthermore, the CL interferes with the ocular defence systems by hindering access of the tear film to the site of corneal damage.

While much is known about the adhesin–receptor interactions involved in the adhesion of *Ps. aeruginosa* to the cornea, far less is known about the adhesion of members of the ocular microbiota. One of the adhesins mediating binding of *Staph. aureus* to the cornea is a collagen-binding protein (Cna). An isogenic mutant of *Staph. aureus* lacking the adhesin (Cna$^-$) was found to not bind collagen, whereas the parent strain (Cna$^+$) and the isogenic mutant complemented with an intact version of the gene (*cna*) did bind collagen. In an animal model of keratitis, contact lenses infected with the parent strain and the mutant complemented with the *cna* gene-induced keratitis, whereas the Cna$^-$ strain did not (Figure 3.6). When the collagen-binding adhesin from *Staph. aureus* was applied to corneal epithelium and then a suspension of the organism was applied, the number of adherent bacteria was substantially reduced compared with controls which had not been administered the adhesin. These experiments show the importance of this adhesin in the pathogenesis of CL-induced keratitis. While a number of the toxins secreted by *Staph. aureus* (Section 4.4.1.5) may result in damage to the cornea, the α-toxin is likely to be of particular importance. This toxin has been shown to induce inflammation of the conjunctiva and cause necrosis and apoptosis in corneal

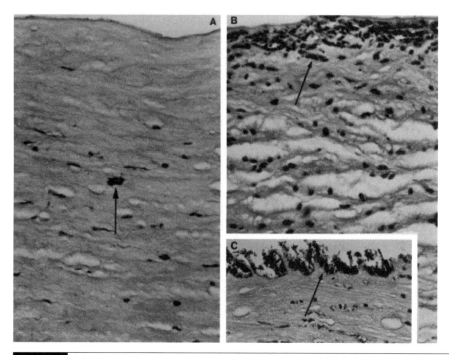

Figure 3.6 Differences in experimental keratitis in the rabbit produced by different strains of *Staph. aureus*. (A) Minimal infiltration of neutrophils (arrow) in a cornea exposed to the Cna^- mutant of the organism (magnification, ×240). (B) Extensive neutrophil infiltration (arrow) arising from infection with the parental Cna^+ strain (magnification, ×240). (C) Adherent cocci (arrow) on the surface of a cornea infected with the parental Cna^+ *S. aureus* strain (magnification, ×360). Reproduced with permission from: Rhem, M.N., *et al. Infection and Immunity* 2000;68:3776–3779.

epithelial cells, thereby enabling penetration of the organism into the deeper layers of the cornea. It also causes severe damage to the iris. The β-toxin, in contrast, causes severe inflammation of the sclera. Together, the activities of these two toxins account for most of the bacterially induced damage accompanying CL-induced keratitis.

3.5.4 Endophthalmitis

Endophthalmitis is a serious infection of the posterior sections of the interior of the eye that generally results in either partial or complete blindness. Fortunately, the outer defences of the eye are very effective at preventing microbial invasion so that the disease is uncommon. There are three pathways by which microbes can gain entry to the internal tissues of the eye: (1) following ocular surgery, (2) during a penetrating trauma, and (3) via the bloodstream from a distant site. The most commonly performed type of ocular surgery is the removal of cataracts. Post-operative endophthalmitis is a rare complication of such operations and in the United States, for example, the infection occurs in approximately 0.1% of cases. The organisms involved are members of the resident ocular microbiota, with up to one-third of infections being polymicrobial in nature. CNS account for approximately 70% of cases, and the majority (>80%) of these are *Staph. epidermidis*. Genetic analysis of the isolates responsible for the infections has shown that, in most cases (82%), the indigenous ocular or nasal microbiota is the source of the *Staph. epidermidis* strain responsible for the infection. Other causative agents, in decreasing order of frequency, are *Staph. aureus*, viridans group streptococci, other Gram-positive bacteria (including enterococci), and Gram-negative bacteria (*Proteus* spp. and *E. coli*).

As many as 17% of penetrating injuries of the eye result in endophthalmitis. Although the main causative agents are, again, members of the indigenous ocular microbiota (particularly CNS), as one might expect, environmental organisms such as *Bacillus* spp. and *Pseudomonas* spp. are also frequently involved. Indeed, *Bacillus cereus* is the second most frequent cause of post-traumatic endophthalmitis. The third form of the disease, in which bacteria gain access to the ocular tissue via the bloodstream, is rare – less than 8% of endophthalmitis cases are attributable to infection via this route. Individuals affected usually include those who are on immunosuppressive therapy, are immunocompromised, are intravenous drug abusers, or have prolonged indwelling devices. The most common aetiological agent is *Can. albicans*, but a wide range of other species is found, including members of the cutaneous and intestinal microbiotas, as well as environmental organisms.

3.5.5 Orbital cellulitis

This is a serious infection of the orbital contents that can result in blindness and death. It is an uncommon infection that sometimes follows injury to, or surgery on, the eye. In some patients, there is evidence of a prior infection of the paranasal sinus, the middle ear, or the teeth. The causative agents include members of the ocular microbiota (*Staph. aureus* and *Strep. pneumoniae*) and members of the respiratory microbiota (*Strep. pyogenes* and *H. influenzae*).

3.5.6 Dacryocystitis

This is a disease involving one or more of the components of the tear drainage system (i.e., the lacrimal gland, lacrimal sac, and the nasolacrimal duct). This system is involved not only in the drainage of tears from the eye, but also is part of the mucosa-associated lymphoid tissue. The epithelium lining the system produces a range of molecules with antimicrobial activities, including lysozyme, lactoferrin, phospholipase A_2, and a number of antimicrobial peptides. Nevertheless, members of the ocular microbiota can infect the system – in particular, *Staph. aureus*, CNS, *Strep. pneumoniae*, and other streptococci. Other organisms that have been implicated in the disease are enterobacteria. The inflammation produced during the infection often results in blockage of the nasolacrimal duct.

3.6 | Further Reading

Reviews and Papers

Aaberg, T.M., Flynn, H.W., Schiffman, J., and Newton, J. (1998). Nosocomial acute-onset postoperative endophthalmitis survey: a 10-year review of incidence and outcomes. *Ophthalmology* **105**, 1004–1010.

Armstrong, R.A. (1998). The immune system and the eye. *Ophthalmic & Physiological Optics* **18** (Suppl 2), S40–S48.

Armstrong, R.A. (2000). The microbiology of the eye. *Ophthalmic & Physiological Optics* **20**, 29–41.

Benz, M.S., Scott, I.U., Flynn, H.W. Jr., Unonius, N., and Miller, D. (2004). Endophthalmitis isolates and antibiotic sensitivities: a 6-year review of culture-proven cases. *American Journal of Ophthalmology* **137**, 38–42.

Berry, M., Harris, A., Lumb, R., and Powell, K. (2002). Commensal ocular bacteria degrade mucins. *The British Journal of Ophthalmology* **86**, 1412–1416.

Brook, I. (2001). Ocular infections due to anaerobic bacteria. *International Ophthalmology* **24**, 269–277.

Callegan, M.C., Engelbert, M., Parke, D.W., Jett, B.D., and Gilmore, M.S. (2002). Bacterial endophthalmitis: epidemiology, therapeutics, and bacterium-host interactions. *Clinical Microbiology Reviews* **15**, 111–124.

Chern, K.C., Shrestha, S.K., Cevallos, V., Dhami, H.L., Tiwari, P., Chern, L., Whitcher, J.P., and Lietman, T.M. (1999). Alterations in the conjunctival bacterial flora following a single dose of azithromycin in a trachoma endemic area. *British Journal of Ophthalmology* **83**, 1332–1335.

Chisari, G., Cavallaro, G., Reibaldi, M., and Biondi, S. (2004). Presurgical antimicrobial prophylaxis: effect on ocular flora in healthy patients. *International Journal of Clinical Pharmacology and Therapeutics* **42**, 35–8.

DeAngelis, D., Hurwitz, J., and Mazzulli, T. (2001). The role of bacteriologic infection in the etiology of nasolacrimal duct obstruction. *Canadian Journal of Ophthalmology* **36**, 134–139.

Erdogan, H., Kemal, M., Toker, M.I., Topalkara, A., and Bakici, Z. (2002). Effect of frequent-replacement contact lenses on normal conjunctival flora. *The CLAO Journal* **28**, 94–95.

Goldstein, M.H., Kowalski, R.P., and Gordon, Y.J. (1999). Emerging fluoroquinolone resistance in bacterial keratitis: a 5-year review. *Ophthalmology* **106**, 1313–1318.

Haynes, R.J., Tighe, P.J., and Dua, H.S. (1999). Antimicrobial defensin peptides of the human ocular surface. *British Journal of Ophthalmology* **83**, 737–741.

Jackson, T.L., Eykyn, S.J., Graham, E.M., and Stanford, M.R. (2003). Endogenous bacterial endophthalmitis: a 17-year prospective series and review of 267 reported cases. *Survey of Ophthalmology* **48**, 403–423.

Jett, B.D. and Gilmore, M.S. (2002). Host-parasite interactions in *Staphylococcus aureus* keratitis. *DNA and Cell Biology* **21**, 397–404.

Knop, N. and Knop, E. (2000). Conjunctiva-associated lymphoid tissue in the human eye. *Investigative Ophthalmology & Visual Science* **41**, 1270–1279.

Kodjikian, L, Burillon, C., Chanloy, C., Bostvironnois, V., Pellon, G., Mari, E., Freney, J., and Roger, T. (2002). *In vivo* study of bacterial adhesion to five types of intraocular lenses. *Investigative Ophthalmology & Visual Science.* **43**, 3717–3721.

Leid, J.G., Costerton, J.W., Shirtliff, M.E., Gilmore, M.S., and Engelbert, M. (2002). Immunology of staphylococcal biofilm infections in the eye: new tools to study biofilm endophthalmitis. *DNA and Cell Biology* **21**, 405–413.

Levine, J. and Snyder, R.W. (1999). Practical ophthalmic microbiology. *Journal of Ophthalmic Nursing and Technology* **18**, 50–59.

Marone, P., Monzillo, V., Carretto, E., Haeusler, E. and Antoniazzi, E. (2002). *In vitro* activity of sagamicin against ocular bacterial isolates. *Ophthalmologica* **216**, 133–138.

McClellan, K.A. (1997). Mucosal defense of the outer eye. *Survey of Ophthalmology* **42**, 233–246.

McClellan, K.A., Cripps, A.W., Clancy, R.L., and Billson, F.A. (1998). The effect of successful contact lens wear on mucosal immunity of the eye. *Ophthalmology* **105**, 1471–1477.

Okhravi, N., Adamson, P., Carroll, N., Dunlop, A., Matheson, M.M., Towler, H.M.A., and Lightman, S. (2000). PCR-based evidence of bacterial involvement in eyes with suspected intraocular infection. *Investigative Ophthalmology & Visual Science* **41**, 3474–3479.

Paulsen, F.P., Pufe, T., Schaudig, U., Held-Feindt, J., Lehmann, J., Schröder, J.M., and Tillmann, B.N. (2001). Detection of natural peptide antibiotics in human nasolacrimal ducts. *Investigative Ophthalmology & Visual Science* **42**, 2157–2163.

Pleyer, U. and Baatz, H. (1997). Antibacterial protection of the ocular surface. *Ophthalmologica* **211** (Suppl 1), 2–8.

Qu, X. and Lehrer, R.I. (1998). Secretory phospholipase A2 is the principal bactericide for staphylococci and other Gram-positive bacteria in human tears. *Infection and Immunity* **66**, 2791–2797.

Ramachandran, L., Sharma, S., Sankaridurg, P.R., Vajdic, C.M., Chuck, J.A., Holden, B.A., Sweeney, D.F., and Rao, G.N. (1995). Examination of the conjunctival microbiota after 8 hours of eye closure. *The CLAO Journal* **21**, 195–199.

Schabereiter-Gurtner, C., Maca, S., Rolleke, S., Nigl, K., Lukas, J., Hirschl, A., Lubitz, W., and Barisani-Asenbauer, T. (2001). 16S rDNA-based identification of bacteria from conjunctival swabs by PCR and DGGE fingerprinting. *Investigative Ophthalmology & Visual Science* **42**, 1164–1171.

Schaefer, F., Bruttin, O., Zografos, L., and Guex-Crosier, Y. (2001). Bacterial keratitis: a prospective clinical and microbiological study. *British Journal of Ophthalmology* **85**, 842–847.

Sechi, L.A., Pinna, A., Pusceddu, C., Fadda, D., Carta, F., and Zanetti, S. (1999). Molecular characterization and antibiotic susceptibilities of ocular isolates of *Staphylococcus epidermidis*. *Journal of Clinical Microbiology* **37**, 3031–3033.

Sotozono, C., Inagaki, K., Fujita, A., Koizumi, N., Sano, Y., Inatomi, T., and Kinoshita, S. (2002). Methicillin-resistant *Staphylococcus aureus* and methicillin-resistant *Staphylococcus epidermidis* infections in the cornea. *Cornea* **21** (7 Suppl), S94–S101.

Stapleton, F., Willcox, M.D.P., Fleming, C.M., Hickson, S., Sweeney, D.F., and Holden, B.A. (1995). Changes to the ocular biota with time in extended- and daily-wear disposable contact lens use. *Infection and Immunity* **63**, 4501–4505.

Tan, K.O., Sack, R.A., Holden, B.A., and Swarbrick, H.A. (1993). Temporal sequence of changes in tear film composition during sleep. *Current Eye Research* **12**, 1001–1007.

Thielen, T.L., Castle, S.S., and Terry, J.E. (2000). Anterior ocular infections: an overview of pathophysiology and treatment. *Annals of Pharmacotherapy* **34**, 235–245.

von Graevenitz, A., Schumacher, U., and Bernauer, W. (2001). The corynebacterial flora of the normal human conjunctiva is lipophilic. *Current Microbiology* **42**, 372–374.

Watanabe, K., Numata-Watanabe, K., and Hayasaka, S. (2001). Methicillin-resistant staphylococci and ofloxacin-resistant bacteria from clinically healthy conjunctivas. *Ophthalmic Research* **33**, 136–139.

Willcox, M.D. and Holden, B.A. (2001). Contact-lens–related corneal infections. *Bioscience Reports* **21**, 445–461.

Willcox, M. and Stapleton, F. (1996). Ocular bacteriology. *Medical Microbiology Reviews* **7**, 123–131.

Willcox, M.D., Harmis, N.Y., and Holden, B.A. (2002). Bacterial populations on high-Dk silicone hydrogel contact lenses: effect of length of wear in asymptomatic patients. *Clinical and Experimental Ophthalmology* **85**, 172–175.

Zegans, M.E., Becker, H.I., Budzik, J., and O'Toole, G. (2002). The role of bacterial biofilms in ocular infections. *DNA and Cell Biology* **21**, 415–420.

Zierhut, M., Elson, C.O., Forrester, J.V., Killstra, A., Kraehenbuhl, J.P., and Sullivan, D.A. (1998). Mucosal immunology and the eye. *Immunology Today* **19**, 148–150.

4

The respiratory system and its indigenous microbiota

The respiratory system consists of a series of tubes (the respiratory tract) for conducting air to the respiratory membrane where oxygen and carbon-dioxide exchange occurs with the bloodstream, as well as a ventilatory system consisting of the lungs, diaphragm, and associated muscles. The primary function of the respiratory system is to supply oxygen to, and remove carbon dioxide from, the blood. The blood is transported to all cells of the body by the cardiovascular system. The site of interaction between the two systems is the lung – this organ being responsible for the exchange of gases between them. Other functions of the respiratory system include (1) regulating blood pH, (2) producing sounds, (3) eliminating water and heat, and (4) housing the receptors responsible for the sense of smell. Only the respiratory tract of the system is colonised by microbes and the ventilatory system is not described further.

4.1 | Anatomy and physiology of the respiratory tract

The respiratory tract, which is also referred to as the "conducting portion" of the respiratory system, delivers air to the lungs (known as the "respiratory portion"), where gas exchange takes place. The conducting portion consists of the nose, pharynx, larynx, trachea, and bronchi, whereas the respiratory portion consists of the bronchioles, alveolar ducts, alveolar sacs, and alveoli (Figure 4.1). The conducting portion, apart from the bronchi, is heavily colonised by microbes, whereas the respiratory portion is generally sterile. It is also sometimes convenient to divide the tract into the upper respiratory tract (nose and pharynx) and the lower respiratory tract (larynx, trachea, bronchi, and lungs).

4.1.1 Nose

This organ has four main functions: (1) to filter particulate matter from incoming air, (2) to warm and moisten the air to speed up the gaseous exchange process in the lungs, (3) to house olfactory receptors, and (4) to aid speech. It consists of an external, visible portion and a large internal cavity within the skull (Figure 4.2). Each of these is separated into a left side and a right side by the nasal septum. Air is taken in via the external nares (nostrils), which contain hairs (vibrissae) to filter out large particles. The nose retains 70–80% of particles with a diameter of 3–5 μm and 60% of particles with a diameter of 2 μm, but is unable to filter out particles with a diameter less than 1 μm. The air then passes over three bony protrusions (conchae) within

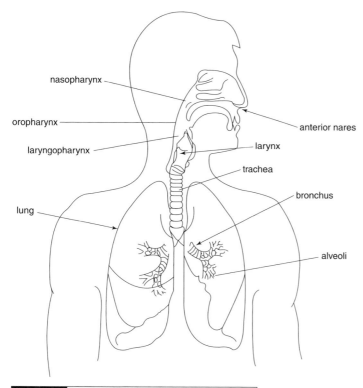

nasopharynx

oropharynx

laryngopharynx

lung

anterior nares

larynx

trachea

bronchus

alveoli

Figure 4.1 The main regions of the respiratory system.

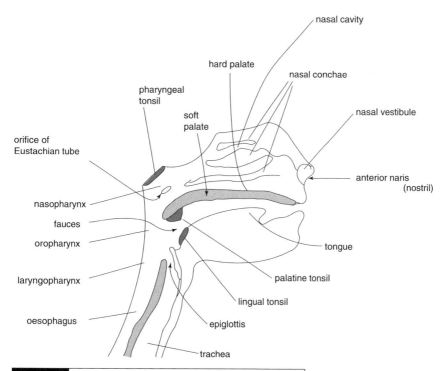

nasal cavity

hard palate

nasal conchae

pharyngeal tonsil

soft palate

nasal vestibule

orifice of Eustachian tube

anterior naris (nostril)

nasopharynx

fauces

oropharynx

tongue

laryngopharynx

palatine tonsil

oesophagus

lingual tonsil

epiglottis

trachea

Figure 4.2 The main anatomical features of the nose and pharynx.

Figure 4.3 Scanning electron micrograph of human nasopharyngeal epithelium showing abundant cilia. Magnification, ×3000. Reproduced with permission from: Shuter, J., *et al. Infection and Immunity* 1996;64:310–318.

the nasal cavity. The conchae provide a large surface area for warming and moisturising the air and are covered with mucus, which traps any small particles that may be present. The transit time of air through the nose is approximately 0.01 second, and during this time the inspired air is warmed to approximately 30°C.

Each naris opens into a vestibule which constitutes one of the anterior chambers of the nasal cavity. The anterior zone of the vestibule is lined by a keratinised, stratified, squamous epithelium which has hairs (but without erector muscles), sebaceous glands, and sweat glands. This extends between 1 and 2 cm inwards to join the non-keratinised squamous epithelium of the posterior region. This forms a transitional zone between the keratinised epithelium of the anterior region and the respiratory mucosa of the rest of the nasal cavity. The respiratory mucosa is a pseudo-stratified, ciliated, columnar epithelium containing numerous mucus-secreting goblet cells. The cilia move the mucus and entrapped particles (at a rate of approximately 1 cm/minute) backwards through the internal nares and into the pharynx, where it is swallowed or expelled by coughing. Approximately 1 litre of fluid is secreted (and swallowed) by the nasal epithelium each day.

4.1.2 Pharynx

This is a tube that extends from the internal nares down to the larynx and consists of three main regions: the nasopharynx, the oropharynx, and the laryngopharynx (Figure 4.2). The nasopharynx has a number of openings: (1) the internal nares through which air and mucus enter from the nasal cavity, (2) two openings that lead into the two Eustachian tubes, and (3) one opening into the oropharynx. At the back of the nasopharynx is a mass of tissue known as the pharyngeal tonsil which, when inflamed, is also known as the adenoids. The nasopharynx is lined with a pseudo-stratified, ciliated, columnar epithelium, the cilia of which move mucus towards the oral cavity (Figure 4.3). The oropharynx, as well as having connections to the nasopharynx and

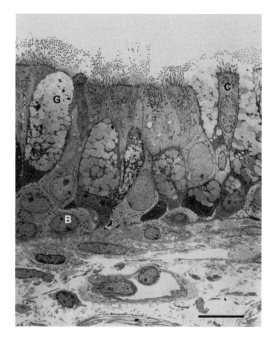

Figure 4.4 Transmission electron micrograph of a cross-section through the respiratory epithelium showing goblet cells (G), ciliated cells (C), and basal cells (B). Bar = 10 μm. Reproduced with permission from: Airway mucosa: secretory cells, mucus, and mucin genes. Jeffery, P.K. and Li, D. *European Respiratory Journal* 1997;10:1655–1662.

laryngopharynx, has an opening to the oral cavity known as the fauces through which food, drink, and air pass. It also contains the palatine and lingual tonsils. The oropharynx is lined with a non-keratinised, stratified, squamous epithelium because this provides greater protection from damage by food. The oropharynx opens into the laryngopharynx which is also lined with a non-keratinised, stratified, squamous epithelium, and this leads into the larynx and the oesophagus. The pharynx acts as a passageway for food, drink, and air, but also harbours the tonsils which are important lymphoid tissues.

4.1.3 Larynx

The larynx connects the pharynx with the trachea, and its main function is to act as a sphincter to prevent the entrance into the lungs of anything other than air. Its upper opening is protected by the epiglottis which, during swallowing, closes like a trap door to prevent the entrance of food and drink. It is lined mainly with a ciliated, columnar epithelium. The laryngeal mucosa has a number of folds which constitute the vocal cords. These are covered by a stratified, columnar epithelium and are responsible for sound production.

4.1.4 Trachea

The trachea connects the larynx to the lungs and branches at its lower end to form the two bronchi. It is lined with a pseudo-stratified, ciliated, columnar epithelium containing numerous goblet cells (Figures 1.14 and 4.4). The cilia move mucus and trapped particles upwards towards the pharynx.

4.1.5 Bronchi and bronchioles

The first two branches from the base of the trachea are known as the left and right primary bronchi. Each of these divides successively to form secondary bronchi, then

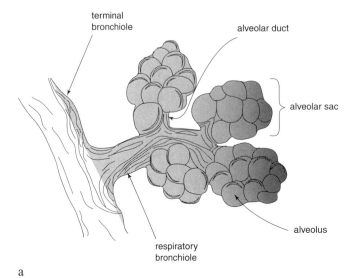

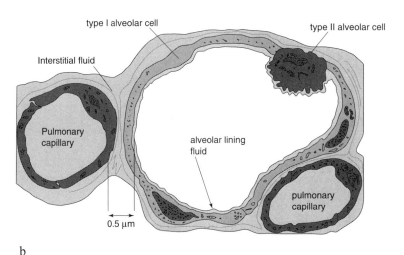

Figure 4.5 (a) The gross structure of an alveolar sac. (b) Cross-section through an alveolus.

tertiary bronchi, then bronchioles, and finally terminal bronchioles. The epithelium lining these structures changes as they get narrower. The bronchi are lined with a pseudo-stratified, ciliated, columnar epithelium containing goblet cells: larger bronchioles have a ciliated, simple, columnar epithelium with a few goblet cells; smaller bronchioles have a ciliated, simple, cuboidal epithelium with fewer goblet cells, whereas the terminal bronchioles have a non-ciliated, simple, cuboidal epithelium without goblet cells.

4.1.6 Alveolus

The terminal bronchioles divide further to form respiratory bronchioles, each of which subdivides into several (between two and eleven) alveolar ducts. Both of these structures are lined with a non-ciliated, simple, cuboidal epithelium. Each alveolar duct opens into an alveolar sac, which consists of a number of alveoli (Figure 4.5). An alveolus is a cup-shaped structure lined by two types of cells. The most numerous (approximately 95%) are the type I alveolar cells, which are simple squamous epithelial cells that form an almost continuous lining of the alveolus – this is the site of gaseous exchange and is known as the respiratory membrane. These cells are extremely flattened (approximately 0.2 μm in thickness) and have long cytoplasmic extensions, with each cell contributing to the lining of more than one alveolus. Interspersed among these are the type II alveolar cells, which are cuboidal epithelial cells with numerous microvilli – these secrete alveolar fluid and are also stem cells and so can replace damaged type I cells. The alveolar fluid provides a warm, moist interface between the air and the type I cells, thereby facilitating gaseous exchange. The fluid also contains surfactants (phospholipids and proteins) which prevent the alveoli from collapsing in on themselves. The alveoli are surrounded by a network of capillaries, and the air in the alveoli is separated from the blood in the capillaries by only four layers: the alveolar wall (i.e., a single layer of type I alveolar cells), the basement membrane underlying the alveolar wall, the basement membrane of the capillary, and the endothelium of the capillary. This four-layered respiratory membrane is only approximately 0.5 μm thick, thereby facilitating gaseous diffusion. Approximately 300 million alveoli are present in the lungs, and these have a total surface area of approximately 70 m^2, thus enabling the rapid exchange of oxygen and carbon dioxide between blood and the atmosphere.

4.2 | Antimicrobial defence mechanisms of the respiratory tract

The average adult inhales and exhales approximately 10,000 litres of air per day, of which 70% reaches the respiratory membrane, the rest remaining in the conducting portion of the tract. Given that the inspired air contains microbes and microbe-laden particulate matter which could potentially cause physical damage to, and infection of, the delicate respiratory membrane, it is not surprising that elaborate and effective mechanisms have evolved to prevent these eventualities. Before outlining the protective mechanisms operating at each region of the respiratory tract, one of these – mucociliary clearance – is described because it is not only one of the most important host defence mechanism, but also it operates in many parts of the conducting portion of the tract.

Mucociliary clearance is one of the first lines of defence of the respiratory tract and is largely responsible for maintaining the respiratory portion virtually free of microbes and particulate matter. Basically, the system involves the entrapment of microbes and particulate matter in a layer of mucus which is propelled by ciliated epithelial cells towards the oropharynx, where it is either swallowed or expectorated. This system (often termed the "mucociliary escalator") is found in the posterior two-thirds of the nasal cavity, the nasopharynx, and from the larynx up to (but not including) the terminal bronchioles. The ciliated cells are the most numerous cells in these regions, and each has approximately 200 cilia on its outer surface (Figures 4.3 and 4.4). Each cilium is approximately 6 μm long and between 0.1 and 0.2 μm in diameter and is surrounded by six microvilli which are much shorter (1 μm) but slightly thicker (0.1–0.3 μm). Each cilium has between three and seven 30-nm hooks projecting from its tip. The ciliated epithelium is covered by a two-layered film of liquid (i.e., airway surface liquid – ASL), the lower of which (the periciliary fluid) is watery and is in the form of a sol, while the upper layer (mucus gel layer) is more viscous and forms a gel; the layers are separated by a thin layer of surfactant (Figure 4.6).

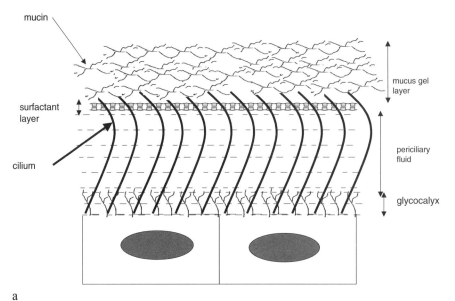

a

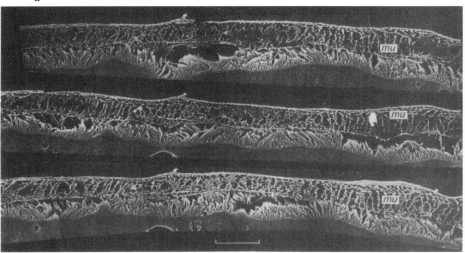

b

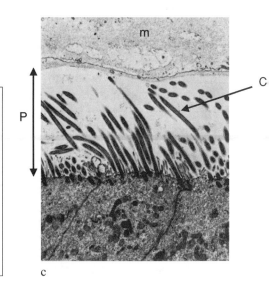

c

Figure 4.6 Figure 4.6. (a) Diagrammatic representation of the main components of the mucociliary escalator. (b) The two layers of airway surface liquid can be clearly seen in this cross-section through the tracheal epithelium of a rabbit (mu = mucus layer). Reproduced with the permission of the Company of Biologists Ltd. from: Ciliary activity of cultured rabbit tracheal epithelium: beat pattern and metachrony. Sanderson, M.J. and Sleigh, M.A. *Journal of Cell Science* 1981;47:331–347. (c) Transmission electron micrograph of a cross-section through the trachea of a rat showing the periciliary layer (P), the mucus layer (m), and cilia (C). Reproduced with permission from: Mucous blanket of rat bronchus: ultrastructural study. Yoneda, K. *American Review of Respiratory Disease* 1976;114:837–842.

Figure 4.7 Transmission electron micrograph of a Clara cell with its apex protruding into the airway lumen. Bar = 2.0 μm. Reproduced with permission from: Airway mucosa: secretory cells, mucus, and mucin genes. Jeffery, P.K. and Li, D. *European Respiratory Journal* 1997;10:1655–1662.

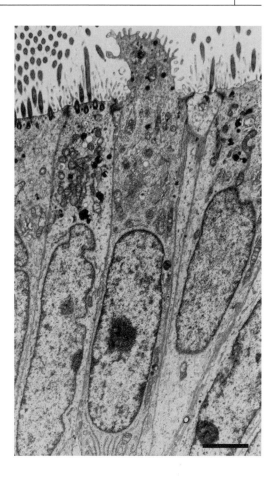

Estimates of the thickness of the ASL film vary widely and range from 5 to 120 μm. Recent determinations using more sophisticated, non-destructive techniques have indicated that, at least in the bronchi, the thickness of the ASL film is between 50 and 60 μm. The periciliary fluid has a depth similar to that of the length of the cilia (i.e., between 5 and 10 μm). Interspersed among the ciliated cells are a variety of secretory cells – goblet, Clara, and serous – as well as the openings of a number of different types of submucosal glands. Collectively, these cells and glands produce the periciliary fluid and mucus of the ASL. The goblet cells secrete mainly acidic mucins, whereas the serous cells produce neutral mucins and lipids. The Clara cells (Figure 4.7) are present only in the terminal bronchioles and produce surfactant (which is described later) and are stem cells which can give rise to both ciliated and mucus cells. The microvilli of the ciliated cells, together with numerous pinocytic vesicles in these cells, are involved in the absorption of fluid (and ions) and so are important in regulating the depth of the periciliary layer. The length of each cilium is such that it spans the periciliary fluid and contacts the mucus layer via its terminal hooks. The proper functioning of the mucociliary escalator requires that the periciliary layer is maintained at a certain depth – if it is too deep, the cilia will not touch the mucus layer and cannot move it forward; if it is too shallow, the mucus prevents the cilia from beating. The cilia of the epithelium beat in a sequential wave-like manner, with each cilium being at a slightly different stage in the beat cycle from its neighbour. In this way, the tips of

Table 4.1.	Antimicrobial compounds present in nasal fluid
Antimicrobial compound	Concentration
lysozyme	250–500 μg/ml
lactoferrin	80–200 μg/ml
secretory leukocyte proteinase inhibitor	10–80 μg/ml
secretory phospholipase A$_2$	–
human β-defensin-1	–
human β-defensin-2	0.3–0.4 μg/ml
human neutrophil peptide-1	–
elafin	–
statherin	–
peroxidase	–
glandulin	–
IgA	49–218 mg/100ml
IgG	14–136 mg/100ml
IgM	<10 mg/100ml

Note: – = concentration not known.

the cilia propel the mucus layer along on their forward stroke at a rate of between 5 and 20 mm/minute. The cilium then rests and, during its return, it bends to the right to avoid touching the mucus layer which would impede the forward movement of the mucus. During its return, it pushes against a neighbouring cilium in its resting phase which stimulates the latter to begin its recovery stroke. The frequency of ciliary beating varies with the anatomical location, but can be as high as 800 strokes/minute. As described later, several pathogens can interfere with ciliary function, thereby preventing mucociliary clearance.

The composition of the two layers of ASL differ mainly in terms of their content of macromolecules – the mucus layer having a number of mucins and other macromolecules, while the periciliary layer is largely devoid of such compounds. ASL contains a large variety of compounds, some of which have antimicrobial activities while others can serve as nutrients for microbes – its composition will be described in Section 4.3.2.

4.2.1 Nasal cavity

As described in Section 4.1.1, large particles present in inhaled air are removed by hairs in the nostrils, while smaller particles and suspended bacteria become trapped in the mucus covering the nasal mucosa. In the posterior two-thirds of the nasal mucosa, the mucociliary escalator propels the mucus-entrapped particles into the pharynx, and it has been shown that the whole of this mucosal layer is replaced every 10–20 minutes. However, this particular physical expulsion mechanism does not operate in the anterior region of the nasal mucosa, and the disposal of entrapped microbes is largely dependent on microbicidal substances present in nasal fluid. A wide range of antimicrobial agents have been detected in this fluid, although their presence and concentration vary widely among individuals (Table 4.1). Lysozyme and lactoferrin are major antimicrobial components of nasal fluid, accounting for 10–30% and 2–4% of its total protein content, respectively. Approximately 75% of plasma cells in the nasal mucosa are IgA-producing, the remainder being mainly producers of IgG. IgA accounts for 10–15% of the total protein content of nasal fluid and causes agglutination of bacteria resulting in high molecular mass complexes that are more easily expelled by the mucociliary escalator. Its other main function is to block microbial adhesion to epithelial cells. IgA also has some complement-activating ability. IgG is present in lower amounts (2–4% of the total protein content) in nasal fluid, but is found in higher concentrations in tissue fluid and is able to limit microbial invasion. Most of the other antimicrobial compounds found in nasal secretions are present at other body sites (their antimicrobial properties have been described in Sections 2.2 and 3.2). Statherin, however, has previously been identified only in saliva. This is a phosphoprotein which is active against many species found in the respiratory tract, including members of the anaerobic genera *Peptostreptococcus* spp., *Fusobacterium* spp., *Prevotella* spp., and *Bacteroides* spp. Elafin is a low molecular mass peptide (9.9 kDa) secreted by respiratory epithelial cells which inhibits human neutrophil elastase and neutralises some of the pro-inflammatory activities of lipopolysaccharide (LPS). It also has antimicrobial properties and can kill *Staph. aureus* and *Ps. aeruginosa*. Glandulin is a small molecule (<1 kDa) which is active mainly against Gram-negative species, including *Pseudomonas* spp. The peroxidase converts the hydrogen peroxide continually produced by the epithelium to reactive oxygen species, which have potent broad-spectrum antimicrobial properties.

The ability of nasal secretions to kill or inhibit the growth of a range of microbes has been demonstrated in a number of studies. Hence, *Staph. aureus, E. coli*, and *Ps. aeruginosa* have all been shown to be susceptible to the fluid, although the potency of the secretions from different individuals shows considerable variation. The antimicrobial effectiveness of the secretions is also markedly affected by the concentration and type of ionic constituents – high ionic strengths decrease antimicrobial activity (mainly because of their effect on antimicrobial peptides) as does a high concentration of Ca^{2+} ions.

Although the anterior portion of the nasal cavity does not have a functioning mucociliary clearance system, sneezing provides an alternative means of physically expelling mucus-entrapped microbes – and provides an excellent means of dispersal of such microbes.

4.2.2 Conducting portion of respiratory tract other than the nasal cavity

In the rest of the conducting portion of the respiratory tract (i.e., pharynx, larynx, trachea, bronchi, and bronchioles), mucociliary clearance is one of the main defence mechanisms against microbes. In the bronchi and bronchioles, this is aided by the branching nature of the system which results in microbes and particles in inhaled air impacting onto the mucus layer. Both sneezing and coughing are additional methods of physically expelling mucus-entrapped microbes from the respiratory tract. The cough reflex is activated whenever mucociliary clearance is either impaired or overwhelmed and results in the explosive release of air at a speed approaching that of sound. This results in the expulsion of mucus with its entrapped microbes and particulate matter.

These physical methods of expelling microbes are supplemented by the presence of antimicrobial substances in the ASL. These are produced mainly by submucosal glands, but epithelial cells, neutrophils, and macrophages are additional sources of

Table 4.2. Antimicrobial compounds detected in airway surface liquid

Antimicrobial compound	Concentration in ASL	Source
lysozyme	0.1–1.0 mg/ml	submucosal glands, epithelium, PMNs, macrophages
lactoferrin	0.1–1.0 mg/ml	submucosal glands, PMNs, macrophages
secretory leukoprotease inhibitor	0.01–0.10 mg/ml	submucosal glands, epithelium, macrophages, PMNs
neutrophil defensins	10 μg/ml	PMNs, epithelium
human β-defensin-1	1 μg/ml	epithelium, PMNs
human β-defensin-2	1 μg/ml	epithelium, PMNs
human β-defensin-3	–	epithelium, PMNs
human β-defensin-4	–	epithelium, PMNs
LL-37	–	epithelium, PMNs, macrophages
α-defensin HD-5	–	epithelium
phospholipase A_2	–	epithelium, PMNs
lactoperoxidase	0.65 mg/mg secreted protein	epithelium
BPI	–	PMNs
IgA	–	plasma cells
IgG	–	plasma cells
peroxidase	–	epithelium
anionic peptide	0.8–1.3 mM	epithelium
nitric oxide and reactive nitrogen species	–	epithelium, macrophages

Notes: – = concentration not known; ASL = airway surface liquid; BPI = bacterial permeability-inducing protein; PMNs = polymorphonuclear leukocytes.

these compounds. Again, a wide range of antimicrobial compounds has been detected in this fluid, and these are summarised in Table 4.2. Again, lysozyme and lactoferrin are major effector molecules of the host defence system in ASL which also contains many of the antimicrobial compounds found at other mucosal surfaces. Uniquely, ASL contains an anionic antimicrobial peptide which is active against Gram-positive and Gram-negative bacteria, including *E. coli*, *Ps. aeruginosa*, *K. pneumoniae*, *Ser. marcescens*, *Staph. aureus*, and *Ent. faecalis*. All of the components of the lactoperoxidase system (i.e., lactoperoxidase, thiocyanate, and hydrogen peroxide) have been detected in ASL, which has been found to exhibit lactoperoxidase-dependent killing of *H. influenzae* and *Ps. aeruginosa*.

Many of the compounds listed in Table 4.2 are expressed constitutively (e.g., human beta-defensin [HBD]-1, lysozyme, lactoferrin), while others (e.g., HBD-2, HBD-3, HBD-4) are induced in response to some infectious stimulus. Hence, levels of both HBD-2 and HBD-3 are increased dramatically by LPS, mucoid *Ps. aeruginosa*, interleukin-1β, and tumor necrosis factor-α. Furthermore, the concentrations of HBD-2, but not HBD-1, in the ASL of patients with inflammatory lung diseases – such as cystic fibrosis and pneumonia – are much higher than those found in disease-free individuals. Levels of HBD-4 produced by human respiratory epithelial cells are increased dramatically by the presence of *Strep. pneumoniae* or *Ps. aeruginosa*. HBD-4 inhibits the growth of *Staph. aureus, E. coli, Strep. pneumoniae*, and *Burkholderia cenocepacia* (an environmental organism responsible for lung infections in patients with cystic fibrosis) at high concentrations, but is active against *Staph. carnosus* and *Ps. aeruginosa* at much lower concentrations. Its activity is diminished by high concentrations of NaCl, but it displays synergy with other antimicrobial peptides, including lysozyme and HBD-3. Although the respiratory epithelium produces nitric acid (which has antimicrobial activity) constitutively, increased amounts of the compound are released in response to the presence of LPS and pro-inflammatory cytokines. Some of the antimicrobial compounds present in ASL can act synergistically. For example, lactoferrin, secretory leukocyte proteinase inhibitor (SLPI), HBD-4, and LL-37 all demonstrate synergy with lysozyme resulting not only in a greater potency than that displayed by the sum of the individual activities of the two compounds, but also a broader antimicrobial spectrum. HBD-2 is synergistic with lysozyme and with lactoferrin.

The antimicrobial activity of ASL is markedly affected by pH, ionic strength, and the presence of divalent cations. In general, high ionic strengths, high concentrations of divalent cations, and deviations from a neutral pH result in decreased antimicrobial potency. Lysozyme, lactoferrin, SLPI, LL-37, and HBDs are all inhibited by high ionic concentrations.

4.2.3 Respiratory portion

The epithelia lining the alveolar and the respiratory bronchioles are not ciliated and, therefore, the mucociliary escalator does not operate in this region of the respiratory tract. These epithelial surfaces are covered by alveolar lining fluid (ALF) which consists of a plasma ultrafiltrate together with secretions produced mainly by type II alveolar cells. ALF contains a number of antimicrobial components, including lysozyme, free fatty acids, immunoglobulins, iron-binding proteins, and surfactant proteins. The main immunoglobulin present in ALF is IgG rather than the IgA predominating in ASL, and this is not locally produced but is present as a result of transudation from serum. It functions as an opsonin and facilitates the phagocytosis of any microbes reaching this part of the respiratory tract. It can also activate complement. Four surfactant proteins (SPs) have been detected in ALF (SP-A, SP-B, SP-C, and SP-D), and all of these are produced by the type II alveolar cells while three of the proteins (SP-A, SP-B, and SP-D) are also produced by Clara cells found in the terminal bronchioles. SP-B and SP-C are primarily involved in maintaining a low surface tension, thus preventing collapse of the alveoli, whereas SP-A and SP-D are important in defending this section of the respiratory tract from microbes. SP-A and SP-D belong to the collectin family of proteins – so-named because they have both a collagen-like domain and a lectin domain – which form large multimeric complexes. SP-A binds to macrophages and to a number of bacteria, including *Strep. pneumoniae, Staph. aureus, H. influenzae, E. coli, Strep. agalactiae*, and *Ps. aeruginosa*. It can, therefore, function as an opsonin, thus helping the phagocytosis of microbes by macrophages. In addition to its opsonic activity, SP-A also stimulates phagocytosis of microbes by macrophages and by polymorphonuclear leukocytes (PMNs), enhances PMN migration, and stimulates the oxidative burst in PMNs. SP-D has similar activities to SP-A, but binds to a slightly different range of microbes, including *Sal. minnesota, E. coli, K. pneumoniae, Ps. aeruginosa, Pneumocystis carinii, Mycob. tuberculosis, Strep. agalactiae,* and *Cryptococcus neoformans.* This difference in microbial recognition arises primarily because SP-A binds mainly to di-mannose repeating units that are found in some capsular polysaccharides and to the lipid A domain of LPS, whereas SP-D preferentially binds to core oligosaccharides of LPS. However, other, as yet uncharacterised, bacterial constituents are recognized by each of the collectins.

The alveoli also have a substantial population of macrophages, together with smaller numbers of lymphocytes and PMNs. Macrophages can generally deal effectively with the small numbers of microbes that arrive in the alveoli either in inhaled air or in small quantities of oropharyngeal secretions that are regularly aspirated (particularly during sleep) into the lungs. The presence of larger numbers of microbes usually induces the macrophages to recruit PMNs from the pulmonary capillaries, which then help to clear microbes from the alveoli.

4.3 | Environmental determinants at different regions of the respiratory tract

The oxygen and carbon dioxide content of the air in the lumen of the respiratory tract vary with the anatomical location. In the nares, the air contains approximately 21% oxygen and 0.04% carbon dioxide, whereas in the alveoli the oxygen content drops to 14% and the carbon dioxide content increases to 5% as a consequence of gas exchange with the bloodstream. Because the film of liquid (whether nasal fluid, ASL, or ALF) covering the respiratory mucosa is so thin, rapid exchange of gases occurs between it and the air in the lumen. Studies have shown that even though oxygen is being consumed by the underlying epithelial cells, there is no gradient in the oxygen content through the liquid film. The respiratory tract, therefore, is predominantly an aerobic environment, thus providing atmospheric conditions suitable for the growth of obligate aerobes and facultative anaerobes. Nevertheless, obligate anaerobes can be isolated from some regions of the respiratory tract. Anaerobic microhabitats can be generated as a result of oxygen utilisation by aerobes and facultative species, and this can be exacerbated by local anatomical features that hinder oxygen replenishment (e.g., the convoluted surfaces of some epithelial cells and the crypts of the tonsils).

The pH of ASL is generally slightly acidic throughout the respiratory tract, with a mean value of 6.78. In the nasal cavity, the pH of the mucosa gradually increases from the anterior nares (pH 5.5) to almost neutral (pH 6.95), at a distance of 6 cm from the tip of the nose. The temperature of the nasopharynx in adults is 34°C.

The main sources of nutrients derived from the host depend on the anatomical region (Table 4.3). In all regions, the fluid present will contain a transudate from plasma, which is the origin of, for example, the albumin detected in these fluids. In addition, we encounter for the first time the possibility that food ingested by the host could function as a source of nutrients for the resident microbes of at least some regions of the respiratory tract – this could occur during the passage of food and saliva through the pharynx. However, the transit time of food and saliva in this region is so rapid (between 1 and 2 seconds/swallow) that it is unlikely that significant quantities of dietary constituents would be transferred to the ASL.

Table 4.3. Sources of host-derived nutrients for microbial residents of the respiratory tract

Region	Sources of nutrients
nose	secretions from airway epithelial cells (especially goblet cells) and submucosal glands; transudate from plasma; tears from nasolacrimal ducts
nasopharynx, larynx, trachea, bronchi	secretions from airway epithelial cells (especially goblet cells) and submucosal glands; transudate from plasma
oropharynx, laryngopharynx	secretions from airway epithelial cells (especially goblet cells) and submucosal glands; transudate from plasma; food ingested by host
respiratory bronchioles	secretions from epithelial and Clara cells, transudate from plasma
alveoli	secretions from type 1 and type 2 alveolar cells; transudate from plasma

4.3.1 Composition of nasal fluid

This fluid contains the secretions of the nasal mucosa (epithelial cells and interspersed goblet cells) and submucosal glands; the transudate from nasal blood vessels; products of cells resident in the mucosa, such as plasma cells, lymphocytes, etc.; and, finally, tears which enter via the nasolacrimal ducts. As described previously, approximately 1 litre of fluid is produced by the nasal epithelium each day, and its main constituents are listed in Table 4.4. The fluid contains approximately 30 mg/g of dissolved and suspended solids.

Table 4.4. Main constituents of nasal fluid

Constituent	Concentration
proteins	414–895 mg/100 ml
mucins	52–112 mg/100 ml
albumin	31–105 mg/100 ml
lipocalin-1	8–18 mg/100 ml
DNA	40 μg/ml
uric acid	5–16 μM
urea	3.3 mM
cystatin S	–
neutral endopeptidase	–
aminopeptidase	–
carboxypeptidase N	–
angiotensin-converting enzyme	–
kallikrein	–
Na^+	98–225 mM
K^+	23–68 mM
Ca^{2+}	3–14 mM
HPO_4^{2-}	3–7 mM

Notes: For details of constituents with antimicrobial properties, see Table 4.1.

– = data not available.

Table 4.5.	Main constituents of airway surface liquid
Constituent	**Concentration**
total protein	3.0%
albumin	48–73 mg/100 ml
mucins	0.5–1.0%
hyaluronic acid	3 mg/100 ml
heparin	–
chondroitin sulphate	–
IgA	50–2,000 (μg/ml)
IgG	90–2,000 (μg/ml)
DNA	0.028%
carbohydrate	0.95%
lipids	0.84%
glutathione	429 μM
Na^+	80–85 mM
Cl^-	75–80 mM
K^+	15 mM
antimicrobial compounds	see Table 4.2

Note: – = concentration not known.

The most abundant proteins in nasal secretions are albumin and lipocalin-1, which constitute approximately 10% and 2% of the total protein content, respectively. Lipocalin-1 has a high affinity for small hydrophobic molecules and is thought to be involved in odour and taste reception, and may also function as a scavenger of toxic and pro-inflammatory lipids. A variety of enzymes has also been detected in nasal fluid, together with cystatin S which is an inhibitor of cysteine proteases. Other proteins identified include transthyretin and immunoglobulin-binding factor. While the Na^+ content of nasal fluid is approximately 25% lower than that of plasma, the K^+ ion concentration is more than five times greater than that of plasma. The Cl^- concentration is slightly higher than it is in plasma. The osmolarity of nasal fluid is 277 mOsm/litre which is very similar to that of plasma (285 mOsm/litre). Its pH is 5.5 in the anterior nares and gradually increases to 6.95 at a distance of 6 cm from the tip of the nose.

4.3.2 Composition of airway surface liquid

A total of between 20 and 100 ml of ASL is produced each day, and its composition varies slightly depending on its location within the respiratory tract. It consists mainly of water, which accounts for between 90% and 95% of its mass, mucins, and proteins (Table 4.5). The principal mucins in ASL are MUC5AC (produced mainly by surface goblet cells) and MUC5B (secreted mainly by submucosal glands). ASL also has a high albumin content ranging from 48 mg/100 ml in the trachea to 73 mg/100 ml in the bronchi. A wide range of proteins in addition to albumin has been detected, and these include immunoglobulins, α1-antitrypsin, α1-antichymotrypsin, α2-macroglobulin, prealbumin, haptoglobin, lipocalin-1, cystatin S, transthyretin, and immunoglobulin-binding factor. High levels of the tripeptide glutathione are present (more than 100-fold greater than the concentration found in plasma), which protects the mucosa against oxidative damage. Lipids are present at a concentration of approximately 1%, most of which are phospholipids, and the most abundant of these is phosphatidylcholine. A range of glycosaminoglycans is present, including heparin sulphate, heparin, chondroitin sulphate, and hyaluronate. The main ions present are sodium, potassium, and chloride. Its osmolarity is 210 mOsm/litre, which is 74% of that of plasma.

4.3.3 Composition of alveolar lining fluid

ALF is a complex mixture of proteins and lipids. The total protein content is approximately 9.0 mg/ml, of which approximately half is albumin. A major constituent of ALF is a mixture of surface-active compounds known as pulmonary surfactant, which lowers the surface tension of the fluid. In the absence of surfactant, the high surface tension of water would cause the alveoli to collapse during expiration. Pulmonary surfactant consists mainly of phospholipids, with smaller quantities of proteins. Its approximate composition is: 40% saturated phosphatidylcholines, 30% unsaturated phosphatidylcholines, 10% phosphatidylglycerol, 4% phosphatidylethanolamine, 2% phosphatidylinositol, 2% sphingomyelin, 4% other lipids, and 8% surfactant proteins. The main component of the mixture is 1,2 dipalmitoyl-glycero-3-phosphocholine, a saturated phosphatidylcholine, which constitutes approximately 30% of its total mass. ALF also contains vitamin C, vitamin E, reduced glutathione, transferrin (324.2 μg/ml), and ceruloplasmin (22.2 μg/ml), which act as antioxidants to protect the epithelium from oxidative damage during its continual exposure to high oxygen concentrations. Transferrin appears to be the most effective of these antioxidants and is able to prevent lipid peroxidation, the most important reaction involved in oxygen-induced tissue damage.

4.3.4 Contribution of microbial residents of the respiratory tract to nutrient availability

In all of the regions of the respiratory tract, the pool of available nutrients in respiratory secretions will be increased by the activities of some members of the microbiota. Hence, macromolecules (i.e. proteins, glycoproteins, lipids, nucleic acids) in the respiratory secretions will be converted by proteases, sialidases, glycosidases, lipases, etc., to carbohydrates, amino acids, and fatty acids for use as carbon, nitrogen, and energy sources (Figure 4.8). Some of the relevant enzymes and the organisms producing them are listed in Table 4.6.

4.4 | Indigenous microbiota of the respiratory tract

4.4.1 Main characteristics of key members of the respiratory microbiota

The various regions of the respiratory tract are colonised by a wide range of microbes. Those most frequently detected include viridans streptococci, *Strep. pyogenes, Strep. pneumoniae, Neisseria* spp., *Haemophilus* spp., *Moraxella* spp., *Staph. aureus,* coagulase-negative staphylococci (CNS), *Corynebacterium* spp., *Propionibacterium* spp., *Prevotella* spp., *Mollicutes* (*Mycoplasma* spp. and *Ureaplasma* spp.), and *Porphyromonas* spp. *Corynebacterium* spp., CNS, and *Propionibacterium* spp. are described in Section 2.4.1, while *Prevotella* spp.

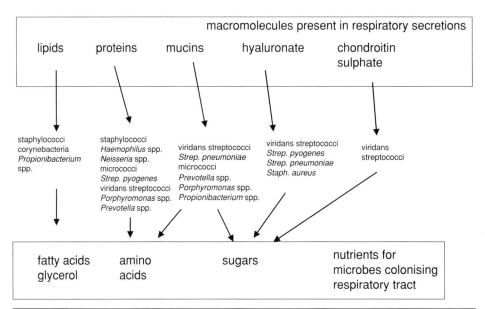

Figure 4.8 Degradation of polymers by, and the provision of nutrients for, members of the indigenous respiratory microbiota.

and *Porphyromonas* spp. are major constituents of the oral microbiota and are more appropriately described in Chapter 9. The other residents of the respiratory tract are described herein.

4.4.1.1 *Neisseria* spp.

These are aerobic, non-sporing, non-motile, Gram-negative cocci that frequently occur as pairs. Sixteen species are recognized in the genus and the G+C content of their DNA ranges from 47 to 52 mol%. They can grow over the temperature range of 22–40°C, but most species grow optimally at 35–38°C. Their growth is stimulated by carbon dioxide, and they can grow anaerobically in the presence of nitrite as a terminal electron

Table 4.6. | Extracellular enzymes produced by microbes colonising the respiratory tract

Enzyme	Produced by
IgA protease	*H. influenzae*, *Strep. pneumoniae*, viridans streptococci, *N. meningitidis*
protease	*Strep. pyogenes*, *Staph. aureus*, *P. acnes*, CNS, viridans streptococci, *Prevotella* spp., *Porphyromonas* spp.
hyaluronidase	*Strep. pneumoniae*, *Strep. pyogenes*, *P. acnes*, *P. granulosum*, *Staph. aureus*, viridans streptococci, *Bacteroides* spp., *Can. albicans*
sialidase	*Strep. pneumoniae*, viridans streptococci, *P. acnes*, *Prevotella* spp., *Porphyromonas* spp.
glycosidase	viridans streptococci, *Prevotella* spp., *Porphyromonas* spp.
DNase	*Mor. catarrhalis*, *Strep. pyogenes*, *Staph. aureus*, *P. acnes*, CNS, *Corynebacterium* spp.
lipase	*Staph. aureus*, *P. acnes*, CNS, *Corynebacterium* spp., *Mor. catarrhalis*

Note: CNS = coagulase-negative staphylococci.

Table 4.7. | Virulence factors of *Neisseria meningitidis*

Virulence factor	Function and/or effect
pili	mediate adhesion to epithelial cells; undergo antigenic variation, thereby interfering with immune response
Opa and Opc	mediate adhesion to epithelial cells; undergo phase variation, thereby interfering with immune response
PorB (outer membrane protein)	interferes with maturation of phagosomes, so enabling survival in vacuole; induces apoptosis
PorA (outer membrane protein)	undergoes phase variation, thereby interfering with immune response
IgA protease	hydrolyses IgA antibodies, thereby interfering with host defence
capsule	inhibits phagocytosis and complement-mediated bacteriolysis; down-regulates immune response because it mimics human neural cell adhesion molecule
transferrin-binding proteins	acquire iron from host transferrin
haemoglobin-binding outer membrane protein	can extract haem from haemoglobin
outer membrane proteins	down-regulate $Fc\gamma$, C1, and C3 receptors, thereby impeding ingestion by PMNs
lipo-oligosaccharide	down-regulates immune response because it mimics human glycosphingolipids; undergoes phase variation, thereby undermining immune response; inhibits serum bactericidal activity; down-regulates complement activation; induces release of pro-inflammatory cytokines

Note: PMNs = polymorphonuclear leukocytes.

acceptor. They are catalase-positive and oxidase-positive and oxidise carbohydrates to produce acid. Apart from *N. meningitidis* and *N. gonorrhoeae*, members of the genus are not nutritionally fastidious. They can grow over the pH range of 6.0–8.0, but growth is optimal at a pH of 7.4–7.6.

Neisseria meningitidis, also known as the meningococcus, is a member of the indigenous microbiota of the nasopharynx and is an important pathogen. Thirteen serogroups (A, B, C, D, etc.) of the organism are recognized on the basis of the antigenicity of the capsular polysaccharide. Further differentiation into twenty serotypes and ten subtypes is possible by analysis of the antigenic nature of porin proteins and other outer membrane proteins, respectively. In addition, thirteen immunotypes can be distinguished on the basis of the antigenicity of the lipo-oligosaccharide of the organism – these are prefixed by the letter L. Although serotyping has been widely used for epidemiological studies of the organism, this is now being replaced by molecular approaches, such as multilocus enzyme electrophoresis, DNA fingerprinting, and PCR.

N. meningitidis has a range of virulence factors that enable it to subvert host defence systems and to cause damage to the host – these are summarised in Table 4.7. The main diseases caused by the organism are meningitis and septicaemia (Section 4.5.1), but it may also be responsible for septic arthritis, pneumonia, conjunctivitis, pericarditis, otitis, and sinusitis.

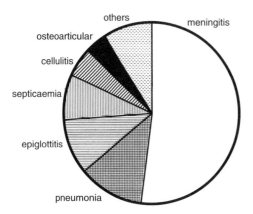

Figure 4.9 Spectrum of diseases due to *H. influenzae* type b and their relative incidences based on data from studies carried out in various parts of the world.

4.4.1.2 *Haemophilus* spp.

These are non-motile, non-sporing, aerobic Gram-negative rods. Depending on the growth conditions, they may appear as coccobacilli or filamentous rods. The G+C content of their DNA ranges from 37 to 44 mol%. Their growth is generally enhanced by elevated levels (5–10%) of carbon dioxide. They can grow over the temperature range of 20–40°C, with growth being optimal at 35–37°C. The optimum pH for growth is 7.6. They are nutritionally exacting, and all species require haemin or nicotinamide adenine dinucleotide or both for growth *in vitro*. They ferment carbohydrates to produce succinic, lactic, and acetic acids.

    H. influenzae is an important respiratory pathogen, and seven major groups of the organism are recognized. Capsulated strains constitute six of these groups and are designated as serotypes a to f on the basis of the structure and antigenicity of their capsules. The most well known of these is *H. influenzae* type b (Hib), which is responsible for many of the severe, invasive diseases due to *H. influenzae* (Figure 4.9). The seventh group of strains is those which do not have a capsule, and these are usually referred to as non-typeable *H. influenzae* (NTHi), although this is a misnomer and they should be termed "non-encapsulated *H. influenzae*". The main virulence factors of *H. influenzae* are summarised in Table 4.8. The main diseases for which the organism is responsible include meningitis, pneumonia, epiglottitis, otitis media, sinusitis, bronchitis, conjunctivitis, and arthritis. These are described in Sections 3.5.1 and 4.5.1–4.5.5.

4.4.1.3 *Streptococcus* spp.

Streptococci are facultatively anaerobic, non-sporing, catalase-negative, Gram-positive spherical, or ovoid cocci which usually occur in pairs or chains. The G+C content of their DNA is between 34 mol% and 46 mol%. The genus consists of at least 39 species, all of which are nutritionally fastidious and are incapable of respiratory metabolism, but ferment carbohydrates to produce mainly lactate. Some species require high carbon-dioxide levels (5%) for growth, and the growth of most is stimulated by increased concentrations of the gas. They are able to grow over the temperature range of 20–42°C, but optimum growth usually occurs at approximately 37°C. Some species are aciduric and can survive at pHs as low as 4.1. The older classification systems based on the haemolytic reactions (α-, β- and non-haemolytic) and immunochemical properties (Lancefield groups A, B, C, D, etc.) of the different species are still widely used to divide the genus into a number of major groups.

Table 4.8.	Virulence factors of *Haemophilus influenzae*
Virulence factor	**Function and/or effect**
LOS	cytotoxic to ciliated epithelial cells; inhibits ciliary activity; induces release of pro-inflammatory cytokines
peptidoglycan fragments	inhibit ciliary activity
IgA1 protease	hydrolyses IgA1, thereby interfering with the mucosal defence system
pili	involved in adhesion to epithelial cells
P2, P5 (outer membrane proteins)	mediate adhesion to mucin
opacity-associated protein A; P5; *Haemophilus* adhesion and penetration protein; *H. influenzae* adhesin; HMW1 and HMW2 proteins	all mediate adhesion to epithelial cells
antigenic and phase variation of LOS	interferes with immune response
antigenic variation of outer membrane	interferes with immune response
molecular mimicry of LOS	sialylated and non-sialylated forms of LOS resemble human glycosphingolipids, therefore not recognized by the host defence system
transferrin-binding proteins	can acquire iron from transferrin
can invade epithelial cells and pass between tight junctions	enables penetration of respiratory mucosa and access to underlying tissues

Notes: As well as having the virulence factors listed, *H. influenzae* type b also has an antiphagocytic capsule.

LOS = lipo-oligosaccharide.

4.4.1.3.1 *Streptococcus pyogenes*

This is a β-haemolytic streptococcus (i.e., its colonies are surrounded by clear zones of haemolysis on blood agar) and has a Lancefield group A cell-wall antigen – it is also frequently known as the "Group A streptococcus". *Strep. pyogenes* can bind to mucin, one of the main components of ASL, by means of two proteinaceous adhesins. One of these is the M protein, while the identity of the other has not been determined. $\alpha2$–6-linked sialic-acid residues of the mucin molecule are the complementary receptors for these bacterial adhesins. Sialic-acid residues of membrane proteins of pharyngeal cells are also the receptors for M protein-mediated adhesion to these epithelial cells. On the basis of the antigenicity of the N-terminal region of its M protein, more than eighty serotypes of *Strep. pyogenes* have been recognised. Another scheme (based on the genes encoding M and M-like proteins – *emm* genes) has resulted in the recognition of 5 *emm* patterns – from A to E. Other adhesins that may be involved in the adhesion of the organism to epithelial cells are listed in Table 4.9. *Strep. pyogenes* is responsible for a very wide range of infections in humans (Section 4.5.7) and has an impressive array of virulence factors (Table 4.10), in addition to the adhesins listed in Table 4.9. The hyaluronidase can hydrolyse hyaluronate present in ASL and connective tissue to release glucuronic acid and *N*-acetylglucosamine which it, and other organisms, can use as carbon and energy sources. The proteases produced by the organism will provide amino acids, which many organisms can use as carbon, nitrogen, and energy sources.

Table 4.9. Adhesins of *Streptococcus pyogenes* and their complementary receptors

Adhesin	Receptor
lipoteichoic acid	fibronectin
M protein	fibrinogen, fibronectin, laminin, sialic acid-containing molecules, galactose
Sfb I	fibronectin
protein F2	fibronectin
glyceraldehyde-3-phosphate dehydrogenase	fibronectin, fibrinogen
hyaluronic acid	CD44
serum opacity factor	fibronectin
collagen-binding protein	collagen
SpeB	integrins, laminin
vitronectin-binding protein	vitronectin
laminin-binding protein	laminin

Table 4.10. Virulence factors of *Streptococcus pyogenes* (see also adhesins listed in Table 4.9)

Virulence factor	Function and/or effect
M protein	protects against phagocytosis; mediates adhesion to keratinocytes; promotes aggregation on epithelial cell surface; involved in invasion of epithelial cells
hyaluronic-acid capsule	protects against phagocytosis; poor immunogen because of its similarity to host connective-tissue components; mediates adhesion to host cells expressing CD44 (e.g., keratinocytes)
lipoteichoic acid	mediates adhesion to fibronectin on epithelial cells; induces release of pro-inflammatory cytokines
fibronectin-binding proteins	mediate adhesion to fibronectin on epithelial cells; some also function as invasins (e.g., FI protein)
streptococcal fibronectin binding protein-I	an invasin; involved in invasion of epithelial cells
streptolysins	cytotoxic to many cells, including PMNs
hyaluronidase	degrades hyaluronic acid in connective tissues
streptokinase	converts plasminogen to plasmin, thereby disrupting clots
C5a peptidase	interferes with PMN chemotaxis
streptococcal inhibitor of complement	prevents complement-mediated bacteriolysis; interferes with phagocytic killing
pyrogenic exotoxins	comprise at least 16 proteins; superantigens, activate large numbers of T cells which induces release of pro-inflammatory cytokines resulting in toxic shock
SpeB	a cysteine protease; cleaves host proteins, including antibodies; superantigen, induces release of pro-inflammatory cytokines; mediates binding to laminin and integrins

Note: PMN = polymorphonuclear leukocyte.

Table 4.11. | Virulence factors of *Streptococcus pneumoniae*

Virulence factor	Function and/or effect
capsule	antiphagocytic
IgA protease	hydrolyses IgA, thereby inactivating a major effector molecule of the host defence system
pneumococcal surface protein A	inhibits complement activation
pneumococcal surface adhesin A	an adhesin; mediates attachment to epithelial cells
lipoteichoic acid	induces cytokine release from host cells
hyaluronate lyase	facilitates tissue invasion by breaking down hyaluronan, an important extracellular matrix component
pneumolysin	released on cell lysis; cytotoxic to ciliated bronchial epithelial cells; slows ciliary beating; disrupts tight junctions and integrity of respiratory epithelium; induces release of pro-inflammatory cytokines; inhibits lymphocyte proliferation; inhibits antibody synthesis; induces apoptosis in neutrophils, macrophages, and neuronal cells
choline-binding protein A	an adhesin; mediates attachment to epithelial cells
sialidase	cleaves terminal sialic acid residues from host cell surface glycans, thereby exposing receptors for bacterial adhesins; degrades protective mucus layer
autolysin A	hydrolyses bacterial cell wall, releasing inflammatory peptidoglycan degradation products and pneumolysin
hydrogen peroxide	toxic to epithelial cells

4.4.1.3.2 *Streptococcus pneumoniae*

This is an α-haemolytic streptococcus (i.e., its colonies are surrounded by a greenish zone of partial haemolysis on blood agar) which is distinguished from other members of this group by its susceptibility to ethylhydrocupreine (optochin). The cocci are usually arranged in pairs, which are enclosed within a polysaccharide capsule. On the basis of the antigenicity of the capsule, more than ninety serotypes can be distinguished. Choline-binding protein A is considered to be an important adhesin for maintaining the organism in the nasopharynx. *Strep. pneumoniae* is an important human pathogen, being responsible for a range of diseases, including pneumonia, meningitis, otitis media, and sinusitis (Sections 4.5.1–4.5.4). Its main virulence factors are listed in Table 4.11. The hyaluronidase liberates glucuronic acid and *N*-acetylglucosamine from hyaluronate in ASL and connective tissue, thus providing molecules that can be used as carbon and energy sources. The organism also produces sialidases (neuraminidases), which can remove sialic-acid residues from the mucins in ASL and attach to the respiratory mucosa, thus, contributing to the degradation of these complex molecules and the release of carbohydrates and amino acids for use as carbon, nitrogen, and energy sources (Section 4.3.4). The IgA protease cleaves IgA molecules at the hinge region, thereby releasing fragments that can be further degraded to provide amino acids such as carbon, nitrogen, and energy sources.

Table 4.12. | Virulence factors of *Moraxella catarrhalis*

Virulence factor	Function/effect
CD protein (an outer membrane protein)	an adhesin; mediates adhesion of the organism to mucins present in the nasopharynx and middle ear
carbohydrate adhesin	enables adhesion to epithelial cells
ubiquitous surface protein A1	inhibits complement-mediated killing; mediates adhesion to epithelial cells
ubiquitous surface protein A2	mediates adhesion to fibronectin and vitronectin; protects against killing by human serum
CopB (an outer membrane protein)	protects against killing by human serum; can acquire iron from transferrin and lactoferrin
phospholipase B	can degrade pulmonary surfactants; induces tissue damage
lipo-oligosaccharide	immunochemically similar to the human Pk antigen, thereby interfering with host immune response; induces release of pro-inflammatory cyokines

4.4.1.3.3 Viridans group streptococci

These include the α-haemolytic streptococci other than *Strep. pneumoniae*. Some viridans streptococci produce a sialidase, and many produce a range of glycosidases and consequently are able to liberate sugars from respiratory mucins. Some also produce a hyaluronidase, chondroitin sulphatase, protease, and urease, and therefore can produce sugars, amino acids, and ammonium ions from the constituents of respiratory secretions. Because they are among the predominant members of the oral microbiota, further information on viridans streptococci is provided in Section 8.4.1.1.

4.4.1.4 *Moraxella catarrhalis*

Mor. catarrhalis is an aerobic, non-motile, Gram-negative coccus which is usually seen as pairs of cocci, each with a diameter of 0.5–1.5 μm. Although its optimum growth temperature is 37°C, it can grow over the range of 20–42°C. It is catalase- and oxidase-positive. It does not produce acid from glucose or other carbohydrates. It produces a DNAse, phospholipase, and esterase. The main virulence factors of the organism are summarised in Table 4.12. *Mor. catarrhalis* is a frequent cause of sinusitis and otitis media in children (Sections 4.5.3 and 4.5.4) and can also cause lower respiratory tract infections, particularly in adults with chronic obstructive pulmonary disease. It has also been responsible for outbreaks of respiratory-tract infections in hospitalized patients.

4.4.1.5 *Staphylococcus aureus*

The main characteristics of the genus *Staphylococcus* are described in Section 2.4.1.3. *Staph. aureus* is distinguished from other members of the genus by its ability to produce the enzyme coagulase. The organism is found mainly in the anterior nares, but is also present in the nasopharynx and the oropharynx. *Staph. aureus* has a number of adhesins, but little is known with regard to which are important in mediating the attachment of the organism within the nasal cavity or to the skin during its involvement in cutaneous infections. In an *in vitro* study, a battery of mutants – each deficient in one of the organism's main adhesins – was tested for its ability to adhere

Table 4.13.	Virulence factors of *Staphylococcus aureus*
Virulence factor	Function/effect
capsule	anti-phagocytic
protein A	inhibits phagocytosis by binding to Fc region of IgG, therefore the IgG cannot interact with Fc receptors on phagocytes
coagulase	converts fibrinogen to fibrin, thereby protecting against host defences
fatty acid-metabolizing enzyme	neutralizes antibacterial effects of fatty acids
leukocidin	kills leukocytes
proteases	destroy host tissues
hyaluronate lyase	destroys extracellular matrix of tissues
staphylokinase	hydrolyses fibrin; may contribute to spreading of the organism
toxic shock syndrome toxin	pyrogenic; superantigenic; induces massive cytokine release
enterotoxins	pyrogenic; superantigenic; induce massive cytokine release; emetic; induce diarrhoeae
α-haemolysin	cytotoxic, haemolytic, dermonecrotic, neurotoxic
β-haemolysin	a sphingomyelinase; haemolytic
γ-haemolysin	haemolytic; kills neutrophils and macrophages
δ-haemolysin	haemolytic; cytotoxic
exfoliative toxins	damage epidermidis

to human keratinocytes. Mutants deficient in expressing protein A, fibronectin-binding proteins A and B, clumping factor, and coagulase exhibited significantly reduced binding to the keratinocytes. Hence, this organism can utilise a variety of adhesins to enable colonisation of the anterior nares and skin. Furthermore, the study revealed that adhesion of the parent strain to keratinocytes was markedly affected by pH – the number of cells adhering at pHs between 3 and 6 was less than half of that at a pH of between 7 and 10. This could be of relevance to diseases such as atopic dermatitis in which the pH of the skin rises to alkaline levels – adhesion of the organism would be enhanced at such pHs, thus, increasing the risk of infection in such patients. Interestingly, the skin lesions of more than 90% of patients with atopic dermatitis are colonised with *Staph. aureus*.

The main virulence factors of the organism are summarised in Table 4.13. *Staph. aureus* produces a number of bacteriocins with broad spectrum activity, including staphylococcin BacR1, which is active against some strains of *Staph. aureus*, some CNS, some viridans streptococci, *Corynebacterium* spp., *H. parasuis*, *Bordetella* spp., *Neisseria* spp., and *Bacillus* spp. Bac1829 is active against some strains of *Staph. aureus*, *Strep. suis*, *Corynebacterium* spp., *H. parasuis*, *Bordetella* spp., *Moraxella bovis*, and *Pasteurella multocida*. The organism is responsible for a wide range of infections, and these are described in Section 4.5.6.

4.4.1.6 *Mollicutes*

The *Mollicutes* are a class of a bacteria whose distinguishing characteristic is the absence of a cell wall. They are the smallest free-living organisms known, some species having a diameter of only 0.2 μm, and they have a very small genome (<600 kb in some

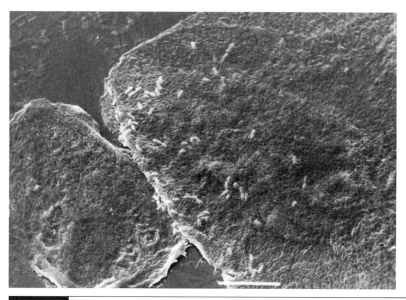

Figure 4.10 Scanning electron micrograph showing bacteria (mainly diplococci) attached to epithelial cells from the oropharynx. Bar = 10 μm. Photomicrograph kindly supplied by Professor Lars-Eric Stenfors, University of Tromso. Reprinted from: Bacterial attachment to oropharyngeal epithelial cells in breastfed newborns. Bjerkli, I.H., Myklebust, R., Raisanen, S., Telimaa, S., and Stenfors, L-E. *International Journal of Pediatric Otorhinolaryngology* 1996;36:205–213. Copyright © 1996, with permission from Elsevier Ireland Ltd.

species). They are pleomorphic, nutritionally fastidious, slow-growing, and exacting in their environmental requirements. The class consists of eight genera, and three of these (*Mycoplasma, Ureaplasma,* and *Acholeplasma*) have species which are members of the indigenous microbiota of humans – they are found mainly on the mucosal surfaces of the respiratory and genito-urinary tracts. The three genera are facultative anaerobes and are distinguished on the basis of their requirement for sterols and their ability to ferment glucose, utilise arginine, and hydrolyse urea. *Ureaplasma* spp. (G+C content = 27–30 mol%) hydrolyse urea, but are unable to metabolise glucose or arginine, whereas *Mycoplasma* spp. (G+C content = 23–41 mol%) cannot hydrolyse urea, whereas most species can utilize arginine and many can metabolise glucose. Unlike the other two genera, *Acholeplasma* spp. (G+C content = 27–36 mol%) do not require sterols for growth.

Because of their exacting nutritional and environmental requirements, mollicutes are difficult to grow, and few surveys of the indigenous microbiota of any body site have used the special media and incubation conditions necessary for their isolation. Consequently, little is known regarding their presence in the various microbial communities inhabiting humans and whether or not they are residents or transients of those body sites from which they can be isolated. Their main habitats appear to be the mucosal surfaces of the respiratory and genito-urinary tracts. Some species considered to be residents of the respiratory tract (e.g., *Mycoplasma fermentans*) are able to act as opportunistic respiratory pathogens.

4.4.2 Acquisition of the respiratory microbiota

The respiratory tract is generally sterile at birth, although in one study of 128 neonates, a small proportion was found to harbour *Staph. epidermidis* (15%), *Bac. subtilis* (7%), or *E. coli* (2%) in their nose and/or throat. In approximately half of the neonates colonised by

Table 4.14. Nasopharyngeal microbiota of 72 infants (38 males, 34 females) during the first 6 months of life

Organism	Age (months)						
	<1	1	2	3	4	5	6
Staph. aureus	64	44	48	34	33	24	28
Staph. epidermidis	62	44	35	25	23	23	18
Strep. pneumoniae	6	11	10	23	23	21	28
Strep. mitis	15	16	19	25	29	31	31
C. pseudodiphtheriticum	20	30	40	45	51	45	44
Mor. catarrhalis	6	9	24	19	33	27	39
Neisseria spp.	1	0	2	8	3	18	26
H. influenzae	0	5	6	11	14	15	13
Enterobacteriaceae	3	2	0	2	4	2	3

Note: Numbers denote the proportion (%) of infants colonised with each organism.

Staph. epidermidis, the same organism was also present in the birth canal of the mother, implying that this was the likely source. Other studies have also reported the presence of CNS and/or E. coli from the mother's birth canal in the anterior nares of neonates. Within 2 days of birth, the oropharynx is colonised almost exclusively by α-haemolytic streptococci, with approximately two-thirds of the epithelial cells having more than fifty bacteria attached, the remaining having between ten and fifty adherent bacteria (Figure 4.10). Surprisingly, few studies have characterised the respiratory microbiota of infants. Most of the microbiological investigations that have been carried out have focused on the nasopharynx and its colonisation by specific pathogens, such as Strep. pneumoniae, N. meningitidis, Mor. catarrhalis, and H. influenzae. In a study of the nasopharyngeal flora of seventy-two infants, staphylococci were the most frequently isolated organisms from the nasopharynx, but their prevalence gradually decreased with age (Table 4.14). The frequency of isolation of streptococci, corynebacteria, Neisseria spp., H. influenzae, and Mor. catarrhalis all gradually increased during the first 6 months of life. After 6 months, the frequency of isolation of Staph. aureus continues to decrease,

Table 4.15. Nasopharyngeal microbiota of infants

Microbe	Age (months)			
	1–3 (n = 451)	4–7 (n = 402)	8–12 (n = 352)	18 (n = 85)
H. influenzae	5	6	13	24
Strep. pneumoniae	12	30	32	32
Mor. catarrhalis	20	26	40	36
Staph. aureus	39	11	4	4
Staph. epidermidis	15	14	9	7
α-haemolytic streptococci	13	16	20	8
coryneforms	39	38	35	40
Moraxella spp.	3	4	7	14
E. coli	1	0	<1	0
H. parainfluenzae	1	<1	1	1

Note: Numbers represent the proportion (%) of individuals harbouring the particular microbe.

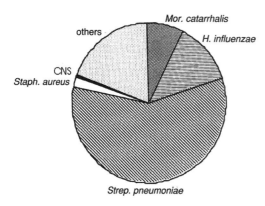

Figure 4.11 Proportions of the various species comprising the cultivable microbiota of the nasopharynx of 18-month-old children. CNS = coagulase-negative staphylococci.

whereas colonisation by other pathogens (*Strep. pneumoniae, Mor. catarrhalis,* and *H. influenzae*) increases (Table 4.15). By the age of 18 months, *Strep. pneumoniae, Mor. catarrhalis,* and *H. influenzae* are, along with coryneforms, the most frequently isolated members of the nasopharyngeal microbiota. Potentially pathogenic organisms, particularly *Mor. catarrhalis H. influenzae* and *Strep. pneumoniae,* also comprise a considerable proportion of the cultivable nasopharyngeal microbiota in this age group (Figure 4.11).

The pattern of acquisition of *N. meningitidis* is very different from that of other pathogens found in the nasopharynx. This organism is rarely isolated during the first 2 years of life, but the carriage rate progressively increases throughout childhood and adolescence (Figure 4.12). Interestingly, the frequency of isolation of the closely related organism, *N. lactamica,* increases during the first 2 years of life and then progressively decreases. The protection afforded by *N. lactamica* against colonisation by *N. meningitidis* is probably due to the induction of cross-reactive antibodies in those colonised by *N. lactamica.* Many of the children were found to have antibodies cross-reactive with *N. meningitidis* serogroups A, B, and C within 2 months of acquiring *N. lactamica.*

Carriage of *Mor. catarrhalis* remains at a high level up to the age of 4 years and then progressively declines to the low levels (1–3%) characteristically found in adults.

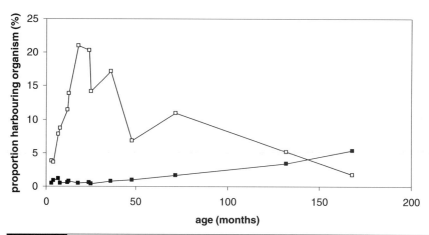

Figure 4.12 Colonisation of the nasopharynx of infants by *N. meningitidis* (■) and *N. lactamica* (□).

Table 4.16. Adhesins and receptors involved in the colonisation of respiratory mucus by important members of the respiratory microbiota

Organism	Adhesin	Receptor
Strep. pyogenes	M protein	$\alpha2$–6-linked sialic-acid residues
H. influenzae	P2 and P5	$\alpha2$–3-linked sialic-acid residues
Strep. pneumoniae	not known	$\alpha2$–3- and $\alpha2$–6-linked sialic-acid residues, GlcNAc($\beta1$–3)Gal
Mor. catarrhalis	CD protein	sialic-acid residues

Similarly, carriage of *Strep. pneumoniae* and *H. influenzae* is lower in schoolchildren than in pre-schoolchildren and then continues to decrease until adulthood.

One of the main determinants of the presence of an organism in the respiratory tract is its ability to adhere to some site within this system. Adhesion is also, of course, important in the pathogenesis of infections of the respiratory tract. Apart from the anterior regions of the nasal vestibules, which have an epithelium different from that present in the rest of the respiratory tract, the problems encountered by an organism attempting to adhere to most regions of the respiratory tract are very similar and will now be described in general terms. Adhesion of bacteria, particularly *Staph. aureus*, to the epithelium of the nasal vestibules is described in Section 4.4.3.1. Unfortunately, there is little information concerning the adhesion of the vast majority of organisms comprising the respiratory microbiota, most investigations having concentrated on those members of the respiratory microbiota that are frequently associated with infections. The epithelial surfaces of most of the respiratory tract are covered by ASL; the outermost layer of this, the mucus layer, is therefore the first possible site of adhesion for those organisms gaining access to the respiratory tract. Many members of the respiratory microbiota have been shown to adhere to respiratory mucus; these include *H. influenzae* (both type b and non-typeable strains), *Strep. pneumoniae, Staph. aureus, Mor. catarrhalis, Strep. pyogenes*, and viridans streptococci. However, other members of the respiratory microbiota, including *N. meningitidis*, do not demonstrate a high affinity for respiratory mucus. Interestingly, the main adhesin of *Mor. catarrhalis* (the CD protein) responsible for binding of the organism to mucins, binds to mucins from the nasopharynx and middle ear, but not to those from the oral cavity or the lower respiratory tract. This may account, in part, for the preferred sites of colonisation of this organism – the nasopharynx and oropharynx – and its ability to cause middle-ear infections. Adhesion may involve specific ligand-receptor and/or hydrophobic interactions. The adhesins and receptors involved in the adhesion to mucus of other respiratory species are listed in Table 4.16. From the point of view of the bacterium and the host, adhesion to the mucus layer of ASL has both advantages and disadvantages. While adhesion to mucus prevents the organism being expelled from the respiratory tract during expiration, it also renders it susceptible to expulsion by means of the mucociliary escalator. From the point of view of the host, adhesion to mucus means that the organism can be expelled by the mucociliary escalator, but it also offers a platform (no matter how temporary) to the organism from where it may be able to gain access to the underlying epithelium to which it may be able to adhere. Alternatively, the mucus-adherent organism may be able to interfere with the operation of the mucociliary escalator in some way, thus preventing or reducing the risks of its expulsion. Hence, a number of organisms are

Table 4.17. | Means by which members of the indigenous respiratory microbiota can damage the mucociliary escalator

Organism	Compound	Effect on mucociliary escalator
H. influenzae	lipo-oligosaccharide	cytotoxic to ciliated epithelial cells; inhibits ciliary activity
	peptidoglycan fragments	inhibit ciliary activity
	unidentified	stimulates secretion of mucus, thereby providing a greater "load" which the cilia cannot transport effectively; leads to blockage of smaller airways
Strep. pneumoniae	pneumolysin	cytotoxic to ciliated cells; slows ciliary beating
	sialidase	cleaves terminal sialic-acid residues from mucins, thereby altering viscosity of mucus which interferes with its transport
	unidentified	stimulates secretion of mucus, thereby providing a greater "load" which the cilia cannot transport effectively; leads to blockage of smaller airways
Staph. aureus	unidentified	stimulates secretion of mucus, thereby providing a greater "load" which the cilia cannot transport effectively; leads to blockage of smaller airways
N. meningitidis	unidentified, possibly peptidoglycan fragments	inhibits ciliary activity; kills ciliated cells

able to damage ciliary cells, inhibit ciliary activity, alter mucus viscosity, or affect the amount of mucus produced. Any of these will have an adverse effect on the performance of the escalator, thereby reducing the effectiveness of expulsion of mucus-entrapped organisms (Table 4.17). Underneath the mucus layer lies the periciliary layer which, because it is less frequently replaced than the mucus layer, offers a more permanent site for microbial colonisation. There is little evidence, however, of microbial colonisation of this region, although this may reflect the absence of appropriate studies and/or the inherent technical difficulties of such investigations. Finally, the epithelium itself constitutes a more permanent site for microbial colonisation. Here, a number of potential sites of colonisation are available – ciliated cells (apart from in the oropharynx and alveoli), non-ciliated cells, goblet cells, serous cells, Clara cells, as well as the mucins and extracellular matrix molecules expressed by many of these cells. Electron microscopy of healthy respiratory mucosae has revealed that ciliated cells rarely have attached microbes and, of the organisms investigated, few appear to be able to adhere to healthy ciliated cells. Hence, *H. influenzae, Strep. pneumoniae, N. meningitidis,* and *Staph. aureus* are rarely observed to adhere to such cells either *in vitro* or *in vivo*. However, once ciliated cells have been damaged in some way, they become susceptible to microbial colonisation, possibly as a result of the uncovering of other receptors for bacterial adhesins. Furthermore, if damage is extensive, molecules of the underlying extracellular

Figure 4.13 Invasion of epithelial cells of adenoidal tissue in culture by *H. influenzae*. Lamellipodia can be seen engulfing the attached cells which are then internalized within membrane-bound vacuoles. Reprinted with permission of Blackwell Publishing from: Molecular and cellular determinants of non-typeable *Haemophilus influenzae* adherence and invasion. St. Geme, J.W. *Cellular Microbiology* 2002;4:191–200.

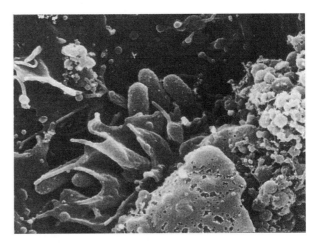

matrix may be exposed, thereby offering additional binding sites. Respiratory organisms display a wide range of adhesins which can mediate binding to a variety of cells and host macromolecules – these are described in Section 4.4.1.

Finally, persistence of bacteria in the respiratory tract can be achieved by invasion of the mucosa. It has been found, for example, that *Strep. pyogenes* is able to invade respiratory epithelial cells, and it has been suggested that this may be the mechanism underlying the ability of the organism to persist in up to 30% of individuals treated with antibiotics for tonsillopharyngitis. *N. meningitidis*, *Strep. pneumoniae*, and *H. influenzae* (including non-typeable strains) can also invade respiratory epithelial cells *in vitro* (Figure 4.13).

4.4.3 Community composition at different sites within the respiratory tract

4.4.3.1 External nares

This is predominantly an aerobic region with the main source of nutrients being nasal fluid (Section 4.3.1), desquamated cells, and secretory products of the nasal microbiota. Within this region, there exists a number of sites for microbial colonisation, including hair follicles, the skin surface, and vibrissae. *Corynebacterium* spp., CNS, and *Propionibacterium* spp. dominate the microbial community of the nares, which usually comprise between 10^6 and 10^7 viable bacteria per nostril (Table 4.18). Gram-negative bacteria are not frequent colonisers of this region and, when present, usually comprise only a small proportion of the microbiota.

There has always been great interest in this region of the respiratory tract because it is the main site of carriage of that most versatile and intractable of pathogens – *Staph. aureus*. However, while innumerable investigations have been undertaken to ascertain whether or not this organism is present in the nasal vestibule, relatively few have involved an analysis of the other members of the microbial community of this site. The importance of the nares as a source of the *Staph. aureus* either found at other body sites or causing infections at such sites has been deduced from the following lines of evidence: (1) elimination of the organism from the nares by topical antimicrobial agents results in its disappearance from other body sites and also

Table 4.18. | Cultivable microbiota of the external nares

Microbe	Proportion of population colonised (%)	Bacteria per nostril (cfu)
CLC group	30	2.4×10^5
C. jeikeum	50	1.5×10^6
C. minutissimum	27	1.3×10^6
C. xerosis	7	0.3×10^3
P. acnes	79	3.0×10^4
P. granulosum	18	2.3×10^3
P. avidum	26	1.2×10^4
all coagulase-negative staphylococci	100	1.9×10^6
Staph. epidermidis	95	NA
Staph. hominis	40	NA
Staph. haemolyticus	48	NA
Staph. cohnii	28	NA
Staph. saprophyticus	18	NA
Micrococcus spp.	48	NA
Staph. aureus	30	2.1×10^6
Neisseria spp.	19	NA
Acinetobacter spp.	8	NA

Notes: CLC = cutaneous lipophilic corynebacteria (see Section 2.4.1.1);
NA = data not available.

reduces nosocomial infection in patients undergoing surgery; (2) rates of infection are higher in carriers of the organism than in non-carriers; and (3) the strain responsible for an infection in an individual is usually identical to that carried in the nose of that person. Epidemiological studies have delineated three types of carriage of *Staph. aureus* in the population: (1) persistent carriers (approximately 20% of the population) – such individuals always carry a particular strain; (2) non-carriers – these individuals (approximately 20% of the population) almost never carry the organism; and (3) intermittent carriers – such individuals (about 60% of the population) carry the organism intermittently and the strain carried changes with time. The number of viable *Staph. aureus* carried in the anterior nares is higher in persistent than in intermittent carriers. Estimates of carriage rates for the general population range from 19.0% to 55.1%, with a mean value of 37.2%. However, the frequency of carriage is affected by many factors, including age, race, gender, immune status, and genetic makeup. It is generally higher in males than in females and is higher in infants than in the elderly. Hospitalisation is also an important risk factor for nasal carriage. In general, higher carriage rates are found in young children, injection drug users, patients with insulin-dependent diabetes, those affected by dermatologic conditions, patients with indwelling intravascular catheters, and health-care workers.

Carriage of the organism in the nasal vestibule is obviously dependent on, among other factors, its ability to adhere to some structure in this rather complex habitat. Possibilities include epithelial cells, extracellular matrix molecules, nasal fluid, vibrissae, and other members of the nasal microbiota. *Staph. aureus* has been shown to adhere to epithelial cells, matrix molecules, and nasal mucin, but its ability to adhere to vibrissae

Table 4.19. Microbiota of the nasal cavity of 10 healthy adults

Organism	Proportion of subjects harbouring the organism (%)	Proportion of microbiota (%)
Corynebacterium spp.	100	0.4–97.7
Staphylococcus spp.	80	0.3–78.1
Streptococcus spp.	60	<0.01–0.1
Aureobacterium (*Microbacterium*) spp.	50	0.8–13.2
Rhodococcus spp.	40	0.8–86.4
Haemophilus spp.	30	<0.01–0.4

and other microbes has received little attention. Evidence suggests, however, that the precise site of colonisation of the organism in the vestibule is the moist squamous epithelium on the septum adjacent to the nasal ostium (i.e., the opening) which is devoid of hairs. It appears to be maintained at this site by the proteinaceous adhesin clumping factor B, the receptor for which is cytokeratin K10. This cytokeratin is expressed on the surface of nasal epithelial cells and keratinocytes. Other adhesin–receptor interactions may also be involved in the adhesion process.

4.4.3.2 Nasal cavity

Apart from the anterior regions (i.e., the vestibules), the nasal cavity is lined by a mucus-covered ciliated epithelium (see Section 4.1.1). This is an aerobic environment, with the main source of nutrients being nasal fluid (Section 4.3.1), dead epithelial cells, and compounds secreted by the resident microbiota. Surprisingly, few studies of the microbiota of the nasal cavity have been carried out. In a study of the aerobic microbiota of ten adults, the most frequently isolated organisms were members of the genera *Corynebacterium, Staphylococcus, Streptococcus*, and *Aureobacterium* (now included in the genus *Microbacterium*). Those organisms, which generally comprised the highest proportions of the microbiota, belonged to the genera *Corynebacterium, Aureobacterium* (*Microbacterium*), and *Rhodococcus* (Table 4.19). *Microbacterium* spp. are aerobic, Gram-positive, non-sporing, rod-shaped bacteria that produce acid from glucose. Gram-negative bacteria were not frequently isolated and, apart from *Haemophilus* spp., the only other Gram-negative species found (in only one of the subjects) was *Moraxella nonliquefaciens*. Although staphylococci were frequently present in the samples, *Staph. aureus* was detected in only one of the subjects, and even then the organism comprised only 0.01% of the microbiota. The most frequently detected staphylococcus was *Staph. epidermidis*. Other studies of the microbiota of the nasal cavity of adults have also found it to consist predominantly of Gram-positive organisms, with Gram-negative species being isolated only occasionally. The presence in the nasal cavity of species that are either rarely detected in other regions of the upper respiratory tract or only comprise a small proportion of the microbiota (e.g., *Aureobacterium* spp., *Rhodococcus* spp.), together with the rare isolation of organisms frequently present in such regions (e.g., *Haemophilus* spp., *Neisseria* spp., *Branhamella* spp., and *Moraxella* spp.), indicates that the nasal cavity constitutes a habitat that is uniquely different from other regions of the upper respiratory tract. Although *Strep. pneumoniae* and *H. influenzae* are rarely present in the nasal cavity of adults, the situation is different in infants, who frequently harbour these organisms.

Table 4.20. Carriage rates of the main groups of organisms found in the nasopharynx of adults

Microbial group/species	Proportion harbouring the organism(s) (%)
Staph. aureus	21
coagulase-negative staphylococci	10
α-haemolytic streptococci	100
non-haemolytic streptococci	100
Strep. pneumoniae	6
Neisseria spp.	100
N. meningitidis	5
H. influenzae	34
H. parainfluenzae	77
other *Haemophilus* spp.	11
Mor. catarrhalis	4
Gram-negative anaerobes	38

4.4.3.3 Nasopharynx

The nasopharynx is an aerobic region with the main sources of nutrients being ASL, dead epithelial cells, and microbial products. Because the mucus layer of the ASL is continually being expelled into the oral cavity, the main site available for permanent colonisation is the mucosal surface. One other possible site of "colonisation" is the periciliary layer, which is replenished less rapidly than the mucus layer. The nasopharynx supports a large and varied microbial community. The number of viable microbes present ranges from 3×10^4 to 4×10^8 cfu, with a median value of approximately 6×10^6 cfu. Most of these organisms are either present in the ASL or are attached to non-ciliated epithelial cells, each of which usually has between ten and fifty adherent bacteria with some having microcolonies. Ciliated cells are rarely colonised by members of the indigenous microbiota. Most individuals harbour streptococci, *Neisseria* spp., and *Haemophilus* spp. (predominantly *H. parainfluenzae*) in the nasopharynx (Table 4.20). Other organisms frequently present include *Staph. aureus* and CNS (Table 4.21).

Table 4.21. Organisms frequently present in the nasopharyngeal microbiota (not an exhaustive list)

Aerobes	Facultative anaerobes	Obligate anaerobes
H. influenzae	*Strep. pneumoniae*	*Fusobacterium* spp.
H. parainfluenzae	*Strep. intermedius*	*Prevotella* spp.
H. parahaemolyticus	*Strep. anginosus*	*Porphyromonas* spp.
H. segnis	*Strep. constellatus*	*Peptostreptococcus* spp.
H. aphrophilus	*Step. sanguis*	
N. meningitidis	*Strep. gordonii*	
N. cinerea	*Strep. mitis*	
N. sicca	*Strep. parasanguis*	
N. subflava	*Strep. crista*	
N. mucosa	*Staph. aureus*	
N. lactamica	*Staph. epidermidis*	
N. flavescens	*Staph. cohnii*	
Mor. catarrhalis	*Kingella kingae*	

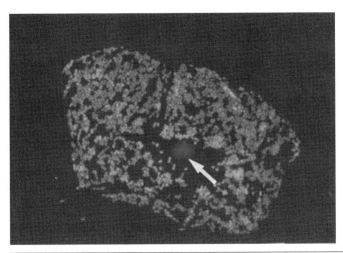

Figure 4.14 Nasopharyngeal epithelial cell with attached bacteria. Several microcolonies can be seen above the cell nucleus (arrow). Photomicrograph kindly supplied by Professor Lars-Eric Stenfors, University of Tromso. Reproduced with permission of the University of Chicago Press from Stenfors, L. and Raisanen, S. *Journal of Infectious Diseases* 1992;165:1148–1150. Copyright © 1992 by The Infectious Diseases Society of America.

Obligate anaerobes, particularly Gram-negative species, may also be isolated. The existence of anaerobes in this aerobic environment is possible because of oxygen utilisation by the aerobic members of the microbiota, which results in the formation of anaerobic microhabitats, particularly within the microcolonies that have been observed on epithelial cells (Figures 4.14 and 4.15) and within the invaginations of epithelial cells (Figure 4.16). An important feature of the nasopharyngeal microbiota is that a number of potential pathogens of the respiratory tract and other sites, including *Strep. pneumoniae*, *H. influenzae*, *Mor. catarrhalis*, and *N. meningitidis*, inhabit this region – this is particularly the case in infants.

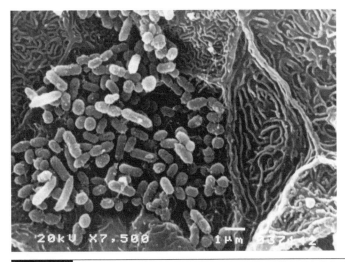

Figure 4.15 Microcolonies of bacteria on the surface of tonsillar epithelial cells. Note also the pattern of microridges on adjacent epithelial cells. Photomicrograph kindly supplied by Professor Lars-Eric Stenfors, University of Tromso, and reproduced with permission from: *Acta Otolaryngologica* 1996;116:620–626.

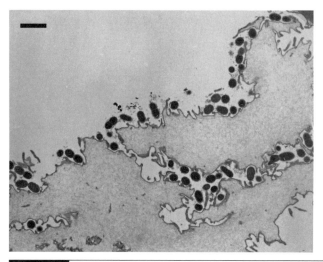

Figure 4.16 Transmission electron micrograph showing bacteria within involutions of the mucosal epithelium. Bar = 2 μm. Photomicrograph kindly supplied by Professor Lars-Eric Stenfors, University of Tromso, and reproduced with permission from: *Acta Otolaryngologica* 1996;116:620–626.

Because of the complexity of the nasopharyngeal microbiota and the enormous interest in its potentially pathogenic members, few studies have attempted to define its composition – most have been concerned with establishing the presence of *Strep. pneumoniae, H. influenzae, Mor. catarrhalis*, and *N. meningitidis*. The presence in the nasopharynx of these organisms, as well as *Staph. aureus*, is affected by many factors, including age, gender, climate, and social factors (Table 4.22).

Table 4.22. Influence of various factors on the carriage of key members of the indigenous microbiota of the nasopharynx

Factor	Effect
age	(1) prevalence of *Strep. pneumoniae, H. influenzae*, and *Mor. catarrhalis* decreases in the order infants > children > adults
	(2) prevalence of *N. meningitidis* is low in infants and the elderly and is highest in teenagers and young adults
season	(1) prevalence of *Strep. pneumoniae* in schoolchildren is higher in summer than in winter
	(2) prevalence of *Mor. catarrhalis* is higher in winter and autumn than in spring and summer
	(3) prevalence of *Staph. aureus* in infants is greater in winter than in summer
gender	density of colonisation of *Staph. aureus* in infants is greater in males than females
social factors	(1) prevalence of *N. meningitidis* is highest in low socioeconomic classes and in institutionalised individuals (e.g., prisoners, military recruits)
	(2) prevalence of *Strep. pneumoniae* and *H. influenzae* is highest in low socioeconomic classes
	(3) children with siblings have increased carriage of *Strep. pneumoniae, H. influenzae*, and *Mor. catarrhalis*

The mean age of acquisition of *Strep. pneumoniae* in the nasopharynx is approximately 6 months and, by the age of 2 years, the nasopharynx of 95% of children will have been colonised by the organism on at least one occasion. In developed countries, carriage reaches a maximum at the age of 2 years when approximately 50% of infants harbour the organism. In developing countries, carriage rates are much higher and can reach almost 100% during the first few months of life. Colonisation is usually by one serotype (only 4% of infants being colonised by more than one serotype at any given time), which may persist for several months and is then replaced by another serotype. The period of carriage of the second and subsequent strains gradually decreases so that carriage rates decrease with increasing age – probably as a consequence of the gradual acquisition of immunity to the dominant serotypes present in the community. The reported carriage rate in adults ranges from 0.8% to 20%. In children, the most common serotypes found in the nasopharynx are, in order of decreasing frequency of isolation, 6, 19, 23, and 14 and these serotypes are also carried for prolonged periods of time – approximately twice as long as other serotypes. These serotypes are also responsible for as many as 87% of *Strep. pneumoniae* infections in children, with serotypes 6 and 14 being the most frequent causative agents. A broader range of serotypes is found in adults, and these include 1, 3, 4–7, 9, 14, 18, 19, and 23.

The presence of *N. meningitidis* in the nasopharynx is also affected by age, but shows a different pattern from that found with *Strep. pneumoniae*. Hence, carriage is low in infants and increases steadily until it reaches a maximum in teenagers and young adults, and then decreases with increasing age. Studies in the United Kingdom have revealed carriage rates of <3% in infants, 24–37% in 15- to 24-year-olds, and <10% in older age groups. During non-epidemic periods, approximately 10% of the population carries the organism in the nasopharynx. However, carriage rates are greater among military personnel, boarding school students, prisoners, and smokers. In approximately one-quarter of the population, colonisation of the nasopharynx by *N. meningitidis* results in prolonged carriage (i.e., several months) of the organism, whereas in one-third of the population, carriage will last for only days or weeks. In the rest of the population, carriage is only transient or infrequent. Strains of *N. meningitidis* isolated from the blood or cerebrospinal fluid of patients suffering from an infection due to the organism most commonly belong to serogroups A, B, C, Y, and, to a lesser extent, W-135. In contrast, the strains isolated from asymptomatic carriers of the organism are usually non-groupable or belong to serogroups B, Y, X, Z, or 29E. Approximately 90% of the strains isolated from carriers are regarded as being "nonpathogenic" because they do not belong to clones that are highly invasive such as ET-5 (ET = electrophoretic type), lineage III, cluster A4, and ET-37. It has been estimated that 1% of individuals infected with such clones will develop meningitis.

Mor. catarrhalis is a frequent coloniser of the nasopharynx of infants, but carriage of the organism is less common in adults. In infants, the reported carriage rate ranges from 28% to 100%, depending on geographical location, socioeconomic status, and at what time of year the samples are obtained. On average, a child will be colonised by 3–4 strains of the organism before the age of 2 years. The carriage rate in adults is considerably lower than that in children – approximately 4% in healthy individuals. However, in adults with chronic respiratory diseases, the carriage rate may be as high as 43%. In both infants and adults, colonisation by a particular strain of the organism lasts for several months, but this strain is then lost and replaced by a different strain and the pattern is then repeated.

With regard to *H. influenzae*, it is important to distinguish between the presence of capsulated serotype b strains of the organism (Hib) and the non-typeable (NTHi) strains (Section 4.4.1.2). Hib strains are highly invasive and can cause infections such as meningitis and other blood-borne diseases, whereas NTHi are generally less invasive and are associated with infections of the respiratory tract. Unfortunately, many studies of the prevalence of *H. influenzae* have failed to establish whether the strains present were Hib or NTHi. Since the introduction of an effective vaccine against Hib in the late 1980s, carriage of this organism in infants has fallen to very low levels (0.1–1.5%) in many developed countries. However, in developing countries that cannot afford to introduce effective vaccination programmes, the prevalence of Hib in the nasopharynx of infants remains high (3–6%). The frequency of carriage of the organism (i.e., Hib + NTHi) in the nasopharynx during the first 2 years of life ranges from 19% to 81%. Carriage rates then decrease with age to between 3% and 19% in adults, but then increase again in those over 65 years of age.

Antagonistic behaviour among members of the nasopharyngeal microbiota has frequently been reported. α-haemolytic streptococci have been shown to be able to inhibit 92% of strains of *Strep. pneumoniae*, 74% of NTHi, and 89% of *Mor. catarrhalis*, and several studies have reported an inverse relationship in the nasopharynx between the presence of α-haemolytic streptococci and NTHi. These observations have prompted clinical trials of the effectiveness of the implantation in the pharynx of α-haemolytic streptococci for re-establishing a normal microbiota in individuals colonised by pathogenic organisms. These have met with some success and are described in greater detail in Chapter 10. *Strep. pneumoniae* has been shown to be antagonistic towards *H. influenzae*, *Mor. catarrhalis*, *Staph. aureus*, and *N. meningitidis*. This has been attributed to the production of hydrogen peroxide by the organism.

4.4.3.4 Oropharynx

The environment of the oropharynx differs from that of the nasal cavity and the nasopharynx in several respects. Firstly, the mucosal surface consists of a non-keratinised, stratified, squamous epithelium which, because it does not have ciliated cells, is not part of the mucociliary escalator. It is, nevertheless, coated in ASL. Secondly, there is the possibility that food taken in by the host could provide additional nutrients for the microbes colonising this region. The very short residence time of food in this region makes it very unlikely that nutrients could be absorbed by the ASL during its transit. However, the continuous swallowing of saliva (containing nutrients derived from the host's diet) could possibly contribute to the nutrient content of ASL – but there is no evidence that this actually occurs. The microbiota of the oropharynx is very complex and variable, not least because it is continually being seeded by nasopharyngeal secretions and saliva – both of which have very dense, as well as very different microbial populations. α-haemolytic streptococci, non-haemolytic streptococci, *Haemophilus* spp., and *Neisseria* spp. are invariably present in the oropharynx. These organisms account for approximately 80% of the microbiota, the proportions of each of the four groups being approximately equal. The species diversity is considerable and at least fifteen different *Streptococcus* spp. have been isolated from the oropharynx, although *Strep. salivarius*, *Strep. mitis* biovars 1 and 2, and *Strep. anginosus* generally comprise more than

75% of the streptococci present. The predominant *Haemophilus* species is *H. parainfluenzae*, which comprises approximately 75% of all the *Haemophilus* spp. present. Other frequently isolated species include *H. segnis* and *H. paraphrophilus*. *H. influenzae* is present in approximately 40% of adults, but comprises only a very small proportion (0.2%) of the total cultivable microbiota. Ten different *Neisseria* spp. have been detected, including *N. perflava*, *N. lactamica*, *N. sicca*, *N. flavescens*, *N. subflava*, *N. cinerea*, and *N. mucosa*. CNS and anaerobes (particularly Gram-negative species) are also frequently present. The existence of anaerobes in this aerobic environment is possible because of oxygen utilisation by the large numbers of aerobes and facultative anaerobes in the microbiota – this results in the generation of anaerobic microhabitats, particularly within the microcolonies that are known to be present on epithelial cells. Furthermore, the convoluted nature of the mucosal surface also disposes toward the creation of anaerobic microhabitats (Figure 4.16). The tonsils of the oropharynx also have many crypts within which anaerobic conditions are likely to be generated as a result of microbial consumption of oxygen exacerbated by the reduced penetration of oxygen-rich fluids into the crypts. The oropharynx appears to be one of the main habitats of *Mollicutes* and at least 11 species have been detected at this site. Frequently isolated species include *Myc. salivarium*, *Myc. orale*, *Myc. buccale*, *Myc. faucium*, *Myc. lipophilum*, and *Acholeplasma laidlawii*. Unfortunately, because of the paucity of studies directed at detecting these organisms in the oropharynx and the difficulties associated with their isolation and quantification, little is known regarding their prevalence or population density.

Like the nasopharynx, the oropharynx frequently harbours potentially pathogenic organisms, such as *N. meningitidis*, *Strep. pneumoniae*, *H. influenzae*, and *Mor. catarrhalis*. In addition, the oropharynx is the principal habitat of *Strep. pyogenes*, which is one of the most ubiquitous and versatile pathogens of humans. Carriage of this organism in healthy adults is usually between 5% and 10%, but is generally higher in schoolchildren. During winter and spring, as many as 20% of schoolchildren may carry the organism in their oropharynx. However, in healthy individuals, the organism rarely comprises more than 8% of the cultivable oropharyngeal microbiota.

Within the oropharynx, lie the tonsils which are important lymphoid tissues. The microbiota of the surface of the tonsils is similar to that of the rest of the oropharynx, with α-haemolytic streptococci, *Neisseria* spp., *H. influenzae*, *H. parainfluenzae*, and *Staph. aureus* being the most frequently isolated organisms. Microscopic studies have revealed the presence of bacteria in three main locations on the tonsillar epithelium: in the mucus layer (Figure 4.17a), attached to the epithelial cells (Figure 4.17b), and within the epithelial cells (Figure 4.17b).

The distribution of bacteria attached to epithelial cells is irregular, with some cells having no adherent bacteria, some a few, while others have large bacterial colonies on their surface. Most of the bacteria present are cocci, often in pairs. The epithelial cells have a distinctive pattern of microridges (see Figure 4.15) and, in many cases, the bacteria appear to be "gripped" by these ridges (see Figure 4.19). The microbiota of the tonsillar crypts, however, is dominated by anaerobic species. In a study of seventeen healthy adults without any symptoms of tonsillitis, the mean number of anaerobic and aerobic/facultative isolates per person was 4.7 and 4.3, respectively. *Prevotella* spp. were the most frequently isolated organisms and constituted 26% of the total isolates; these were followed by viridans streptococci (20% of isolates), staphylococci (6%), *Corynebacterium*

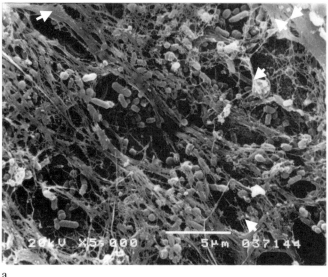

a

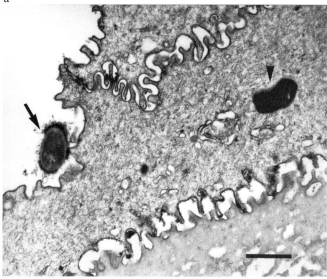

b

Figure 4.17 (a) Scanning electron micrograph showing bacteria surrounded by mucus which has contracted to form fibrils (arrows) during sample processing. (b) Transmission electron micrograph showing bacteria attached to (arrow) and within (arrowhead) the tonsillar epithelium. Bar = 1 μm. Photomicrographs kindly supplied by Professor Lars-Eric Stenfors, University of Tromso, and reproduced with permission from: *Acta Otolaryngologica* 1996;116:620–626.

spp. (5%), and *Porphyromonas* spp. (5%). Other anaerobes isolated less frequently included *Fusobacterium* spp. and *Veillonella* spp.

4.4.3.5 Lower respiratory tract

The larynx, trachea, bronchi, bronchioles, and alveoli of healthy individuals are not usually colonised by microbes. Small numbers of bacteria can be isolated from these regions, particularly the larynx and trachea, but they are only intermittently present and are usually disposed of by the extensive defence systems operating in the lower respiratory tract. These organisms are present as a result of the aspiration of bacteria-laden secretions from the upper respiratory tract. This is a frequent occurrence, particularly during sleep when it has been shown that approximately 50% of healthy individuals aspirate nasopharyngeal secretions into the lungs. Although the quantity of fluid involved is very small, ranging from 0.01 to 0.2 ml/night, the high concentration of bacteria present in these secretions (up to 10^8 cfu/ml) means that the number of bacteria entering the lower respiratory tract can be appreciable.

4.4.4 Interactions among members of the respiratory microbiota

The main beneficial interactions among microbes colonising the respiratory tract involve the provision of nutrients. Among the most abundant nutrient sources in respiratory secretions are mucins, the degradation of which requires the concerted action of microbes expressing sialidases, glycosidases, and proteases. Carbohydrates, amino acids, and fatty acids are also released by the action of hyaluronidase, chondroitin sulphatase, proteases, and lipases by some members of the respiratory microbiota (Figure 4.8). Metabolic end-products, such as lactate produced by streptococci, staphylococci, and *Haemophilus* spp., can also be utilised as carbon and energy sources by *Neisseria* spp. and *Veillonella* spp. Oxygen consumption by *Neisseria* spp., *Haemophilus* spp., and *Mor. catarrhalis* will also contribute to the establishment of microhabitats with low oxygen concentrations, thus enabling the growth of microaerophiles and anaerobes.

Although many *in vitro* studies have demonstrated the ability of respiratory tract microbes to inhibit or kill other members of the respiratory microbiota, the identity of the compounds involved has often not been determined. Antagonistic substances produced by members of the respiratory microbiota include bacteriocins, fatty acids, and hydrogen peroxide. In addition, microenvironments with a low pH and/or low oxygen content may be produced as a result of acid generation and oxygen utilisation by respiratory microbes, and this will restrict or prevent the growth of some species. Although for many of these antagonistic interactions there is little evidence that they operate *in vivo*, in some cases the *in vitro* and *in vivo* data do support the hypothesis that they are involved in controlling the composition of the respiratory-tract microbiota. This has prompted trials of the ability of indigenous species to prevent respiratory-tract infections (Sections 10.4.2 and 10.4.4). A number of *in vitro* studies have shown that viridans streptococci are able to inhibit the growth of respiratory pathogens, such as *Strep. pyogenes, Strep. pneumoniae, H. influenzae*, and *Staph. aureus*. For example, viridans streptococci isolated from the nasopharynx of children who were not prone to otitis media have been shown to inhibit the growth of NTHi *in vitro*. Furthermore, these children were found to have high proportions of viridans streptococci, but low proportions of NTHi in their nasopharynx. In contrast, children who were otitis-prone had low concentrations of viridans streptococci, but high proportions of NTHi. An inverse association between the presence of viridans streptococci and *Strep. pyogenes* in the oropharynx has also been reported and is described further in Section 10.4.2. *Strep. pneumoniae* has been reported to produce hydrogen peroxide at levels that can kill *H. influenzae in vitro*, and it has been suggested that this might account for the observed decrease in nasopharyngeal carriage of *H. influenzae* in individuals colonised by *Strep. pneumoniae*.

4.4.5 Dissemination of organisms from the respiratory tract

Microbes inhabiting the respiratory tract are disseminated in respiratory secretions mainly during breathing, talking, coughing, and sneezing. Microbe-laden secretions will also adhere to the fingers when these are placed in the mouth, a habit particularly common among young children, from where they can be transferred to a variety of other locations. Kissing and spitting constitute other means by which microbes in respiratory secretions can be transmitted to other sites. In all of the aforementioned situations, the secretions dispersed are mainly those present in the oral cavity, and this is the case even during sneezing, although, additionally, strings of mucus may be released from the nose. These secretions consist mainly of saliva, together with mucus-laden organisms from the nasal cavity and other regions of the respiratory tract that have been deposited in the mouth by the mucociliary escalator. Coughing and sneezing also expel respiratory secretions directly via the nose. Microbe-laden nasal secretions are also transferred onto handkerchiefs from which they can contaminate hands, clothing, and objects.

Large numbers of microbe-laden droplets are expelled from the mouth even during normal speech. Hence, it has been shown that speaking 2,000 words out loud results in the production of between 10^3 and 10^4 droplets – similar numbers are produced by a vigorous cough. Sneezing produces considerably more droplets – approximately 10^6. The fate of the droplets expelled depends on their size. Large droplets (>100 μm in diameter) rapidly fall to the ground, but the distance they travel will depend on their velocity during expulsion (i.e., whether by sneezing, coughing, or talking). Nevertheless, they are rarely propelled more than 2 m. Smaller droplets dry out before they reach the ground, resulting in the production of "droplet nuclei" with a diameter approximately one-fifth of the original droplets. Ordinary air movements will prevent these from settling to the ground, and they may remain suspended for considerable periods of time and are dispersed over large distances. Given that saliva has approximately 10^8 cfu/ml, many of the droplets produced will contain viable microbes, and it has been calculated that the minimum droplet size necessary to contain one microbe is 10–15 μm. The number of oral streptococci in a given volume of air has been used to estimate the dissemination of microbes from the respiratory tract; in an office environment, this has been found to be between 7 and 12 cfu/m^3, although the total concentration of microbes (which will include those from other sources such as the skin) was much higher than this. Measures to reduce the microbial content of air, such as ultraviolet irradiation and increased ventilation, generally achieve reductions no greater than 75%. Studies have shown that very few *Staph. aureus* are dispersed via the nose in comparison with the numbers dispersed on skin scales. Approximately 0.14 cfu are expelled from the respiratory tract by nasal carriers of the organism during 15 minutes of talking.

Although organisms colonising the respiratory tract are undoubtedly widely disseminated as airborne particles, their ability to survive in the droplet varies enormously, depending on the particular species with Gram-positive species generally surviving longer than Gram-negative species (Table 4.23). Furthermore, their ability to survive dessication (once the droplet has dried out) and exposure to sunlight and temperature extremes will also vary from species to species. Delicate organisms, therefore, will die rapidly on surfaces contaminated by droplets once these have dried out and cannot

Table 4.23.	Survival of members of the respiratory microbiota (and other organisms) in droplets

Organism	Survival time (hours)
H. influenzae	0.5–1.0
Strep. pneumoniae	>48
Strep. pyogenes	>48
Staph. aureus	54–70
E. coli	6–8
Ps. aeruginosa	8–24

Note: Suspensions of the organism were aerosolized to produce droplets with a volume of approximately 1 nl, exposed to air at a temperature of 21–27°C and a relative humidity of 70%, and samples taken for viable counting at various time intervals.

easily be transferred from such surfaces to other individuals. For example, *N. meningitidis* does not survive well in air or in the environment, and droplets that have traveled more than 3 feet from an infected individual generally do not contain viable organisms. The organism, therefore, can be transferred to others only by direct contact or between individuals who are in close proximity. Although organisms such as *Strep. pyogenes* and *Strep. pneumoniae* survive better in the environment, their major routes of transmission to other individuals are via droplets and direct contact.

4.4.6 Effects of antibiotics and other interventions on the respiratory microbiota

4.4.6.1 Antibiotics

The administration of various antibiotics has been shown to have a pronounced effect on the nasopharyngeal carriage of *Strep. pneumoniae, Mor. catarrhalis*, and *H. influenzae*. In a study involving seven groups of children (approximately 100 in each group), each given one of seven antibiotics, the main findings were: (1) amoxycillin, cefpodoxime, amoxicillin/clavulanate, and erythromycin/sulfisoxazole induced a decrease in the carriage of penicillin-susceptible strains, thereby increasing the proportion of penicillin-resistant *Strep. pneumoniae* carried after treatment; (2) β-lactam antibiotics increased the proportion of *Strep. pneumoniae* resistant to macrolides; (3) nasopharyngeal carriage of *Mor. catarrhalis* was significantly decreased by amoxycillin, amoxycillin/clavulanate, cefixime, cefpodoxime, and erythromycin/sulfisoxazole; and (4) carriage of *H. influenzae* was decreased by erythromycin/sulfisoxazole and cefaclor. However, in none of the children were bacteria detected with a decreased susceptibility to any of the antibiotics.

The administration of cephalexin orally for 12 days has been shown to result in a 99.9% reduction in the total viable count in the anterior nares after 6 days, but levels returned to pre-treatment values approximately 1 month after the end of treatment. However, lipophilic and non-lipophilic coryneforms and *Staph. aureus* did not return to their former levels and were replaced by CNS. Before treatment, all of the organisms were susceptible to 10–20 μg/ml cephalexin, but 9 days after the end of treatment, 53% of the ten individuals had CNS and Gram-negative rods resistant to the antibiotic.

Rifampicin is widely used for the prophylaxis of meningitis due to *N. meningitidis* and *H. influenzae*. Oral rifampicin given for 2 days to adult males has been shown to reduce the nasopharyngeal carriage rate of *N. meningitidis* from 37% to 6% and of

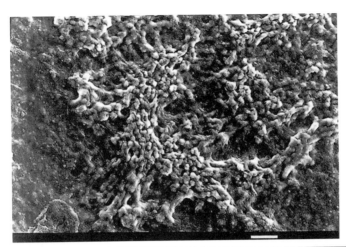

Figure 4.18 Scanning electron micrograph showing a biofilm on an endotracheal tube. Bar = 10 μm. Reproduced with the permission of Springer-Verlag GmbH & Co. from: Eradication of endotracheal tube biofilm by nebulised gentamicin. Adair, C.G., Gorman, S.P., Byers, L.M., Jones, D.S., Feron, B., Crowe, M., Webb, H.C., McCarthy, G.J., and Milligan, K.R. *Intensive Care Medicine* 2002;28:426–431.

H. influenzae from 34% to 7%. However, 25% of nasopharyngeal isolates were resistant to rifampicin after adminisration of the antibiotic. Twenty-seven percent of streptococci, 33% of neisseria (mainly *N. sicca-subflava* and some *N. meningitidis*), and 10% of *Haemophilus* spp. (all were *H. parainfluenzae*) were found to be resistant to the antibiotic.

4.4.6.2 Vaccination

Immunisation against a number of important pathogens has had a dramatic effect on their prevalence in the nasopharynx. The use of a number of different vaccines against *H. influenzae* type b has resulted in a decrease in nasopharyngeal carriage of the organism. Similarly, vaccination against *Strep. pneumoniae* using a polysaccharide-protein conjugate results in decreased carriage and acquisition of the vaccine serotypes. However, the nasopharynx of vaccinated children becomes colonised with other non-vaccine serotypes of *Strep. pneumoniae*. In studies of the effects on the carriage of *N. meningitidis* serogroup C strains following vaccination with meningococcal group A and C polysaccharide vaccine, carriage of the organism was found to decrease by between 47% and 65%. This was accompanied by large increases in the carriage of non-groupable strains of the organism.

4.4.6.3 Intubation

Nasogastric and endotracheal intubation are frequently used medical procedures. Nasogastric tubes are used for a variety of purposes, including feeding, decompression of the stomach, and flushing of the stomach to remove toxic substances. Endotracheal tubes are used in patients requiring mechanical ventilation. Such devices interfere with anatomical barriers to aspiration, thus allowing direct access of microbe-containing secretions to the lower airways, and they also provide a substratum for the growth of biofilms which can act as a reservoir of potential pathogens (Figure 4.18). The non-shedding substratum facilitates colonisation by organisms such as facultatively anaerobic Gram-negative rods that are not usually able to colonise the mucosal surfaces of the respiratory tract. The presence of such devices, therefore, predisposes patients to nosocomial pneumonia due to predominantly facultatively anaerobic Gram-negative species – ventilator-associated pneumonia is a major cause of death in intensive-care patients. A number of studies have shown that the presence of a nasogastric tube has a pronounced effect on the oropharyngeal microbiota. In general, the presence of the tube for 2–3 days results in increased levels of Gram-negative bacilli, including *Pseudomonas* spp., *Klebsiella* spp., *Proteus* spp., and *E. coli* in the oropharynx, and this is usually accompanied by decreased proportions of members of the indigenous micro-biota. Biofilm formation takes place rapidly on the external surface of the tube, and a substantial biofilm is usually present after only 24 hours. Similarly, biofilm formation rapidly occurs on the inner surface of endotracheal tubes, and facultatively anaerobic Gram-negative organisms can be isolated from the oropharynx within 36 hours and from the lower respiratory tract within 60–84 hours.

4.4.6.4 Radiation therapy

Post-operative sepsis is a frequently encountered consequence of head and neck surgery, particularly in those administered radiation therapy for malignant lesions. Infections account for between 28% and 47% of deaths in such patients. In a study of eighty patients suffering from head and neck cancers who received tumouricidal doses of Cobalt[60] (2,000–6,600 cGy) over a period of between 2 and 7 weeks, dramatic changes in the oropharyngeal microbiota were found. Hence, the proportions of *Staph. aureus*, *Strep. pyogenes*, *Can. albicans*, *Proteus* spp., and *Ps. aeruginosa* all showed a statistically significant increase by the end of the treatment period. Such changes are likely to have resulted from a combination of the effects of the radiation itself (different organisms being affected to varying extents), as well as the immunological and physiological changes induced in the host.

4.5 | Diseases caused by members of the respiratory microbiota

4.5.1 Meningitis

The nasopharynx is the primary habitat of the three main causative agents of meningitis: *N. meningitidis*, *Strep. pneumoniae*, and *H. influenzae*.

N. meningitidis is the main cause of meningitis worldwide and is responsible for both epidemic and endemic outbreaks of the disease – the total number of cases being more than 500,000 per year worldwide. The only reservoir of the organism is the nasopharynx of humans, and carriage is highest in teenagers and young adults. Al-though this is a life-threatening disease in adults, the attack rate and case/fatality ratio are much greater in children. The highest incidence of the disease is in the 6- to 24-month age group when protective maternal antibodies have disappeared and the infant has not developed specific immunity to the organism. This immunity arises from repeated exposure to the organism and/or to the antigenically related commensal organism *N. lactamica*. In industrialised countries, most infections are caused by strains belonging to serogroups B and C, while in developing countries, infections are due mainly to serogroup A strains and, to a lesser extent, strains belonging to serogroup C. Approximately 90% of infections worldwide are due to strains belonging to serogroups A, B and C. During the last 30 years, a limited number of serogroup B strains (belonging to the ET-5, lineage III, cluster A4, and ET-37 clonal complexes) have been the main cause of meningitis in industrialised countries. In sub-Saharan Africa (known as the "meningitis belt"), serogroup A strains are responsible for annual waves of meningitis which begin at the end of the dry season. During epidemic peaks, as many as 1% of the inhabitants of this region may become infected. The case fatality rate for meningitis due to *N. meningitidis* is between 10% and 20%, and a high proportion of survivors suffer from some permanent disability, such as deafness or mental retardation. Penicillin is effective in the treatment of the disease, but strains demonstrating intermediate resistance to the antibiotic are increasingly being isolated. There is, therefore, great interest in disease prevention by vaccination. A polysaccharide vaccine has been shown to be effective against infections caused by serogroups A, C, Y, and W135 in adults and children, but is not effective in infants younger than 2 years. Protein-polysaccharide conjugate vaccines are, therefore, being developed to overcome this problem, and a serogroup C conjugate vaccine is now used routinely in the United Kingdom. The vaccine has been shown to be effective at preventing invasive meningococcal disease in infants, children, and young adults.

Strep. pneumoniae is the second leading cause of meningitis and can occur in all age groups, particularly young children, adults with underlying diseases, the elderly, and immunocompromised individuals. The case fatality rate is high and ranges from 10% to 60%, and tends to be higher in adults than in children. Fifty percent of those who survive the disease develop some neurological dysfunction which may be short-lived or life-long. Because approximately 24% of strains of *Strep. pneumoniae* are now resistant to penicillin, the main antibiotic used in the treatment of the disease, there is increasing emphasis on disease prevention by vaccination. Currently, a polyvalent vaccine effective against 23 of the 90 serotypes of the organism is widely used. The twenty-three serotypes include 90% of those responsible for invasive infections due to the organism, and the effectiveness of the vaccine in preventing all invasive diseases due to the organism has been reported to range from 50–70%.

Meningitis due to *H. influenzae* serotype b affects mainly children younger than 4 years of age – particularly those in developing countries. Prior to the introduction of conjugate vaccines, the number of deaths per year due to *H. influenzae* serotype b was approximately 104,200, with most of these occurring in infants younger than 4 years. It has been estimated that the use of these vaccines prevents 21,000 cases of meningitis per year in infants in developed regions (i.e., 78% of cases in this age group). Unfortunately, its use in developing countries has been negligible because of its high cost, so that the number of cases of meningitis prevented by vaccination represents only 6% of the total number of cases worldwide in this age group. From a global perspective, therefore, the introduction of these highly effective vaccines has had little impact on this life-threatening disease.

4.5.2 Pneumonia

Pneumonia is a leading cause of morbidity and mortality, and is the sixth most common cause of death in the United Kingdom and the United States. Various types of pneumonia are recognised, the most common of which are (1) lobar pneumonia – a primary infection affecting one or more of the five lobes of the lungs; (2) broncho-pneumonia – this is usually a secondary infection following viral infection, some other lung disease, or heart disease; (3) aspiration pneumonia – this results from the in-halation into the lungs of food, drink, vomit, respiratory secretions, or saliva; and (4) nosocomial pneumonia – pneumonia acquired during hospitalisation. Although pneumonia can be caused by a variety of organisms, the most frequent aetiologi-cal agents include members of the respiratory microbiota – mainly *Strep. pneumoniae*, *H. influenzae*, and *Staph. aureus*. *Strep. pneumoniae* is responsible for approximately 95% of all cases of lobar pneumonia, a disease that affects mainly infants younger than 2 years of age and those over 65 years. Worldwide, more than 1 million children under 5 years of age die of the disease each year. Bronchopneumonia is also caused by *Strep. pneumo-niae*, but other residents of the respiratory tract are frequently involved, including *H. influenzae*, *Mor. catarrhalis*, and *Staph. aureus*.

Aspiration pneumonia is frequently a polymicrobial infection involving organisms from the upper respiratory tract. The organisms usually associated with community-acquired aspiration pneumonia include *Strep. pneumoniae*, *Staph. aureus*, *H. influenzae*, and members of the *Enterobacteriaceae*. In contrast, aspiration pneumonia acquired in hospitals is usually caused by *Enterobacteriaceae* and *Pseudomonas* spp.

Although one of the main causative agents of nosocomial pneumonia is *Staph. au-reus*, most of the organisms responsible for the disease are not normally residents of the respiratory tract and are facultatively anaerobic Gram-negative rods (mainly *K. pneumo-niae*, *Ps. aeruginosa*, and *Acinetobacter* spp.) *H. influenzae* and *Strep. pneumoniae* may also be involved, but less frequently.

4.5.3 Sinusitis

This is an inflammation of the mucosa of the paranasal sinuses which, in its chronic form, can lead to irreversible tissue damage. Acute sinusitis is a common disease in infants and accounts for between 5% and 10% of upper respiratory tract infections in this population group. *Strep. pneumoniae* is the causative agent in approximately 40% of cases, while *H. influenzae* and *Mor. catarrhalis* are each responsible for approximately 20% of cases. In chronic sinusitis, anaerobic species are often present, as well as the previously described organisms, and these are mainly *Prevotella* spp., *Porphyromonas* spp., and *Fusobacterium* spp., which are found in the oral cavity and oropharynx.

4.5.4 Otitis media

Otitis media is a very frequent infection in infants, with 50% and 70% of children expe-riencing at least one episode of the disease by the age of 1 year and 3 years, respectively. In developed countries, it is one of the most frequently diagnosed infections of chil-dren younger than 15 years. Approximately 40% of cases are attributable to infection by *Strep. pneumoniae*, with non-typeable *H. influenzae* and *Mor. catarrhalis* accounting for 30% and 25% of cases, respectively. Other causative agents include *Strep. pyogenes* and *Staph. aureus*. Often, the disease follows, or is accompanied by, a viral infection. The most common serotypes of *Strep. pneumoniae* responsible are 1, 3, 4, 6, 7, 9, 14, 15, 18, 19, and 23.

Table 4.24.	Diseases caused by *Staphylococcus aureus*
localised	folliculitis, furuncles, carbuncles, pyomyositis, botryomycosis, secondary infections of eczema lesions, abscesses, conjunctivitis, blepharitis, dacrocystitis
invasive	bacteraemia, endocarditis, osteomyelitis, septic arthritis, pneumonia, empyema, keratitis, orbital cellulitis
toxin-mediated	impetigo, septic shock, toxic shock syndrome, scalded skin syndrome, food-borne gastroenteritis

Interestingly, the nasopharyngeal orifice of the Eustachian tube of children who are otitis-prone is more densely colonised with bacteria than that of children who are not prone to the infection. Seventy-six percent of the epithelial cells from this region in otitis-prone children have more than fifty attached bacteria, whereas such densely colonised cells are found in only 8% of children who are not otitis-prone.

4.5.5 Epiglottitis

This is an infection of the epiglottis and neighbouring tissues, which can result in obstruction of the airways and is, therefore, life-threatening. It occurs mainly in children between the ages of 2 and 7 years. The infection is invariably caused by *H. influenzae* type b, and it is the second most common disease due to this organism in industrialised countries. Prior to introduction of the conjugate Hib vaccine, the annual number of cases in developed countries was 24,000, but this has fallen dramatically due to vaccination. The disease is far less prevalent in developing countries, with approximately 2,000 cases/year.

4.5.6 Diseases due to *Staph. aureus*

This organism is responsible for a wide range of infections, including skin, soft tissue, respiratory, bone, joint, and endovascular diseases (Table 4.24). Approximately half of all soft-tissue infections are attributable to *Staph. aureus*. The number of infections due to the organism is relentlessly increasing, and there is also a disturbing increase in the proportion of infections due to methicillin-resistant (MRSA) and multi-resistant strains of the organism. Hence, both the number of infections due to *Staph. aureus* and the number due to MRSA both approximately doubled over the period from 1987 to 1997. Nasal carriage of *Staph. aureus* is a major risk factor for the development of infections due to the organism, particularly in individuals undergoing surgery or haemodialysis and those on continuous ambulatory peritoneal dialysis. Efforts to control or eliminate nasal carriage of the organism have generally focussed on the local application of antibiotics and disinfectants, of which mupirocin has proved to be one of the most effective. Systemic antibiotics have been less successful and are of limited use because of the rapid development of resistance to them. Another possible approach is one based on the phenomenon of bacterial interference. That such a strategy may be effective has been prompted by a number of observations, which imply that other members of the nasal microbiota exert an inhibitory effect on *Staph. aureus*. For example, a negative correlation has been found between the presence of *Staph. aureus* in the nasal vestibule and either *Staph. epidermidis* or *Corynebacterium* spp. Whether this stems from the production of some inhibitory compound (e.g., a bacteriocin, toxic end-product of metabolism) or competition for adhesion sites remains to be established. *Staph. epidermidis*, but not corynebacteria, are known to produce compounds able to inhibit *Staph. aureus*, and corynebacteria display a greater affinity for mucus than does *Staph. aureus*. A recent study has shown that implantation of a *Corynebacterium* sp. into the nasal cavity of *Staph. aureus* carriers was able to eliminate *Staph. aureus* from this site – this is described in greater detail in Section 10.4.1.

The mortality rate from bacteraemia due to *Staph. aureus* is high and ranges from 11% to 43%, while that from endocarditis caused by the organism is as high as 56%. Staphylococcal toxic shock syndrome (STSS) is a disease that may follow a localized, non-invasive infection by the organism. It is characterised by vasodilation, disseminated intravascular coagulation, renal failure, acute respiratory distress syndrome, and multiple-organ failure. These symptoms are a consequence of the massive release of pro-inflammatory cytokines induced by one or more exotoxins produced by the organism. The exotoxins responsible are toxic shock syndrome (TSS) toxin-1 (TSST-1), staphylococcal exotoxin B (SEB), and staphylococcal exotoxin C (SEC). These are superantigens which, at picomolar concentrations, are able to stimulate the proliferation of approximately 20% of T cells, which in turn results in the production of huge quantities of pro-inflammatory cytokines. The case fatality rate is between 3% and 5%. Two forms of STSS are recognized – menstrual and non-menstrual. Menstrual TSST occurs within 1–2 days of the beginning or end of menses and is associated with the use of high-absorbancy tampons. In this form of the disease, only TSST-1 is involved, and a single clone of *Staph. aureus* is responsible – electrophoretic type ET 41. In non-menstrual TSS, three toxins are involved – TSST-1, SEB, and SEC – and the disease may follow any infection with *Staph. aureus*. The number of cases of menstrual STSS has declined during the last 20 years due to a decrease in tampon absorbancy and greater awareness of TSS among women. However, STSS is still a significant cause of morbidity worldwide.

Sudden infant death syndrome (SIDS) is an important cause of death in infants between 2 and 3 months of age. Because many of the risk factors for the syndrome are similar to those for respiratory tract infections, it has been suggested that it may have a microbial aetiology. Studies of the nasopharyngeal microbiota of infants who have died from the syndrome have revealed a significantly greater frequency of isolation of *Staph. aureus* than from a healthy control group. Furthermore, the strains isolated were toxigenic, and exotoxins have been identified in the tissues of most infants who have died from SIDS. The toxins detected include TSST and SEC_1, which act as superantigens and can induce toxic shock and death in otherwise healthy individuals.

4.5.7 Diseases due to *Strep. pyogenes*

Like *Staph. aureus*, *Strep. pyogenes* is responsible for a wide spectrum of infections, ranging from the trivial to the life-threatening, and these are summarized in Table 4.25.

Pharyngitis is one of the most commonly encountered infections due to the organism. The predominant M protein serotypes responsible are 1, 3, 5, 6, 12, 14, 18, 19, and 24. The *emm* patterns (Section 4.4.1.3.1) associated with pharyngitis are A, B, and C, while *emm* pattern E strains may cause pharyngitis or skin infections. If the infecting

Table 4.25. | Diseases caused by *Streptococcus pyogenes*

superficial infections	pharyngitis, tonsillitis, impetigo, erysipelas
invasive infections	bacteraemia, meningitis, pneumonia, necrotising fasciitis, puerperal sepsis, myositis, cellulitis, pericarditis, septic arthritis
toxin-mediated	scarlatina, streptococcal toxic shock syndrome
immunologically mediated	acute rheumatic fever, acute post-streptococcal glomerulonephritis, reactive arthritis

strain produces one or more pyrogenic exotoxins (SpeA, SpeB, SpeC), pharyngitis may be accompanied by a rash and skin desquamation – a condition known as scarlet fever. In infants, pharyngitis occurs more frequently in the autumn/winter than in the spring/summer months and involves mainly the nasopharynx, but can then spread to the middle ear and even to the meninges. In older children and adults, the tonsils are frequently involved. The preferred sites of colonisation on the tonsils are the microridges of the epithelial cells (Figures 4.15 and 4.19). Long chains of bacteria have been seen attached to the tonsillar epithelium during cases of acute tonsillitis (Figure 4.20). Although pharyngitis itself is rarely life-threatening, as many as 10% of infected individuals then suffer an attack of rheumatic fever, and up to 40% may develop acute glomerulonephritis.

Acute rheumatic fever is a life-threatening inflammatory disease affecting the heart and joints and is the main cause of acquired heart disease in children. It follows approximately 3 weeks after an upper respiratory tract infection by certain strains of

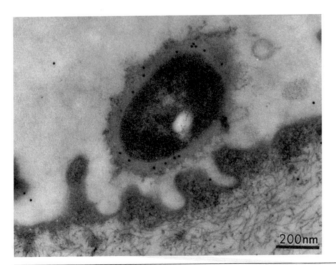

Figure 4.19 Transmission electron micrograph showing *Streptococcus pyogenes* attached to the crests of cellular projections (microridges) of the tonsillar epithelium. The bacteria are coated with rabbit antiserum to *Strep. pyogenes* labelled with gold particles. Photomicrograph kindly supplied by Professor Lars-Eric Stenfors, University of Tromso. Reprinted from: Where are the receptors for *Streptococcus pyogenes* located on the tonsillar surface epithelium? Lilja, M., Silvola, J., Raisanen, S., and Stenfors, L.-E. *International Journal of Pediatric Otorhinolaryngology* 1999;50:37–43. Copyright © 1999, with permission from Elsevier Ireland Ltd.

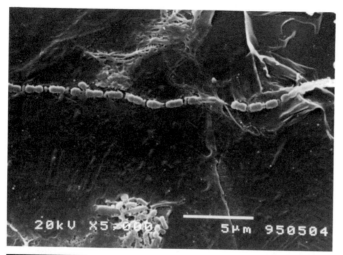

Figure 4.20 Scanning electron micrograph of the tonsillar epithelium with a long chain of cocci. Photomicrograph kindly supplied by Professor Lars-Eric Stenfors, University of Tromso. Reprinted from: Initial events in the pathogenesis of acute tonsillitis caused by *Streptococcus pyogenes*, Lilja, M., Raisanen, S., and Stenfors, L.-E. *International Journal of Pediatric Otorhinolaryngology* 1998;45:15–20. Copyright © 1998, with permission from Elsevier Ireland Ltd.

Strep. pyogenes – particularly M types 1, 3, 5, 6, 14, 18, 19, and 24. The exact mechanism by which these rheumatogenic strains cause the disease is uncertain, but it is thought to be a consequence of the similarity of certain surface molecules of the organism to molecules present in the heart, synovium, and neurons of humans. Antibodies raised by the host to these bacterial antigens then cross-react with the host molecules causing tissue inflammation and damage. Individuals who survive one attack of rheumatic fever are vulnerable to subsequent episodes of the disease following further respiratory-tract infections.

Acute post-streptococcal glomerulonephritis occurs mainly in children and young adults, and follows 1–4 weeks after a skin or respiratory-tract infection with a nephritogenic strain of *Strep. pyogenes*. The nephritogenic strains most often associated with skin infections are M types 2, 49, 42, 56, 57, and 60, while those causing throat infections are M types 1, 4, 12, and 25. Again, the exact mechanism remains uncertain but, like rheumatic fever, it is probably a consequence of the antigenic similarity between certain bacterial constituents and those present in human kidneys. Antibodies raised against human glomeruli are known to cross-react with certain streptococcal M proteins. Unlike the situation with rheumatic fever, individuals who have had one episode of acute glomerulonephritis are unlikely to suffer further attacks – this is probably because of the limited number of nephritogenic strains of *Strep. pyogenes*.

Necrotising fasciitis is an invasive disease in which the organism spreads through the tissues above the deep fascia, and this is accompanied by thrombosis of the blood vessels. This results in gangrene and the thrombosis prevents antibiotics from penetrating into the tissues so that the disease spreads rapidly despite antimicrobial chemotherapy. Affected individuals often also suffer from a bacteraemia and toxic shock syndrome and in approximately one-third of cases the disease is fatal. Risk factors for the disease include diabetes, minor or major trauma, and immunocompromisation. Fortunately, this life-threatening condition is rare, with an incidence of 0.4 per 1,000,000. The M types most frequently associated with the disease are 1, 3, 11, 12, and 28.

Streptococcal toxic shock syndrome accompanies a high proportion of invasive infections with *Strep. pyogenes*, particularly those due to M types 1 and 3. As with staphylococcal TSS, the affected individual suffers from vasodilation, disseminated intravascular coagulation, renal failure, acute respiratory distress syndrome, and multiple organ failure due to the release of pro-inflammatory cytokines induced by one or more superantigens produced by the organism. The superantigens responsible are streptococcal pyrogenic exotoxin A and streptococcal pyrogenic exotoxin C. The case fatality rate is much higher than that encountered in staphylococcal TSS and ranges from 30% to 80%.

4.6 | Further Reading

Books

Ellis, M.E. (ed.) (1998). Infectious diseases of the respiratory tract. Cambridge: Cambridge University Press.

Niederman, M.S., Sarosi, G.A., and Glassroth, J. (2001). *Respiratory infections*. Philadelphia: Lippincott Williams & Wilkins.

Reviews and Papers

Adamsson, I., Edlund, C., Sjostedt, S., and Nord, C.E. (1997). Comparative effects of cefadroxil and phenoxymethylpenicillin on the normal oropharyngeal and intestinal microflora. *Infection* **25**, 154–158.

Alouf, J.E. and Muller-Alouf, H. (2003). Staphylococcal and streptococcal superantigens: molecular, biological, and clinical aspects. *International Journal of Medical Microbiology* **292**, 429–440.

Archer, G.L. (1998). *Staphylococcus aureus*: a well-armed pathogen. *Clinical Infectious Diseases* **26**, 1179–1181.

Aubrey, R. and Tang, C. (2003). The pathogenesis of disease due to type b *Haemophilus influenzae*. *Methods in Molecular Medicine* **71**, 29–50.

Blackwell, C.C., MacKenzie, D.A., James, V.S., Elton, R.A., Zorgani, A.A., Weir, D.M., and Busuttil, A. (1999). Toxigenic bacteria and sudden infant death syndrome (SIDS): nasopharyngeal flora during the first year of life. *FEMS Immunology and Medical Microbiology* **25**, 51–8.

Bogaert, D., De Groot, R., and Hermans, P.W. (2004). *Streptococcus pneumoniae* colonisation: the key to pneumococcal disease. *Lancet Infectious Diseases* **4**, 144–154.

Boucher, R.C. (1999). Molecular insights into the physiology of the thin film of airway surface liquid. *Journal of Physiology* **516**, 631–638.

Brook, I. and Gober, A.E. (2002). Effect of amoxicillin and co-amoxiclav on the aerobic and anaerobic nasopharyngeal flora. *Journal of Antimicrobial Chemotherapy* **49**, 689–692.

Brook, I. and Shah, K. (2001). Effect of amoxycillin with or without clavulanate on adenoid bacterial flora. *Journal of Antimicrobial Chemotherapy* **48**, 269–273.

Cantin, A.M. (2001). Biology of respiratory epithelial cells: role in defense against infections. *Pediatric Pulmonology* (Suppl 23), 167–169.

Cockeran, R., Anderson, R., and Feldman, C. (2002). The role of pneumolysin in the pathogenesis of *Streptococcus pneumoniae* infection. *Current Opinion in Infectious Diseases* **15**, 235–239.

Cole, A.M., Dewan, P., and Ganz, T. (1999). Innate antimicrobial activity of nasal secretions. *Infection and Immunity* **67**, 3267–3275.

Cole, A.M., Tahk, S., Oren, A., Yoshioka, D., Kim, Y.-H., Park, A., and Ganz, T. (2001). Determinants of *Staphylococcus aureus* nasal carriage. *Clinical and Diagnostic Laboratory Immunology* **8**, 1064–1069.

Courtney, H.S., Hasty, D.L., and Dale, J.B. (2002). Molecular mechanisms of adhesion, colonization, and invasion of group A streptococci. *Annals of Medicine* **34**, 77–87.

Cunningham, M.W. (2000). Pathogenesis of group A streptococcal infections. *Clinical Microbiology Reviews* **13**, 470–511.

De Lencastre, H. and Tomasz, A. (2002). From ecological reservoir to disease: the nasopharynx, day-care centres and drug-resistant clones of *Streptococcus pneumoniae*. *Journal of Antimicrobial Chemotherapy* **50** (Suppl S2), 75–81.

Devine, D.A. (2003). Antimicrobial peptides in defence of the oral and respiratory tracts. *Molecular Immunology* **40**, 431–443.

Diamond, G., Legarda, D., and Ryan, L.K. (2000). The innate immune response of the respiratory epithelium. *Immunological Reviews* **173**, 27–38.

Emonts, M., Hazelzet, J.A., de Groot, R., and Hermans, P.W. (2003). Host genetic determinants of *Neisseria meningitidis* infections. *Lancet Infectious Diseases* **3**, 565–577.

Gagliardi, D., Makihara, S., Corsi, P.R., Viana Ade, T., Wiczer, M.V., Nakakubo, S., and Mimica, L.M. (1998). Microbial flora of the normal esophagus. *Diseases of the Esophagus* **11**, 248–250.

Ganz, T. (2002). Antimicrobial polypeptides in host defense of the respiratory tract. *Journal of Clinical Investigation* **109**, 693–697.

Garcia-Rodriguez, J.A. and Fresnadillo Martinez, M.J. (2002). Dynamics of nasopharyngeal colonization by potential respiratory pathogens. *Journal of Antimicrobial Chemotherapy* **50** (Suppl C), 59–73.

Gomes, G.F., Pisani, J.C., Macedo, E.D., and Campos, A.C. (2003). The nasogastric feeding tube as a risk factor for aspiration and aspiration pneumonia. *Current Opinion in Clinical Nutrition and Metabolic Care* **6**, 327–333.

Gunnarsson, R.K., Holm, S.E., and Soderstrom, M. (1998). The prevalence of potential pathogenic bacteria in nasopharyngeal samples from healthy children and adults. *Scandinavian Journal of Primary Health Care* **16**, 13–17.

Hardy, K.J., Hawkey, P.M., Gao, F., and Oppenheim, B.A. (2004). Methicillin resistant *Staphylococcus aureus* in the critically ill. *British Journal of Anaesthesia* **92**, 121–130.

Hood, D.W. (2003). The genome sequence of *Haemophilus influenzae*. *Methods in Molecular Medicine* **71**, 147–159.

Jedrzejas, M.J. (2001). Pneumococcal virulence factors: structure and function. *Microbiology and Molecular Biology Reviews* **65**, 187–207.

Karalusa, R. and Campagnaria, A. (2000). *Moraxella catarrhalis*: a review of an important human mucosal pathogen. *Microbes and Infection* **2**, 547–559.

Klugman, K.P. and Feldman, C. (2001). *Streptococcus pneumoniae* respiratory tract infections. *Current Opinion in Infectious Diseases* **14**, 173–179.

Kononen, E., Jousimies-Somer, H., Bryk, A., Kilp, T., and Kilian, M. (2002). Establishment of streptococci in the upper respiratory tract: longitudinal changes in the mouth and nasopharynx up to 2 years of age. *Journal of Medical Microbiology* **51**, 723–730.

Kreikemeyer, B., McIver, K.S., and Podbielski, A. (2003). Virulence factor regulation and regulatory networks in *Streptococcus pyogenes* and their impact on pathogen-host interactions. *Trends in Microbiology* **11**, 224–232.

Kvalsvig, A.J. and Unsworth, D.J. (2003). The immunopathogenesis of meningococcal disease. *Journal of Clinical Pathology* **56**, 417–422.

Leiberman, A., Dagan, R., Leibovitz, E., Yagupsky, P., and Fliss, D.M. (1999). The bacteriology of the nasopharynx in childhood. *International Journal of Pediatric Otorhinolaryngology* **49** (Suppl 1), S151–S153.

Leung, A.K. and Kellner, J.D. (2004). Acute sinusitis in children: diagnosis and management. *Journal of Pediatric Health Care* **18**, 72–76.

LeVine, A.M. and Whitsett, J.A. (2001). Pulmonary collectins and innate host defense of the lung. *Microbes and Infection* **3**, 161–166.

Lowy, F.D. (2003). Antimicrobial resistance: the example of *Staphylococcus aureus*. *Journal of Clinical Investigation* **111**, 1265–1273.

Lund, B., Edlund, C., Rynnel-Dagoo, B., Lundgren, Y., Sterner., J., and Nord, C.E. (2001). Ecological effects on the oro- and nasopharyngeal microflora in children after treatment of acute otitis media with cefuroxime axetil or amoxycillin-clavulanate as suspensions. *Clinical Microbiology and Infection* **7**, 230–237.

Mandal, S., Berendt, A.R., and Peacock, S.J. (2002). *Staphylococcus aureus* bone and joint infection. *Journal of Infection* **44**, 143–151.

Marrs, C.F., Krasan, G.P., McCrea, K.W., Clemans, D.L., and Gilsdorf, J.R. (2001). *Haemophilus influenzae* – human specific bacteria. *Frontiers in Bioscience* **6**, 41–60.

McCormick, J.K., Yarwood, J.M., and Schlievert, P.M. (2001). Toxic shock syndrome and bacterial superantigens: an update. *Annual Review of Microbiology* **55**, 77–104.

McDaniel, L.S. and Swiatlo, E. (2004). Pneumococcal disease pathogenesis, treatment, and prevention. *Infectious Diseases in Clinical Practice* **12**, 93–98.

Meats, E., Brueggemann, A.B., Enright, M.C., Sleeman, K., Griffiths, D.T., Crook, D.W., and Spratt, B.G. (2003). Stability of serotypes during nasopharyngeal carriage of *Streptococcus pneumoniae*. *Journal of Clinical Microbiology* **41**, 386–392.

Mehta, R. and Niederman, M. (2002). Nosoconial pneumonia. *Current Opinion in Infectious Diseases* **15**, 387–394.

Meli, D.N., Christen, S., Leib, S.L., and Tauber, M.G. (2002). Current concepts in the pathogenesis of meningitis caused by *Streptococcus pneumoniae*. *Current Opinion in Infectious Disease* **15**, 253–257.

Mitchell, T.J. (2003). The pathogenesis of streptococcal infections: from tooth decay to meningitis. *Nature Reviews Microbiology* **1**, 219–230.

Murphy, T.F. (2003). Respiratory infections caused by non-typeable *Haemophilus influenzae*. *Current Opinion in Infectious Diseases* **16**, 129–134.

Nassif, X. (2002). Genomics of *Neisseria meningitidis*. *International Journal of Medical Microbiology* **291**, 419–423.

Obaro, S. and Adegbola, R. (2002). The pneumococcus: carriage, disease and conjugate vaccines. *Journal of Medical Microbiology* **51**, 98–104.

Palmer, L.B., Albulak, K., Fields, S., Filkin, A.M., Simon, S., and Smaldone, G.C. (2001). Oral clearance and pathogenic oropharyngeal colonization in the elderly. *American Journal of Respiratory and Critical Care Medicine* **164**, 464–468.

Palmu, A.A., Herva, E., Savolainen, H., Karma, P., Makela, P.H., and Kilpi, T.M. (2004). Association of clinical signs and symptoms with bacterial findings in acute otitis media. *Clinical Infectious Diseases* **38**, 234–242.

Pathan, N., Faust, S.N., and Levin, M. (2003). Pathophysiology of meningococcal meningitis and septicaemia. *Archives of Disease in Childhood* **88**, 601–607.

Peacock, S.J., de Silva, I., and Lowy, F.D. (2001). What determines nasal carriage of *Staphylococcus aureus*? *Trends in Microbiology* **9**, 605–610.

Peacock, S.J., Justice, A., Griffiths, D., de Silva, G.D., Kantzanou, M.N., Crook, D., Sleeman, K., and Day, N.P. (2003). Determinants of acquisition and carriage of *Staphylococcus aureus* in infancy. *Journal of Clinical Microbiology* **41**, 5718–5725.

Pericone, C.D., Overweg, K., Hermans, P.W., and Weiser, J.N. (2000). Inhibitory and bactericidal effects of hydrogen-peroxide production by *Streptococcus pneumoniae* on other inhabitants of the upper respiratory tract. *Infection and Immunity* **68**, 3990–3997.

Rao, V.K., Krasan, G.P., Hendrixson, D.R., Dawid, S., and St. Geme III, J.W. (1999). Molecular determinants of the pathogenesis of disease due to non-typable *Haemophilus influenzae*. *FEMS Microbiology Reviews* **23**, 99–129.

Rasmussen, T.T., Kirkeby, L.P., Poulsen, K., Reinholdt, J., and Kilian, M. (2000). Resident aerobic microbiota of the adult human nasal cavity. *APMIS* **108**, 663–675.

Regev-Yochay, G., Raz, M., Dagan, R., Porat, N., Shainberg, B., Pinco, E., Keller, N., and Rubinstein, E. (2004). Nasopharyngeal carriage of *Streptococcus pneumoniae* by adults and children in community and family settings. *Clinical Infectious Diseases* **38**, 632–639.

Rovers, M.M., Schilder, A.G., Zielhuis, G.A., and Rosenfeld, R.M. (2004). Otitis media. *Lancet*. **363**, 465–473.

Seal, D. (2001). Necrotizing fasciitis. *Current Opinion in Infectious Diseases* **14**, 127–132.

Shopsin, B. and Kreiswirth, B.N. (2001). Molecular epidemiology of methicillin-resistant *Staphylococcus aureus*. *Emerging Infectious Diseases* **7**, 323–326.

Stjernquist-Desatnik, A. and Holst, E. (1999). Tonsillar microbial flora: comparison of recurrent tonsillitis and normal tonsils. *Acta Oto-Laryngologica* (Stockholm) **119**, 102–106.

Taha, M.-K., Deghmane, A.-E., Antignac, A., Zarantonelli, M.L., Larribe, M., and Alonso, J.-M. (2002). The duality of virulence and transmissibility in *Neisseria meningitidis. Trends in Microbiology* **10**, 376–382.

Travis, S.M., Singh, P.K., and Welsh, M.J. (2001). Antimicrobial peptides and proteins in the innate defence of the airway surface. *Current Opinion in Immunology* **13**, 89–95.

Tzeng, Y.-L. and Stephens, D.S. (2000). Epidemiology and pathogenesis of *Neisseria meningitidis. Microbes and Infection* **2**, 687–700.

van Deuren, M., Brandtzaeg, P., and van der Meer, J.W. (2000). Update on meningococcal disease with emphasis on pathogenesis and clinical management. *Clinical Microbiology Reviews* **13**, 144–166.

Verduin, C.M., Hol, C., Fleer, A., van Dijk, H., and van Belkum, A. (2002). *Moraxella catarrhalis*: from emerging to established pathogen. *Clinical Microbiology Reviews* **15**, 125–144.

Wildes, S.S. and Tunkel, A.R. (2002). Meningococcal vaccines: a progress report. *BioDrugs* **16**, 321–329.

The urinary system and its indigenous microbiota

The urinary system consists of two kidneys, two ureters, a bladder, and a urethra. When considering the microbiota of this system, it is convenient to deal with the male and female urinary systems separately for a number of reasons. Firstly, the anatomy of the system in males and females differs significantly. Secondly, in males, the terminal portion of the urinary system, the urethra, also constitutes part of the reproductive system, resulting in important functional differences from the urethra of females. Thirdly, the urethral opening in females is closer to the anus than in males and is also close to the vaginal introitus, and these heavily colonised sites provide important additional sources of potential microbial colonisers. These factors combine to generate significant differences not only in the types of microbes colonising the urinary system of males and females, but also their relative susceptibility to infection.

5.1 | The urinary system of females

5.1.1 Anatomy and physiology

Whereas the kidneys, ureters, and bladder are normally sterile, the urethra of females is usually colonised by microbes along its whole length. Further discussion of the anatomy of the urinary system is, therefore, confined to the urethra. In females, the urethra is a short tube (approximately 3.8 cm in length) leading from the floor of the bladder to the external environment via the external urethral orifice which lies in front of the vaginal opening (Figure 5.1). However, far from being a simple tube, the urethra has a complex structure that includes many glands (paraurethral and mucous) and pit-like recesses which open into the lumen. The urethra is lined by a non-keratinised, stratified, squamous epithelium, similar to that of the vagina, in contrast to the transitional epithelium which lines the bladder. However, the epithelium becomes keratinised near the external orifice.

5.1.2 Antimicrobial defence mechanisms of the urinary system

One of the main antimicrobial defence mechanisms of mucosal surfaces is the shedding of the outermost cells with their adherent microbial populations (Table 5.1). This occurs in the female urethra, but is under hormonal control. Hence, the urethra of menstruating women and post-menopausal women on hormone replacement therapy (HRT) has abundant exfoliating cells. In contrast, exfoliation occurs at a slower

Table 5.1. Antimicrobial defence mechanisms of the female urinary system

Defence mechanism	Mode of action
desquamation	outermost cells are shed with their attached microbes
urine flow	mechanically removes bacteria
high-pressure zone in mid-urethra	restricts ascent of bacteria into bladder
pH of urine	low pH can be microbicidal or microbistatic
osmolality of urine	high osmolality can be microbicidal or microbistatic
urea	high concentration can be microbicidal or microbistatic
antimicrobial peptides	inhibit microbial growth or microbicidal
Tamm-Horsfall protein	prevents bacterial adhesion to epithelium
manno-oligosaccharides	prevent bacterial adhesion to epithelium
GP51	prevents bacterial adhesion to epithelium
sIgA antibodies	prevent bacterial adhesion
epithelial cell-dependent bacterial growth inhibition	inhibits growth of attached bacteria

Note: sIgA = secretory IgA.

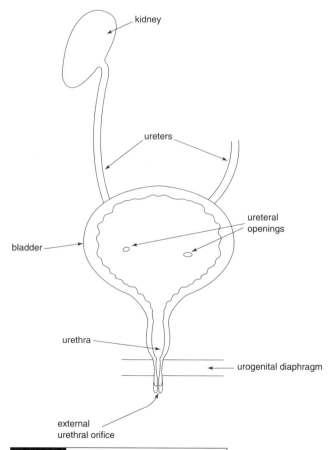

Figure 5.1 Organs of the female urinary system.

rate in the urethra of pre-menarchal girls and post-menopausal women not on HRT. Another defence mechanism is the secretion of mucus by the paraurethral glands – this forms a layer on the epithelial surface and entraps microbes, as well as preventing their adhesion to the urethral epithelium. Removal of this layer results in significantly increased adhesion of bacteria to the urethral mucosa. Colonisation of the urethra by microbes is hindered to a great extent by the flushing action of urine. Between 1 and 2 litres of urine are produced each day, and this is expelled in 200–400 ml quantities during urination at a flow rate of between 40 and 80 ml/hour – urinary flow, therefore, exerts a considerable flushing action. Furthermore, a high-pressure zone exists approximately mid-way along the urethra (where the urethra passes through the urogenital diaphragm), and this acts as a barrier to microbial ascent from the urethral orifice to the bladder. Adhesion of microbes to the urethral epithelium is also hindered by the presence of a low-molecular mass protein known as the Tamm-Horsfall protein or uromodulin. This is synthesised in the kidney and secreted into the urine from which it is deposited along the uroepithelium, forming a continuous coating. The protein contains receptors for the adhesins of various enterobacteria, including the type 1 and S fimbriae of E. coli – the organism most frequently responsible for urinary tract infections (UTIs). Binding of bacteria to these adhesins protects the underlying epithelial cells and facilitates the removal of the bacteria by the flushing action of urine. The protein also binds to polymorphonuclear leukocytes, thus enhancing phagocytosis of the bacteria-protein complex. Urine also contains manno-oligosaccharides which bind to the adhesins on type 1 fimbriae of E. coli, thereby preventing the organism adhering to the uroepithelium. Interestingly, none of the constituents of urine appear able to bind to adhesins on the P fimbriae of E. coli. A major component of the mucus layer coating the bladder is a glycoprotein, GP51, which is produced by bladder epithelial cells and is also secreted into the urine. GP51 binds to a range of organisms, including E. coli, Enterobacter cloacae, K. pneumoniae, Proteus spp., Ps. aeruginosa, Ser. marcescens, Staph. aureus, Staph. epidermidis, and Ent. faecalis, thereby preventing their adhesion to underlying epithelial cells. The bacteria form aggregates, and this facilitates their removal by the flushing action of urine. The concentration of GP51 is significantly higher in the urine of patients with a UTI than in uninfected controls.

The pH of urine in healthy individuals ranges from 4.6 to 7.5, with a mean value of 6.0 (Table 5.1). Vegetarians tend to produce urine with a higher pH, while those on a high-protein diet have a more acidic urine. The acidic pH of the urine of the average individual can prevent the growth of some microbes and can be microbicidal. The high urea content of urine (200–400 mmoles/litre) can also exert a microbicidal or microbistatic effect, as can the high osmolality (up to 1,300 mOsm/kg). The antibacterial effect of urine is greatest at low pHs.

Urine contains 10–100 μg/litre of human β-defensin (HBD)-1, but E. coli is affected only by higher concentrations. It is possible that local concentrations (e.g., at epithelial surfaces) are higher than those detected in urine, so that E. coli may be inhibited or killed when attached to the epithelium. Another antimicrobial peptide present in urine is hepcidin 20, which is active against E. coli, Staph. epidermidis, Staph. aureus, Can. albicans, and Strep. agalactiae. However, again, a microbicidal effect is seen only at concentrations much higher than those found in urine. It has also been shown that uroepithelial cells are able to inhibit the growth of bacteria adhering to them. The mechanism

| Table 5.2. | Composition of urine |

Solute	Concentration (mmol/L)
urea	200–400
sodium	50–130
chloride	50–130
potassium	20–70
ammonium	30–50
phosphate	25–60
calcium	10–24
creatinine	6–20
uric acid	0.7–8.7
bicarbonate	0–2
protein	trace
glucose	trace

involved has not been elucidated, but appears to be independent of the receptor–ligand interaction mediating adhesion and is induced by transmembrane signalling involving calcium flux triggered by bacterial contact. The effector molecules of this response are likely to be antimicrobial peptides.

Secretory IgA (sIgA), together with smaller quantities of IgG and IgM, are present in urine, but their role in the prevention of UTIs is not clear. For example, it has been shown that individuals with hypogammaglobulinaemia are not more susceptible to UTIs. In contrast, other studies have shown that individuals with low urinary concentrations of sIgA are more susceptible to UTIs. sIgA has been shown to block adhesion of *E. coli* and other bacteria to epithelial cells.

5.1.3 Environmental determinants within the urethra

Little information is available regarding the nature of the environment in the lumen of the urethra. However, the oxygen content of urine is high, with a mean partial pressure of approximately 81 mm Hg, implying that the habitat is largely aerobic. The pH ranges from 5.8 in the distal urethra to 6.1 in the proximal regions, but will be affected by the intermittent voiding of urine which generally has a pH of approximately 6. With regard to nutrient availability, urine would provide several nitrogen sources (urea, NH_4^+, and creatinine), phosphate, and a number of trace elements (Table 5.2), but, except for individuals suffering from diabetes, no carbohydrates. However, nutrients would also be available from dead and dying epithelial cells, as well as from the mucus and other substances secreted by the numerous glands lining the urethra. Sugars and amino acids can be liberated from mucins by a number of species resident in the urethra, including coagulase-negative staphylococci (CNS), lactobacilli, viridans streptococci, Gram-positive anaerobic cocci (GPAC), and *Bacteroides* spp. Furthermore, many of these species are also proteolytic and produce amino acids from host proteins for use as a carbon and/or energy source. As in many host secretions, the concentration of free iron is low, and there is evidence that bacterial growth in urine *in vivo* is iron-restricted.

Surprisingly, there is no consensus regarding the ability of urine to support bacterial growth. Different studies have reached the following conclusions: (1) a variety of bacteria can grow in urine; (2) only some species can grow in urine; (3) bacterial growth is inhibited by urine; (4) urine can kill many bacteria; (5) *E. coli* and other facultative Gram-negative bacilli can survive and grow in urine better than staphylococci and enterococci; and (6) uropathogenic species can grow well in urine, whereas other species cannot. The confusion arises mainly because of differences in experimental methodology, such as whether single, several, or pooled urine samples are used, the characteristics (age, sex, etc.) of the individual(s) from whom the urine was obtained, the size of the bacterial inoculum, the range of species tested, the origin of the strains used, and the pH of the urine. In some of the better-designed studies in which small inocula (corresponding to those which would be expected to exist *in vivo*) of strains isolated from the human urethra have been used, the ability of urine to support the growth of bacteria has, not surprisingly, been found to depend on the nature of the bacterial strain and the particular urine sample. Often, urine is inhibitory for many strains of bacteria that can be isolated from the healthy and infected urinary tract, and this inhibitory effect is generally related to the urea content of the sample. However, it must be remembered that the urethra is, of course, not filled with urine for most of the time. During the periods between urination, traces of urine will be present, which may be unable to exert an antibacterial effect (because of dilution by mucus and glandular secretions), but could supply essential nutrients (particularly nitrogen sources) to support bacterial growth.

In order for microbes to survive in the urethra, they must, of course, be able to withstand the numerous antimicrobial mechanisms protecting this region (see Table 5.1).

5.1.4 The indigenous microbiota of the female urethra

As previously mentioned, only the urethra is colonised by microbes – the kidneys, ureters, and bladder are sterile in healthy individuals. The repeated flushing of the urethra during urination is a strong environmental selecting factor, making the ability to adhere to the urethral epithelium a key requirement for any successful coloniser of this region.

5.1.4.1 Main characteristics of key members of the urethral microbiota

The main organisms colonising the female urethra are *Lactobacillus* spp., CNS, *Corynebacterium* spp., viridans streptococci, *Bacteroides* spp., GPAC, and *Mollicutes*. *Corynebacterium* spp. and CNS are described in Chapter 2, while viridans streptococci are described in Chapter 4. *Mollicutes* are also described in Chapter 4 and are discussed further in Chapter 6. Because *Lactobacillus* spp. and *Bacteroides* spp. are the predominant organisms of the vagina and colon, respectively, it is more appropriate to describe these organisms in Chapters 6 and 7.

The GPAC comprise the following genera: *Peptococcus, Peptostreptococcus, Anaerococcus, Peptoniphilus, Gallicola, Finegoldia, Micromonas, Ruminococcus, Schleiferella, Coprococcus*, and *Sarcina*. Species belonging to these genera may be present as members of the indigenous microbiota of the urinary tract, oral cavity, vagina, skin, and intestinal tract. Until recently, many species of GPAC found in the female urethra were assigned to the genera *Peptostreptococcus* or *Peptococcus*. However, the nomenclature of these organisms has undergone dramatic changes and now only one species belonging to the genus *Peptococcus* is recognised – *Peptococcus niger* – while members of the genus *Peptostreptococcus* have

Table 5.3.	Nomenclature of Gram-positive anaerobic cocci previously considered to belong to the genus *Peptostreptococcus*
Previous nomenclature	**Current or proposed nomenclature**
Pep. anaerobius	unchanged
Pep. asaccharolyticus	*Schleiferella asaccharolytica*
Pep. indolicus	*Schleiferella indolica*
Pep. magnus	*Finegoldia magna*
Pep. micros	*Micromonas micros*
Pep. prevotii	*Anaerococcus prevotii*
Pep. productus	unchanged
Pep. vaginalis	*Anaerococcus vaginalis*

been reclassified into five genera. The new nomenclature of species mentioned in this and subsequent chapters is given in Table 5.3. All eleven genera of GPAC are obligately anaerobic, non-sporing cocci which may occur in pairs, tetrads, clusters, or chains. They have complex growth requirements and often need to be supplied with a number of vitamins and amino acids. Species belonging to the genera *Peptococcus, Peptostreptococcus, Anaerococcus, Peptoniphilus, Gallicola, Schleiferella, Finegoldia*, and *Micromonas* use peptones as their major energy sources and do not need carbohydrates for growth. The main end-products of metabolism are acetic and butyric acid, although a few species produce mainly isovaleric or caproic acids. Many species of these genera are also able to ferment a number of carbohydrates and secrete proteases and glycosidases. Species belonging to the genera *Ruminococcus, Coprococcus*, and *Sarcina*, however, do require carbohydrates for growth and produce mainly lactate, acetate, formate, and succinate. *Ruminococcus* spp. are predominant members of the intestinal microbiota and are described in more detail in Chapter 8.

Although they are associated with a number of infections (Section 5.1.5.1), little is known about the virulence factors of GPAC. Many species are proteolytic and produce enzymes such as collagenase, gelatinase, and immunoglobulin proteases which can damage host tissues and undermine host defence systems. Other exoenzymes produced include coagulase, hyaluronidase, DNase, RNase, and haemolysins.

5.1.4.2 Acquisition of the urethral microbiota

The main groups of organisms found in the urethra depend very much on the age and sexual maturity of the individual (Section 5.1.4.3). Nevertheless, they invariably include members of the genera *Lactobacillus, Corynebacterium, Staphylococcus, Bacteroides, Streptococcus, Fusobacterium*, and *Veillonella*, as well as GPAC. These organisms are also members of the indigenous microbiota of either the skin, vagina, or intestinal tract and are presumably derived from these regions. Studies have shown that strains of uropathogens – such as *E. coli* and *Staph. saprophyticus* that can colonise the urethra – are identical to those present in the intestinal tract of the same individual.

Unfortunately, few studies of the adhesion to uroepithelial cells of microbes indigenous to the urethra have been carried out; most studies of this type have involved uropathogenic organisms. Electron microscopy has revealed that bacteria are not uniformly distributed over the urethral epithelium and that most cells have no attached bacteria. Those cells that are colonised by bacteria generally have at least ten adherent

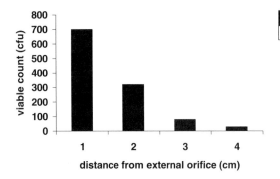

Figure 5.2 Number of viable bacteria detected in 1-cm sections of the urethra in 52 adult females.

bacteria which are evenly distributed over the cell surface. The adherent bacteria are usually single or in pairs – microcolonies are only occasionally present.

5.1.4.3 Community composition within the female urethra

Unlike the urethra of males in which only the distal portion is colonised by microbes (Section 5.2.4.3), bacteria have been detected along the entire length of the female urethra – although their population densities vary at different points. In a study involving fifty-two healthy females, all of the women harboured bacteria in the first 1-cm segment of the urethra, whereas fewer (54%) had bacteria in the 1-cm segment adjacent to the bladder. The number of bacteria detected was found to decrease with increasing distance from the external urethral orifice (Figure 5.2). The normal urethra is colonised by between two and eleven (median = 8) different organisms, with a total population of between 10^5 and 10^6 (mean = $2-10^5$) viable microbes. Lactobacilli dominate the microbiota, with CNS and corynebacteria also comprising substantial proportions of the urethral microbial community (Figure 5.3). *Mollicutes* are also frequently present and include *Myc. hominis, Myc. genitalium, Myc. fermentans*, and *U. urealyticum. Bacteroides* spp. and *Peptostreptococcus* spp. (including former members of the genus) are the most frequently encountered anaerobes in the urethra, but a variety of other anaerobic species may also be present (Table 5.4). Unfortunately, little is known regarding the distribution of the various species within the urethra.

Age and/or sexual maturity have a profound effect on the urethral microbiota, both qualitatively and quantitatively (Table 5.5). Hence, the diversity of the community and the number of organisms present are greater in the older age groups. Also, whereas facultative anaerobes dominate the urethral microbiota of pre-menarchal girls and pre-menopausal women, obligate anaerobes are dominant in post-menopausal women. Furthermore, facultative Gram-negative bacilli are generally absent from the urethras of pre-menarchal girls and pre-menopausal women, but are frequently present in post-menopausal women. These organisms, however, are frequently present in the vagina and/or the vulva of pre-menarchal women. Other noticeable shifts in the urethral microbiota include (1) a decreased isolation frequency of lactobacilli in pre-menarchal girls and post-menopausal women, (2) decreased proportions of CNS and *Corynebacterium* spp. in post-menopausal women, and (3) an increase in the proportion of black-pigmented Gram-negative anaerobic bacilli in post-menopausal women. Many of these differences will be a consequence of age-related, hormone-induced changes in the urethra similar to those which occur in the vagina

Table 5.4. Frequency of isolation of obligate anaerobes from the urethra of pre-menopausal women	
Organism	Frequency of isolation (%)
BPGNAB	70
B. capillosus	40
B. fragilis group	30
other Bacteroides spp.	30
Finegoldia magna (formerly Pep. magnus)	30
Anaerococcus prevotii (formerly Pep. prevotii)	30
Veillonella spp.	20
Propionibacterium spp.	20
B. amylophilus	10
Schleiferella asaccharolytica (formerly Pep. asaccharolyticus)	10
F. nucleatum	10
Pep. anaerobius	10
Pep. productus	10
Eggerthella lenta (formerly Eubacterium lentum)	10
Clostridium spp.	10

Notes: BPGNAB = black-pigmented Gram-negative anaerobic bacilli.
Frequency of isolation = proportion of the study population from which the organism was isolated.

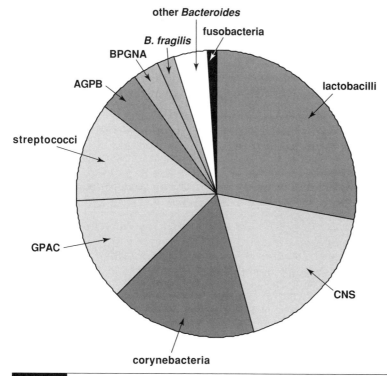

Figure 5.3 Proportions of organisms that comprise the indigenous microbiota of the urethra of healthy pre-menopausal women. AGPB = anaerobic Gram-positive bacilli; BPGNAB = black-pigmented Gram-negative anaerobic bacilli; CNS = coagulase-negative staphylococci; GPAC = Gram-positive anaerobic cocci.

Table 5.5. Main features of the urethral microbiotas of pre-menarchal girls, pre-menopausal, women, and post-menopausal women

| Population group | Number of species present (median) | Mean total viable count (cfu) | Proportion (%) of | | | | Frequency of isolation of | |
			FA	CS	CNS	BPGNAB	Lactobacilli (%)	FGNB (%)
pre-menarchal	7	7×10^4	66	33	13	11	33	0
pre-menopausal	8	2×10^5	74	17	18	3	80	0
post-menopausal	9	1×10^5	36	7	1	36	20	50

Notes: BPGNAB = black-pigmented Gram-negative anaerobic bacilli; CNS = coagulase-negative staphylococci; CS = *Corynebacterium* spp.; FA = facultative anaerobes; FGNB = facultative Gram-negative bacilli.

(Section 6.3.1). Hence, in pre-menopausal women, the high proportions of lactobacilli with their wide-ranging antibacterial and anti-adhesive properties (Section 6.4.3.1.1) will have a profound effect on the composition of the urethral microbiota.

5.1.4.4 Dispersal of organisms from the urethra

Dissemination of microbes from the urethra will occur mainly during urination. The number of viable bacteria per millilitre of urine in individuals without a UTI ranges from 0 to approximately 10^5 cfu, with most individuals having between 10 and 1,000 cfu. The diagnosis of a UTI is based on determining the concentration of viable bacteria in urine. For this purpose, contamination by the indigenous microbiota must be avoided – this is accomplished by taking a "mid-stream" specimen in which the urethral microbiota has been flushed out by the initial urinary flow. If the concentration of viable bacteria in such a sample is greater than 10^5 cfu/ml (known as a "significant bacteriuria"), then the individual probably has a UTI. Urethral microbes may also be transmitted to a partner during sexual intercourse, and can be transferred to the hands and underclothing and subsequently disseminated from these sites.

5.1.4.5 Effect of antibiotics and other interventions on the urethral microbiota

5.1.4.5.1 Antibiotics

Few studies have determined the effect of antibiotic administration on the indigenous microbiota of the urethra and all of these have involved individuals with UTIs. Antibiotic chemotherapy for UTIs has been found to have a profound effect on the urethral microbiota. For example, 4–6 weeks after the administration of amoxycillin or bacampicillin, the urethral microbiotas were found to be dominated by *E. coli* and *Staph. epidermidis*, respectively (Table 5.6). The urethral microbiota of healthy individuals is dominated by lactobacilli, and these organisms are important in preventing UTIs (Section 7.5.7.1). Administration of amoxycillin or bacampicillin, therefore, results in urethral communities that would have a reduced ability to prevent subsequent colonisation by uropathogens and possible re-infection. In fact, the effect of amoxycillin was to increase colonisation by *E. coli*, a major uropathogen. In contrast, those

Table 5.6.	Dominant organisms in the urethral microbiota of women 4–6 weeks after completion of a course of amoxycillin, bacampicillin, or enoxacin

Dominant urethral organisms after administration of					
Amoxycillin		Bacampicillin		Enoxacin	
E. coli	40	Staph. epidermidis	36	lactobacilli	57
lactobacilli	33	lactobacilli	29	Staph. epidermidis	29
Proteus spp.	13	E. coli	21	Gram-positive cocci	14
Staph. epidermidis	7	Strep. agalactiae	7		
Klebsiella spp.	7	yeasts	7		

Note: Numbers represent the percentage of women in each group who were colonised by the organism.

patients administered enoxacin had a post-treatment urethral microbiota dominated by lactobacilli. Interestingly, 4–6 weeks after treatment, 29% of the amoxycillin-treated patients and 20% of those treated with bacampicillin had another UTI, whereas re-infection did not occur in the enoxacin-treated patients. E. coli was the causative agent in 87% of the patients and it was found that, in the group treated with amoxycillin, 10% of the E. coli strains isolated from the urethra, introitus, and rectum were resistant to the antibiotic prior to treatment, but 72% were resistant after treatment. In the bacampicillin group, the corresponding figures were 15% and 40%. In contrast, none of the E. coli strains were resistant to enoxacin before or after treatment with this antibiotic.

Administration of co-trimoxazole appears to have little effect on the urethral microbiota. Ps. aeruginosa, an organism not often detected in the urethra, may be isolated from the urethra of individuals with recurrent UTIs who are frequently treated with antibiotics.

5.1.4.5.2 Catheterisation

A urinary catheter is a tube inserted along the urethra into the bladder where it is held in place by a small saline- or air-containing balloon. Urine continuously drains from the bladder via the catheter into a bag and consequently there is no need for the catheterised individual to use a toilet or bedpan for urination. Catheters are used in those patients who have lost the ability to control urination. Approximately 25% of hospital patients and 5% of those in nursing homes have an indwelling urinary catheter. Many studies have reported that the outer surface of the catheter in contact with the urethra is frequently colonised with a variety of organisms. The extent of colonisation and the type of organisms isolated depend on a number of factors, among which the duration of catheterisation is particularly important. Within 24 hours of catheterisation, bacteria can be isolated from the urethra-contacting surface of the catheter in more than 50% of cases. While some of these appear to be members of the normal urethral microbiota (CNS, Corynebacterium spp., viridans streptococci, etc.), E. coli is the most frequently detected species and has been isolated in approximately 75% of cases. Catheter-associated UTIs (Section 7.5.7.1) are a significant health problem and are, in fact, the most common type of nosocomial infection – affecting between 10% and 20% of patients with indwelling urethral catheters. The likelihood of such infections grows with increasing duration of catheterisation, and bacteria colonising the urethra-contacting surface of the catheter are a major source of the organisms responsible for the infection – patients who are colonised in this way are twice as likely to have a UTI as those who are not.

5.1.5 Diseases caused by members of the urethral microbiota of females

Many of the organisms found in the urethra are also present at other body sites, and the infections with which they are associated are dealt with in other chapters. Conversely, infections of the urinary tract are caused by microbes that are normally resident in the gastrointestinal tract and these will be described in Chapter 8.

5.1.5.1 Infections due to Gram-positive anaerobic cocci

Although these organisms are members of the indigenous microbiota of the female urethra, they are also found on all other mucosal surfaces, as well as on the skin. GPAC are associated with infections at a variety of body sites, and the organism involved will usually have been derived from the site affected rather than from the urethra. Nevertheless, it is convenient to describe infections due to GPAC in this section. Recent surveys of the organisms responsible for anaerobic infections have shown that GPAC comprise 25–30% of the isolates from such infections. GPAC have been isolated from abscesses and from infections of the mouth, skin and soft tissues, bone and joints, respiratory tract, and female genito-urinary tract – the most frequently encountered GPAC in such cases being *Pep. anaerobius, Schleiferella asaccharolytica, Finegoldia magna* (all of which are members of the microbiota of the female urethra), and *Micromonas micros*. Although most of these infections are polymicrobial (involving also microaerophilic streptococci and *Fusobacterium* spp.), some are due solely to GPAC. In mono-infections, *Finegoldia magna* is the most frequent isolate, but *Pep. anaerobius, Schleiferella indolica, Micromonas micros*, and *Anaerococcus vaginalis* have also been isolated from such infections.

5.1.5.2 Urethral syndrome

Urethral syndrome is a disease characterised by increased frequency of and painful urination, but without a significant bacteriuria. It is thought to have a prevalence similar to that of acute UTI in females. However, whether or not the disease has a microbial aetiology is uncertain, and over the years a number of organisms have been suggested as the causative agent, including lactobacilli, corynebacteria, and *Strep. milleri*. In a study of the bladder and urethral microbiota of patients with the disease, a number of obligate anaerobes, including *B. fragilis*, were isolated. The patients responded to therapy with metronidazole, an agent active against obligate anaerobes, suggesting the involvement of these organisms in the disease.

5.2 | The urinary system of males

5.2.1 Anatomy and physiology

In the male, the urethra is not only the terminal duct of the urinary system, but also constitutes part of the reproductive system (Figure 5.4). It is, therefore, a passageway

a

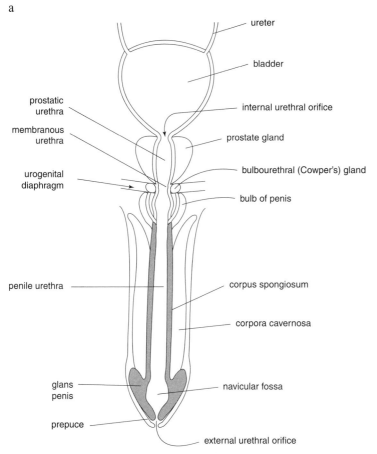

ureter

bladder

prostatic urethra

internal urethral orifice

membranous urethra

prostate gland

urogenital diaphragm

bulbourethral (Cowper's) gland

bulb of penis

penile urethra

corpus spongiosum

corpora cavernosa

glans penis

navicular fossa

prepuce

external urethral orifice

Figure 5.4 (a) Cross-section through the penis and associated structures. (*continued*)

for the excretion of urine and also for the ejaculation of semen. Semen consists of spermatozoa suspended in seminal plasma. The seminal plasma consists of secretions from the male accessory sex glands with most being provided by the seminal vesicles (approximately 60%) and the prostate gland (approximately 25%). Smaller quantities are derived from the Cowper's glands; Littre glands (which line the urethra, particularly the penile urethra); the ampulla (the terminal portion of the vas deferens); and the epididymis. The male urethra is approximately 20 cm long and consists of three regions: (1) the prostatic urethra – this section lies within the prostate and is between 3 and 4 cm in length, (2) the membranous urethra – this section is surrounded by the sphincter urethrae muscle and is approximately 2 cm long; and (3) the penile (or spongiose) urethra – this section lies within the bulb and body of the penis and is approximately 15 cm long. Most of the urethra is lined by a pseudo-stratified or stratified columnar epithelium. The navicular fossa, however, is lined with a stratified, squamous epithelium which becomes keratinised near the external meatus. Numerous glands (urethral glands) are present along the epithelium which also contains many small collections of mucous cells (glands of Littre). The external orifice of the urethra is located on the glans penis which, in uncircumcised individuals, is covered by a retractable fold of skin known as the prepuce or foreskin. On the corona of the glans penis

b

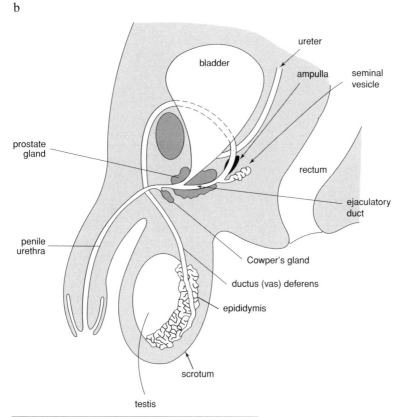

Figure 5.4 (*continued*). (b) Male organs of reproduction.

are sebaceous glands (glands of Tyson) which secrete a white cheesy material known as smegma.

It is important to note that the urethra of males is approximately five times longer than that of females. This anatomical feature explains why ascending infections (i.e., infections due to microbes colonising the urethra and peri-urethral regions) of the bladder and kidneys are much less frequent in males than in females.

5.2.2 Antimicrobial defence mechanisms

Many of the defence mechanisms operating in the male urethra are similar to those in the female urethra and were described in Section 5.1.2. They include desquamation of epithelial cells, urinary flow, and the presence in urine of antimicrobial (urea, peptides, antibodies) and anti-adhesive (TH protein, sIgA) factors. Furthermore, the male urethra also acts as a duct for semen, which contains additional antimicrobial compounds. For example, the prostate secretes zinc, which is present at high concentrations in prostatic secretions and can inhibit the growth of *E. coli, Chlam. trachomatis, Can. albicans*, and *Trichomonas vaginalis* at the levels found *in vivo*. Prostatic fluid also contains phospholipase A_2, which is active against many Gram-positive organisms, including staphylococci, streptococci, micrococci, enterococci, and *Lis. monocytogenes*. Other antimicrobial compounds found in seminal plasma are spermidine, spermine (which has activity against a wide range of organisms, including *Chlam. trachomatis*), lactoferrin, secretory leukocyte protease inhibitor, HBD-4, and lysozyme. The activities of these compounds were previously described.

As in the female genital tract, no organized lymphoid follicles are present in the genital tract of the male. However, it can function as an effector site for the common mucosal immune system and appears to be able to mount local immune responses. IgA- and IgM-producing plasma cells are present along the entire length of the penile urethra, and the epithelium expresses secretory component, the polymeric-IgA and -IgM transport molecule. IgA, IgG, and, to a lesser extent, IgM are present in seminal fluid, but there is uncertainty with regard to the relative quantities of IgA and IgG produced. Most of the IgG present is derived from the circulation, whereas the IgA is produced locally. Immunohistochemistry has revealed that the penile urethra is covered in mucus containing IgA, IgG, and IgM, thereby providing an immunological barrier against bacterial adhesion and invasion.

5.2.3 Environmental determinants within the male urethra

Because it is approximately 20 cm in length, it is likely that the urethra offers different environments within its various regions. However, little information is available regarding any such variations with distance along the urethra. Because the oxygen content of urine is high (approximately 81 mm Hg), the urethra is likely to be predominantly aerobic.

The pH of the urethra will be subject to marked variation, ranging from acidic (when urine is present) to slightly alkaline (pH 7.1 to 7.5) when semen is ejaculated. The pH of the glans penis ranges from 4.8 to 7.2, with a median value of 6.02. The host-derived nutrients available to colonising microbes will be those present in the secretions of the various urethral glands, desquamated epithelial cells, and those supplied (intermittently) by the passage of urine and semen through the urethra. As described in Section 5.1.3, urine provides several nitrogen sources (urea, NH_4^+, and creatinine), phosphate, and a number of trace elements, but no carbohydrates. Human ejaculate contains spermatozoa and secretions of the male accessory glands. Fluids produced by the seminal vesicles and the prostate constitute the bulk of seminal plasma, contributing approximately 60% and 25% of the fluid volume, respectively. Seminal plasma consists predominantly of water (92%), but also contains lipids, proteins, and carbohydrates (Table 5.7) and has a pH ranging from 7.1 to 7.5. A large number of proteins has been identified in seminal plasma, including albumin, prealbumin, insulin, α_1-antitrypsin, α_1-glycoprotein, transferrin, IgA, and IgG. As well as these proteins, the fluid also contains a wide variety of enzymes, including adenosine triphosphatase, lactic dehydrogenase, acid phosphatase, alkaline phosphatase, glucose phosphate isomerase, acetylcholinesterase, lysozyme, α-amylase, pepsin, and glucose-6-phosphatase. Most of the lipids in the fluid are phospholipids – the main ones being phosphatidyl choline, phosphatidyl ethanolamine, and sphingomyelin. The passage of urine and semen along the urethra, therefore, will deposit a range of potential nutrients for microbes colonising this site, and these are listed in Table 5.8. The mucins, proteins, and lipids present can be degraded by a number of species resident in the urethra (including CNS, *Corynebacterium* spp., viridans streptococci, clostridia, bifidobacteria, GPAC, *Prevotella* spp., and *Bacteroides* spp.), thereby providing sugars, amino acids, and fatty acids. The glans penis is a lipid-rich environment

Table 5.7.	Principal constituents of human seminal plasma

Constituent	Concentration
fructose	3.5–28 mM
glucose	—
citric acid	20–39 mM
spermine	—
prostaglandins	—
zinc	0.1 mg/ml
ascorbic acid	—
inositol	—
neutral amino acids	6.4 mg/ml
acidic amino acids	2.8 mg/ml
basic amino acids	3.4 mg/ml
proteins	35–55 mg/ml
enzymes	—
lipids	185 mg/ml

Note: — = concentration not known.

in which the main source of nutrients for microbes is smegma, which contains approximately 27% fat and 13% protein. In uncircumcised individuals, the region between the glans penis and the overlying prepuce (foreskin) is known as the preputial space. The prepuce reduces moisture loss and desquamation, resulting in the preputial space being a warm, moist region with a neutral to slightly alkaline pH and a reduced oxygen content.

5.2.4 The indigenous microbiota of the male urethra

5.2.4.1 Main characteristics of key members of the microbiota

The main organisms colonising the male urethra are CNS, *Corynebacterium* spp., viridans streptococci, GPAC, and *Mollicutes*. *Corynebacterium* spp. and CNS were described in

Table 5.8.	Substances present in the male urethra that can serve as microbial nutrients

Nutrient	Main source
urea	urine
NH_4^+	urine
trace elements (Na^+, K^+, Cl^-, Mg^{2+}, SO_4^{2-}, PO_4^{3-}, Ca^{2+}, Zn^{2+})	urine, seminal fluid
citric acid	prostate
fructose	seminal fluid
glucose	seminal fluid
glycogen	epididymis
citric acid	seminal fluid
amino acids	seminal fluid
proteins	seminal fluid
lipids	epididymis, prostate
various	desquamated epithelial cells

| Table 5.9. | Urethral microbiota of 35 sexually active heterosexual men without urethritis or a urinary tract infection |

Organism	Frequency of isolation (%)
Gram-positive anaerobic cocci	49
Staphylococcus spp.	48
Streptococcus spp.	28
Veillonella spp.	14
Bifidobacterium spp.	14
Clostridium spp.	9
Prevotella spp.	9
Corynebacterium spp.	3
Bacteroides spp.	2
Porphyromonas spp.	2
Fusobacterium spp.	0
Eubacterium spp.	0

Note: Frequency of isolation = proportion of the study population from which the organism was isolated.

Chapter 2, while viridans streptococci are described in Chapters 4 and 8. *Mollicutes* were also described in Chapter 4 and are discussed further in Chapter 6. GPAC were described in Section 5.1.4.1.

5.2.4.2 Acquisition of the microbiota of the male urethra

Many of the organisms most frequently isolated from the male urethra (staphylococci, coryneforms, streptococci) are members of the cutaneous microbiota and are, presumably, acquired from this source (Section 5.2.4.3). The urethra of sexually active males supports a more diverse microbial community than that of non-sexually active individuals, and many of the organisms present will have been acquired from the vagina and cervix of female sexual partners (Sections 6.4.3.1 and 6.4.3.2).

As is found for the female urethra (Section 5.1.4.2), bacteria colonising the male urethra are not evenly distributed over the epithelium, but adhere to only a proportion of the host cells. Likewise, laboratory studies of the adhesion of *Staph. saprophyticus* and *E. coli* to samples of male urethral mucosa in organ culture have shown that the proportion of cells with adherent bacteria is usually less than 20%.

5.2.4.3 Community composition of the microbiota of the male urethra

Only the distal 6 cm of the male urethra is normally colonised by microbes. As might be expected, the types of organisms present are affected by the usual variables (Section 1.1.1) and by factors particular to this site. The latter include the frequency of sexual intercourse; the number of sexual partners; the nature (if any) of the birth control measure used by the sexual partner; whether the individual engages in oral, anal, or vaginal intercourse; whether or not a condom is used; whether the individual is circumcised; and the gender of the sexual partner. In a study of the urethral microbiota of thirty-five heterosexual sexually active men, the most frequently detected organisms were found to be GPAC, staphylococci, coryneforms, streptococci, bifidobacteria, clostridia, *Veillonella* spp., and *Prevotella* spp. (Table 5.9). Among the

Table 5.10. Microbiota of the urethra of sexually active and non-sexually active adolescent males

Organism	Frequency of isolation (%)	
	Non-sexually active (n = 16)	Sexually active (n = 42)
aerobes/facultative anaerobes		
streptococci		
α-haemolytic streptococci	69	38
non-haemolytic streptococci (not group D)	13	33
Strep. agalactiae	0	5
Group D streptococci	0	10
staphylococci		
coagulase-negative staphylococci	75	98
Staph. aureus	13	2
coryneforms	75	69
lactobacilli	19	29
enterococci	6	2
G. vaginalis	6	36
unidentified coryneforms	0	12
Klebsiella spp.	0	2
H. influenzae	0	2
H. parainfluenzae	0	5
Acin. calcoaceticus	0	5
Mycoplasma spp.	0	19
U. urealyticum	0	50
Anaerobes/microaerophiles		
Lactobacillus spp.	0	5
Propionibacterium spp.	6	5
Actinomyces spp.	13	5
Peptostreptococcus spp. (including former members of the genus)	62	67
unidentified Gram-positive cocci	0	36
V. parvula	6	26
Bacteroides spp.	19	33
Fusobacterium spp.	0	7
unidentified Gram-negative rods	0	17

Note: Frequency of isolation = proportion of the study population from which the organism was isolated.

GPAC, *Finegoldia magna* (fomerly *Pep. magnus*) and *Pep. productus* were the predominant isolates.

The effect of engaging in sexual intercourse on the male urethral microbiota can be seen from a study involving adolescent boys (Table 5.10). This study revealed that the complexity of the urethral microbiota was significantly greater in sexually active individuals than in those who were not sexually active – the mean number of isolates being 7.2 in the former and 4.1 in the latter. The other notable differences in the sexually active group, compared with the non-sexually active group, were: (1)

Table 5.11. | Aerobic and facultatively anaerobic bacteria isolated from different regions of the urethra of 30 males

Organism	External orifice		Navicular fossa		Penile urethra	
	F (%)	n (cfu)	F (%)	n (cfu)	F (%)	n (cfu)
CNS	50	9,428	37	1,950	17	720
viridans streptococci	50	28,560	23	10,500	23	3,000
Corynebacterium spp.	40	14,325	20	3,800	3.3	0
Enterococcus spp.	20	24,350	10	1,900	7	750
Staph. aureus	3.3	0	0	0	0	0
Strep. pyogenes	3.3	0	3.3	0	0	0
Enterobacter spp.	3.3	0	0	0	0	0
Strep. agalactiae	3.3	0	0	0	0	0
Micrococcus spp.	3.3	0	0	0	0	0

Notes: The samples were not cultured under anaerobic conditions. CNS = coagulase-negative staphylococci; F = frequency of isolation; n = number of viable bacteria detected.

an increase in the frequency of isolation of *G. vaginalis*, Gram-negative aerobes and facultative anaerobes, *Strep. agalactiae*, Group D streptococci, unidentified anaerobic Gram-positive, non-spore-forming rods and cocci, *V. parvula*, Gram-negative anaerobic rods, *Mycoplasma* spp., and *U. urealyticum*; and (2) a decrease in the frequency of isolation of α-haemolytic streptococci, *Actinomyces* spp., and *Staph. aureus*. Many of the organisms with a higher prevalence in the urethra of sexually active males are members of the vaginal microbiota (e.g., *Strep. agalactiae*, *G. vaginalis*, Gram-negative anaerobic rods and cocci, and *U. urealyticum*) and can be assumed to have been acquired during sexual intercourse. Interestingly, although lactobacilli predominate in the vaginal ecosystem, the frequency of isolation of such organisms from the male urethra is not dramatically increased when males become sexually active. This may be due to the fact that conditions in the male urethra are very different from those of the vagina – particularly with regard to the pH, which is much lower in the latter.

Studies of the urethral microbiota are usually based on samples obtained by insertion of a swab into the urethra. Such an approach will fail to detect possible variations in the microbial communities along the length of the urethra. Most organisms colonising the urethra are, in fact, found in the first centimetre from the urethral orifice (i.e., within the navicular fossa). The numbers present decrease with increasing distance from the orifice until, after approximately 6 cm, no bacteria are generally detectable. Not surprisingly (given the different prevailing environmental conditions at each site), the types of organisms present have been found to vary at different locations along the urethra. In a study of thirty sexually active, uncircumcised males, the most frequently isolated organisms from the external orifice were staphylococci, viridans streptococci, and *Corynebacterium* spp. (Table 5.11). The frequency of isolation of all three bacterial groups decreased with increasing distance from the orifice, as did the complexity of the communities. Within the penile urethra, beyond the navicular fossa, only staphylococci, viridans streptococci, *Corynebacterium* spp., and enterococci were detectable. Unusually for this type of study, a semi-quantitative analysis was also carried out, and viridans streptococci were found to be the dominant species at all three sites. The CNS isolated were *Staph. haemolyticus*, *Staph. auriculares*, *Staph. hominis*, *Staph. epidermidis*,

Table 5.12. | Microbiota of the external urethral orifice and navicular fossa of 97 uncircumcised adults

Organism	Number of isolates from:	
	External orifice	Navicular fossa
CNS	74	72
Pep. magnus (Finegoldia magna)	44	43
Pep. anaerobius	34	24
Ent. faecalis	28	30
Pep. asaccharolyticus (Schleiferella asaccharolytica)	20	15
G. vaginalis	14	12
Prev. bivia	14	12
Corynebacterium spp.	13	10
B. ureolyticus	12	9
α-haemolytic streptococci	10	13
Prev. disiens	9	7
Bacillus spp.	8	4
Prev. melaninogenica	8	4
E. coli	8	6
Total	**334**	**316**

Note: CNS = coagulase-negative staphylococci.

and *Staph. simulans*. Unfortunately, this study did not attempt anaerobic cultivation of the samples, and consequently no information was obtained regarding the presence of anaerobic species in different regions of the urethra. In a similar study which compared the microbiotas of the external orifice and the navicular fossa, anaerobic cultivation of the samples obtained was carried out and the results are shown in Table 5.12. While CNS were, again, frequently isolated, the results showed that obligate anaerobes, particularly GPAC, were also very often present. In fact, organisms belonging to the genera *Peptostreptococcus, Finegoldia*, and *Schleiferella* were among the most frequently isolated organisms from both the navicular fossa and the external orifice. The table also clearly demonstrates the similarity of the indigenous microbiotas of the two regions.

The microbiota of the glans penis shows tremendous variation between individuals, depending on a number of factors, including whether or not the individual is circumcised, the level of hygiene practiced, and the type and frequency of sexual activity (if any) in which the individual engages. In circumcised individuals, the microbiota tends to resemble that of sebaceous gland-rich skin regions and is dominated by *P. acnes*, CNS, and coryneforms. *Malassezia* spp. (mainly *Mal. sympodialis and Mal. globosa*) and *Can. albicans* are also frequently present. In uncircumcised individuals, the population density is higher, but the frequency of isolation of *P. acnes* is lower. Gram-negative anaerobes, enterococci, *Staph. aureus*, and facultatively anaerobic Gram-negative rods are more frequently isolated and comprise higher proportions of the microbiota in uncircumcised individuals. The frequency of isolation of *Malassezia* spp. is greater in circumcised than uncircumcised individuals, being 62% and 49%, respectively. In contrast, the frequency of isolation of *Can. albicans* is higher in uncircumcised (21%) than circumcised (8%) males.

For many years, the existence of microbes in the prostate of healthy individuals has been a controversial issue because of the problem of interpreting the results obtained from the studies carried out. Hence, it is very difficult to obtain samples uncontaminated by the normal urethral microbiota and also, as with any site, a negative result based on culture does not rule out the possibility of the presence of species that cannot be cultivated. Recently, prostate samples from organ donors were obtained under sterile surgical conditions and samples analysed by PCR of 16S rRNA genes. None of the samples revealed the presence of DNA corresponding to bacterial 16S rRNA genes, suggesting that the prostate is a sterile site.

5.2.4.4 Dispersal of organisms from the urethra

The passage of urine through the urethra will invariably dislodge a portion of the urethral microbiota, resulting in bacteria-containing urine, and this constitutes the main means by which urethral organisms are disseminated. In individuals with a UTI (generally defined as more than 10^3 cfu/ml of urine in males as opposed to 10^5 cfu/ml in females), much higher concentrations of bacteria will be present. Urethral microbes will also be transferred to underclothes, to neighbouring body sites, and to the hands, and can then be disseminated from these sites. Although, like urine, semen is sterile when first produced, it will accumulate microbes during its passage through the urethra. It has been shown that the semen of approximately 80% of individuals contains microbes – mainly bacteria. In a study of ninety-seven heterosexual sexually active men, between one and nine different bacterial species were isolated from the semen of each individual, and a total of 253 different species was cultured from the whole group. The most frequently isolated bacteria were, in decreasing order, CNS, *Ent. faecalis*, *Finegoldia magna* (*Pep. magnus*), and *Pep. anaerobius*. However, because semen contains a number of antimicrobial compounds, it is not known how long these organisms would survive. Nevertheless, sexual intercourse is likely to provide another means by which urethral microbes can be disseminated.

5.2.4.5 Effect of antibiotics and other interventions on the urethral microbiota

5.2.4.5.1 Antibiotic use

No studies of the effects of antibiotics on the microbiota of the urethra in males appear to have been published.

5.2.4.5.2 Catheterisation

Catheterisation does not appear to have a dramatic effect on the urethral microbiota in males because the organisms recovered from the urethra-contacting catheter surface are usually members of the indigenous urethral microbiota – mainly CNS, viridans streptococci, and *Ent. faecalis*. *E. coli* and facultative Gram-negative rods are isolated significantly less frequently from catheterised male patients than from catheterised female patients (Section 5.1.4.5.2). Furthermore, the frequency of colonisation in males is less than half that found in females, and this is accompanied by a similar decrease in the incidence of catheter-associated UTI in males.

Table 5.13. | Organisms able to inhibit the growth of *Neisseria gonorrhoeae* in vitro

Most strains are inhibitory	Inhibition depends on the particular strain, some strains being very inhibitory
Pseudomonas spp.	**viridans streptococci**
Can. diversus	**bifidobacteria**
Enterobacter cloacae	*E. coli*
Ser. marcescens	*Providencia stuartii*
	Staph. aureus
	Staph. epidermidis
	Can. albicans
	lactobacilli

Note: Organisms usually found in the male urethra or on the glans penis are boldface.

5.2.5 Diseases caused by members of the urethral microbiota of males

5.2.5.1 Urethritis

The main cause of urethritis in males is *N. gonorrhoeae* but, as this is an exogenous pathogen, it will not be discussed further. However, it is worth pointing out that the urethral microbiota can, like the microbiotas of all sites, protect against exogenous pathogens. This is particularly evident with regard to its ability to prevent infection by *N. gonorrhoeae*. Studies involving humans and other animals have demonstrated that the presence of certain species in the urethra (see Table 5.13) can protect against colonisation by *N. gonorrhoeae*. Inhibition of *N. gonorrhoeae* is thought to be due to the production of bacteriocins and/or fatty acids.

Non-gonococcal urethritis (NGU) in males may be caused by a variety of agents, including the exogenous pathogens *Chlamydia trachomatis*, *Trichomonas vaginalis*, and *Myc. genitalium*. The possible involvement of members of the indigenous urethral microbiota in NGU remains controversial, but some studies have reported a significantly increased frequency of occurrence of *U. urealyticum*, *B. ureolyticus*, and GPAC in males with NGU, but who were not infected with exogenous urethral pathogens. The GPAC showing an association with NGU were *Anaerococcus prevotii* (formerly *Pep. prevotii*), *Pep. productus*, and *Anaerococcus tetradius* (formerly *Pep. tetradius*).

5.2.5.2 Prostatitis

This disease is characterised by pain, together with one or more of the following symptoms: problems with urination, impotency, and ejaculatory dysfunction. Its prevalence worldwide is between 2% and 10%, and it has been estimated that approximately 50% of adult men experience prostatitis at some time in their lives. Three main forms of the disease are recognised – acute bacterial, chronic bacterial, and chronic idiopathic (CIP) or prostatodynia. Only approximately 10% of patients have the acute form of the disease, and this is generally caused by the same organisms as are responsible for UTIs (i.e., *E. coli*, *Proteus* spp., *Providencia* spp., and other *Enterobacteriaceae*), and these are members of the indigenous microbiota of the colon and can frequently be isolated from the glans penis of uncircumcised individuals. Another occasional causative agent is the exogenous pathogen *N. gonorrhoeae*. Chronic bacterial prostatitis is characterised by recurrent UTIs due to enterobacteria or *Ent. faecalis*, the latter being a resident of the urethra as well as of the gastrointestinal tract. The aetiology of CIP – often called

| Table 5.14. | Organisms implicated as the aetiological agents of chronic idiopathic prostatitis (non-bacterial prostatitis) |

Staph. epidermidis, Staph. haemolyticus
C. minitussimum, C. seminale, C. xerosis, C. afermentans
Chlam. trachomatis
U. urealyticum
Tr. vaginalis
Finegoldia magna
Schleiferella asaccharolytica
P. acnes
Aeromonas hydrophila
E. coli
Enterobacter spp.
Klebsiella spp.
Pseudomonas spp.
viridans streptococci

non-bacterial prostatitis – is more controversial and a number of organisms have been implicated (Table 5.14). Many of the aforementioned organisms (e.g., staphylococci, corynebacteria, streptococci, and GPAC) are found in the urethra of healthy individuals. Recently, sequencing of 16S rRNA genes that had been PCR-amplified from DNA extracted from the prostatic fluid of men with chronic prostatitis has revealed the frequent presence of rRNA genes from species belonging to the following genera (in decreasing order of occurrence): *Corynebacterium, Staphylococcus, Peptostreptococcus*, and *Streptococcus*. Although antibiotics do appear to aid the recovery of some patients with CIP, the aetiology of the condition remains controversial, and it has been suggested that the disease has an autoimmune or neuromuscular cause.

5.2.5.3 Balanitis
Balanitis is an infection of the glans penis by *Can. albicans*, which is a member of the indigenous microbiota of this region. It is an uncommon disease, and in the United Kingdom the annual incidence is only twelve cases per 100,000. The main symptoms are inflammation and irritation of the glans penis and, if present, the foreskin. It occurs more frequently in uncircumcised than circumcised individuals, and this is likely to be related to the greater frequency of isolation of the yeast from the glans penis of the former.

5.3 | Further Reading

Brook, I. (2004). Urinary tract and genito-urinary suppurative infections due to anaerobic bacteria. *International Journal of Urology* **11**, 133–141.

Chambers, C.V., Shafer, M.A., Adger, H., Ohm-Smith, M., Millstein, S.G., Irwin, C.E., Jr., Schachter, J., and Sweet, R. (1987). Microflora of the urethra in adolescent boys: relationships to sexual activity and nongonococcal urethritis. *Journal of Pediatrics* **110**, 314–321.

Colleen, S. and Mardh, P.A. (1981). Bacterial colonization of human urethral mucosa. II. Adherence tests using tissue organ cultures. *Scandinavian Journal of Urology and Nephrology* **15**, 181–187.

Domingue, G.J., Sr. and Hellstrom, W.J. (1998). Prostatitis. *Clinical Microbiology Reviews* **11**, 604–613.

Ganz, T. (2001). Defensins in the urinary tract and other tissues. *Journal of Infectious Diseases* **183**, S41–S42.

Hochreiter, W.W., Duncan, J.L., and Schaeffer, A.J. (2000). Evaluation of the bacterial flora of the prostate using a 16S rRNA gene-based polymerase chain reaction. *Journal of Urolology* **163**, 127–130.

Horner, P., Thomas, B., Gilroy, B.C., Egger, M., and Taylor-Robinson, D. (2001). Role of *Mycoplasma genitalium* and *Ureaplasma urealyticum* in acute and chronic nongonococcal urethritis. *Clinical Infectious Diseases* **32**, 995–1003.

Hua, V.N. and Schaeffer, A.J. (2004). Acute and chronic prostatitis. *Medical Clinics of North America* **88**, 483–494.

Krieger, J.N., Ross. S.O., and Riley, D.E. (2002). Chronic prostatitis: epidemiology and role of infection. *Urology* **60** (Suppl 6), 8–12.

Kunin, C.M., Evans, C., Bartholomew, D., and Bates, D.G. (2002). The antimicrobial defense mechanism of the female urethra: a reassessment. *Journal of Urology* **168**, 413–419.

Kunin, C.M. and Steele, C. (1985). Culture of the surfaces of urinary catheters to sample urethral flora and study the effect of antimicrobial therapy. *Journal of Clinical Microbiology* **21**, 902–908.

Lewis, S.A. (2000). Everything you wanted to know about the bladder epithelium but were afraid to ask. *American Journal of Physiology: Renal Physiology* **278**, F867–F874.

Lipsky, B.A. (1999). Prostatitis and urinary tract infection in men: what's new; what's true? *American Journal of Medicine* **106**, 327–334.

Marrie, T.J., Swantee, C.A., and Hartlen, M. (1980). Aerobic and anaerobic urethral flora of healthy females in various physiological age groups and of females with urinary tract infections. *Journal of Clinical Microbiology* **11**, 654–659.

Mazuecos, J., Aznar, J., Rodriguez-Pichardo, A., Marmesat, F., Borobio, M.V., Perea, E.J., and Camacho, F. (1998). Anaerobic bacteria in men with urethritis. *Journal of the European Academy of Dermatology & Venereology* **10**, 237–242.

Montagnini Spaine, D., Mamizuka, E.M., Pereira Cedenho, A., and Srougi, M. (2000). Microbiologic aerobic studies on normal male urethra. *Urology* **56**, 207–210.

Montgomerie, J.Z., McCary, A., Bennett, C.J., Young, M., Matias, B., Diaz, F., Adkins, R., and Anderson, J. (1997). Urethral cultures in female patients with a spinal cord injury. *Spinal Cord* **35**, 282–285.

Morris, N.S., Stickler, D.J., and McLean, R.J. (1999). The development of bacterial biofilms on indwelling urethral catheters. *World Journal of Urology* **17**, 345–350.

Murdoch, D.A. (1998). Gram-positive anaerobic cocci. *Clinical Microbiology Reviews* **11**, 81–120.

Park, C.H., Valore, E.V., Waring, A.J., and Tomas Ganz, T. (2001). Hepcidin, a urinary antimicrobial peptide synthesized in the liver. *Journal of Biological Chemistry* **276**, 7806–7810.

Perra, M.T., Turno, F., and Sirigu, P. (1997). Human urethral epithelium: immunohistochemical demonstration of secretory IgA. *Archives of Andrology* **39**, 45–53.

Pudney, J. and Anderson, D.J. (1995). Immunobiology of the human penile urethra. *American Journal of Pathology* **147**, 155–165.

Riemersma, W.A., van der Schee, C.J., van der Meijden, W.I., Verbrugh, H.A., and van Belkum, A. (2003). Microbial population diversity in the urethras of healthy males and males suffering from nonchlamydial, nongonococcal urethritis. *Journal of Clinical Microbiology* **41**, 1977–1986.

Riley, D.E., Berger, R.E., Miner, D.C., and Krieger, J.N. (1998). Diverse and related 16S rRNA-encoding DNA sequences in prostate tissues of men with chronic prostatitis. *Journal of Clinical Microbiology* **36**, 1646–1652.

Rupp, M.E., Soper, D.E., and Archer, G.L. (1992). Colonization of the female genital tract with *Staphylococcus saprophyticus*. *Journal of Clinical Microbiology* **30**, 2975–2979.

Schwartz, M.A. and Hooton, T.M. (1998). Etiology of nongonococcal nonchlamydial urethritis. *Sexually Transmitted Diseases* **4**, 727–733.

Sirigu, P., Perra, M.T., and Turno, F. (1995). Immunohistochemical study of secretory IgA in the human male reproductive tract. *Andrologia* **27**, 335–339.

Tanner, M.A., Shoskes, D., Shahed, A., and Pace, N.R. (1999). Prevalence of corynebacterial 16S rRNA sequences in patients with bacterial and "nonbacterial" prostatitis. *Journal of Clinical Microbiology* **37**, 1863–1870.

Terai, A., Ishitoya, S., Mitsumori, K., and Ogawa, O. (2000). Molecular epidemiological evidence for ascending urethral infection in acute bacterial prostatitis. *Journal of Urology* **164**, 1945–1947.

Uuskula, A. and Kohl, P.K. (2002). Genital mycoplasmas, including *Mycoplasma genitalium*, as sexually transmitted agents. *International Journal of STD and AIDS* **13**, 79–85.

Willen, M., Holst, E., Myhre, E.B., and Olsson, A.M. (1996).The bacterial flora of the genito-urinary tract in healthy fertile men. *Scandinavian Journal of Urology and Nephrology* **30**, 387–393.

Woolley, P.D. (2000). Anaerobic bacteria and non-gonococcal urethritis. *International Journal of STD and AIDS* **11**, 347–348.

Woolley, P.D., Kinghorn, G.R., Talbot, M.D., and Duerden, B.I. (1990). Microbiological flora in men with non-gonococcal urethritis with particular reference to anaerobic bacteria. *International Journal of STD and AIDS* **1**, 122–125.

Yoshida, T., Deguchi, T., Ito, M., Maeda, S., Tamaki, M., and Ishiko, H. (2002). Quantitative detection of *Mycoplasma genitalium* from first-pass urine of men with urethritis and asymptomatic men by real-time PCR. *Journal of Clinical Microbiology* **40**, 1451–1455.

The reproductive system and its indigenous microbiota

The only region of the male reproductive system that is colonised by microbes is the urethra, and its indigenous microbiota has already been described in Chapter 5. This chapter, therefore, is devoted to a discussion of the indigenous microbiota of the reproductive system in females.

6.1 | Anatomy and physiology of the female reproductive system

The female organs of reproduction include the ovaries, Fallopian tubes, uterus, cervix, vagina, and vulva (Figure 6.1). Of these, only the vagina, vulva, and cervix of the uterus are normally colonised by microbes. While a large number of studies have been devoted to elucidating the indigenous microbiota of the vagina and, to a lesser extent, the cervix, relatively little is known with regard to which microbes inhabit the various regions of the vulva.

The vulva is a term used to denote the external female genitalia, which include the mons pubis, labia majora and minora, clitoris, and vestibule (Figure 6.2). The vestibule is lined with a stratified squamous epithelium and located within it are the vaginal orifice, the external urethral orifice, and the openings of the mucus-secreting paraurethral and Bartholins glands. The mons pubis consists of a layer of adipose tissue covered by skin and pubic hair (Table 6.1). The labia majora are longitudinal folds of skin which contain adipose tissue, have sebaceous and sudoriferous glands, and are covered in pubic hair on their outer surfaces. In contrast, the two folds of skin known as the labia minora do not contain adipose tissue and are free of hair. They are covered with a stratified squamous epithelium, which may have a thin keratinised layer on its surface. Numerous sebaceous glands are present, but few sudoriferous glands. From the medial (i.e., innermost) surfaces of the labia majora to the vagina, there is a gradual transition in the structure of the epidermis as it changes from a keratinised epithelium typical of the outer body surface to the mucosal epithelium found in the vagina. The clitoris consists of erectile tissue and nerves, and is covered with a thin, stratified squamous epithelium – it has a hood of skin known as the prepuce or foreskin.

The vagina is a muscular, tubular organ situated between the bladder and the rectum. It is approximately 8 cm in length and extends from the cervix to the external genitalia (Figure 6.3). It serves as a passageway during childbirth, is an outlet for the menstrual flow, and is a receptacle for the penis during sexual intercourse. It is, therefore, a distendable organ and its wall is composed mainly of smooth muscle and has

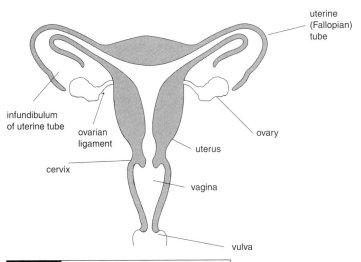

Figure 6.1 The organs of reproduction in females.

a folded lining consisting of a stratified, squamous, non-keratinised epithelium with a total surface area of approximately 360 cm^2. Notably, no glands are present, and the surface is lubricated by mucus produced by the cervix. Three main layers can be distinguished in the vaginal mucosa. The innermost layer comprises basal cells, which are actively engaged in cell division. Above this lies the intermediate layer, which contains cells whose main function is glycogen production – numerous intracytoplasmic glycogen granules are present in intermediate cells, but are absent from basal cells. Cells of the superficial layer have fewer glycogen granules, but are rich in intracytoplasmic microfilaments which provide rigidity and protection to the underlying layers. The outermost cells of this layer are covered in microridges. Migrating lymphocytes and Langerhans cells are also present throughout the mucosa. During the menstrual

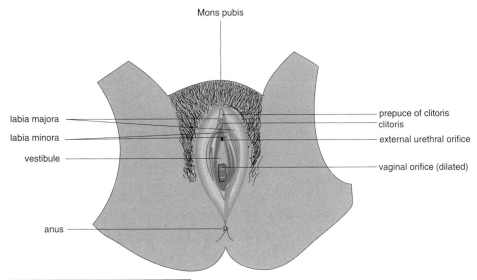

Figure 6.2 Various regions of the vulva.

Table 6.1. Main anatomical features of the various regions of the vulva

Structure	Type of epithelium	Hair	Adipose tissue	Sebaceous glands	Sudoriferous glands
mons pubis	keratinised	+	+	+	+
labia majora	keratinised	+	+	+	+
labia minora	stratified squamous	−	−	+	±
clitoris	stratified squamous	−	−	−	−

cycle (Figure 6.4), there is a constant remodelling of the epithelium which involves the proliferation, maturation, and desquamation of cells. In general, oestrogens (particularly oestradiol) stimulate this process, whereas progesterone inhibits maturation of the cells. The low levels of oestrogens present in individuals before menarche (i.e., the onset of menstruation) and after menopause (i.e., the end of menstruation) results in them having a relatively thin vaginal epithelium.

The cervix is the lower, narrow portion of the uterus that opens into the vagina (Figure 6.3). The upper region of the cervix opens into the uterus via the "internal os". The lumen then widens to form the cervical canal, which narrows to produce an opening (the external os) into the vagina. The portion of the cervix protruding into the vagina is known as the endocervix. The latter is covered by a stratified, squamous, non-keratinised epithelium identical to that found in the vagina. This type of epithelium extends into the cervical canal for a distance which varies, depending on the age of the individual. It then changes into a simple columnar epithelium which lines the rest of the canal. Almost all of the cells of the epithelium of the cervix are mucus-secreting, the

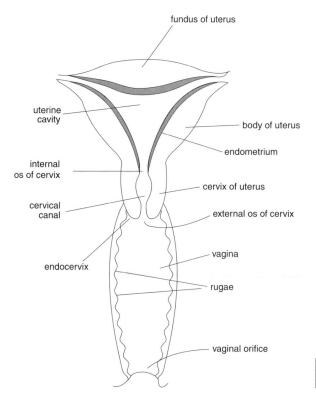

Labels:
fundus of uterus
uterine cavity
body of uterus
endometrium
internal os of cervix
cervix of uterus
cervical canal
external os of cervix
endocervix
vagina
rugae
vaginal orifice

Figure 6.3 Main structural features of the vagina and cervix.

(a) the menstrual (uterine endometrial) cycle

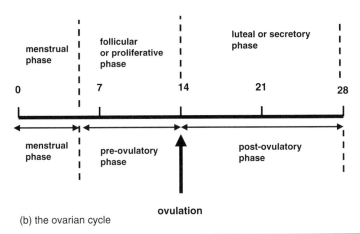

ovulation

(b) the ovarian cycle

Figure 6.4 The term "female reproductive cycle" refers to (a) the "menstrual cycle" – a series of changes in the endometrium of the uterus and (b) the "ovarian cycle" – a monthly series of events involving the maturation of an oocyte.

rest being ciliated. Unlike the mucosa of the uterus, the cervical epithelium is not shed during menstruation. However, it is affected by changes in hormone levels during the menstrual cycle (Figure 6.4). Hence, the quantity of mucus secreted increases 10-fold (to approximately 700 mg/day), when the level of oestrogen peaks at mid-cycle. Following ovulation, the increased level of progesterone causes less mucus to be produced, and it becomes more viscous so that it forms a plug which seals the canal.

6.2 | Antimicrobial defence mechanisms of the female reproductive system

The mons pubis and the labia majora have the characteristic innate and acquired defence mechanisms associated with skin, whereas the labia minora, clitoris, and vestibule have the defence mechanisms typical of those found at other mucosal surfaces. The vagina and cervix have the usual innate and acquired defence mechanisms found in other mucosal regions, although there are two aspects of the acquired immune response of these regions which render them unique – the preponderance of IgG rather than IgA in secretions and the influence exerted by hormonal fluctuations.

6.2.1 Innate defence mechanisms of the female reproductive system

Despite the fact that the vagina and cervix are heavily colonised by a range of microbes, the uterus and upper regions of the genital tract are usually sterile – this is important for the protection of any fertilised ovum. The cervix is the main potential portal of entry for organisms from the heavily colonised lower genital tract and is equipped with a variety of antimicrobial defence mechanisms. The first of these is the presence of cilia on many of the epithelial cells which, together with the production of mucus by other cells of the epithelium, constitute a mucociliary escalator (cf. the respiratory tract; Section 4.2) which continually sweeps incoming organisms back down into the

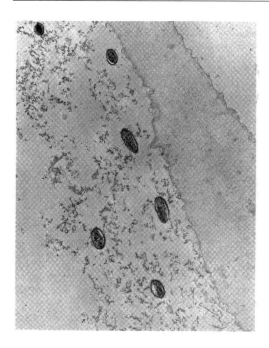

Figure 6.5 Transmission electron micrograph showing Gram-positive, rod-shaped bacteria embedded in the mucus layer on the surface of the vaginal epithelium of a healthy volunteer. Reproduced with permission of Taylor and Francis from: A morphological study of the in situ tissue-associated autochthonous microbiota of the human vagina. Sadhu, K., Domingue, P.A.G., Chow, A.W., et al. *Microbial Ecology in Health and Disease* 1989;2:99–106.

vagina. There is evidence that the mucus covering the mucosa consists of two layers – an inner, hydrophilic layer containing glycoproteins and glycolipids secreted by epithelial cells and a thicker gel consisting mainly of glycoproteins secreted by goblet cells. The mucus itself is protective in a variety of ways. Firstly, it can be extremely viscous, thus preventing penetration of microbes through to the underlying epithelium where they may be able to become permanently attached (Figure 6.5). However, the viscosity of mucus varies dramatically during the course of the menstrual cycle and reaches a minimum just before ovulation. The mucus consists predominantly of glycoproteins (Section 1.5.3) and proteins together with smaller quantities of low molecular mass compounds such as sugars. Some of the carbohydrate moieties in mucus are receptors for the adhesins of many bacteria which are thereby prevented from binding to receptors on underlying epithelial cells. The mucus-bound organisms are then swept back into the vagina. Mucus also contains a variety of antimicrobial compounds, including lysozyme, lactoferrin, secretory leukocyte protease inhibitor (SLPI), calprotectin and a number of antibacterial peptides which can either kill microbes or inhibit their growth. The activities of most of these compounds were described in previous chapters (Sections 2.2, 3.2, and 4.2). However, unlike other body sites, the concentrations of some of these vary during the menstrual cycle (Figure 6.6). Hence, the concentration of lactoferrin in vaginal secretions is at its lowest during the secretory phase of the menstrual cycle (3.8–11.4 μg/mg of protein) in individuals who are not taking oral contraceptives and at its highest (62.9–218 μg/mg of protein) immediately after menses. Interestingly, the lactoferrin content of the vaginal secretions of women taking oral contraceptives is significantly lower (never exceeding 19.8 μ/mg of protein) and shows no monthly variation. The concentration of lysozyme in cervical mucus is lowest during mid-cycle, while that of SLPI is at a maximum at this time, being approximately five times higher than during the follicular and luteal phases. Human β-defensin (HBD)-1 is present in vaginal secretions, and HBD-1 m-RNA has been detected in the epithelia

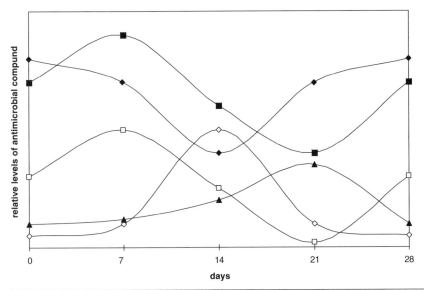

Figure 6.6 Fluctuations in the concentration of antimicrobial compounds in the vagina and cervix during the menstrual cycle. □ = lactoferrin; ♦ = lysozyme; ◇ = secretory leukocyte protease inhibitor; ▲ = human defensin-5; ■ = IgA and IgG.

of the vagina and cervix. Its antimicrobial activities were described previously (Section 2.2). Higher levels of HBD-1 are found in the vaginal secretions of pregnant women, compared with non-pregnant individuals. The vagina and cervix also produce HBD-3 and human cationic antimicrobial protein, which is the precursor of the antimicrobial peptide LL-37. Another defensin, HD-5, is also produced by the endocervix and vagina, and immunohistochemical studies have shown that HD-5 mRNA is present in the upper half of the vaginal epithelium, with its concentration increasing towards the lumen. HD-5 production is at its highest level during the secretory phase of the menstrual cycle and is active against *Sal. typhimurium*, *Lis. monocytogenes*, *E. coli*, and *Can. albicans*. Calprotectin is produced by polymorphonuclear leukocytes (PMNs) and epithelial cells, and inhibits the growth of *E. coli*, *Klebsiella* spp., *Staph. aureus*, *Staph. epidermidis*, and *Can. albicans*. Complement is also present in cervical and vaginal secretions at levels approximately one-tenth of those found in serum, and its concentration varies during the menstrual cycle, being at a minimum during mid-cycle. Another important defence mechanism, of course, is the shedding of the outermost epithelial cells and their attached microbes. This is affected by the cyclic fluctuations in the levels of various hormones that occur during the menstrual cycle. Oestrogen stimulates the rate of cell maturation and thereby the rate of desquamation of epithelial cells.

Between menarche and menopause, the vaginal mucosa contains large stores of glycogen which, when metabolised anaerobically, results in the release of organic acids, including lactic and acetic acids. These organic acids result in a very acidic environment (pH 4), which inhibits the growth of many bacterial species and can also exert a direct antimicrobial effect. It has been shown that a low vaginal pH correlates with a decreased risk of genital infection by chlamydia, trichomonads, and mycoplasmas, as well as urinary tract infections (UTIs). In contrast, when the vaginal pH is greater than 4.5, there are increased risks for the development of bacterial vaginosis, infection with human immunodeficiency virus (HIV), and pelvic inflammatory disease.

Table 6.2. | Antimicrobial compounds present in the cervical mucus plug

Compound	Concentration (mg/g)
lysozyme	660
lactoferrin	100
secretory leukocyte protease inhibitor	750
calprotectin	38
human β-defensin 1	1
HNP-1	12
HNP-2	12
HNP-3	12

Note: HNP = human neutrophil peptide.

If conception occurs, an additional antimicrobial defence mechanism is activated – the cervical mucus thickens and forms a large plug blocking the cervical canal. This "plug" not only acts as a physical barrier preventing microbial ingress, but also displays broad-spectrum antimicrobial activity; it therefore provides protection to the developing foetus from the enormous number of microbes present in the lower genital tract. Importantly, the plug is active against *Strep. agalactiae*, the leading cause of mortality and morbidity in neonates. A range of antimicrobial compounds have been detected in the plug, and these are listed in Table 6.2.

6.2.2 Acquired immune defence mechanisms of the female reproductive system

The acquired immune system of this region has certain characteristics which distinguish it from that operating at other mucosal sites. Firstly, no lymphoid follicles appear to be present. These are characteristic of most mucosal surfaces (and are known as mucosa-associated lymphoid tissue) and may occur singly (e.g., in the appendix and bronchi) or in organized clusters (e.g., the Peyers patches of the large intestine and the tonsils). Such follicles are responsible for the primary mucosal immune response and contain dendritic cells (involved in antigen-presentation), as well as immature B and T cells. The second distinguishing characteristic is that the antibodies present in genital-tract secretions are not only produced locally, but are also derived from serum by transudation. Finally, many aspects of the acquired immune response of the genital mucosa are under hormonal control. Although no lymphoid follicles are evident in the genital mucosa, lymphoid aggregates containing B and T cells have been detected. Also, antigen-presenting cells (dendritic cells and macrophages) are present, and it has been shown that uterine epithelial cells can also function as antigen-presenting cells. It would appear, therefore, that the genital mucosa is capable of mounting at least a local immune response. However, this is certainly under hormonal control because the levels of both IgA and IgG are higher in the pre-ovulatory than the post-ovulatory phase (Figure 6.6). Furthermore, vaginal immunisation has been shown to elicit a substantial production of IgA only during the first half of the menstrual cycle (when oestrogen levels are highest), but not during the second half (when progesterone levels are high). Interestingly, intranasal immunisation has been shown to be an effective means of inducing an IgA response in the female genital tract.

| Table 6.3. | Main host defence mechanisms operating in the female reproductive system |

Defence mechanism	Function
desquamation	removes adherent microbes
mucus	prevents access of microbes to underlying epithelium
mucociliary escalator	expels microbes from cervix into vagina
acidity of the vagina	inhibits or kills microbes
antimicrobial peptides (HBD-1, HBD-3, HD-5, LL-37, HNP-1, HNP-2, HNP-3)	inhibit or kill microbes; anti-inflammatory
lysozyme	antimicrobial
lactoferrin	antimicrobial
SLPI	antimicrobial
calprotectin	antimicrobial
antibodies (IgG, IgA, and IgM)	inhibit microbial adhesion; antimicrobial; neutralise toxins

Note: SLPI = secretory leukocyte protease inhibitor.

Unlike other mucosal secretions, the predominant class of antibody in genital fluids is IgG rather than IgA. Although both antibody classes are produced locally by plasma cells in the genital tissues, they are also derived from serum – but the means by which they are transported from the bloodstream to the lumen (i.e., whether by active transport or passive diffusion) has not been established. The density of immunoglobulin-producing and immunoglobulin-containing cells in the cervix is almost four times greater than in the vagina, and the class of antibody produced varies slightly in these organs. Hence, the proportions of IgA-, IgG-, and IgM-producing cells in the cervix are 73%, 15%, and 12%, respectively, while in the vagina the proportions are 79%, 14%, and 7%, respectively. The main functions of these antibodies are to prevent microbial adhesion to, and invasion of, the epithelial surface and to neutralise toxins and other potentially harmful antigenic substances. They may also be involved in antibody-dependent, cell-mediated cytotoxicity and the promotion of phagocytosis by neutrophils. Their role in determining the composition of the vaginal and cervical microbiotas has not been established. Furthermore, despite the perceived importance of antibodies in the defence of the genital mucosa, there is little direct evidence that they can protect against infection in humans. However, animal studies have shown that secreted antibodies do provide protection against vaginal infections.

The main host defence mechanisms operating in the female reproductive system are summarised in Table 6.3.

6.3 | Environmental determinants at different regions of the reproductive system

6.3.1 Vagina

The concentrations of a number of hormones (oestrogens and progesterone) are known to influence the anatomy and physiology of the vagina and, thereby, the ecology of this

Table 6.4. Conditions in the vagina at various stages of sexual maturity

Characteristic	Neonate	Pre-menarch	Post-menarch	Post-menopause
oestrogen level	high	low	high	low
pH	acidic	neutral	acidic	neutral
glycogen content	high	low	high	low
redox potential	high	low	variable	low

organ. Differences, therefore, exist in the environmental selecting factors operating in the vagina before menarche, once menarche has occurred, and after menopause. Furthermore, hormonal variations during the monthly reproductive cycle in post-menarchal/pre-menopausal females also result in changes within the vaginal ecosystem. Such changes exert a profound effect on the vaginal microbiota. Unfortunately, while differences in the vaginal microbiota have been detected between pre-menarchal, post-menarchal/pre-menopausal, and post-menopausal individuals, as well as during the reproductive cycle of post-menarchal/pre-menopausal women, the environmental determinants responsible for inducing such changes have not always been elucidated.

The partial pressure of oxygen in the vagina of healthy individuals during menstruation has been reported to be approximately 4 mm Hg, which is 2% of that present in air and 10% of that found in tissue cells. Oxygen levels in the vagina have been shown to vary during the menstrual cycle, with lower levels being found in mid-cycle. Consequently, the vaginal ecosystem can be regarded as predominantly microaerophilic. However, dramatic fluctuations in the oxygen content of the vagina do occur. Hence, the insertion of a contraceptive diaphragm results in an increase in the partial pressure of oxygen to approximately 82% of that present in air. This slowly declines to "normal" levels over a 2-hour period. The insertion of a tampon increases the oxygen content of the vagina to a level similar to that found in air – the concentration then decreases slowly and returns to "normal" after approximately 8 hours. Because many women change tampons every 3–6 hours, this could result in the oxygen content of the vagina being maintained at atmospheric concentrations for between 3 and 5 days (the normal duration of menstruation). This could have a profound effect on the vaginal microbiota and may encourage the growth of organisms such as *Staph. aureus* – the causative agent of toxic shock syndrome (Section 4.5.6).

The pH of the vagina is strongly affected by age and hormone levels (Table 6.4). There is considerable controversy as to whether the pH of the vagina is dictated primarily by the host or by the indigenous microbiota. Strong evidence in support of the former hypothesis is that the vagina is acidic at birth (a pH of approximately 4.5), despite the fact that it is free of bacteria. This suggests that acids, such as lactic acid produced by metabolism of the glycogen present in epithelial cells, may be the prime factor governing vaginal pH. The high oestrogen levels in neonates are derived from the mother and result in a thick epithelium with a high glycogen content. As the oestrogen level falls (after approximately 3 weeks), the vaginal epithelium becomes thinner and, therefore, has a lower glycogen content. This results in a decrease in the amount of lactic acid being produced and a consequent increase in pH to a value of around 7.0. Dramatic changes in pH also take place during the menstrual cycle. As can be seen in Figure 6.7, at the onset of menstrual flow, the pH is almost neutral. Towards the end of menstruation, the pH has generally decreased to approximately 5.3, and this decrease

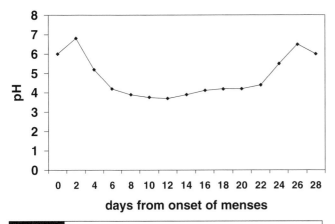

Figure 6.7 Variation in the pH of the vagina during the menstrual cycle.

continues to a minimum (<4) at mid-cycle. The pH remains below 4.5 for approximately another week, then steadily increases until the onset of menstruation. Pregnancy does not appear to affect the pH of the vagina, which remains less than 4.0.

The main source of host-derived nutrients for the vaginal microbiota is vaginal fluid. This is a mixture of fluids derived from a number of sources and includes a transudate from the vaginal mucosa, cervical mucus, endometrial fluid, secretions of the Bartholins glands, desquamated vaginal epithelial cells, and leukocytes. In post-menarchal/pre-menopausal women, an additional source of nutrients would be provided once monthly during menstruation, which is considered later in this section. In those engaging in sexual intercourse, yet another possible source of nutrients would be seminal fluid (Section 5.2.3). Between 1 and 3 g of vaginal fluid are produced each day by women of reproductive age, and a crude analysis has shown that it contains carbohydrates (approximately 48 mg/g), proteins (approximately 4 mg/g), and amino acids (0.2 mmoles/g) and has a pH of less than 4.5. The transudate is derived from the underlying tissues by passage through the gap junctions in the vaginal epithelium and contains a number of proteins (including immunoglobulins – mainly IgA and IgG), urea, amino acids, other organic molecules, and a range of inorganic ions. Like many other aspects of vaginal physiology, transudation is controlled by oestrogen levels and is reduced after mid-cycle when oestrogen levels decrease. Cervical mucus is a major component of the fluid (between 20 and 600 mg is produced per day, depending on the stage of the menstrual cycle), and this contains mucins, proteins (including immunoglobulins), low-molecular mass organic compounds, and inorganic ions (Section 6.3.2). Glycogen is particularly abundant in the epithelial cells of neonates and post-menarchal/pre-menopausal females, but not in pre-menarchal or post-menopausal individuals (Table 6.4). Oligosaccharides and lower molecular mass degradation products of glycogen are also present. These important carbon and energy sources would be available to bacteria from dead or dying epithelial cells. The main constituents of vaginal fluid are listed in Table 6.5.

In post-menarchal/pre-menopausal women, menstruation would provide an additional source of nutrients. Menstrual fluid is produced for approximately 3–5 days and although the total volume produced can range from 10 to 400 ml, most individuals produce approximately 80 ml of fluid. This consists of 30–50% whole blood, with the

Table 6.5.	Principal constituents of vaginal fluid in post-menarchal/ pre-menopausal women

Type of molecule/ion	Predominant examples
protein	mucins, albumin, immunoglobulins (IgG, IgA, IgM), transferrin, lactoferrin, α_2-haptoglobin, α_1-antitrypsin, α_2-macroglobulin
carbohydrate	glucose (0.62 g/100 g), glycogen (1.5 g/100 g), oligosaccharides (80–160 mM), mannose, glucosamine, fucose
lipid	neutral lipids and phospholipids
low-molecular mass organic compounds	urea (49 mg/100 ml), lactic acid, acetic acid, butanoic acid, propanoic acid
amino acids	fourteen identified, including alanine, glycine, histidine, leucine, trytophan
inorganic ions	Na^+ (23 mmol/L), Cl^- (62 mmol/L), K^+ (61 mmol/L)

Note: Approximate concentrations of the constituents are given where these are known.

remainder being an endometrial transudate, and provides a range of additional nutrients, including carbohydrates, amino acids, proteins, urea, lipids, and fatty acids. Furthermore, its high haemoglobin content would be a valuable source of iron, as many organisms are able to extract the element from this compound.

Although there is a plentiful and varied supply of nutrients available to microbes colonising the vagina, certain features of the environment would be unfavourable to many organisms (e.g., the low pH and oxygen content). Furthermore, the range of antimicrobial mechanisms described in Section 6.2 would also exert a selective effect and influence the composition of the vaginal microbiota.

6.3.2 Cervix

As for the vagina, the physiology and, ultimately, the environment of the cervix, are markedly dependent on the levels of oestrogen and progesterone. The cervix, like the vagina, has a low oxygen content, with its partial pressure being approximately 22 mm Hg (which is only 14% of that present in air) for women in the ovulatory phase of the menstrual cycle. The partial pressure of oxygen decreases slightly during the secretory phase, but falls to approximately 12 mm Hg during the proliferative phase. The presence of an intrauterine device has been shown to have little effect on the partial pressure of oxygen in the cervix. The pH of the cervix is generally higher than that of the vagina and ranges from 5.4 to 8.2, with a median pH of 7.0. The pH varies through the menstrual cycle, being more alkaline preceding and during menstruation and at ovulation (Figure 6.8). The pH of the cervix during pregnancy is approximately 6.5.

The main host-derived sources of nutrients for bacteria colonising the cervix are cervical mucus and, in the case of post-menarchal/pre-menopausal women, menstrual fluid. As previously mentioned, copious quantities of mucus are produced in the cervix. Both the amount of mucus produced and its water content gradually increase until midcycle and then decrease. While predominantly consisting of water (90–95%), the mucus

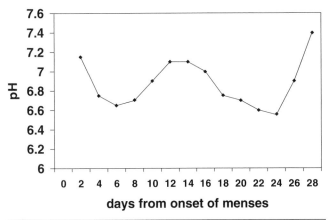

Figure 6.8 Variations in the pH of the cervix during the menstrual cycle.

contains a variety of substances which could serve as nutrients for microbes inhabiting this body site, and these are listed in Table 6.6. Specific proteins present include albumin, transferrin, lactoferrin, α_2-haptoglobin, α_2-macroglobulin, β-lipoprotein, elastase, and α_1-antitrypsin. The albumin content of the mucus decreases significantly prior to, and during, ovulation and then rises again. The viscosity of the mucus is also at a minimum during ovulation, presumably to enable easier sperm penetration. The concentration of immunoglobulins (mainly IgA and IgG) is highest during the follicular phase (first half of the cycle) than during the luteal phase (second half of the cycle).

An important feature of the cervical environment is, of course, the wide range of antimicrobial mechanisms described in Section 6.2.

6.3.3 Vulva

Very little is known about the environments of the various regions of the vulva. The labia majora appear to have a greater level of hydration in the stratum corneum and a higher pH (6.05) than those found on the skin of the forearm. The oxygen content of the labia minora is low – approximately 12% of that found in air. This lack of basic information regarding the environmental conditions that exist in these regions makes it very difficult to gain an adequate understanding of the development and composition of their indigenous microbiotas.

Table 6.6. | Main constituents of cervical mucus

mucins
proteins – including enzymes and immunoglobulins
amino acids – mainly alanine, glutamate, threonine, taurine
sugars – mainly glucose and fucose
lipids – including cholesterol
fatty acids – mainly lactate, acetate
urea
electrolytes – mainly Na^+ and Cl^-

Table 6.7. Members of the vaginal microbiota that produce enzymes able to contribute to the hydrolysis of mucins

Organism	Enzyme
Can. albicans	N-acetylglucosaminidase, aspartyl proteinase
Bifidobacterium spp.	sialidase, α- and β-glycosidases, α- and β-D-glucosidases, α- and β-D-galactosidase, β-D-fucosidase, proteases
Strep. agalactiae	peptidase and proteinase
Prevotella spp.	sialidase, sulphatase, α-fucosidase, β-galactosidase, N-acetyl-β-glucosaminidase, α-galactosidase, protease
B. fragilis	sialidase, α-fucosidase, β-galactosidase, N-acetyl-β-glucosaminidase, N-acetyl-α-galactosidase
Porphyromonas spp.	sialidase, glycosidases, proteases
G. vaginalis	sialidase
E. coli	sialidase, glycosidases
Ent. faecalis	sialidase, glycosidases, proteases
P. acnes	sialidase, protease
Actinomyces spp.	sialidase, protease
viridans streptococci	sialidase, proteases, glycosidases
lactobacilli	α- and β-galactosidases, α- and β-D-glucosidases, β-glucuronidase
Clostridium spp.	sialidase, α- and β-galactosidases, α- and β-D-glucosidases, β-glucuronidase

6.3.4 Contribution of microbial residents of the reproductive system to nutrient availability

Many of the organisms present in the vagina of healthy women produce enzymes that are able to degrade mucins, even if each species is able to achieve only partial degradation (Section 1.5.3). Some of the relevant enzymes produced by these organisms are listed in Table 6.7. Furthermore, enzymes relevant to the degradation of mucins have been detected in anaerobic Gram-negative rods isolated from the vagina of high proportions of healthy individuals (Table 6.8). Proteins are, of course, another valuable

Table 6.8. Proportion of women (n = 68) harbouring vaginal anaerobic Gram-negative rods that produce enzymes involved in the degradation of mucins

Enzyme	Proportion (%) of women with detected activity
N-acetylglucosaminidase	53
β-galactosidase	53
sialidase	50
α-fucosidase	50
α-galactosidase	40
glycine aminopeptidase	35
arginine aminopeptidase	34

source of nutrients and can be degraded by a wide range of vaginal microbes, including *Bacteroides* spp., *Porphyromonas* spp., *Prevotella* spp., *Fusobacterium* spp., enterococci, *P. acnes*, *Ent. faecalis*, staphylococci, *Can. albicans*, *Strep. agalactiae*, and viridans streptococci.

6.4 | The indigenous microbiota of the female reproductive system

6.4.1 Main characteristics of key members of the microbiota

The main organisms found in the female reproductive system are *Lactobacillus* spp., *Staphylococcus* spp., *Corynebacterium* spp., *Streptococcus* spp., *Enterococcus* spp., *Can. albicans*, *Bifidobacterium* spp., *Gardnerella vaginalis*, *Propionibacterium* spp., Gram-positive anaerobic cocci (GPAC), *Bacteroides* spp., *Porphyromonas* spp., *Prevotella* spp., *Clostridium* spp., *Fusobacterium* spp., *Veillonella* spp., *Ureaplasma* spp., and *Mycoplasma* spp.

Staphylococcus spp., *Propionibacterium* spp., and *Corynebacterium* spp. were described in Section 2.4.1, streptococci and *Mycoplasma* spp. in Section 4.4.1, and GPAC in Section 5.1.4.1. Gram-negative anaerobic rods, *Bifidobacterium* spp., *Bacteroides* spp., and *Clostridum* spp. are predominant members of the microbiota of the colon and are more appropriately described in Chapter 7. *Veillonella* spp., *Porphyromonas* spp., *Prevotella* spp., and *Fusobacterium* spp. are predominant members of the oral microbiota and are described in Chapter 8.

6.4.1.1 *Lactobacillus* spp.

Lactobacilli are non-sporing, non-motile Gram-positive bacilli which are usually long and slender. At least thirty-four species are currently recognised and most are micro-aerophillic, although some are obligate anaerobes. The G+C content of their DNA is 37–53 mol%. They are catalase-negative and obtain their energy by the fermentation of sugars, producing a variety of acids, alcohol, and carbon dioxide. Species which produce mainly lactic acid from glucose are termed "homofermentative," while those that produce smaller proportions of lactic acid – together with acetic acid, formic acid, and alcohol – are termed "heterofermentative". They have little proteolytic activity. Lactobacilli are aciduric and acidophilic, and are able to grow over the pH range of 3.5 to 6.8. Optimum growth occurs at approximately pH 6.0. Their temperature range for growth is from 15° to 45°C. They have complex nutritional requirements and may need to be supplied with amino acids, peptides, fatty-acid esters, salts, nucleic acid derivatives, or vitamins.

Lactobacilli display a wide range of adhesins, which enable them to adhere to the vaginal epithelium (Figure 6.9). These include lipoteichoic acids, as well as ill-defined proteinaceous and non-proteinaceous adhesins, some of which are located on fimbriae. *L. acidophilus* and *L. gasseri* appear to utilise glycoproteins as adhesins when binding to vaginal epithelial cells, the receptors being glycolipids. In contrast, *L. jensenii* binds to epithelial cells by means of a carbohydrate adhesin.

Lactobacilli very rarely cause infections in humans.

6.4.1.2 *Ureaplasma urealyticum*

This organism belongs to the *Mollicutes*, a class of bacteria that do not have a cell wall and which constitute the smallest free-living organisms (Section 4.4.1.6). It is a small

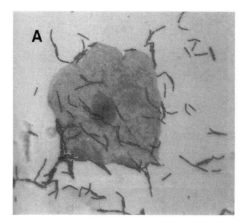

Figure 6.9 Micrographs showing (A) adhesion of *L. acidophilus* to a vaginal epithelial cell and (B) the lack of adhesion of a dairy strain of *L. plantarum*. Reproduced with permission from: Boris, S., *et al. Infection and Immunity* 1998;66:1985–1989.

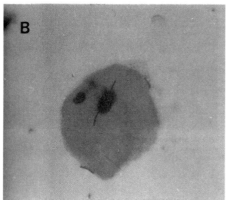

(0.2–0.3 μm in diameter), pleomorphic, facultative anaerobe which is nutritionally fastidious and requires sterols and an enriched medium for growth. It can grow over a broad pH range of pH 5–9. The organism is unable to metabolise glucose or arginine, but produces a urease which is involved in energy generation. The urease hydrolyses urea to ammonia, some of which accumulates intracellularly, thereby creating a chemiosmotic potential which is used to generate ATP – approximately 95% of its ATP is produced in this way. The organism secretes a number of enzymes that have been implicated as virulence factors, including elastase, several phospholipases, and an IgA protease. The organism's haemolytic activity has been attributed to its production of hydrogen peroxide but may, in fact, be due to the production of a haemolysin encoded by the *hlyA* gene identified in the organism's genome (see herein). Ammonia liberated by the urease is also thought to be involved in host pathology because of its cytotoxicity.

The genome of *U. urealyticum* has recently been sequenced and is the third smallest genome to be sequenced to date. It has 613 genes encoding proteins and 39 which encode RNA. Of the protein-encoding genes, 53% have been assigned biological functions, 19% are similar to genes with no known function, and the remaining genes have no significant similarity to genes from other organisms. In keeping with its known limited biosynthetic capacity, genes for the synthesis of purines and pyrimidines are absent and a large number of transporters are available for the uptake of a range of essential nutrients. Surprisingly, genes encoding an IgA protease and phospholipases A and C could not be identified. These *U. urealyticum* enzymes may have diverged so far from orthologues in other bacteria that they have become unrecognisable, or else they may have convergently evolved with little sequence similarity to other known enzymes with these activities. A haemolysin-encoding gene, *hlyA*, has been identified in the genome and could be a new virulence factor because haemolytic activity in the organism had previously been attributed to its production of reactive oxygen species.

6.4.1.3 *Gardnerella vaginalis*

G. vaginalis is the only member of the genus *Gardnerella*. It is a non-motile, non-sporing, facultatively anaerobic Gram-positive bacillus. The G+C content of its DNA is 42–44 mol%. Gram-stained smears show characteristic slender Gram-variable rods or coccobacilli, and it produces an exopolysaccharide and pili. The organism grows best in carbon-dioxide-enriched air and is limited to growing over the pH range of 6–7; the optimum pH for its growth is between 6.0 and 6.5. It can grow over the temperature range 25°–42°C and grows optimally at 35°–37°C. It is nutritionally fastidious and has a slow, fermentative metabolism producing mainly acetate and lactate. It does not produce a catalase or a urease, but is able to ferment glucose, maltose, and sucrose. It also produces a sialidase, implying that it can partially degrade mucins to obtain sugars that could be used as carbon and energy sources.

With regard to virulence factors, the pili and surface polysaccharide are involved in adhesion of the organism to epithelial and red blood cells. It produces a 60 kDa haemolysin that can lyse neutrophils and endothelial cells, as well as red blood cells. Studies of iron acquisition by the organism have shown that it produces siderophores and can directly bind a number of iron-containing compounds, including haem, haemoglobin, catalase, and lactoferrin – all of these compounds, including a number of inorganic iron salts, can be used as an iron source by the organism. In contrast, it is unable to bind or utilise human transferrin as a source of iron.

6.4.1.4 *Candida albicans*

This is a yeast with oval-shaped cells 3–6 μm in diameter. It reproduces by forming buds (blastoconidia), but these often fail to detach, resulting in a chain of cells known as a pseudohypha. It is a dimorphic fungus and can also produce true septate hyphae – this often occurs when the concentration of oxygen is depleted. Hyphal production is also induced by incubating the organism in serum at 37°C; after 3 hours, short hyphae, known as germ tubes, are produced and this is an important, rapid test for the presumptive identification of the organism. It is a facultative anaerobe, but grows best under aerobic conditions. It can grow over the temperature range of 20°–40°C and the pH range of 2–8. The optimum pH for its growth is 5.1–6.9. It can utilise a wide range of sugars as a carbon and energy source, including glucose, maltose, sucrose, galactose, and xylose. It produces an N-acetylglucosaminidase and a number of aspartyl proteinases, which implies that it may be able to partially degrade mucins as well as oligosaccharides and proteins present in the vagina and obtain sugars and amino acids for use as carbon, nitrogen, and energy sources.

A large number of adhesins has been identified in *Can. albicans*, including a number of mannoproteins and integrin analogues which mediate adhesion to epithelial cells and extracellular matrix molecules, such as collagen, fibronectin, and laminin. With the exception of Hwp1 (hyphal wall protein 1), these adhesins bind to their receptors on host cells or matrix molecules by receptor-ligand and/or hydrophobic interactions. Hwp1, however, mediates adhesion by a unique mechanism. This mannoprotein has an amino-terminal region that resembles substrates for host transglutaminase which is involved in cross-linking epithelial cell proteins, and the enzyme has been shown to link Hwp1 to proteins on epithelial cells. This results in the yeast being covalently bound to epithelial cells and, therefore, very difficult to dislodge. Evidence that this occurs *in vivo* comes from studies showing that the yeast remains attached to epithelial cells when samples of yeast-infected tissues are boiled with a detergent. Interestingly, the protein is expressed only by the hyphal form of the organism and may provide an effective means of counteracting the removal of the organism from mucosal surfaces by desquamation. Hence, the production of hyphae enables penetration into the deeper layers of the mucosa and, if this is followed by covalent linkage to these underlying epithelial cells, then shedding of the outer layers of the mucosa would not dislodge the organism from its site of colonisation.

The yeast secretes two large families of enzymes which are thought to be involved in damage to host tissues – four phospholipases and nine secreted aspartyl proteinases. Studies have shown that deletion of the genes encoding many of these proteins results in strains with diminished virulence in animal models. It also produces a haemolysin which may be important in ensuring its access to an adequate supply of iron. Many strains of the organism produce a gliotoxin, which has antibacterial and immunosuppressive activities. Hence, it is able to kill *Micrococcus* spp. and can inhibit chemotaxis and phagocytosis by PMNs. Vaginal samples of individuals with vaginal candidiasis contain significant levels of gliotoxin, whereas no toxin is detectable in control individuals not colonised by the yeast.

6.4.1.5 *Streptococcus agalactiae* (Group B streptococcus)

This is a β-haemolytic streptococcus (Section 4.4.1.3), which contains the Lancefield's Group B antigen. A number of serotypes can be distinguished, one of these – serotype III – is the most frequently associated with neonatal invasive diseases. On the basis of its DNA restriction digest pattern (RDP), a high virulence clone belonging to this serotype has been identified: serotype III RDP type III. The organism produces a hyaluronidase and proteases, which means it could obtain sugars and amino acids for use as carbon, nitrogen, and energy sources from the hyaluronic acid and proteins present in vaginal fluid.

Remarkably few virulence factors have been identified in *Strep. agalactiae*, and most of these are involved in undermining host defence systems. Hence, it is able to resist phagocytosis because of its polysaccharide capsule, it produces a C5a-peptidase, and it has a surface protein (β-antigen) which can bind IgA via the Fc region of the molecule. However, it also produces a hyaluronidase and can induce cytokine release from host cells – both of which could lead to damage to host tissues. The organism can adhere to epithelial cells of the vagina and cervix *in vitro* by means of a number of adhesins, including lipoteichoic acid and unidentified proteins. It can also bind to extracellular matrix components, including fibronectin and fibrinogen. Interestingly, a proteinaceous adhesin, Lmb, has been detected on the surface of the organism which enables it to bind to human placental laminin.

The organism is able to inhibit the growth of many microbial species found in the vagina, including lactobacilli, *G. vaginalis*, coryneforms, viridans streptococci, enterococci, and peptostreptococci. It is unable to inhibit the growth of staphylococci, but, interestingly, it is itself inhibited by coagulase-negative staphylococci (CNS).

6.4.2 Acquisition of the microbiota of the female reproductive system

The vagina of a newly born infant has a pH of approximately 4.5 and, because of the transplacental transfer of oestrogen from the mother, the vaginal epithelium is similar to that of the post-menarchal/pre-menopausal female. This state is maintained for approximately 3 weeks. Passage through the birth canal results in the infant being exposed to its mother's vaginal microbiota, as well as to organisms from other anatomical regions and from the environment. Initially, the vaginal microbiota of the infant is similar to that of the post-menarchal female, with a predominance of lactobacilli. However, as the oestrogen level falls, the vaginal epithelium becomes thinner and the glycogen content decreases, resulting in a gradual rise in pH to approximately 7.0. The microbiota changes accordingly and becomes highly mixed with no particular species predominating. In a study involving twenty-five pre-menarchal girls, the mean number of species isolated from the vagina was found to be nine – slightly more than half of which were obligate anaerobes. The most frequently isolated organisms were *Staph. epidermidis* and coryneforms among the facultative organisms and species from the genera *Peptostreptococcus, Bacteroides, Clostridium*, and *Eubacterium* among the anaerobes. Because these organisms are commonly found on the skin and in the intestinal tract, respectively, data suggest that these two regions are the main sources of the bacteria that colonise the pre-menarchal vagina. In adults, additional possible sources of vaginal colonisers include the skin and oral cavity during sexual intercourse and foreplay.

The microbiota of the cervix is derived from microbes colonising the vagina, while the vulva may be colonised by organisms from the vagina, skin, and intestinal tract.

In order to successfully colonise the vagina or cervix, an organism must be able to adhere to the epithelium underlying the protective mucus layer. Members of the indigenous microbiotas of these sites can adhere to vaginal and cervical epithelial cells *in vitro*; the mechanisms involved, where known, are described for the individual organisms in Section 6.4.1.

6.4.3 Community composition at different sites within the female reproductive system

6.4.3.1 Vagina

While, for reasons discussed in Chapter 1, it is always difficult to describe and define the indigenous microbiota of a particular site, the problems are exacerbated when it comes to defining the microbiota of the vagina. This is because its composition is markedly affected by a number of host factors (Table 6.9), including (1) the sexual maturity of the individual, (2) whether or not the individual is pregnant, (3) the stage of the reproductive cycle in post-menarchal/pre-menopausal individuals, (4) whether or not the individual is sexually active, and (5) whether or not the individual is using contraceptives and, if so, what method is being used.

Although many studies of the vaginal microbiota have been undertaken, a large proportion of these have used populations in which the aforementioned factors have

Table 6.9. | Host factors that affect the composition of the vaginal microbiota

sexual maturity (i.e., whether pre- or post-menarchal, whether pre- or
  post-menopausal)
pregnancy
stage of the reproductive cycle in menarchal individuals
sexual activity – type, frequency, etc.
use of contraceptive devices/treatments
use of tampons
hormone replacement therapy in post-menopausal individuals
use of vaginal douches
antibiotic use

Note: These are in addition to the usual factors – outlined in Chapter 1 – that affect the
composition of the microbiota at other sites.

not been taken into account – this has resulted in considerable problems in trying to
define the microbiotas characteristic of these different groups of females. Additional
difficulties arise because of the variety of sampling methods that have been used. These
include swabbing, the use of bacteriological loops, direct aspiration, and aspiration
following introduction of saline into the vagina. The use of a swabbing method is
advantageous in that it ensures that organisms adherent to the vaginal epithelium
will be sampled, but has the disadvantage that, depending on how it is carried out, it
may be localised and, therefore, may not collect organisms from all of the different
regions of the vagina. Aspiration should ensure that microbes from all regions of the
vagina are being collected, but may not dislodge firmly adherent organisms from the
vaginal epithelium.

Initially, in this section, the vaginal microbiota of healthy, post-menarchal/pre-
menopausal, sexually active, non-pregnant individuals is described. Deviations from
this "norm" are referred to in this and subsequent sections. However, it is important
to bear in mind that, for unknown reasons, the vaginal microbiota of disease-free individuals fluctuates on a daily basis between one that is regarded as being compatible
with health and one that is considered "abnormal". It has been estimated that the majority of women (78%) do not have what has come to be regarded as a "healthy" vaginal
microbiota (i.e., one dominated by lactobacilli) throughout their menstrual cycle, even
though they show no symptoms of vaginal infection.

6.4.3.1.1 Microbiota of post-menarchal/pre-menopausal women

That the vaginal microbiota is a complex community is somewhat of an understate-
ment. For example, one study of the vaginal microbiota of fifty-four healthy women
reported nearly 5,000 isolates belonging to ninety-four species of forty genera. Furthermore, in most studies, between ten and fifteen isolates (approximately 60% of which are
obligate anaerobes) are usually cultured from each vaginal sample, and the number of
organisms detected can be two or three times this when special culture techniques or
culture-independent detection methods are used. Because of the difficulties referred
to previously and in Chapter 1, it is not easy to reach a concensus regarding which
organisms should and should not be regarded as being members of the indigenous microbiota of the vagina. Table 6.10 shows the results of some of the studies that have been
carried out and illustrates the enormous variation in the proportion of the community

Table 6.10.	Cultivable vaginal microbiota of post-menarchal/ pre-menopausal women		
Organism	Percentage of microbiota
facultative anaerobes	
Lactobacillus spp.	50–90
Staphylococcus spp.	0–65
Corynebacterium spp.	0–60
Streptococcus spp.	10–59
Enterococcus spp.	0–27
G. vaginalis	17–43
yeasts	13–16
Enterobacteriaceae	6–15
Ureaplasma spp.	0–54
obligate anaerobes	
Peptostreptococcus spp.	14–28
(including former members of the genus)	
Lactobacillus spp.	29–60
Eubacterium spp.	0–36
(including former members of the genus)	
Bacteroides spp.	4–80
Fusobacterium spp.	0–23
Veillonella spp.	9–29
Propionibacterium spp.	0–14
Bifidobacterium spp.	5–15
Clostridium spp.	5–18

Note: Data indicate the range of values obtained from six studies.

that a particular genus can constitute. Another perspective on the vaginal microbiota can be gained by examining the frequency with which a particular organism occurs in the normal population – such data are given in Table 6.11, from which the variability in isolation rates of all vaginal organisms can be appreciated.

Although the designation of lactobacilli and other organisms that occur frequently and in high proportions in vaginal samples as members of the vaginal microbiota is fairly uncontroversial, the situation with regard to other organisms is less clear. What should be the status of those organisms that are fairly regularly isolated (e.g., in 20% of individuals) and in not insignificant (say between 5% and 10%) proportions? Should these organisms be classed as members of the indigenous microbiota; are they the agents of as yet unknown, undiagnosed, or incipient infections; or are they merely transients? Opinions differ widely with regard to these issues. This situation applies to organisms such as *G. vaginalis, Myc. hominis, Mobiluncus* spp., and many others.

The application of culture-independent approaches to the analysis of the vaginal microbiota has resulted in some interesting findings. Hence, in a study involving nineteen pre-menopausal women, PCR-DGGE analysis (Section 1.4.3) of the DNA extracted from the vaginal communities revealed that, in most subjects, the profiles obtained were dominated by between one and three DNA bands. In 79% of the women, at least one of these dominant bands had a sequence corresponding to that of a *Lactobacillus* species, with *L. iners* being the most frequently detected. Another DGGE analysis of the vaginal microbiota also found *L. iners* to be the most frequently detected *Lactobacillus* species.

Table 6.11.	Frequency of isolation of microbes from the vagina of healthy post-menarchal/pre-menopausal women

Organism	Proportion of women harbouring the organism (%)
aerobes/facultative anaerobes	
Lactobacillus spp.	45–90
Staphylococcus spp.	10–62
Corynebacterium spp.	8–72
Can. albicans	5–30
Streptococcus spp.	15–68
G. vaginalis	15–52
U. urealyticum	13–80
E. coli	5–26
obligate anaerobes	
Peptostreptococcus spp.	10–74
(and former members of the genus)	
Bacteroides spp.	4–50
Prevotella spp.	7–68
Propionibacterium spp.	5–15
Veillonella spp.	9–20
Eubacterium spp.	
(and former members of the genus)	4–36
Bifidobacterium spp.	1–10
Fusobacterium spp.	7–23

However, in most culture-based studies of the vaginal microbiota, *L. iners* has not featured in the list of organisms isolated – this is not surprising, given that it is unable to grow on the main selective media used for the isolation of lactobacilli (i.e., MRS and Rogosa-Sharp). This illustrates one of the drawbacks of relying solely on culture-based approaches for the analysis of the microbiota of a site. Another culture-independent approach to analysing the vaginal microbiota is one based on microscopic examination of a Gram stain of the vaginal specimen and counting the various morphotypes (compare dark-field microscopy used in analysing the microbiota of subgingival plaque – Section 8.5.2.3). This approach is often used for the diagnosis of vaginal infections (particularly bacterial vaginosis) and involves counting three different morphotypes: large Gram-positive rods (i.e., lactobacilli), small Gram-negative or Gram-variable rods (i.e., *G. vaginalis, Bacteroides* spp., *Prevotella* spp., and *Porphyromonas* spp.), and curved rods (i.e., *Mobiluncus* spp.). The genus *Mobiluncus* consists of motile, anaerobic Gram-positive curved rods which produce mainly succinic and lactic acids as metabolic end-products. Lactobacilli are regarded as "beneficial", whereas *Gardnerella*-like species and *Mobiluncus* spp. are associated with disease (Section 6.5.1.1). Each of these morphotypes is given a point score based on the numbers present in five microscopic fields. However, whereas the point score increases with increasing numbers of *Gardnerella*-like species and *Mobiluncus* spp., in the case of lactobacilli, the score decreases with increasing number. The three values are summed to give the "Nugent score" for the patient, which crudely reflects the relative proportions of these three groups of organisms – a high score indicates few lactobacilli are present and/or high numbers of disease-associated organisms, whereas a low score is indicative of the opposite. The scores

Table 6.12. Characteristics of lactobacilli that contribute to their dominance within the vaginal ecosystem

Characteristic	Role in vaginal colonisation
acidophilic	enables survival in acidic environment
acidogenic	inhibits competing organisms
facultative or obligate anaerobes	enables survival in the (generally) oxygen-limited environment
hydrogen peroxide production	kills competing organisms
variety of adhesins	enable adhesion to vaginal epithelium
bacteriocin production	inhibit growth of competing organisms
biosurfactant production	prevent attachment of other species to vaginal epithelium
competitive exclusion	compete for receptors on epithelial cells

are interpreted as follows: 0–3 (normal vaginal microbiota), 4–6 (intermediate altered microbiota), and 7–10 (altered microbiota, indicative of disease or disease potential).

One feature of the vaginal microbiota of post-menarchal/pre-menopausal women that appears to be uncontroversial and generally accepted is that lactobacilli are the dominant organisms. The average number of lactobacilli in the vagina is between 10^8 and 10^9 and, in most women, only a single *Lactobacillus* sp. is present. In 15% to 20% of women, small proportions of one or two additional species of lactobacilli are also found. Because of failure to speciate the lactobacilli isolated, difficulties with the identification of *Lactobacillus* spp., and revision of their taxonomy, there is confusion regarding which actual species are present. Many early studies reported the predominant species as being *L. acidophilus*. However, genotypic analysis has revealed that this "species" is comprised of six different species, although their differentiation by phenotypic methods is very difficult, if not impossible. In a recent study, 272 strains of lactobacilli were isolated from 101 women and were identified using whole DNA genomic probes from known strains. *L. jensenii* and *L. crispatus* (a member of the *L. acidophilus* complex) were found to comprise 41% and 38% of these isolates, respectively. Other frequently isolated strains include *L. gasseri* (a member of the *L. acidophilus* complex), *L. cellobiosus*, *L. fermentum*, and *L. iners*. Less frequently isolated species include *L. plantarum*, *L. rhamnosus*, *L. brevis*, *L. casei*, *L. vaginalis*, *L. delbrueckii*, and *L. salivarius*. Not surprisingly, therefore, the results from studies based on the cultivation of samples, and those from culture-independent approaches appear to be at variance, with the most frequently detected species being *L. jensenii* and *L. crispatus* in the former and *L. iners* in the latter.

How, then, do lactobacilli manage to dominate the vaginal microbiota? First of all, they are ideally suited to colonisation of the vagina – they are either facultative or obligate anaerobes – and, consequently, can survive in this generally oxygen-restricted environment (Table 6.12). Furthermore, they are aciduric and acidophilic and contribute to, or may even be responsible for, maintaining the low pH of this region, resulting in an advantage over less acid-tolerant organisms competing for a foothold in this environment. Apart from having an effect on the vaginal pH, some of the acids produced by lactobacilli (lactic and acetic acids) are microbicidal. Some lactobacilli also produce hydrogen peroxide, which exerts an antibacterial effect on a wide range

Table 6.13.	Possible means by which lactobacilli reduce the risk of acquiring HIV-1 infections

killing of HIV-1 via the peroxidase-halide system
inactivation of HIV-1 by lactic acid
acidic environment produced by lactobacilli decreases activation of T
    lymphocytes, therefore decreasing their susceptibility to infection with the
    virus
inhibition of other organisms by lactobacilli decreases production of substances
    that can interfere with host defences: e.g., production of succinate by
    Gram-negative anaerobes inhibits PMNs, degradation of mucins removes
    protective barrier of the epithelium, production of IgA proteases reduces
    effectiveness of mucosal immunity

Note: HIV = human immunodeficiency virus; PMNs = polymorphonuclear leukocytes.

of competing species. Indeed, 95% of strains of two of the predominant species of lactobacilli (*L. jensenii* and *L. crispatus*) from healthy individuals are known to produce hydrogen peroxide. Such strains contain elevated levels of manganese ions, which are thought to provide protection against damaging superoxide radicals. Another way in which bacterially produced hydrogen peroxide can exert an antimicrobial effect is by oxidising chloride ions to hypochloric acid and/or chlorine, which are potent microbicides. This reaction is catalysed by peroxidase, which is present in vaginal secretions. A number of studies have revealed the importance of peroxide-producing strains of lactobacilli in preventing genito-urinary infections – women harbouring such strains have a significantly lower risk of acquiring HIV infection, gonorrhoea, and bacterial vaginosis. There is also evidence that certain lactobacilli can produce a bacteriocin effective against other organisms. For example, in a study of twenty-two strains of vaginal lactobacilli, 80% were found to produce bacteriocins active against *G. vaginalis*. Bacteriocins have been identified and characterized from *L. gasseri, L. plantarum, L. fermentum, L. casei, L. delbrueckii*, and *L. salivarius*. The bacteriocin of *L. salivarius* is of particular interest because it is active against *N. gonorrhoeae* and *Ent. faecalis*, which are important pathogens of the reproductive system and urinary tract, respectively. Many lactobacilli have also been shown to secrete biosurfactants (i.e., compounds which tend to accumulate at interfaces particularly liquid–air interfaces). These biosurfactants not only prevent the adhesion of bacteria to surfaces, but can also induce the detachment of already adherent cells. That this phenomenon may be of importance in the vaginal ecosystem is implied by the finding that biosurfactants from at least two vaginal species (*L. fermentum* and *L. rhamnosus*) are able to induce the detachment of bacteria adhering to vaginal epithelial cells.

As well as being able to exert control over the proportions of potentially pathogenic members of the indigenous vaginal microbiota, lactobacilli have also been shown to be important in preventing colonisation of the genital tract by exogenous pathogens such as HIV-1 and *N. gonorrhoeae*. Hence, the risk of acquiring HIV-1 infection is highest in women lacking vaginal lactobacilli and lowest in those colonised by peroxide-producing lactobacilli. Women with non-peroxide-producing strains of lactobacilli have a risk intermediate between the two. The means by which lactobacilli can reduce the risk of HIV-1 infection remain to be established. However, some of the likely factors involved are listed in Table 6.13. Equal protection against infection with

Table 6.14.	Obligate anaerobes frequently isolated from the vagina

B. ureolyticus, B. fragilis group
Prev. bivia, Prev. disiens, Prev. buccalis, Prev. corporis
Por. asaccharolytica
Pep. anaerobius, Pep. productus
Micromonas micros (formerly *Pep. micros*)
Schleiferella asaccharolytica (formerly *Pep. asaccharoyticus*)
Anaerococcus prevotii (formerly *Pep. prevotii*)
Finegoldia magna (formerly *Pep. magnus*)
Cl. perfringens, Cl. innocuum, Cl. indolis, Cl. ramosum
Bif. adolescentis, Bif. breve
Eg. lenta (formerly *Eub. lentum*)
Eub. limosum, Eub. alactolyticum, Eub. rectale

N. gonorrhoeae, however, is provided by both peroxide- and non-peroxide-producing strains of lactobacilli. This is probably attributable to the ability of all lactobacilli to kill this pathogen at low pHs. Other factors involved may include some of those listed in Table 6.13, as well as direct competition for nutrients between lactobacilli and the gonococcus.

As well as the numerically dominant lactobacilli, high proportions of coryneforms, CNS, viridans streptococci, and obligate anaerobes are usually present in the vagina. With regard to the obligate anaerobes, few studies have adequately identified the various organisms present, although they have been shown to include species from the following genera: *Bacteroides, Prevotella, Peptostreptococcus* (including former members of this genus), *Propionibacterium, Eubacterium* (including former members of this genus), *Clostridium, Bifidobacterium, Fusobacterium, Porphyromonas*, and *Veillonella*. The main species detected are listed in Table 6.14.

Although antagonistic interactions between members of the vaginal microbiota have been identified (and previously discussed), there is little information concerning the positive interactions (e.g., cross-feeding) that occur in this complex community. Some likely possibilities include: (1) the utilisation of oxygen by aerobes or facultative anaerobes (e.g., staphylococci) to provide an anaerobic environment for obligate anaerobes; (2) the secretion of proteases by proteolytic organisms (e.g., *Bifidobacterium* spp., *Bacteroides* spp., *Prevotella* spp., *Porphyromonas* spp., and viridans streptococci), which result in the production of amino acids that can be used as nitrogen and/or energy sources by others; (3) the degradation of mucins by bacterial consortia producing sialidases, glycosidases, and proteases – thereby releasing sugars and amino acids; (4) the release of metabolic end-products (e.g., lactic acid by lactobacilli, succinic acid by *Bacteroides* spp.) that can be used as a carbon and/or energy source by others (e.g., *Veillonella* spp. and *Propionibacterium* spp.); and (5) the production of an acidic environment (e.g., lactobacilli, streptococci) to create conditions suitable for the growth of acidophiles (e.g., lactobacilli, streptococci).

While considerable attention has been devoted to the lactobacilli found in the vagina, relatively little is known about the other members of the microbiota other than those species that are able to cause infections. The latter include *Can. albicans, G. vaginalis, U. urealyticum, Strep. agalactiae, Mobiluncus* spp., and *Bacteroides* spp.; these organisms – together with the diseases they cause – are discussed in Section 6.5.

Table 6.15. Vaginal microbiota of 50 individuals at three points during the menstrual cycle

Finding	Days from onset of menses		
	1–5	7–12	19–24
Lactobacillus			
any *Lactobacillus*	82	98	88
high levels	70	92	84
any H_2O_2-positive *Lactobacillus*	60	64	64
high levels	54	60	62
any H_2O_2-negative *Lactobacillus*	28	44	38
high levels	20	38	34
any non-*Lactobacillus* isolate	96	96	90
high levels	72	40	40
high levels, excluding mycoplasmas	68	36	38
Strep. agalactiae	30	16	18
E. coli	14	16	6
other Gram-negative rods	0	6	2
any *G. vaginalis*	26	24	18
high levels	18	6	8
U. urealyticum	30	32	38
Mycoplasma hominis	4	4	6
Can. albicans	6	8	12
Prevotella spp.	56	26	28
GPAC	40	50	42
black-pigmented anaerobic Gram-negative rods	10	10	12
B. fragilis group	2	6	10

Notes: Data refer to the percentage of subjects. High levels denote $\geq 10^6$ cfu/ml. GPAC = Gram-positive anaerobic cocci; H_2O_2 = hydrogen peroxide. Reproduced with permission of the University of Chicago Press from: *Clinical Infectious Diseases.* Vol. 30, pp. 901–907. Eschenbach, D.A., Thwin, S.S., Patton, D.L., Hooton, T.M., Stapleton, A.E., Agnew, K., Winter, C., Meier, A., and Stamm, W.E. Copyright © 2000, by The Infectious Diseases Society of America.

Surprisingly few studies have investigated the influence of the menstrual cycle on the vaginal microbiota. The results of the most extensive of these are shown in Table 6.15. In this study, the vaginal microbiota of fifty women aged 18–40, who had regular monthly menstrual cycles, only one (or no) sexual partner, and no use of contraception other than condoms was investigated. Samples were taken during menses (days 1–5) and at two subsequent time points (7–12 days and 19–24 days) prior to onset of the next menses. The main changes in the microbiota detected were: (1) fewer women had high levels of lactobacilli during menses, (2) recovery of high levels of non-*Lactobacillus* spp. was greatest during menses, (3) the proportion of women with *Prevotella* spp. was greatest during menses, and (4) the proportion of women with *B. fragilis* was at its lowest during menses. It is interesting to note that the composition of the microbiota was markedly altered during menses. This is not really surprising, given the profound alterations in the vaginal ecosystem during this event. Hence, the discharge of blood could supply a range of nutrients, trace elements, and oxygen required for the growth of organisms other than lactobacilli. Furthermore, the vaginal pH is higher during menses than at other times within the menstrual cycle, and this

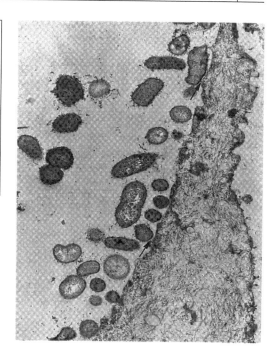

Figure 6.10 Transmission electron micrograph of epithelial cells and adherent microcolonies from the vaginal epithelium of a healthy volunteer. The sample was subjected to vortex mixing, centrifugation, and sonication, yet still had microcolonies attached to epithelial cells. Arrows indicate the condensed exopolysaccharide glycocalyx on the bacterial surface which appear to mediate bacterial adhesion to the epithelium. Reproduced with permission of Taylor and Francis from: A morphological study of the in situ tissue-associated autochthonous microbiota of the human vagina. Sadhu, K., Domingue, P.A.G., Chow, A.W., *et al. Microbial Ecology in Health and Disease* 1989;2:99–106.

would reduce the competitive advantage enjoyed by lactobacilli during the rest of the cycle.

While many studies have revealed the complexity of the vaginal microbiota, little information is available with regard to its structural organisation. Electron microscopy of vaginal scrapings from healthy pre-menopausal individuals obtained at different time points during the menstrual cycle has revealed the presence of microcolonies, predominantly of lactobacilli, attached to epithelial cells (Figure 6.10). Approximately 200 lactobacilli were present on each epithelial cell. The lactobacilli were encased in a polysaccharide matrix and were difficult to detach (by vortexing or sonication) from the epithelial cells and constituted a biofilm. As well as microcolonies adherent to host cells, some of the colonies were also embedded in the viscous material associated with the epithelial surface. Washing with phosphate-buffered saline removed some of the adherent organisms, and analysis revealed that this loosely adherent population consisted mainly of CNS. Three key members of the vaginal microbiota – *L. acidophilus, L. gasseri,* and *L. jensenii* – have all been shown to be capable of self-aggregation via proteinaceous surface components (Figure 6.11). This is likely to be important in the formation of an adherent biofilm.

6.4.3.1.2 Microbiota of pre-menarchal girls

As can be seen from a comparison of Table 6.11 and Table 6.16, the vaginal microbiota of pre-menarchal girls is strikingly different from that found in post-menarchal/pre-menopausal individuals. The most notable differences being that, in girls, anaerobes are very common; coryneforms and CNS are also frequent isolates, whereas lactobacilli are less common. *E. coli* is also more frequently detected, and *G. vaginalis* is usually absent. A wide range of anaerobes is present – the most frequently detected being those listed in Table 6.17. The skin and intestinal tract would appear to be the

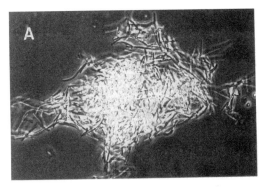

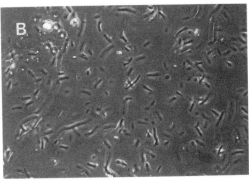

Figure 6.11 Micrographs showing (A) autoaggregation of *L. acidophilus* and (B) the disruption of aggregation by pre-treatment of the organism with proteinase K. Reprinted with permission from: Boris, S., *et al. Infection and Immunity* 1998;66:1985–1989.

main sources of organisms colonising the vagina of pre-menarchal girls because many of the organisms present are members of the indigenous microbiota of these sites.

6.4.3.1.3 Microbiota of post-menopausal women

It has been recognised for some time that onset of the menopause leads to changes in the vaginal microbiota. In a study of the vaginal microbiota of seventy-three post-menopausal women who were not receiving oestrogen replacement therapy, lacto-bacilli, *G. vaginalis*, yeasts, and mycoplasmas were recovered less frequently than from

Table 6.16. | Vaginal microbiota of pre-menarchal girls

Organism	Proportion (%) of girls harbouring the organism
anaerobes	76–92
coryneforms	7–80
coagulase-negative staphylococci	36–84
lactobacilli	2–40
E. coli	16–34
viridans streptococci	13–40
enterococci	9–29
H. influenzae	4–7
Strep. agalactiae	2–11
Proteus spp.	2–5
Klebsiella spp.	8–15
Staph. aureus	2–4

Table 6.17. Frequency of isolation of obligate anaerobes from pre-menarchal girls

Organism	Frequency of isolation (%)
Anaerococcus prevotii (formerly Pep. prevotii)	60
Pep. anaerobius	56
Cl. perfringens	32
Cl. incocuum	12
B. fragilis	24
Schleiferella asaccharolytica (formerly Pep. asaccharolyticus)	24
B. vulgatus	20
Porphyromonas spp.	14–20
Prevotella spp.	16
B. ovatus	16
Eg. lenta (formerly Eub. lentum)	8
Bifidobacterium spp.	8

Note: Frequency of isolation = proportion of the study population from which the organism was isolated.

pre-menopausal women, whereas viridans streptococci were among the most frequently isolated species (Table 6.18). However, the mean count of lactobacilli was the highest of all the organisms isolated, whereas viridans streptococci, although frequently isolated, were present in only small numbers (Figure 6.12). These changes may be attributable to a number of factors, including: (1) a decrease in the glycogen content of the vaginal epithelium due to the low oestrogen levels, thereby resulting in

Table 6.18. Cultivable vaginal microbiota of post-menopausal women

Organism	Proportion (%) of women harbouring the organism
viridans streptococci	74
Schleiferella asaccharolytica (formerly Pep. asaccharolyticus)	66
coagulase-negative staphylococci	59
coryneforms	58
lactobacilli	49
Finegoldia magna (formerly Pep. magnus)	45
E. coli	40
Enterococcus spp.	38
Anaerococcus prevotii (formerly Pep. prevotii)	36
Prev. bivia	33
B. fragilis group	30
B. ureolyticus	29
G. vaginalis	27
Prev. buccalis	25
Strep. agalactiae	23
Por. asaccharolytica	21
Actinomyces spp.	15
U. urealyticum	13
yeasts	1

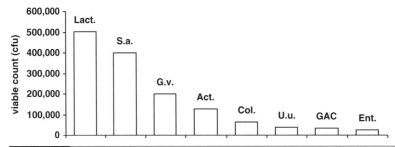

Figure 6.12 Quantitative analysis of the cultivable vaginal microbiota of seventy-three post-menopausal women. Lact. = lactobacilli; S.a. = *Strep. agalactiae*; G.v. = *G. vaginalis*; Act. = *Actinomyces* spp.; Col. = coliforms; U.u. = *U. urealyticum*; GAC = Gram-positive anaerobic cocci; Ent. = enterococci.

an increased pH; (2) thinning of the vaginal epithelium; (3) absence of, or changes in, the receptors available for bacterial adhesins; (4) changes in the innate and/or adaptive immune response; and (5) alterations in the production of vaginal fluid, thereby affecting the availability of bacterial nutrients. The decreased frequency of isolation of organisms associated with vaginal infections (*G. vaginalis, U. urealyticum,* and *Can. albicans*) may account for the decreased incidence of bacterial vaginosis and candidal vaginitis observed among these women. In post-menopausal women who receive oestrogen replacement therapy, the vaginal pH is lower than in those not receiving the hormone, and there is a corresponding decrease in the proportions of obligate anaerobes in the vaginal microbiota.

6.4.3.2 Cervix

Like the vagina, the cervix is colonised by a diverse microbiota, which appears to consist, on average, of approximately six cultivable microbial species. The population density on the epithelium is high, and most microbes are present as microcolonies attached to the epithelial surface (Figure 6.13). The results of a typical study of the cervical microbiota are shown in Table 6.19, from which it can be seen that it is similar to that of the vagina in that lactobacilli are the most frequently isolated species, and coryneforms, staphylococci, streptococci, and GPAC are also usually present. The composition of the cervical microbiota is of interest because several of its members (notably the obligate anaerobes) are able to cause infections of the genital tract and infections associated with childbirth (Section 6.5.2). As is the case for the vagina, a wide range of factors (see Table 6.9) is likely to influence the composition of the microbial community in the cervix, but few of these have been investigated in detail. The effects of contraceptive use and pregnancy on the cervical microbiota are described in Sections 6.4.5.2 and 6.4.5.4.

6.4.3.3 Vulva

Because of its close proximity to the anus and because of the presence of the external openings of the vagina and urethra, the vulva will invariably be contaminated with members of the microbiotas of these sites. The extent to which this happens will depend on the particular anatomy of the individual, the type of clothing worn, personal hygiene, etc. Only a very limited number of studies of the various regions comprising the vulva have been undertaken. Two of these were limited to the labia majora, which

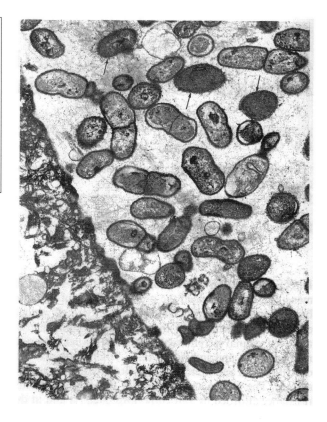

Figure 6.13 Transmission electron micrograph of material scraped from the cervix of a healthy volunteer. The bacteria are present as a microcolony on the epithelial surface. Most of the organisms in the microcolony are Gram-positive, but some Gram-negative cells (arrowed) can be seen. Reproduced with the permission of Taylor and Francis from: A morphological study of the in situ tissue-associated autochthonous microbiota of the human vagina. Sadhu, K., Domingue, P.A.G., Chow, A.W., *et al.* *Microbial Ecology in Health and Disease* 1989;2:99–106.

is basically a specialised skin region. In the first of these studies, the composition of the microbiota of the labia majora was compared with that of the skin of the forearm of the same individual (Table 6.20). Although coryneforms predominated in the vulval microbiota, there were also high proportions of staphylococci, micrococci, and lactobacilli. Apart from the lactobacilli, these microbial groups also dominated the microbiota of the forearm, although the density of colonisation was considerably lower (by at least 3 $\log_{10}$ units) on the forearm in all cases. Notable differences between the microbiotas of the two regions were the vastly greater density of colonisation of the vulva and the presence of much higher numbers (but not greater proportions) of lactobacilli, *Staph. aureus*, streptococci, and Gram-negative rods. With regard to the prevalence of the various microbes in the vulvas of the eighteen women, all were colonised by coryneforms and CNS, more than half harboured *Staph. aureus* and micrococci, while approximately 40% harboured streptococci, lactobacilli, Gram-negative rods, and yeasts. The high carriage rate of *Staph. aureus* (67%) is particularly interesting because it is considerably higher than that reported for other carriage sites, such as the anterior nares (37%) and perineum (20%). This finding implies that the labia majora should also be sampled when screening for carriers of the organism. Interestingly, in an *in vitro* study of the adhesion of *Staph. aureus* to epithelial cells from various body regions, the organism was found to adhere to cells from the labia majora as effectively as to nasal epithelial cells. The numbers of bacteria adhering to vaginal epithelial cells and to cells from the labia minora were considerably lower. Adhesion of *Staph. aureus* to the labia majora cells could be blocked by ribitol teichoic acid, suggesting the involvement of this

Table 6.19. Microbiota of the cervix of 100 non-pregnant, pre-menopausal women

Organism	Frequency of isolation (%)
lactobacilli	75
Schleiferella asaccharolytica (formerly *Pep. asaccharolyticus*)	48
coagulase-negative staphylococci	41
coryneforms	38
Group D streptococci	36
Pep. anaerobius	34
non-haemolytic streptococci (not Group D)	33
E. coli	28
β-haemolytic streptococci	22
unidentified anaerobic Gram-positive cocci	21
Anaerococcus prevotii (formerly *Pep. prevotii*)	17
α-haemolytic streptococci (not Group D)	17
unidentified anaerobic Gram-positive rods	16
Bacteroides spp.	12
Veillonella spp.	11
Finegoldia magna (formerly *Pep. magnus*)	11
G. vaginalis	8
Eubacterium spp.	7
Micromonas micros (formerly *Pep. micros*)	7
Strep. intermedius	5
Proteus spp.	5
yeasts	4
Klebsiella spp.	4
Cl. perfringens	4
B. fragilis	4
N. gonorrhoeae	1

Note: Frequency of isolation = proportion of the study population from which the organism was isolated.

adhesin in the process. The composition of the microbiota of the labia majora was also found to change during the menstrual cycle (Table 6.21). The population densities of lactobacilli, β-haemolytic streptococci, other streptococci, Gram-positive rods, and Gram-negative rods all decreased during menstruation (i.e., day 2 to day 4). In contrast, the numbers of *G. vaginalis* increased substantially during this period. There were few dramatic differences between the vulval microbiota during menstruation compared with that at day 21, other than the numbers of non-pathogenic *Neisseria* increased while the number of micrococci decreased.

The microbiota of the vestibular region surrounding the urethra has also been investigated in healthy pre-menarchal girls. The mean population density varied widely and ranged from 10^4 to 10^7 cfu/cm^2 with a mean of 1.4×10^6 cfu/cm^2. The mean number of species isolated per subject was 5.5, and 95% of the isolates were obligate anaerobes. In most cases, the microbiota was dominated by anaerobic Gram-positive cocci or rods and included organisms belonging to the genera *Lactobacillus*, *Propionibacterium*, *Eubacterium*, and *Bifidobacterium*. Anaerobic Gram-negative rods were often isolated in low numbers and included *Bacteroides* spp., *Fusobacterium* spp., and *Leptotrichia* spp. The most frequently isolated facultative anaerobes were non-haemolytic

Table 6.20. Cultivable microbiota of the labia majora and the forearm of 18 adults

Organism	Mean number present (cfu/cm^2)	
	Labia majora	Forearm
total count	2.8×10^6	6.4×10^2
coryneforms	1.3×10^6	1.2×10^2
coagulase-negative staphylococci	5.7×10^5	1.8×10^2
micrococci	5.1×10^5	2.9×10^2
lactobacilli	4.6×10^5	10
Staph. aureus	4.1×10^4	14
Gram-negative rods	1.8×10^3	1
streptococci	3.7×10^2	5
yeasts	82	8
Bacillus spp.	0	12

Note: Reproduced with permission of Blackwell Publishing from: Quantitative microbiology of human vulva. Aly, R., Britz, M.B., Maibach, H.I. *British Journal of Dermatology* 1979;101: 445–448.

streptococci, coryneforms, and CNS. Facultative Gram-negative rods and *Enterococcus* spp. were only infrequently present.

The vulva is not a uniform body site, but consists of a variety of environments which differ with respect to the nature of the epithelium, moisture content, type and density of glands, degree of occlusion, proximity to other densely populated sites, etc. (Sections 6.1 and 6.3.3). Unfortunately, few studies of the different microbial communities within the vulva have been carried out. Nevertheless, the studies undertaken so far indicate that the microbiotas of the various regions of the vulva are distinct from those found on either the skin or in the vagina, although they resemble the microbiota of the

Table 6.21. Microbiota of the labia majora of 20 women at various stages of the menstrual cycle

Organism	Population density (cfu/cm^2) on:		
	Day 2	Day 4	Day 21
coryneforms	1.2×10^6	4.8×10^5	4.6×10^5
lactobacilli	1.8×10^5	2.9×10^3	3.4×10^5
coagulase-negative staphylococci	2.2×10^5	1.2×10^5	6.9×10^5
Staph. aureus	5.6×10^3	4.0×10^3	6.1×10^3
micrococci	5.7×10^4	2.0×10^4	6.5×10^3
β-haemolytic streptococci	1.0×10^2	0	65
other streptococci	3.1×10^5	8.5×10^2	3.7×10^3
Gram-positive rods	1.0×10^4	55	8.5×10^3
Gram-negative rods	1.9×10^2	0	3.5×10^2
G. vaginalis	5.7×10^2	2.2×10^5	8.0×10^4
non-pathogenic *Neisseria*	0	0	1.9×10^3
yeasts	0	10	0
total organisms	2.0×10^6	8.9×10^5	1.6×10^6

Note: Reprinted from *Seminars in Dermatology*. Vol. 9, pp. 300–304. Elsner, P. and Maibach, H.I. Copyright © 1990, with permission from Elsevier.

Table 6.22.	Possible routes for the dissemination of microbes from the female reproductive tract

shedding of epithelial cells with attached organisms
vaginal and cervical secretions
menstrual fluid
transfer to partner during sexual foreplay and intercourse
transfer to neonate during birth

latter more than that of the former. This is likely to be attributable mainly to the high moisture content of the region and its proximity to the vagina, from which it will be continually supplied with densely populated secretions.

6.4.4 Dissemination of organisms from the female reproductive system

There are a number of means by which microbes colonising the female reproductive system can be dispersed into the environment or transferred directly to other individuals (Table 6.22). Epithelial cells from the vulva, vagina, and cervix are regularly shed with their attached organisms, and these will be deposited on undergarments or directly into the environment, depending on the clothing habits of the individual. Fluids from the vagina (particularly during menstruation) and cervix will also contain large numbers of microbes, and these can be deposited onto clothing, tampons, etc., and so gain access to the environment. Transfer of organisms from the female urogenital tract to a sexual partner can occur during sexual foreplay and intercourse. A variety of organisms may be transferred in this way to the penis, skin, oral cavity, vagina, or other regions of the partner, depending on the individual's sexual proclivities. Passage of a baby through the birth canal results in the transfer of a variety of bacteria to various sites on the neonate. Hence, it has been shown that vaginal organisms can be isolated from the skin, ear, and conjunctiva of neonates. However, little is known with regard to how long such organisms persist on the neonates.

6.4.5 Effect of antibiotics and other interventions and events on the indigenous microbiota of the female reproductive system

6.4.5.1 Antibiotics

Antibiotic use can have a dramatic effect on the vulval microbiota. For example, amoxycillin use for 5–7 days by a group of eight girls has been shown to result in dramatic reductions in both the numbers and variety of anaerobic organisms in the periurethral region, while the number of enterobacteria (mainly *E. coli* and *Klebsiella* spp.) increased markedly. Three weeks after the end of the treatment course, the numbers of anaerobes had returned to levels similar to those found prior to antibiotic treatment, while enterobacteria were undetectable. Because colonisation of the periurethral region with enterobacteria is a risk factor for UTIs, the administration of amoxycillin to females could predispose them to such infections. In contrast, administration of trimethoprin-sulphamethoxazole for 5–7 days to a group of ten girls had no detectable effect on the periurethral microbiota, and no enterobacteria were detectable before, during, or after the treatment course.

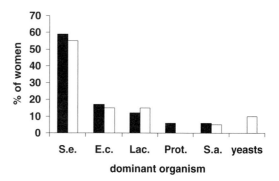

Figure 6.14 Dominant organisms in the vaginal introitus of women 4–6 weeks after completion of a course of amoxycillin or bacampicillin. Each bar represents the proportion of women in each group whose microbiota was dominated by each of the organisms shown. ■ = amoxycillin; □ = bacampicillin; S.e. = *Staph. epidermidis*; E.c. = *E. coli*; Lac. = lactobacilli; Prot. = *Proteus* spp.; S.a. = *Strep. agalactiae*.

Antibiotics administered for the treatment of urinary tract infections (UTIs) can have a profound effect on the microbiota of the vaginal introitus (opening). For example, 4–6 weeks after the administration of either amoxycillin or bacampicillin to thirty-five and twenty-five women, respectively, the microbiota of the vaginal introitus was dominated by *Staph. epidermidis* in most of the women (Figure 6.14). In only 12% and 15% of the women who had taken amoxycillin or bacampicillin, respectively, was the vaginal microbiota dominated by lactobacilli. Because lactobacilli are important in preventing colonisation of the vagina by uropathogens (Section 7.5.7.1), the absence of these organisms in an individual would predispose her to a UTI. In fact, 29% of the patients given amoxycillin and 20% of those given bacampicillin had another UTI within 6 weeks of completing their course of antibiotics.

One of the main side effects associated with the administration of antibiotics to females is an alteration in the vaginal microbiota. A number of antibiotics increase the frequency of recovery of *Can. albicans* and/or decrease the recovery of lactobacilli, and such changes increase the risk of vaginal candidiasis and UTIs, respectively. Studies have also shown that antibiotic use is associated with a loss of vaginal colonisation by hydrogen-peroxide-producing species of lactobacilli. Table 6.23 summarises some of the reported effects of antibiotics on the vaginal microbiota. In a study of nearly 32,000 women in the United Kingdom, a significant increase in the risk of developing vaginal candidiasis was found to be associated with the use of any of the following antibiotics: ciprofloxacin, ofloxacin, norfloxacin, cefixime, azithromycin, and fosfomycin.

Table 6.23. Effects of antibiotic administration on the vaginal microbiota

Antibiotic	Main effects on vaginal microbiota
phenoxymethylpenicillin	increase in frequency of recovery of *Can. albicans* and facultative Gram-negative rods; decrease in frequency of recovery of lactobacilli
minocycline	increase in frequency of recovery of *Can. albicans*
tetracycline	increase in frequency of recovery of *Can. albicans*
clarithromycin	increase in frequency of recovery of *Can. albicans*; decrease in frequency of recovery of lactobacilli
ciprofloxacin	increase in frequency of recovery of *Can. albicans* and increase in vaginal yeast infections
cefprozil	increase in vaginal yeast infections

6.4.5.2 Contraceptives

A variety of contraceptive practices are in common use, and many of these (spermicides, oral contraceptive pills, diaphragm, cervical cap) have the potential to alter the vaginal microbiota. Whereas a number of studies have shown that the use of oral contraceptive pills has little effect on the vaginal microbiota, the other contraceptive methods mentioned have all been shown to alter the microbial community of the vagina in some way. A large study involving groups of women using oral contraceptive pills, a cervical cap, or a diaphragm plus spermicide has recently been carried out. In the latter two groups, there was a significant increase in the proportion of women harbouring the uropathogens *E. coli* and *Enterococcus* spp. in their vagina after 1 and 4 weeks of use. In contrast, there was a significant decrease in vaginal colonisation by *E. coli* among those women using oral contraceptive pills. In none of the groups were there any changes in the proportions of women harbouring *Strep. agalactiae, G. vaginalis, Staph. aureus*, or anaerobic Gram-negative rods. What are the explanations for these changes in the composition of the vaginal microbial communities? In the case of the group using the diaphragm plus spermicide, one contributing factor is likely to be the differential antibacterial properties of the widely used spermicide (nonoxynol-9). While nonoxynol-9 has little effect on the viability of many strains of *E. coli*, it is bactericidal for many lactobacilli at concentrations likely to be present in the vagina. Hence, the use of this spermicide, by reducing the effectiveness of the "protective" lactobacilli, would encourage vaginal colonisation by *E. coli*. Also, as previously described (Section 6.3.1), the insertion of a diaphragm is known to lead to a dramatic increase in the oxygen content of the vagina, and this is likely to encourage the growth of aerobic and facultative organisms, such as *E. coli*. Just as the insertion of a diaphragm increases the vaginal oxygen content, the insertion of a cervical cap may have a similar effect and encourage the growth of organisms such as *E. coli*.

The use of a copper-releasing intra-uterine device (IUD) as a contraceptive for at least 3 years has been found to increase the diversity of the cervical microbiota. This increased diversity was attributable mainly to an increase in the number of obligately anaerobic species isolated – particularly *Peptostreptococcus* spp. (including former members of this genus), *Bacteroides* spp., and *Fusobacterium* spp. This finding is of concern because these organisms are associated with pelvic inflammatory disease (Section 6.5.2.1). A similar shift in the cervical microbiota was seen in women using oral contraceptives, whereas no such changes were observed in women whose partners used condoms – the cervical microbiota of the latter remained dominated by lactobacilli.

6.4.5.3 Sexual intercourse

In view of the mechanical abrasion of the vaginal mucosa that takes place during sexual intercourse and the transfer of semen with its range of microbes, nutrients, and antimicrobial constituents (Sections 5.2.2 and 5.2.3), a change in the vaginal microbiota would be expected following intercourse. Any changes would most likely be temporary and would be followed by a reconstitution of the indigenous microbiota, although this is likely to be dependent on the frequency of intercourse. Few studies have investigated the effect of sexual intercourse on the microbiotas of the female genital tract. However, in one study, a number of sexual behaviour patterns were associated with changes in the vaginal microbiota when assessed by analysis of Gram-stained smears and use of the Nugent score (Section 6.4.3.1.1). A Nugent score indicative of an "unstable" microbiota showed a significant correlation with those who (1) engaged in frequent vaginal intercourse, (2) had a large number of sexual partners, or (3) engaged in frequent receptive oral sex. An increased frequency of sexual intercourse (unaccompanied by use of a contraceptive device or spermicide) is also associated with an increased risk of vaginal colonisation by *E. coli*. This is an interesting finding because it has long been known that sexual intercourse predisposes young women to UTIs, the most frequent causative agent of such infections being *E. coli*. This is most likely due to mechanical effects which result in the transfer of organisms from the vagina to the urethral mucosa.

6.4.5.4 Pregnancy

Pregnancy appears to have a significant effect on the cervical microbiota. The frequency of isolation of *E. coli*, peptococci, peptostreptococci (including former members of this genus), *B. fragilis*, and other *Bacteroides* spp. decreases progressively from the first trimester through the third trimester. The pronounced decrease in the isolation of anaerobes may be attributable to the increased vascularity of the genital tract during pregnancy. The tissues would, therefore, be provided with more oxygen which would result in an increase in the partial pressure of oxygen and the redox potential in the cervix, thereby inhibiting the growth of obligate anaerobes. In addition, hormonal changes during pregnancy would affect the physiology of the epithelium and result in an altered microenvironment, which could affect the resident microbiota. Hence, during pregnancy, there is an increase in the concentration of both glycogen and lactic acid, and this encourages the proliferation of lactobacilli, which are generally found to be more prevalent in the vaginal microbiota of pregnant women than in non-pregnant individuals. For example, in a study involving 85 pregnant women, lactobacilli were isolated from 96% of the subjects and comprised approximately 90% of the vaginal microbiota (Table 6.24).

Dramatic differences are found between the microbiota during the third trimester and that present 3 days post-partum. Hence, the frequency of isolation of anaerobes, aerobic Gram-negative rods, GPAC, anaerobic Gram-negative rods, and *B. fragilis* is greater post-partum, whereas the isolation frequency of yeasts and aerobic Gram-positive rods declines. Such changes are not surprising in view of the dramatic alterations that would have occurred in the cervical environment because of the delivery of the baby, the discharge of lochia, and the presence of dead and dying tissues from the uterus. Lochia is a fluid produced by the uterus for as long as 8 weeks following childbirth, and contains cells, mucins, carbohydrates, proteins (including enzymes and immunoglobulins), and low-molecular mass compounds and ions. The presence of this nutrient-rich fluid would be expected to have a considerable effect on the cervical microbiota. Significantly, many of the organisms found more frequently post-partum are those associated with endometritis – particularly the obligate anaerobes (Section 6.5.2.2). At 6 weeks post-partum, nevertheless, the cervical microbiota returns to that found in the first trimester, with a few exceptions – *E. coli* and some *Bacteroides* spp. are isolated more frequently, while lactobacilli are less frequently present.

6.4.5.5 Douching

The results of studies of the effect of douching on the indigenous microbes of the genital tract are often difficult to evaluate because of variables such as the frequency

Table 6.24. The vaginal microbiota of 85 pregnant women – proportion of individuals harbouring a particular microbe in their vagina

Organism	Proportion (%) harbouring the organism
lactobacilli	96
coagulase-negative staphylococci	89
Schleiferella asaccharolytica (formerly *Pep. asaccharolyticus*)	88
U. urealyticum	78
coryneforms	72
anaerobic Gram-negative rods	91
Prev. bivia/disiens	61
viridans streptococci	55
Finegoldia magna (formerly *Pep. magnus*)	53
G. vaginalis	46
Enterococcus spp.	39
B. ureolyticus	36
Anaerococcus prevotii (formerly *Pep. prevotii*)	32
Can. albicans	31
Por. asaccharolytica	31
P. acnes	22
Anaerococcus tetradius (formerly *Pep. tetradius*)	21
Peptococcus niger	20
E. coli	17
Strep. agalactiae	15
Veillonella spp.	14

Note: Reproduced with permission of the University of Chicago Press from: Hillier, S.L., Krohn, M.A., Rabe, L.K., Klebanoff, S.J., and Eschenbach, D.A. *Clinical Infectious Diseases* 1993;16 (Suppl 4): S273–S281. Copyright © 1993 by The Infectious Diseases Society of America.

of douching and the composition of the fluid used. However, a recent study involving 1,200 women who used a similar fluid has shown that douching at least once a month results in increased vaginal colonisation by *G. vaginalis, Myc. hominis*, and non-pigmented Gram-negative anaerobic rods, but decreased numbers of hydrogen-peroxide-producing lactobacilli. There was also an increased risk of vaginosis. The disruption to the vaginal microbiota was probably due to the differential effect on the various members of the microbiota of the variety of antimicrobial compounds in the douching fluid. A number of other studies have also shown that douching is associated with an increased risk of vaginosis. Furthermore, the practice has also been linked to an increased risk of HIV acquisition, pelvic inflammatory disease, pre-term delivery, and cervical cancer.

6.5 | Diseases caused by members of the indigenous microbiota of the female reproductive system

6.5.1 Infections of the vagina

The most common infections of the vagina are bacterial vaginosis, candidiasis, and trichomoniasis. Of these, only the first two are considered further because they are caused by organisms frequently isolated from the vaginal ecosystem. Trichomoniasis is a sexually transmitted disease caused by the protoctist *Trichomonas vaginalis*.

6.5.1.1 Bacterial vaginosis

The prevalence of bacterial vaginosis varies widely and ranges from 5% to 51% in different populations. It is the most common vaginal infection in women of child-bearing age and is characterised by three or more of the following signs: (1) a grey vaginal discharge, (2) a vaginal pH > 4.5, (3) the presence of epithelial cells coated with small bacteria (known as clue cells), and (4) a fishy odour when potassium hydroxide is added to the discharge. Early studies implicated a small, rod-shaped organism as the aetiological agent of the disease, and this has been variously named as *Haemophilus vaginalis, Corynebacterium vaginale*, and, more recently, as *Gardnerella vaginalis*. More recent investigations, however, have revealed that vaginosis has a polymicrobial aetiology in which the normally dominant lactobacilli of the vagina are replaced by a mixture of bacteria which generally includes *G. vaginalis, Prevotella* spp. (and/or other Gram-negative obligate anaerobes), *Peptostreptococcus* spp., *Mobiluncus* spp., *Myc. hominis*, and *U. urealyticum*. Many of these are considered to be members of the indigenous vaginal microbiota, although the status of some (e.g., *G. vaginalis, Myc. hominis, Mobiluncus* spp.) is unresolved. Although the disease may be diagnosed on the basis of the symptoms previously described, a Gram-stained vaginal smear has proved useful in aiding diagnosis and in detecting individuals at risk of developing the disease – this is described in Section 6.4.3.1.1. A characteristic feature of the vaginal ecosystem in patients with bacterial vaginosis is that the pH is raised, usually to between 5.0 and 5.5. While the aetiology of the disease is becoming clearer, we still know little about what induces the shift in the microbiota away from one dominated by lactobacilli or what virulence factors of the postulated aetiological agents are responsible for the observed pathological changes in the vagina. It is easy to appreciate that any decrease in the numbers of lactobacilli would result in changes to the vaginal community. Hence, the production of lactic acid, hydrogen peroxide, biosurfactants, and bacteriocins would enable the proliferation of previously inhibited species, while competition for adhesion sites would be less intense (Table 6.25). Hydrogen-peroxide production by lactobacilli appears to be particularly important for maintaining vaginal health because several studies have shown that women colonised by peroxide-producing strains have a decreased risk of vaginosis, compared with those colonised by strains unable to produce peroxide. A number of *in vitro* studies have implied the nutritional interdependence of members of the vaginosis-associated microbiota. *G. vaginalis*, for example, produces amino and keto acids (e.g., pyruvate) which serve as nutrients for a number of anaerobes. These organisms, in turn, can produce volatile amines (e.g., putrescine and cadaverine) which would contribute to a rise in pH within the ecosystem and can stimulate the growth of *Myc. hominis*. Furthermore, these amines are known to be tissue irritants and could induce sloughing of the vaginal epithelium. *Prev. bivia* appears to be able to supply nutrients to both *G. vaginalis* (ammonia) and *Pep. anaerobius* (amino acids).

One possible factor that could precipitate a change in the vaginal microbiota to one characteristic of vaginosis is the rise in pH associated with the onset of menstruation. During maximal blood flow (days 2–3), the vaginal pH is approximately 7.0, and this high pH, if appropriate bacterial species were present, could induce changes in the community composition to one that is associated with vaginosis. Hence, growth of

Table 6.25. | Effects of a decrease in the proportion of lactobacilli in the vagina

Effect on vaginal ecosystem	Possible consequences
decreased lactic-acid production	rise in pH, encourages growth of *G. vaginalis, Mobiluncus* spp., *Can. albicans, E. coli*
decreased hydrogen-peroxide production	encourages growth of *G. vaginalis* and *Prev. bivia*
decreased production of bacteriocins	encourages growth of *G. vaginalis, Mobiluncus* spp., *Peptococcus* spp., *Pep. anaerobius*
decrease in number of lactobacilli adhering to epithelium	enables increased numbers of *G. vaginalis* and *Can. albicans* to adhere to epithelium
decreased production of biosurfactants	enables increased numbers of enterococci, staphylococci, *Can. albicans,* and *E. coli* to adhere to epithelium

Note: Possible consequences of this decrease are based on extrapolations from the results of *in vitro* studies or else have been observed *in vivo.*

G. vaginalis is inhibited below pH 4.5, but the organism grows well between pH 6.0 and 6.5. Furthermore, the activity of bacteriocins produced by lactobacilli decrease dramatically as the pH rises. Alternatively, it has been suggested that killing of lactobacilli by bacteriophages could be responsible for the population shift. More than 30% of vaginal lactobacilli have been shown to have lysogenic bacteriophages, which were able to induce lysis in a wide range of lactobacilli. Factors that are associated with an increased risk of bacterial vaginosis are shown in Table 6.26. As previously mentioned, little is known with regard to the pathogenesis of vaginosis. Table 6.27 lists some virulence factors that may be involved in the pathology associated with the infection.

6.5.1.2 Vaginal candidiasis

It has been estimated that 40–75% of sexually active women have experienced this disease, which is characterised by a white, cottage-cheese–like discharge and itching of the vulva – hence, it is often referred to as vulvovaginitis. A number of factors pre-dispose to the infection, and these are summarised in Table 6.28. The causative agent of the

Table 6.26. | Risk factors for bacterial vaginosis

increased vaginal pH
decreased proportion of lactobacilli
decreased proportion of peroxide-producing lactobacilli
previous history of bacterial vaginosis
onset of menses
use of vaginal medications
use of spermicides
increased number of sexual partners
increased frequency of vaginal intercourse
increased number of episodes of receptive oral sex

Table 6.27. Microbial virulence factors that may be involved in the pathogenesis of bacterial vaginosis

Virulence factor	Organism	Effect
succinic acid	*Prevotella, Bacteroides, Mobiluncus*	inhibits neutrophil chemotaxis
sialidases	*Prev. bivia, Prev. disiens, G. vaginalis*	cleavage of mucin – contributes to vaginal discharge
haemolysin	*G. vaginalis*	lyses red blood cells – liberates nutrients
putrescine	anaerobes	sloughing of vaginal epithelium
malic acid	*Mobiluncus*	vaginal irritation

infection is *Can. albicans*, and treatment involves the application of a topical antifungal agent such as an imidazole. Many studies have reported that *Can. albicans* is a member of the indigenous vaginal microbiota, and the organism has been isolated from up to 30% of pre-menopausal, non-pregnant women. However, its frequency of occurrence and the numbers present in the vagina are influenced by a variety of factors, including those listed in Table 6.28. Vaginal candidiasis is a frequent consequence of antibiotic therapy. For example, administration of penicillins has been shown to eliminate vaginal lactobacilli and increase the proportion of *Can. albicans*. Lactobacilli control the proportions of *Can. albicans* by the production of hydrogen peroxide, which can kill the organism, and by interfering with its adhesion to the vaginal epithelium. Elimination of lactobacilli by antibiotics removes such controlling factors, resulting in yeast overgrowth within the vagina.

The situation with non-antibiotic-induced candidiasis, however, is not so straightforward because this condition does not appear to be associated with a dramatic alteration in the vaginal microbiota. For example, in a study of the vaginal microbiota of twenty healthy women and twenty-four women with non-antibiotic-induced vaginal candidiasis, there were no significant differences other than the frequency of isolation of *Can. albicans* – this being 25% and 92%, respectively. However, this study did not determine whether the lactobacilli isolated from each group were peroxide-producing, and it has been suggested that vaginitis may be associated with a change to a microbiota dominated by non-peroxide-producing lactobacilli. However, it must be said that, for the present, little is known regarding the changes in the vaginal microbiota that result

Table 6.28. Risk factors for vaginal candidiasis

recent (previous 15–30 days) antibiotic therapy
pregnancy
use of oral contraceptives containing high levels of oestrogen
oestrogen therapy
diabetes
use of condoms
prior history of gonorrhoea
frequent vaginal intercourse

in the overgrowth of *Can. albicans* in patients with non-antibiotic-associated vaginal candidiasis.

6.5.2 Infections at sites other than the vagina

6.5.2.1 Pelvic inflammatory disease

Pelvic inflammatory disease (PID), also known as salpingitis, is a serious condition which involves inflammation of the Fallopian tubes and uterus. PID is one of the most frequently occurring gynaecological problems and is a disease with high morbidity – 20% of women with PID become infertile, 20% develop chronic pelvic pain, and 10% of those who conceive have an ectopic pregnancy. A number of organisms have been isolated from the upper genital tract of women with PID, including several members of the cervical and vaginal microbiotas. While the involvement of the exogenous pathogens *N. gonorrhoeae* and *Chlam. trachomatis* in the pathogenesis of PID is well established, the possible role of members of the indigenous genital microbiota remains uncertain. Nevertheless, there is some evidence suggesting that such organisms, particularly obligate anaerobes, facilitate infection by exogenous pathogens and are capable of damaging tissues of the Fallopian tubes *in vitro* and in animal models. The complex of organisms responsible for bacterial vaginosis has also been implicated in the disease.

The possibility of an association between PID and the use of IUDs remains controversial. Some studies have implied that biofilm formation on the device is a risk factor for PID. The organisms present in the biofilms include β-haemolytic streptococci, *Staph. aureus*, *Bacteroides* spp., *Peptostreptococcus* spp., and *E. coli*.

6.5.2.2 Endometritis

The uterus is generally sterile in pregnant and non-pregnant women. However, after rupture of the amniotic sac during delivery, organisms from the lower genital tract gain access to the uterus and may cause endometritis – a potentially life-threatening infection of the lining of the uterus (endometrium). The amnionic fluid, however, does have a range of antimicrobial defence mechanisms, including lysozyme and IgA. Furthermore, the amnionic and choriamnionic membranes themselves are able to inhibit the growth of a range of bacteria, including *Strep. pyogenes*, *Staph. saprophyticus*, and *Staph. aureus*. Endometritis is uncommon following vaginal delivery (less than 3% of births), but is one of the most frequent complications of caesarean delivery and occurs in 11–23% of such cases. The causative organisms are usually members of the indigenous microbiota of the lower genital tract, and the infection is invariably polymicrobial (Table 6.29).

6.5.2.3 Chorioamnionitis

This is an infection of the amnionic cavity and the chorioamnionic membranes which affects between 0.5% and 10.5% of pregnancies. It is associated with increased maternal and perinatal morbidity and mortality, and accounts for up to 40% of cases of febrile morbidity in the mother and 20–40% of cases of neonatal sepsis and pneumonia. It is a polymicrobial infection and invariably involves members of the indigenous microbiota of the lower reproductive tract, with the mean number of organisms being 2.2.

Table 6.29.	Organisms frequently associated with chorioamnionitis or endometritis	
Organism	Chorioamnionitis	Endometritis
E. coli	+	+
Strep. agalactiae	+	+
Ent. faecalis	+	+
Staph. aureus	+	+
Streptococcus spp.	+	+
Fusobacterium spp.	+	+
G. vaginalis	+	+
B. fragilis	+	+
B. bivius	−	+
other *Bacteroides* spp.	+	+
Peptostreptococcus spp.	+	+
Peptococcus spp.	+	+
Clostridium spp.	+	+
U. urealyticum	+	−

Although a wide range of organisms are associated with the infection, (Table 6.29), those most frequently responsible are *Strep. agalactiae* and *E. coli*.

6.5.2.4 Pre-term birth

Between 6% and 10% of all births are pre-term, and such births account for approximately 70% of perinatal mortality. Although many causes of pre-term birth have been identified, it is thought that infections are responsible for about 50% of cases, and antibiotic administration following the pre-term rupture of membranes has been shown to reduce neonatal morbidity. The causative organism(s) involved include *Strep. agalactiae*, *U. urealyticum*, and the complex of organisms associated with bacterial vaginosis. As many as 78% of women harbour *U. urealyticum* in the lower genital tract where, under suitable circumstances, it can become a member of the complex of organisms associated with bacterial vaginosis (Section 6.5.1.1). A very large study (involving 13,000 women) in the United States demonstrated that its presence in the lower genital tract is not associated with an increased risk of pre-term delivery. However, when it gains access to the uterus of pregnant women, it can induce pre-term labour. The mechanism involved is possibly related to the ability of the organism to produce phospholipase A_2. This enzyme can generate prostaglandins, which are responsible for inducing labour.

There is also evidence of a link between the organisms associated with periodontitis and pre-term birth (Section 8.5.3).

6.5.2.5 Neonatal infections

As previously mentioned, *U. urealyticum* is frequently present in the lower genital tract, which can result in colonisation of neonates during delivery. Approximately half of normal-term infants born to *U. urealyticum*-positive mothers are colonised by the organism at skin and mucosal sites; but, in pre-term infants, the frequency of colonisation can be as high as 90%. The organism has been associated with a number of infections in neonates, particularly in very low birth weight infants, and these are listed in

| Table 6.30. | Neonatal infections due to *Ureaplasma urealyticum* |

Condition	Frequency
sepsis	0–34% depending on population – most common in pre-term infants, less common in normal-term infants
meningitis	0–8%
pneumonia	63% of low birth weight infants colonised by the organism
wheezing	1.7%
chronic lung disease	23–100% in very low birth weight infants

Table 6.30. The virulence factors of the organism that are responsible for these conditions have not been established. However, *U. urealyticum* is known to produce phospholipases A_1, A_2, and C; elastase; and urease. Of these, the elastase may induce abnormal development of the developing lung, while phospholipase A_2 can inhibit pulmonary surfactant. Both of these factors could contribute to chronic lung disease.

The principal habitat of *Strep. agalactiae* is the gastrointestinal tract from which site colonisation of the vagina and cervix occurs. The organism is frequently isolated from the vagina and/or cervix of both non-pregnant and pregnant individuals. It can adhere to epithelial cells of the vagina and cervix *in vitro*, and can also bind to extracellular matrix components, including fibronectin and fibrinogen. The organism can also bind to human placental laminin, which is a major component of the placental basement membrane, and is known to colonise, and even invade, intact membranes resulting in septic abortion. Between 5% and 40% of women have been shown to harbour *Strep. agalactiae* in the lower genital tract, and between 50–70% of the children of these individuals are colonised by the organism during delivery. In 1–2% of such infants, the organism causes severe invasive diseases such as sepsis, pneumonia, and meningitis. Most infections occur during the first 24 hours of birth (early-onset disease) and involve sepsis and pneumonia. Late-onset disease characteristically involves meningitis, which occurs between 1 week and 3 months after birth. Although modern intensive neonatal care has reduced the fatality rate in infants infected with the organism, this rate is still approximately 5%. The administration of an antibiotic (usually penicillin) to pregnant women colonised by *Strep. agalactiae* immediately prior to delivery has resulted in a substantial decrease in the incidence of neonatal infections due to the organism.

6.5.2.6 Bartholinitis

Infections of the Bartholin's glands often result in the formation of abscesses or cysts, and the exogenous pathogen *N. gonorrhoeae* is a frequent cause of such infections. However, members of the vaginal microbiota are also responsible for a substantial proportion of infections and, in such cases, the aetiology is usually polymicrobial. Gram-negative anaerobic rods and GPAC are often isolated from the abscesses, together with *Staph. epidermidis* and streptococci.

6.6 | Further Reading

Aroutcheva, A., Gariti, D., Simon, M., Shott, S., Faro, J., Simoes, J.A., Gurguis, A., and Faro, S. (2001). Defense factors of vaginal lactobacilli. *American Journal of Obstetrics and Gynecology* **185**, 375–379.

Arya, O.P., Tong, C.Y., Hart, C.A., Pratt, B.C., Hughes, S., Roberts, P., Kirby, P., Howel, J., McCormick, A., and Goddard, A.D. (2001). Is *Mycoplasma hominis* a vaginal pathogen? *Sexually Transmitted Infections* **77**, 58–62.

Balu, R.B., Savitz, D.A., Ananth, C.V., Hartmann, K.E., Miller, W.C., Thorp, J.M., and Heine, R.P. (2002). Bacterial vaginosis and vaginal fluid defensins during pregnancy. *American Journal of Obstetrics and Gynecology* **187**, 1267–1271.

Bayo, M., Berlanga, M., and Agut, M. (2002). Vaginal microbiota in healthy pregnant women and prenatal screening of group B streptococci (GBS). *International Microbiology* **5**, 87–90.

Berner, R. (2002). Group B streptococci during pregnancy and infancy. *Current Opinion in Infectious Diseases* **15**, 307–313.

Boris, S., Suarez, J.E., Vazquez, F., and Barbes, C. (1998). Adherence of human vaginal lactobacilli to vaginal epithelial cells and interaction with uropathogens. *Infection and Immunity* **66**, 1985–1989.

Boris, S.B. (2000). Role played by lactobacilli in controlling the population of vaginal pathogens. *Microbes and Infection* **2**, 543–546.

Boskey, E.R., Telsch, K.M., Whaley, K.J., Moench, T.R. and Cone, R.A. (1999). Acid production by vaginal flora *in vitro* is consistent with the rate and extent of vaginal acidification. *Infection and Immunity* **67**, 5170–5175.

Brandtzaeg, P. (1997). Mucosal immunity in the female genital tract. *Journal of Reproductive Immunology* **36**, 23–50.

Burton, J.P., Cadieux, P.A., and Reid, G. (2003). Improved understanding of the bacterial vaginal microbiota of women before and after probiotic instillation. *Applied and Environmental Microbiology* **69**, 97–101.

Burton, J.P. and Reid, G. (2002). Evaluation of the bacterial vaginal flora of 20 postmenopausal women by direct (Nugent score) and molecular (polymerase chain reaction and denaturing gradient gel electrophoresis) techniques. *Journal of Infectious Disease* **186**, 1770–1780.

Calderone, R.A. and Fonzi, W.A. (2001). Virulence factors of *Candida albicans*. *Trends in Microbiology* **9**, 327–335.

Cauci, S., Driussi, S., De Santo, D., Penacchioni, P., Lannicelli, T., Lanzafame, P., De Seta, F., Quadrifoglio, F., de Aloysio, D., and Guaschino, S. (2002). Prevalence of bacterial vaginosis and vaginal flora changes in peri- and postmenopausal women. *Journal of Clinical Microbiology* **40**, 2147–2152.

Clarke, J.G., Peipert, J.F., Hillier, S.L., Heber, W., Boardman, L., Moench, T.R., and Mayer, K. (2002). Microflora changes with the use of a vaginal microbicide. *Sexually Transmitted Diseases* **29**, 288–293.

Eckert, L.O., Hawes, S.E., Stevens, C.E., Koutsky, L.A., Eschenbach, D.A., and Holmes, K.K. (1998). Vulvovaginal candidiasis: clinical manifestations, risk factors, management algorithm. *Obstetrics & Gynecology* **92**, 757–765.

Eggert-Kruse, W., Botz, I., Pohl, S., Rohr, G., and Strowitzki, T. (2000). Antimicrobial activity of human cervical mucus. *Human Reproduction* **15**, 778–784.

Eschenbach, D.A., Thwin, S.S., Patton, D.L., Hooton, T.M., Stapleton, A.E., Agnew, K., Winter, C., Meier, A., and Stamm, W.E. (2000). Influence of the normal menstrual cycle on vaginal tissue, discharge, and microflora. *Clinical Infectious Diseases* **30**, 901–907.

Garland, S.M., Ni Chuileannain, F., Satzke, C., and Robins-Browne, R. (2002). Mechanisms, organisms, and markers of infection in pregnancy. *Journal of Reproductive Immunology* **57**, 169–183.

Gupta, K., Hillier, S.L., Hooton, T.M., Roberts, P.L., and Stamm, W.E. (2000). Effects of contraceptive method on the vaginal microbial flora: a prospective evaluation. *Journal of Infectious Diseases* **181**, 595–601.

Hein, M., Valore, E.V., Helmig, R.B., Uldbjerg, N., and Ganz, T. (2002). Antimicrobial factors in the cervical mucus plug. *American Journal of Obstetrics and Gynecology* **187**, 137–144.

Hillier, S.L. and Lau, R.J. (1997). Vaginal microflora in postmenopausal women who have not received estrogen replacement therapy. *Clinical Infectious Diseases* **25** (Suppl 2), S123–S126.

Jaquiery, A., Stylianopoulos, A., Hogg, G., and Grover, S. (1999). Vulvovaginitis: clinical features, aetiology, and microbiology of the genital tract. *Archives of Disease in Childhood* **81**, 64–67.

Johansson, M. and Lycke, N. (2003). Immunology of the human genital tract. *Current Opinion in Infectious Diseases* **16**, 43–49.

Kjaergaard. N., Hein, M., Hyttel, L., Helmig, R.B., Schøheyder, H.C., Uldbjerg, N., and Madsen, H. (2001). Antibacterial properties of human amnion and chorion in vitro. *European Journal of Obstetrics & Gynecology and Reproductive Biology* **94**, 224–229.

Kubota, T., Nojima, M., and Itoh, S. (2002). Vaginal bacterial flora of pregnant women colonized with group B streptococcus. *Journal of Infection and Chemotherapy* **8**, 326–330.

Larsen, B. and Monif, G.R. (2001). Understanding the bacterial flora of the female genital tract. *Clinical Infectious Diseases* **32**, 69–77.

Loudon, I. (2000). The cause and prevention of puerperal sepsis. *Journal of the Royal Society of Medicine* **93**, 394–395.

Mandar, R. and Mikelsar, M. (1996). Transmission of the mother's microflora to the newborn at birth. *Biology of the Neonate* **69**, 30–35.

Marrazzo, J.M., Koutsky, L.A., Eschenbach, D.A., Agnew, K., Stine, K., and Hillier, S.L. (2002). Characterization of vaginal flora and bacterial vaginosis in women who have sex with women. *Journal of Infectious Diseases* **185**, 1307–1313.

Mikamo, H., Sato, Y., Hayasaki, Y., Hua, Y.X., and Tamaya, T. (2000). Vaginal microflora in healthy women with *Gardnerella vaginalis*. *Journal of Infection and Chemotherapy* **6**, 173–177.

Navarro-Garcia, F., Sanchez, M., Nombela, C., and Pla, J. (2001). Virulence genes in the pathogenic yeast *Candida albicans*. *FEMS Microbiology Reviews* **25**, 245–268.

Newton, E.R., Piper, J.M., Shain, R.N., Perdue, S.T., and Peairs, W. (2001). Predictors of the vaginal microflora. *American Journal of Obstetrics and Gynecology* **184**, 845–853.

Ocana, V. and Nader-Macias, M.E. (2001). Adhesion of *Lactobacillus* vaginal strains with probiotic properties to vaginal epithelial cells. *Biocell* **25**, 265–273.

Paavonen, J. (1983). Physiology and ecology of the vagina. *Scandinavian Journal of Infectious Diseases* **40** (Suppl), 31–35.

Parks, D.K., Yetman, R.J., Moyer, V., and Kennedy, K. (2000). Early-onset neonatal group B streptococcal infection: implications for practice. *Journal of Pediatric Health Care* **14**, 264–269.

Pybus, V. and Onderdonk, A.B. (1999). Microbial interactions in the vaginal ecosystem, with emphasis on the pathogenesis of bacterial vaginosis. *Microbes and Infection* **1**, 285–292.

Rajan, N., Cao, Q., Anderson, B.E., Pruden, D.L., Sensibar, J., Duncan, J.L., and Schaeffer, A.J. (1999). Roles of glycoproteins and oligosaccharides found in human vaginal fluid in bacterial adherence. *Infection and Immunity* **67**, 5027–5032.

Regan, J.A. and Greenberg, E.M. (2001). Perinatal ureaplasma urealyticum infection and colonisation: the association with pre-term delivery and the spectrum of disease in neonates. *Reviews in Medical Microbiology* **12**, 97–107.

Ross, J. (2001). Pelvic inflammatory disease. *British Medical Journal* **322**, 658–659.

Ross, J.D. (2002). An update on pelvic inflammatory disease. *Sexually Transmitted Infections* **78**, 18–19.

Russell, M.W. and Mestecky, J. (2002). Humoral immune responses to microbial infections in the genital tract. *Microbes and Infection* **4**, 667–677.

Schwebke, J.R. (2001). Role of vaginal flora as a barrier to HIV acquisition. *Current Infectious Diseases Reports* **3**, 152–155.

Schwebke, J.R., Richey, C.M., and Weiss, H.L. (1999). Correlation of behaviors with microbiological changes in vaginal flora. *Journal of Infectious Diseases* **180**, 1632–1636.

Schwebke, J.R. and Weiss, H. (2001). Influence of the normal menstrual cycle on vaginal microflora. *Clinical Infectious Diseases* **32**, 325.

Sherman, D., Lurie, S., Betzer, M., Pinhasi, Y., Arieli, S., and Boldur, I. (1999). Uterine flora at caesarean and its relationship to postpartum endometritis. *Obstetrics and Gynecology* **94**, 787–791.

Spellerberg, B. (2000). Pathogenesis of neonatal *Streptococcus agalactiae* infections. *Microbes and Infection* **2**, 1733–1742.

Spiegel, C.A. (2002). Bacterial vaginosis. *Reviews in Medical Microbiology* **13**, 43–51.

Sundstrom, P. (1999). Adhesins in *Candida albicans*. *Current Opinion in Microbiology* **2**, 353–357.

Vallor, A.C., Antonio, M.A., Hawes, S.E., and Hillier, S.L. (2001). Factors associated with acquisition of, or persistent colonisation by, vaginal lactobacilli: role of hydrogen-peroxide production. *Journal of Infectious Diseases* **184**, 1431–1436.

Wilson, J. (2004). Managing recurrent bacterial vaginosis. *Sexually Transmitted Infections* **80**, 8–11.

The gastrointestinal tract and its indigenous microbiota

The gastrointestinal tract (GIT) and the accessory digestive organs (teeth, tongue, salivary glands, liver, gallbladder, and pancreas) together constitute the digestive system whose function is to break down dietary constituents into small molecules and then to absorb these molecules for subsequent distribution throughout the body. The GIT is, basically, a continuous tube extending from the mouth to the anus (Figure 7.1). The main regions of the GIT include the oral cavity, the oropharynx, and laryngopharynx (which are also part of the respiratory tract), oesophagus, stomach, small intestine (duodenum, jejunum, and ileum), and the large intestine (caecum, colon, and rectum). Because of differences in their anatomy, physiology, organisation, and location, each of these regions provides a different set of environmental conditions for potential microbial colonisers and has, therefore, a distinctive microbiota. Although each of these regions is distinctly different, the oral cavity has some unique features (e.g., a complex anatomy leading to a great variety of habitats, the presence of non-shedding surfaces; i.e., teeth, which enable biofilm formation and a high oxygen content), which render it very different from the rest of the GIT. In contrast, the other regions of the GIT have a far simpler structure (basically, tubular in nature), which results in a lower habitat diversity within a particular region, non-shedding surfaces are absent, most regions (below the stomach) are anaerobic, and many regions (within the large intestine) have a lumen which is usually filled with material. Because of such differences, the oral cavity and its microbiota are described in a separate chapter (Chapter 8). The microbiota of the pharynx was described in Chapter 4. This chapter, therefore, is concerned only with that part of the GIT which extends from the oesophagus to the anus.

7.1 | Anatomy and physiology of the gastrointestinal tract

Apart from the oral cavity (see Chapter 8) and pharynx (see Chapter 4), the GIT consists basically of a tube comprising four layers of tissue (Figure 7.2). The innermost layer is the mucosa, which consists of an epithelium surrounded by connective tissue and a thin layer of muscle – the muscularis mucosae. Muscular contractions result in folding of the mucosa, which increases its surface area, thereby aiding digestion and absorption. Mucosa-associated lymphoid tissue (MALT) is also present. The mucosa is surrounded by connective tissue (the submucosa), which contains blood and lymphatic vessels, as well as nerves for controlling GIT secretions and motility. The submucosa also binds the mucosa to the muscularis, which consists of an inner sheet

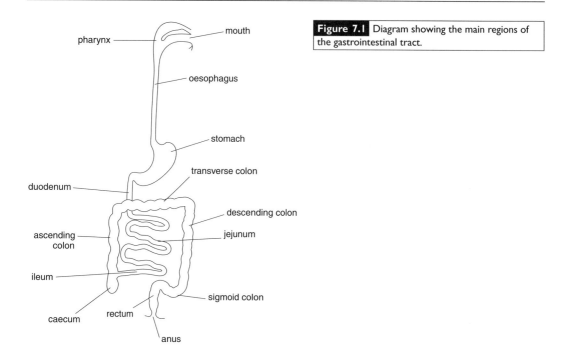

Figure 7.1 Diagram showing the main regions of the gastrointestinal tract.

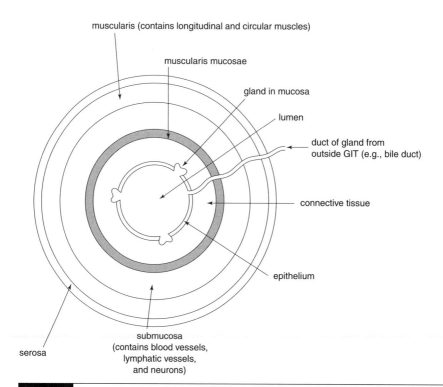

Figure 7.2 Cross-section through the wall of the gastrointestinal tract (GIT) showing the arrangement of the four basic layers of tissues.

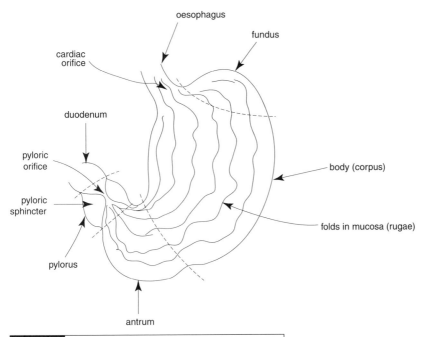

oesophagus

fundus

cardiac
orifice

duodenum

pyloric
orifice

body (corpus)

pyloric
sphincter

folds in mucosa (rugae)

pylorus

antrum

Figure 7.3 Diagram showing the main regions of the stomach.

of circular muscle and an outer sheet of longitudinal muscle. Contraction of these muscles contributes to the physical breakdown of food, mixes the food with digestive secretions, and propels it along the GIT. The outermost layer, the serosa (or visceral peritoneum), consists of connective tissue covered by squamous epithelium. It secretes a watery fluid which enables the various regions of the GIT to move easily over one another, over other organs in the abdomen, and against the parietal peritoneum which lines the wall of the abdominal cavity. Blood and lymphatic vessels are also present, as well as nerves. Although the aforementioned general description applies to those regions of the GIT from the oesophagus to the anus, each region shows characteristic variations on this theme; these will now be briefly described.

The oesophagus is a tubular structure approximately 25 cm long that connects the laryngopharynx to the stomach, and its function is to transport food between these regions. It is lined by a stratified, squamous epithelium and has numerous mucus glands. Food entering the oesophagus is propelled towards the stomach by peristalsis, which is initiated by contraction of the circular muscles above the bolus of food – this constricts the oesophagus and pushes the bolus downwards. At the same time, the longitudinal muscles in front of the bolus contract, thereby expanding the oesophagus and enabling the bolus to move into this region. The whole cycle is then repeated sequentially along the oesophagus until the food is propelled into the stomach. Movement of the bolus is aided by the mucus secreted by the oesophagus and also by the presence of saliva. Swallowing occurs 70 times/hour during fasting, >200 times/hour during eating, and <10 times/hour when sleeping.

The stomach is a J-shaped structure (Figure 7.3) linking the oesophagus to the duodenum and has a total capacity of approximately 1,500 ml. Food enters the stomach from the oesophagus via the cardiac orifice, which is situated below the uppermost

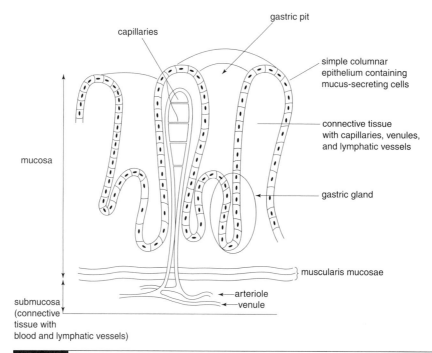

Figure 7.4 Diagram of a cross-section through the gastric mucosa showing gastric pits and glands.

region of the stomach known as the fundus. The central portion of the stomach is known as the body (or corpus), which narrows to form the pylorus (or antrum). The pyloric sphincter separates the stomach from the duodenum. When the stomach is empty, its mucosa forms folds known as rugae. The gastric epithelium consists of a layer of nonciliated simple columnar epithelial cells and has a large number of shallow involutions known as gastric pits, into the base of which tubular glands (gastric glands) open (Figure 7.4). Three types of secretory cells may be present in gastric glands – mucus neck cells, chief cells, and parietal cells – and these secrete mucus, pepsinogen, and hydrochloric acid, respectively. Collectively, these secretions are known as gastric juice. Two types of gastric glands are recognised: oxyntic and pyloric. Oxyntic glands secrete HCl, pepsinogen, intrinsic factor (Section 7.5.10), and mucus, and are located mainly in the body region. Pyloric glands are located predominantly in the antrum and secrete mainly mucus, together with some pepsinogen and gastrin. In addition, mucus glands are distributed over the entire gastric mucosa. Gastrin stimulates secretion of gastric juice, as well as increasing gut motility. The HCl produced by the parietal cells converts the enzymically inactive pepsinogen to the active enzyme pepsin. The latter converts dietary proteins to peptides and has its optimum activity at the low pH (usually less than 2.0) of the stomach. The epithelium is protected from its action by a thick layer of mucus between 0.2 and 0.6 mm deep. Acid produced by the mucosal glands is thought to pass through the mucus gel via channels so that it does not come into contact with the epithelial surface. Furthermore, the gel is relatively impermeable to the bicarbonate produced by epithelial cells so that the pH of the mucosal surface underlying the gel is maintained at a high pH relative to the gastric lumen. The outer surface of the mucus layer is also covered by a layer of phospholipids (mainly phosphatidylcholine and phosphatidylethanolamine), rendering it hydrophobic so that it repels aqueous acid in the gastric lumen. Peristaltic movements of the stomach result in mixing of the food with gastric juice to produce a watery fluid known as chyme, small quantities of which are continually forced through the pyloric sphincter into the small intestine.

The small intestine is the main site of digestion and absorption, and is a tubular structure approximately 3 m long and 2.5–3.5 cm in diameter. It consists of three regions: the duodenum (approximately 25 cm in length), the jejunum (approximately 1.0 m), and the ileum (approximately 2.0 m). The mucosa is highly folded to produce permanent ridges projecting into the lumen so that the chyme is made to follow a spiral pathway through the small intestine – this increases the surface area of the lumen and thereby enhances digestion and absorption. Furthermore, the mucosa is arranged to form numerous finger-like projections (0.5–1.0 mm in length) known as villi, of which there are between 20 and 40/mm^2 (Figure 7.5). Covering the epithelial surface is a mucus layer between 10 and 250 μm thick. The epithelium consists mainly of simple columnar epithelial cells, absorptive cells, and mucus-secreting goblet cells. The surfaces of the absorptive cells (often called enterocytes) have numerous projections (microvilli), which provide a very large surface area (often known as a "brush border") for the absorption of low-molecular mass digestion products. In the mucosa, between the bases of the villi, are a number of crypts (i.e., crypts of Lieberkühn) – between 6 and 10 per villus. These are the proliferative units of the mucosa and contain undifferentiated stem cells. These cells exit the crypt and move from the base of the villus to its tip, undergoing transformation to an enterocyte, a goblet cell, or an enteroendocrine cell. On reaching the tip of the villus, each cell is shed into the lumen. The whole process takes approximately 3 days. The crypts produce a fourth cell type – the Paneth cell – which remains within the crypts and synthesises a range of antimicrobial compounds (Section 7.2.1). The combined secretions of the cells of the small intestine are known as intestinal juice. MALT is also abundant in the small intestine and is present as solitary lymphatic nodules, and as aggregated lymphatic follicles known as Peyer's patches (Section 7.2.2). The latter are found throughout the small intestine in infants, but in adults are localised predominantly in the ileum.

Approximately 2.0 litres of chyme enters the duodenum after each meal, and localised contractions of the small intestine cause mixing of chyme with enzymes produced by epithelial cells, intestinal juice, pancreatic juice (secreted by the pancreas into the duodenum), and bile (produced by the liver and transported via the gallbladder into the duodenum). This process aids the enzymatic degradation of food macromolecules resulting in low-molecular mass products, which are taken up by the absorptive cells – approximately 90% of all absorption occurs in the small intestine. This absorptive phase is followed by peristalsis, which results in expulsion of the chyme into the large intestine.

The large intestine consists of the caecum, colon (ascending, transverse, descending, and sigmoid), rectum, and anal canal, and is about 1.5 m long, 6.5 cm in diameter, and has a surface area of approximately 1,250 cm^2. Its mucosal surface is very different from that of the small intestine in that it has no permanent folds or villi and has

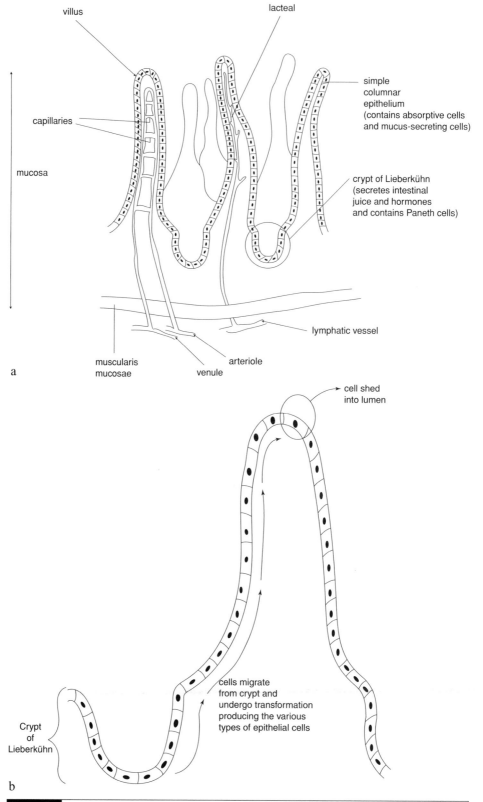

lumen of small intestine

villus

lacteal

capillaries

simple
columnar
epithelium
(contains absorptive cells
and mucus-secreting cells)

mucosa

crypt of Lieberkühn
(secretes intestinal
juice and hormones
and contains Paneth cells)

lymphatic vessel

muscularis
mucosae

arteriole

venule

a

cell shed
into lumen

cells migrate
from crypt and
undergo transformation
producing the various
types of epithelial cells

Crypt
of
Lieberkühn

b

Figure 7.5 Diagram showing (a) the structure of the mucosal surface of the small intestine and (b) the production and transformation of intestinal epithelial cells from stem cells in the crypts of Lieberkühn.

Figure 7.6 Transmission electron micrograph of intestinal epithelium showing a goblet cell (G) and absorptive epithelial cells (AEC) with associated microvilli. Magnification, ×4675. Reproduced with permission from: Mayhew, T.M. *et al. Cell and Tissue Research* 1992;270:577–585. Copyright © 1992, Springer-Verlag.

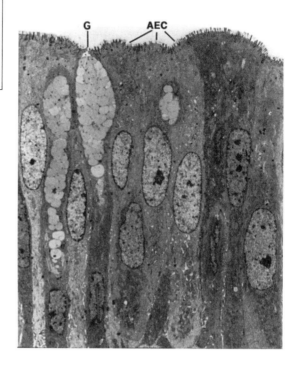

a large number of narrow invaginations (known as intestinal glands) lined by absorptive and goblet cells. The absorptive cells, which have many microvilli, are involved primarily in water absorption, while the goblet cells secrete mucus which provides lubrication to facilitate the peristaltic movement of the luminal contents (Figure 7.6). The mucus layer coating the epithelial surface increases in thickness from the proximal colon (107 ± 48 μm) to the rectum (155 ± 54 μm) – it is continually being synthesised by goblet cells and degraded by microbial action. No digestive enzymes are secreted by the mucosa of the large intestine – further breakdown of dietary constituents in this region is carried out by the resident microbiota. Between 3 and 10 hours after entering the large intestine, the chyme becomes solid as a result of water absorption and is known as faeces. Faeces is transferred into the rectum from the lower, S-shaped region of the colon known as the sigmoid colon.

The main macromolecular constituents of the diet are carbohydrates, proteins, and lipids, and their fate within the digestive system is now briefly outlined. Carbohydrates are present in the diet mainly as starch, lactose, and sucrose. Dietary fibre also contains high levels of plant polysaccharides (e.g., cellulose, hemicellulose, gums), as well as non-carbohydrate polymers (e.g., lignin) that cannot be digested by the host, but are converted by colonic bacterial enzymes mainly to short-chain fatty acids which can then be absorbed. Starch is digested by salivary and pancreatic amylases to dextrins, maltotriose, and maltose. These, as well as dietary sucrose and lactose, are then hydrolysed by carbohydrases present on the surfaces of mature enterocytes. The resulting monosaccharides (glucose, fructose, and galactose) are then absorbed by the gut mucosa. Proteins present in the diet (70–100 g/day) and in the epithelial cells shed into the lumen (2–30 g/day) are digested by proteases produced by the stomach and pancreas.

A number of gastric proteases are produced, although the main one is pepsin. These have a pH optimum less than 5 and lose their activity once they enter the duodenum where the pH is greater than this. The most important phase of protein digestion occurs in the small intestine, with the main source of proteases being the pancreas which produces a number of endopeptidases and exopeptidases. The endopeptidases trypsin, chymotrypsin, and elastase convert proteins to oligopeptides, which are converted by the exopeptidases carboxypeptidase A and B to amino acids. Peptidases are also present on the surface of enterocytes and, consequently, protein digestion also takes place on the mucosal surface, as well as in the lumen. The amino acids produced are absorbed throughout the small intestine, but absorption is greater in proximal regions than in distal regions. Most dietary fats are triglycerides, which are insoluble in water. Little degradation of fats occurs in the stomach; most is accomplished by pancreatic lipases and takes place in the small intestine. On arriving in the duodenum, fats are solubilised by bile salts with which they form micelles. Pancreatic lipases then attach to these micelles and hydrolyse the fats to monoglycerides and fatty acids. These degradation products remain as micelles, in which form they are taken up by the mucosa of the small intestine. Any undigested fats or their hydrolysis products that enter the colon are generally not metabolised by the microbes present and are excreted in faeces.

7.2 | Antimicrobial defence mechanisms of the gastrointestinal tract

7.2.1 Innate defence systems

Peristaltic activity in the stomach and small intestine is an important clearance mechanism and rapidly propels fluid and particulate matter through these regions, thereby hindering microbial colonisation of the mucosal surface. An important innate defence component that is present throughout the GIT is mucus. This consists mainly of water and large glycoproteins (mucins), with a high oligosaccharide content (>80%) which polymerise to form a viscous gel that covers the epithelial surface (Section 1.5.3). As well as being constituents of mucus, some mucins produced by the intestinal mucosa remain anchored to the surface of epithelial cells. The mucin content of mucus varies with its location in the GIT and gradually decreases from the stomach (50 mg/ml) to the colon (20 mg/ml). The secreted mucins are produced mainly by goblet cells and, to a lesser extent, by absorptive cells. The high carbohydrate content of the mucins provides the gel with lectin-binding properties, thus enabling it to bind and trap bacteria (and certain molecules), thereby preventing their access to the underlying epithelium. At least seven different mucins are produced in the GIT, but the principal type present in mucus is MUC2, which is secreted by goblet cells. Another mucin found in the intestinal tract is MUC3, which is produced by both goblet and absorptive cells. Apart from mucins, the mucus layer also contains various effector molecules of the innate and acquired host defence systems (see later). The main functions of the mucus layer are to protect the underlying epithelium from microbial colonisation, to provide lubrication to facilitate movement of the luminal contents through the GIT, to protect the underlying mucosa (mainly in the stomach and duodenum) from acid and digestive enzymes, and to protect the mucosa from the shear forces generated by movement of material through the GIT. Microbes that bind to the carbohydrate receptors of mucins may eventually be removed as a result of peristaltic movement of the luminal contents. If such organisms

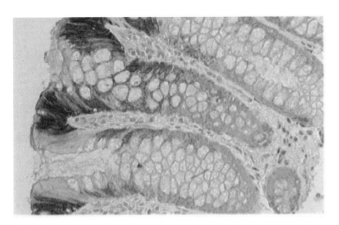

Figure 7.7 Immunohistochemical detection of the cathelicidin LL-37 (darkly stained regions) in the human colon. Biopsies were analysed by indirect immunoperoxidase staining for LL-37 expression. The sections were immunostained with an LL-37-specific antibody. Magnification, ×400. Reproduced with permission from: Hase, K., *et al. Infection and Immunity* 2002;70:953–963.

are members of the indigenous microbiota, they serve to protect the host by occupying the available binding sites, thereby preventing colonisation by pathogens.

The intestinal mucosa, like other epithelia, can produce a range of molecules with antimicrobial activities. These include lysozyme, lactoferrin, lactoperoxidase, secretory phospholipase A_2, bacterial permeability-inducing protein, collectins, and several antibacterial peptides. The antimicrobial activities of these compounds were described in previous chapters (Sections 2.2, 3.2, and 4.2). The α-defensins human defensin-5 (HD-5) and human defensin-6 (HD-6) are secreted by Paneth cells in the crypts of the small intestine. In contrast, the β-defensins human β-defensin-1 (HBD-1) and human β-defensin-2 (HBD-2) are secreted by a variety of epithelial cells in the stomach, small intestine, and colon. HBD-1 is constitutively expressed, although increased quantities are produced in response to interleukin (IL)-1α, whereas HBD-2 is produced by the mucosa only in response to IL-1α, bacterial infection, or during inflammation. HBD-4 is produced constitutively by epithelial cells of the gastric antrum and exhibits synergy with lysozyme. Human cathelicidin LL-37 is produced by cells of the surface and upper regions of crypts in the colon, but not in the small intestine (Figure 7.7). The antimicrobial properties of these peptides were described in previous chapters.

Adrenomedullin is produced constitutively by gastric and colonic epithelial cells and can kill *B. fragilis, E. coli*, and a number of oral and skin bacteria. Its production is up-regulated in gastric epithelial cells by *B. fragilis, Hel. pylori, E. coli, Strep. bovis*, and *Sal. enterica*. Another compound with antimicrobial properties produced by the intestinal mucosa is histone H1, which displays activity against *Salmonella* spp., *Ent. faecalis, Ent. faecium, Staph. aureus*, and *E. coli*. This is present within the cytoplasm of cells from the surface of villi and is released when these cells undergo apoptosis. The presence of H1 in the cytoplasm of epithelial cells may provide protection against invasive microbes, while extracellular H1 may contribute to the control of microbes in the lumen.

Excessive secretion of fluids (i.e., diarrhoea), although often regarded as principally a means of pathogen dispersal, may also be considered to be a host defence mechanism as it flushes out organisms resident in the GIT. Hence, the instillation of up to 2.5 litres of isotonic saline into the small intestine of human volunteers has been shown to result in a considerable reduction in the microbial population of the GIT.

If bacteria do overcome these defence mechanisms and manage to adhere to epithelial cells, the last line of the innate defence system comes into operation – exfoliation. The intestinal epithelium, like all mucosae, is a shedding surface and so bacteria adhering to epithelial cells will be removed when the latter are shed. The lifespan of an intestinal epithelial cell is typically between 2 and 5 days, after which time it is shed from the mucosal surface – it has been estimated that approximately 2×10^{11} epithelial cells are shed into the lumen of the GIT each day.

As previously mentioned (Section 7.1), the secretion of HCl by the parietal cells of the gastric glands maintains a low pH (pH <2) in the stomach. This not only protects the stomach from colonisation by a wide range of microbes unable to tolerate this low pH, but also acts as a barrier preventing access of acid-sensitive organisms to the small and large intestines. Both the small and large intestines contain bile and proteolytic enzymes, which can kill a range of microbes. Bile is a mixture of cholesterol, phospholipids, bile acids, and immunoglobulins, which is produced in the liver, stored in the gallbladder, and secreted into the duodenum. The bile acids (mainly cholic and deoxycholic) are conjugated to amino acids (usually glycine or taurine) and are amphipathic, surface-active compounds with potent, but selective, antimicrobial activity *in vitro*. In general, they are more active against Gram-positive than Gram-negative species, but there may be profound differences in the susceptibility of individual species within a particular genus. The different bile acids vary in their antimicrobial potency, with the unconjugated compounds having greater activity than the conjugated forms. However, many organisms present in the colon (e.g., *Bacteroides* spp., *Bifidobacterium* spp., *Clostridium* spp., *Lactobacillus* spp., *Streptococcus* spp.) are able to convert conjugated bile acids to their unconjugated form. Studies of populations with different faecal bile acid contents have shown that the proportion of *Bacteroides* spp. is higher, and that of Gram-positive non-sporing rods is lower, in the faeces of those with a high faecal bile concentration.

7.2.2 Acquired immune defence system

The epithelium of the GIT, like other mucosae, produces secretory IgA, which is able to block adhesion of microbes to the epithelium and also causes microbial aggregation, thereby generating aggregates that are more easily expelled from the system. An immune response is generated following uptake of an antigen by the M cells of Peyer's patches. The antigenic material is transported to underlying lymphoid cells, where antigen presentation and affinity maturation take place. The activated antigen-specific lymphocytes migrate to the lamina propria where they differentiate into antibody-producing plasma cells. However, the acquired immune response appears to have little effect on the composition of the microbiota of the GIT because individuals with IgA deficiency, a relatively common disorder affecting 0.3% of the population, have an intestinal microbiota similar to that of healthy individuals. The reasons for this have not yet been elucidated. However, it may be that, in the stomach and duodenum, antibodies may not be able to function effectively at the low pH of these environments. In other regions of the GIT (the colon and ileum), the IgA produced may not be able to exert a detectable effect because of the huge numbers of organisms present.

| Table 7.1. | Host defence mechanisms in the gastrointestinal tract |

Mechanism	Effect
production of mucus by goblet and absorptive cells	prevents microbial adhesion to epithelial cells; entrapment of microbes; facilitates movement of gut contents, thereby expelling microbes; provides receptors for adhesion of members of indigenous microbiota, thereby preventing colonisation by pathogens
rapid transit of luminal contents in certain regions	hinders colonisation in oesophagus, stomach, and small intestine
exfoliation	removes microbes attached to exfoliated cells
secretion of antimicrobial peptides and proteins by epithelial cells	kills or inhibits microbes
release of histone H1 from apoptotic epithelial cells	kills or inhibits microbes
release of cytokines by epithelial cells	attracts and activates phagocytes
low pH of stomach	kills or inhibits a wide range of microbes
bile acids	antimicrobial – Gram-positive organisms generally more susceptible than Gram-negative species
proteolytic enzymes	microbicidal
stimulation of excessive fluid secretion	flushes out microbes from GIT
production of IgA	blocks adhesion of microbes to epithelial cells

Note: GIT = gastrointestinal tract.

However, IgA may be expected to play a role in the initial colonisation of the GIT and could exert an effect on the composition of the mucosa-associated microbiota of these regions.

The host defence mechanisms operating in the GIT are summarised in Table 7.1.

7.3 | Environmental determinants within different regions of the gastrointestinal tract

In general, remarkably little is known of the environmental determinants in the various regions of the GIT because, unlike certain other anatomical sites (e.g., the skin, oral cavity), many of these regions are difficult to access and/or involve sampling procedures that are uncomfortable or embarrassing for the individual. Attempts at circumventing these difficulties (e.g., by anaesthetisation) invariably alter the environment of the region due to the resulting changes in gut motility and in the flow of secretions as well as the necessity of prior fasting, etc. One of the distinguishing features of the GIT which sets it apart from all of the body sites described so far is that the microbes colonising these regions are not dependent wholly on host secretions and the metabolic end-products of other organisms for their nutrients – they also have access to nutrients present in food ingested by the host. Epithelial cells, which are shed into the lumen in huge numbers, constitute another important source of nutrients because of their slow transit time through the GIT.

7.3.1 Oesophagus

The environment within this tubular structure is not ideal for microbial colonisation because it is being flushed constantly by saliva and intermittently by food and drinks ingested by the host. Even in the absence of food and fluid intake, saliva is constantly being swallowed at a rate of approximately 20 ml/hour. Furthermore, the average individual in a temperate climate consumes between 1 and 2 litres of fluids per day. The passage of food through the oesophagus (described in Section 7.1) also exerts a cleansing action. The ability of an organism to adhere to the epithelium, therefore, is a prime requisite for any would-be coloniser of this region. The oesophagus is aerobic, has a temperature of 37°C, and the pH of the mucus layer covering its epithelial lining is usually slightly acid – approximately 6.8. Although nutrient-rich, host-ingested food and fluids are regularly passing through the oesophagus, their transit time is short, and they are unlikely to contribute significantly to the nutrients available to colonising microbes. Although saliva is continually being swallowed, this is relatively devoid of nutrients between mealtimes. The only reliable source of nutrients, therefore, is likely to be the mucus layer, which contains mucins, as well as substances excreted and secreted by epithelial cells.

7.3.2 Stomach

The temperature of gastric fluid is normally 37°C, but will be affected temporarily by the intake of food and fluids at different temperatures. It is an aerobic environment, and the partial pressure of oxygen at the lumenal surface of the mucosa is 46.3 ± 15.4 mm of Hg, which corresponds to 29% of the oxygen content of air.

Although the median 24-hour intragastric pH has been reported to be 1.4, the pH is influenced by many factors, including age, diet, and whether food and/or drinks have recently been ingested. For example, food has a strong buffering action, and the pH of gastric juice over a 24-hour period has been shown to range from approximately 1 to 5, with the higher pHs corresponding to mealtimes (Figure 7.8).

Three major sources of nutrients exist for microbes in the gastric lumen: (1) food ingested by the host (which has a transit time of 2–6 hours), (2) host secretions, and (3) interstitial tissue fluid as a consequence of mucosal inflammation. Salivary amylases and gastric proteases produce disaccharides, oligosaccharides, and oligopeptides from carbohydrates and proteins in the host's diet. These can be further degraded to assimilable molecules (monosaccharides and amino acids) by the action of various organisms present in the saliva taken in with the food, as well as by organisms resident in the gastric lumen. The mucins present in the mucus layer coating the gastric epithelium can be degraded by a number of salivary organisms and by *Hel. pylori*, which is frequently present in the stomach. This results in the production of a variety of carbohydrates and amino acids which can serve as nutrients for resident microbes. Finally, damage to the gastric epithelium by *Hel. pylori* (due to the production of its vacuolating cytotoxin and as a consequence of the inflammatory response it induces – see Section

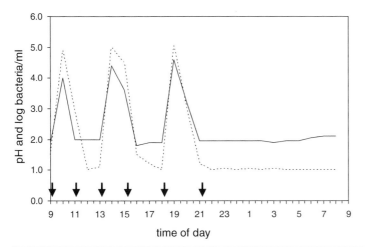

Figure 7.8 Effect of food and drink on the pH (. . . .) and bacterial content (—) of gastric juice over a 24-hour period. The bacterial concentration is expressed as log_{10} of total bacteria per milliliter of gastric juice. Arrows denote when food or a beverage together with a biscuit were taken. Note that the concentration of bacteria in the gastric juice correlates closely with the pH – a rise in pH results in an increase in the bacterial concentration. Reprinted with permission of CRC Press (via the Copyright Clearance Center) from: Factors controlling the microflora of the healthy upper gastrointestinal tract. Hill, M.J. In: *Human microbial ecology*, Hill, M.J. and Marsh, P.D. (eds.). Copyright © 1989, by CRC Press.

7.5.1) increases its permeability, thereby enabling the ingress of interstitial fluid with its plentiful supply of nutrients.

Antimicrobial factors operating in the stomach include IgA, pepsin, antimicrobial peptides, and lysozyme. However, the low pH of the gastric lumen is likely to reduce, or abolish, their effectiveness (apart from pepsin) so that they are likely to exert an antimicrobial effect only at the mucosal surface or during periods of increased pH when food is present.

7.3.3 Small intestine

The duodenum, jejunum, and proximal regions of the ileum are subjected to rapid peristalsis, which results in a short transit time (between 3 and 5 hours) for the lumenal contents of these regions. However, peristalsis in the distal section of the ileum is much slower. Large quantities of fluid enter into, and are produced by, the small intestine, and this exerts a considerable flushing action hindering microbial colonisation, particularly in the upper regions. Hence, in 1 day, the total quantity of fluid entering the small intestine of an adult amounts to approximately 9.0 litres. This consists of 2.0 litres of drinks, 1.5 litres of saliva, 2.5 litres of gastric juices, 0.5 litres of bile, 1.5 litres of pancreatic juice, and 1.0 litres of intestinal fluid. More than 80% of this fluid is absorbed by the small intestine. Other factors limiting microbial colonisation are the high concentrations of bile salts and proteolytic enzymes (entering the duodenum from the gallbladder and pancreas, respectively), other host antimicrobial defences (Section 7.2), and the low pH of the upper regions. Although the chyme entering the small intestine (approximately 2.0 litres after each meal) is very acidic, the pH soon increases because of the alkaline fluids produced by, and delivered to, the small intestine. These include intestinal juice (pH 7.6), pancreatic juice (pH 7.1–8.2), and bile (pH 7.6–8.6). The pH gradually increases along the small intestine and ranges from 5.7 to 6.4 in the duodenum, from 5.9 to 6.8 in the jejunum, and from 7.3 to 7.7 in the ileum. The secretion of bicarbonate by the mucosa also contributes to this change in pH along the small intestine.

The chyme present in the lumen provides an abundance of a variety of nutrients for microbes resident in this region. Mucins and exfoliated intestinal cells constitute additional nutrient sources, as also do molecules produced by the microbes present.

Oxygen is present in the various fluids secreted into the lumen and diffuses into the lumen from the tissues underlying the muosa. However, its concentration in the small intestine is low, and its partial pressure at the lumenal surface of the mucosa has been reported to range from 34 to 36 mm Hg, which corresponds to approximately 22% of the oxygen content of air. There is very little information regarding the oxygen concentration of the lumenal contents. Microbial activity, however, will reduce the oxygen content as the chyme passes along the small intestine, and the redox potential of the ileum has been reported to be as low as −150 mV.

7.3.4 Large intestine

The human colon is approximately 150 cm long, with a volume of approximately 540 ml. It receives approximately 1.5 kg of material/day from the small intestine, but much of this is water and is rapidly absorbed. The quantity of material present within the colon of an adult ranges from 58 to 908 g, with an average of 220 g, of which 35 g is dry matter. The transit time of material through the colon is very long and ranges from 20 to 140 hours, with a mean value of 60 hours; during this time, the water content of the material decreases from 86% in the caecum to 77% in the rectum. Approximately 120 g of faeces is produced each day, and bacteria comprise 55% of the faecal solids. The fluid material entering the caecum is mixed thoroughly and any readily digestible compounds present are rapidly broken down by resident bacteria. It has been estimated that digested matter is retained within the caecum for approximately 18 hours from where it passes into the ascending colon. Portions of this material are periodically transferred to the transverse colon; as it moves through the rest of the large intestine, water is absorbed which reduces its viscosity. Microbial activity continually reduces the nutrient content and alters the composition of the material as it passes along the colon. This, in turn, will dictate which organisms can survive and grow in a particular region, which affects the nature of the bacterial metabolic end-products, thereby altering the local environment.

The large intestine is often described as resembling a continuous culture system in which chyme enters at one end and faeces exits at the other. However, this description applies only to the caecum and ascending colon because these regions are regularly supplied with fresh substrate and the contents are contiguous. In contrast, throughout the rest of the large intestine, there is no further nutrient input and the contents are usually present as isolated masses – these regions are more like a batch or fed-batch system.

The pH varies along the length of the colon, with the pH of the caecum being approximately 5.7. This represents a sharp decrease from the pH found in the ileum and is attributed to the rapid bacterial fermentation of unabsorbed carbohydrates to short-chain fatty acids (SCFAs), which are weak acids. SCFAs (e.g., butyrate, acetate, and

Table 7.2. Nutrients present in the colon and their origins

Nutrient	Amount reaching colon (g per day)
dietary origin	
non-starch polysaccharide	8–18
starch	8–40
oligosaccharides	2–8
unabsorbed sugars	2–10
proteins and peptides	10–15
fats (as fatty acids and glycerol)	6–8
host origin	
digestive enzymes	5–8
bile acids	0.5–1.0
mucins	2–3
epithelial cells	20–30

Note: End-products of microbial metabolism constitute an additional source of a variety of nutrients.

propionate) constitute 75% of the anions present in the colon. The pH remains low in the ascending colon (mean pH 5.6) and transverse colon (mean pH 5.7), but then increases to 6.6 in the descending and sigmoid colons. This increase is due to a combination of SCFA absorption and bicarbonate secretion by the mucosa. The pH in the rectum ranges from 6.6 to 6.8.

The lumen of the colon is an anaerobic region and has a very low redox potential, with values in the range of -200 mV to -300 mV. However, because of oxygen supplied by underlying tissue, the mucosal surface has a high oxygen content relative to that found in the lumen. The partial pressure of oxygen on the lumenal side of the mucosa is approximately 30 mm Hg in the caecum, 39 mm Hg in the transverse colon, 29 mm Hg in the descending colon, and 39 mm Hg in the sigmoid colon. These values correspond to 19, 25, 18, and 25%, respectively, of the oxygen content of air.

A wide variety of nutrients is available to microbes inhabiting the colon, including those present in, or derived from, the individual's diet, host secretions, and cells shed from the intestinal mucosa (Table 7.2). Data shown in this table must be regarded as only approximations because the diet of the individual will have a marked effect on the quantity and composition of material entering the colon. In addition, nutrients will, of course, be available from microbial end-products of metabolism. A wide range of carbohydrates may be present, including unabsorbed monosaccharides (glucose, fructose), oligosaccharides, and polysaccharides, such as starch, cellulose, pectins, xylan, inulin, hemicelluloses, various gums, and mucopolysaccharides (e.g., hyaluronic acid, chondroitin sulphate). Oligosccharides and/or monosaccharides may also be liberated from host glycoproteins and glycosphingolipids by bacterial enzymes. A wide range of bacterial genera is able to degrade the various polysaccharides that may be present in the human colon (Table 7.3). However, some of the more complex molecules (e.g., mucins and glycosphingolipids) often require the concerted actions of a variety of organisms for their complete breakdown (Section 1.5.3). In fact, only three species – *Ruminococcus torques, Ruminococcus gnavus*, and a *Bifidobacterium* sp. – have been reported to be capable of completely degrading mucins on their own. The resulting monosaccharides

Table 7.3. Polysaccharide and glycoprotein degradation by representative colonic genera

Macromolecule	Bacterial genus				
	Bacteroides	*Bifidobacterium*	*Eubacterium*	*Ruminococcus*	*Clostridium*
pectin	+	+	+	−	+
cellulose	+	−	−	+	+
mucins	+	+	−	+	+
heparin	+	−	−	−	−
chondroitin sulphate	+	−	−	−	+
xylan	+	+	−	−	+
guar gum	+	−	−	+	•
amylose	+	+	+	+	+
amylopectin	+	+	+	+	+
arabinogalactan	+	+	−	−	•
galactomannans	+	−	−	+	•
hyaluronic acid	+	•	•	•	+

Notes: + = some species are able to degrade the polysaccharide; − = most strains are unable to degrade the polysaccharide; • = data not available.

and amino acids are then used by many colonic bacteria as carbon and energy sources. Many other organisms can contribute to mucin degradation by producing a sialidase (*Bacteroides* spp., *Bifidobacterium* spp., *Clostridium* spp., *Prevotella* spp., *Bacteroides* spp., *E. coli*, and *Ent. faecalis*) and/or glycosidases (*Bacteroides* spp., *Bifidobacterium* spp., *Clostridium* spp., *Lactobacillus* spp., *Prevotella* spp., *Bacteroides* spp., *E. coli*, and *Ent. faecalis*). The main end-products of metabolism are SCFAs, carbon dioxide, and hydrogen. In fact, SCFAs (mainly acetate, propionate, and butyrate) are the principal products of fermentation in the colon – between 300 and 400 mmoles are produced per day. As well as acting as nutrients for other colonic bacteria, the SCFAs produced by the microbial fermentation of carbohydrates and amino acids are absorbed by the colonic mucosa and are metabolised by the host to provide between 3% and 9% of his/her energy requirements (Section 9.3.1). Because of their utilization by resident bacteria, the carbohydrate concentration of the colonic contents decreases with distance from the caecum and becomes less significant as carbon and energy sources for resident microbes.

Undigested fats, and unabsorbed digestion products of fats, which reach the colon, appear not to be used by colonic microbes and are excreted in faeces.

The concentrations of urea, ammonia, and free amino acids are low in the effluent from the ileum; thus, the main sources of nitrogen for colonic bacteria are proteins and peptides. Proteins account for 48–51% of the nitrogen content of material reaching the colon, peptides account for 34–42%; while urea, ammonia, free amino acids, and nitrate together account for between 10% and 15%. However, as described later, large quantities of ammonia are produced by the microbial fermentation of amino acids released from peptides and proteins, and this constitutes an important nitrogen source for many colonic microbes. The proteins present in the colon include those from the diet, tissue proteins (e.g., collagen), serum albumin, antibodies, pancreatic enzymes, and those derived from exfoliated mucosal cells. Bacterial proteases, rather than host proteases, appear to be responsible for most of the protein degradation occurring in the colon. Many colonic organisms are proteolytic, including *Clostridium* spp.

(e.g., *Cl. perfringens, Cl. bifermentans*), *Bacteroides* spp. (e.g., *B. fragilis* group, *B. splanchnicus*), *Fusobacterium* spp., *Prevotella* spp., *Propionibacterium* spp., enterococci, and staphylococci. The resulting amino acids can serve as sources of nitrogen, carbon, and energy, and become increasingly important nutrient sources as carbohydrate levels in the lumenal contents are depleted during their passage along the colon. Many colonic bacteria are able to carry out amino-acid fermentation; these include *Clostridium* spp., *Eg. lenta*, *Fusobacterium* spp., *Peptostreptococcus* spp. (as well as former members of this genus), *Prev. melaninogenica, Acidaminococcus* spp., and *Peptococcus* spp. Although a diverse range of fermentation products are generated (e.g., ammonia, amines, indoles, organic acids, alcohols, hydrogen), the major products are SCFAs. The main SCFAs produced include acetate, propionate and butyrate, as well as branched chain fatty acids (BCFAs) – isobutyrate, 2-methylbutyrate, and isovalerate.

A number of vitamins are present in the colon and are derived both from the diet and from the colonic microbiota – particularly *Bacteroides* spp., *Bifidobacteria* spp., *Clostridium* spp., and enterobacteria. Vitamins produced by colonic bacteria include vitamin K, nicotinic acid, folate, pyridoxine, vitamin B_{12}, and thiamine (Section 9.3.2).

Although the environment within the large intestine gradually changes from the caecum to the anus, it is convenient to recognise three main environmental regions corresponding to distinct anatomical sites, and these are summarised in Table 7.4.

7.4 | The indigenous microbiota of the gastrointestinal tract

A major problem in determining the composition of the microbiota of the GIT is obtaining samples for analysis. Access to all regions of the GIT is difficult and is either very uncomfortable for the individual or requires some form of anaesthesia, which can itself affect the motility and secretory activity of the GIT. Samples may be obtained during operation, but these are unlikely to be representative because the patient would invariably have been fasting prior to the operation, would probably have been given prophylactic antibiotics, and would probably have been given an enema or purgative to empty his/her bowels. Post-mortem samples have been used, but a major problem with these is that a considerable period of time has usually elapsed between death and sampling – which results in cross-contamination between sites and enables the overgrowth of certain members of the microbiota of each site. Ideally, samples should be taken endoscopically because the patients are not usually anaesthetised, are not fasting, and are not given antibiotics. Furthermore, endoscopic systems have been designed which enable samples to be taken without becoming contaminated by saliva or nasal fluids during the insertion of the device. Another approach that has been used involves swallowing a capsule which is designed to open when it has reached a particular region of the GIT and then to close again to avoid contamination as it continues its passage through the GIT.

Analysis of the microbial communities colonising the human GIT by culture and molecular methodologies have revealed them to be extremely complex, with estimates of the number of different species present ranging from 500 to 1,000. The total number of microbial genes in these communities (known as the microbiome or metagenome) is between 2 million and 4 million (i.e., 70–140 times more than that of their host), which represents an enormous metabolic potential which is far greater than that possessed

Table 7.4. Environmental features of and microbial activities in the three main regions of the colon

Environment	Main characteristics
caecum/ascending colon	contents have a high water content; acidic pH (≤ 5.7); easily fermentable substrates; very active fermentation; high bacterial growth rates; plentiful supply of carbohydrates; large quantities of SCFAs produced; main gases produced are H_2 and CO_2
transverse colon	contents are less fluid; contents move more slowly than in ascending colon; less acidic; reduced content of easily fermentable substrates; bacterial growth rates slower; smaller quantities of SCFAs produced
descending colon/sigmoid-rectum	contents are less fluid; contents move more slowly than in transverse colon; pH approaches neutrality; smaller quantities of SCFAs produced; concentrations of protein fermentation products increase; bacterial growth rates are low; main gases produced are H_2, CO_2, and CH_4; amines, phenols, and ammonia produced

Note: SCFAs = short-chain fatty acids.

by their host. Co-evolution of humans and the communities inhabiting the GIT has resulted in an excellent example of a symbiosis. In the upper regions of the GIT (the stomach, duodenum, and jejunum), the host provides an environment which is not conducive to extensive colonisation, and it is in these regions that the host extracts many of the nutrients from the constituents of its diet. In the terminal ileum and the colon, however, the environment is suitable for the establishment of large and diverse microbiotas. Within these regions, the host takes advantage of the enormous metabolic potential of its indigenous microbiota to degrade dietary constituents that it is itself unable to digest. Many of the end-products of these microbial metabolic processes are then absorbed by the colon, along with vitamins synthesised by some microbes. The microbes, in turn, are supplied with an environment suitable for their growth and a constant supply of nutrients from the host's diet. As well as supplying nutrients for their host, the indigenous microbiota also plays a role in the development of the intestinal mucosa and immune system of the host (Section 9.2).

As described in Section 7.3, the environment of the GIT varies considerably along its length, resulting in distinct microbial communities in each of the main regions. In addition, it is important to appreciate that there is also a horizontal stratification at any point within the tract (Figure 7.9). Hence, the lumen, the mucus layer, and the epithelial surface all offer different environments and may support different microbial communities. In some regions of the GIT, the crypts at the bases of villi, as well as various glands, would also provide a different set of environmental

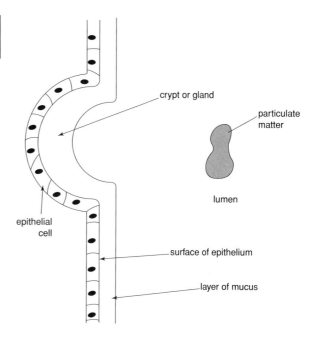

Figure 7.9 The various microhabitats available for microbial colonisation in the gastrointestinal tract.

crypt or gland

particulate matter

lumen

epithelial cell

surface of epithelium

layer of mucus

conditions and may harbour a distinct microbiota. Within the lumen, particulate matter provides a substratum for biofilm formation and enables the establishment of communities whose composition is likely to differ from that present in the fluid phase of the lumenal contents. However, because of the difficulties in obtaining samples from these microhabitats that are not contaminated by microbes from other sites, few studies have been able to provide information regarding the composition of these communities.

7.4.1 Main characteristics of key members of the intestinal microbiota

The organisms most frequently isolated from the GIT of humans belong to the genera *Bacteroides, Eubacterium* (and related genera), *Clostridium, Bifidobacterium, Streptococcus, Lactobacillus, Peptostreptococcus* (and related genera), *Peptococcus, Ruminococcus, Fusobacterium, Veillonella, Enterococcus, Propionibacterium, Actinomyces, Methanobrevibacter, Desulphovibrio, Helicobacter, Porphyromonas, Prevotella, Escherichia, Enterobacter, Citrobacter, Serratia, Candida, Gemella,* and *Proteus*. Of these, the genera *Streptococcus, Lactobacillus, Propionibacterium, Peptostreptococcus, Candida, Gemella,* and *Peptococcus* were described in previous chapters, while the genera *Fusobacterium, Veillonella, Porphyromonas, Prevotella,* and *Actinomyces* are described in Section 8.4.1. The main features of the remaining genera are outlined in this section.

7.4.1.1 *Bacteroides* spp.

These are non-motile, non-sporing, anaerobic Gram-negative pleomorphic rods. Although they are anaerobes, they are among the most aerotolerant of all anaerobic species. The G+C content of their DNA is 40–48 mol%. They have simple nutritional requirements, although most require haemin and vitamin B_{12} for growth. NH_4^+ is the primary source of nitrogen for *Bacteroides* spp. Most species can utilise a variety of carbohydrates as energy and carbon sources, and can degrade a wide range of polysaccharides, including starch, cellulose, xylan, pectin, dextran, arabinogalactan, guar gum, laminarin, chondroitin sulphate, glycosaminoglycans, malto-oligosaccharides, gum arabic, hyaluronic acid, and heparin. Mucins can also be degraded, and some species are proteolytic and can hydrolyse fibrinogen, casein, trypsin, chymotrypsin, transferrin, and ovalbumin. The major end-products of carbohydrate metabolism are succinate, propionate, and acetate. Lactate, formate, and malate are also produced, but in smaller quantities.

One of the most abundant organisms in the lower regions of the GIT is *B. thetaiotaomicron*, and recently its 6.26 Mb genome has been sequenced – one of the largest bacterial genomes sequenced so far. Of the 4,779 predicted proteins in its proteome, 58% showed homology to other known proteins and could be assigned putative functions, 18% were homologous to proteins with no known function, and 24% had little homology to any of the proteins in databases. A large proportion of the proteome appears to be devoted to the acquisition, degradation, and utilisation of complex carbohydrates. Hence, many (163) of the outer membrane proteins encoded by the genome are likely to be involved in the acquisition of a variety of carbohydrates; and twenty putative, sugar-specific transporters and twenty permease subunits of ATP-binding cassette transporters have been identified. Remarkably, a large number (172) and variety (twenty-three different activities) of glycosylhydrolases are present – more than that of any other bacterium sequenced. This shows an adaptation to its role in the degradation of carbohydrates that the host cannot digest. Most (61%) of these glycosylhydrolases are predicted to be in the periplasm, in the outer membrane, or are extracellular – which implies that their degradation products could also be used either by the host or by other members of the intestinal community. Many of the predicted proteins are enzymes which enable the degradation of host-derived polymers, including mucin, chondroitin sulphate, hyaluronate, and heparin. Another remarkable feature of the organism's proteome is the large number of one- and two-component signal transduction systems, which suggests that the organism is well adapted to sense and respond to changes in its environment. A number of mobilisable genetic elements are also present, including a plasmid, sixty-three transposases, and four conjugative transposons. This suggests that the organism is able to engage in the horizontal transfer of genes encoding antibiotic resistance, virulence factors, and so forth.

Bacteroides spp. have a high pathogenic potential and account for approximately two-thirds of all anaerobes isolated from clinical specimens – the most frequently isolated species being *B. fragilis*.

7.4.1.2 *Eubacterium* spp.

Until recently, this was a heterogeneous group of organisms which included anaerobic Gram-positive rods whose taxonomic position was uncertain. Two main groups were recognized – saccharolytic and asaccharolytic. Most of the latter are currently retained within the genus *Eubacterium*, while many of the former have been assigned to new genera. Hence, two species regularly present in the GIT of humans, *Eub. lentum* and *Eub. aerofaciens*, have been re-named *Eggerthella lenta* and *Collinsella aerofaciens*, respectively, and are often referred to as "*Eubacterium*-like organisms". Species currently recognised as belonging to the genus *Eubacterium*, which are present in the GIT, include *Eub. biforme*, *Eub. contortum*, *Eub. rectale*, *Eub. cylindroides*, *Eub. hadrum*, *Eub. ventriosum*, *Eub. barkeri*, *Eub.*

limosum, and *Eub. moniliforme*. All of these ferment sugars to produce a mixture of fatty acids. *Eubacterium* spp. and *Eubacterium*-like organisms are anaerobic, non-motile, non-sporing Gram-positive bacilli. They are not very aerotolerant. Some intestinal species can hydrolyse starch and pectin. Most of these organisms produce one or more fatty acids from glucose, but *Eg. lenta* does not. *Eg. lenta* can use amino acids as an energy source and produces a mixture of lactate, acetate, formate, and succinate. *Coll. aerofaciens* ferments sugars and produces formate, lactate, and hydrogen.

7.4.1.3 *Clostridium* spp.

These are obligately anaerobic, spore-forming, Gram-positive rods. Most species are motile – important exceptions include *Cl. perfringens, Cl. ramosum*, and *Cl. innocuum*. The genus is one of the largest among the prokaryotes and contains more than 100 species. Most species are fermentative and/or proteolytic, but some are asaccharolytic and non-proteolytic. Many produce a number of SCFAs (e.g., acetate, butyrate) when grown in carbohydrate-containing media, as well as a variety of other fermentation products, such as acetone and butanol. Species regularly isolated from the human GIT include *Cl. perfringens, Cl. ramosum, Cl. innocuum, Cl. paraputrificum, Cl. sporogenes, Cl. tertium, Cl. bifermentans*, and *Cl. butyricum*. Of these, *Cl. perfringens* is the most frequent cause of gas gangrene – a life-threatening infection (Section 7.5.4). *Cl. bifermentans* (and possibly *Cl. tertium* and *Cl. sporogenes*) is also able to cause gas gangrene. A number of the clostridia previously mentioned may also be involved in polymicrobial infections, such as peritonitis, intra-abdominal abscesses, and septicaemia.

Cl. perfringens is proteolytic, is able to hydrolyse various polysaccharides and glycoproteins (including starch, mucin, and hyaluronic acid), and can ferment a range of sugars producing mainly acetate and butyrate. It produces numerous extracellular toxins, including α-toxin (phospholipase C), θ-toxin (perfringolysin O), and κ-toxin (collagenase), as well as an enterotoxin.

7.4.1.4 *Bifidobacterium* spp.

These are non-sporing, non-motile, pleomorphic Gram-positive rods which occur singly, in chains, or in clumps. They are generally obligate anaerobes, although some can grow in carbon-dioxide-enriched air. The G+C content of their DNA varies from 45 to 67 mol%, with most species being within the range of 55–67 mol%. Thirty-two species are recognised, most of which have been detected in the human GIT, and many also occur in the vagina and oral cavity. They all produce acid from glucose, the end-products being acetate and lactate in the molar ratio 2:3. They are able to ferment a wide range of sugars and can hydrolyse a variety of polysaccharides, including starch, xylan, pectin, inulin, gum arabic, and dextran. They can also hydrolyse proteins and peptides, and produce a number of enzymes important in the degradation of mucins, including sialidases, α- and β-glycosidases, α- and β-D-glucosidases, α- and β-D-galactosidase, and β-D-fucosidase. They are able to grow over the temperature range of 20°–49.5°C, but their optimum growth temperature is 37°–41°C. Their pH range for growth is 4.0–8.5, but growth is optimal between pH 6.5 and pH 7.0. They are acid-tolerant, but not acidophilic. All of the species inhabiting humans can utilise NH_4^+ as the sole source of nitrogen, and many excrete large quantities of a variety of amino acids. They also produce a number of vitamins, including thiamine, folic acid, nicotinic acid, pyridoxine, cyanocobalamin, and biotin. Growth is stimulated by a number of oligosaccharides, protein hydrolysates, and glycoproteins, and these are often termed "bifidogenic factors". Bifidobacteria can inhibit the growth of a number of microbes by virtue of the low pH generated by the production of acetate and lactate. Some species also produce a bacteriocin, bifidocin B, as well as other unidentified antimicrobial compounds.

The most commonly occurring species in the GIT are *Bif. pseudocatenulatum*, *Bif. longum*, *Bif. angulatum*, *Bif. adolescentis*, *Bif. catenulatum*, *Bif. bifidum*, *Bif. gallicum*, *Bif. infantis*, and *Bif. breve*. Species usually present in the vagina include *Bif. breve* and *Bif. adolescentis*, as well as, to a lesser extent, *Bif. longum* and *Bif. bifidum*. *Bif. denticolens*, *Bif. adolescentis*, *Bif. inopinatum*, and *Bif. dentium* are the most frequently isolated species from the oral cavity.

There is considerable interest in the use of bifidobacteria as probiotics; thus, a number of studies have been carried out to assess their antagonistic effects on other organisms. They have been shown to (1) prevent the attachment of enteropathogens to intestinal epithelial cells; (2) inhibit the growth of intestinal organisms by producing acidic metabolic end-products; (3) produce a broad-spectrum antimicrobial compound that can inhibit the growth of *B. fragilis*, *Cl. perfringens*, and enteropathogens, such as *V. cholerae*, *Lis. monocytogenes*, *Sh. sonnei*, *Campylobacter spp.*, *Salmonella* spp., and *E. coli*; (4) produce a low-molecular mass (3,500 Da), lipophilic compound able to kill *Sal. typhimurium*, *Lis. monocytogenes*, *Y. pseudotuberculosis*, *Staph. aureus*, *Ps. aeruginosa*, *K. pneumoniae*, and *E. coli* – the compound also protected mice against lethal infection by *Sal. typhimurium*; and (5) produce a bacteriocin, bifidocin B, which has a broad antimicrobial spectrum inhibiting the growth of species of *Listeria*, *Enterococcus*, *Bacillus*, *Lactobacillus*, *Leuconostoc*, and *Pediococcus*.

The genome of *Bif. longum* has recently been sequenced, and analysis has revealed many features showing its adaptation to the environment of the colon. These include (1) homologues of enzymes needed to ferment a wide range of sugars; (2) homologues of enzymes for the fermentation of amino acids; (3) more than twenty predicted peptidases; (4) numerous predicted proteins for carbohydrate transport and metabolism, including more than forty glycosyl hydrolases; (5) enzymes with predicted activities which include xylanases, arabinosidases, galactosidases, neopullanase, isomaltase, maltase, and inulinase; and (6) three α-mannosidases and an endo-NAc glucosaminidase. These findings show that the organism has the potential to hydrolyse and utilise the degradation products from a range of complex plant polymers which survive digestion in the small intestine and also host glycoproteins. *Bifidobacterium* spp. are very rarely implicated in human infections.

7.4.1.5 *Enterococcus* spp.

These are facultatively anaerobic, non-sporing, catalase-negative Gram-positive cocci that occur singly, in pairs, or in chains. The genus consists of seventeen species, but only two of these – *Ent. faecalis* and *Ent. faecium* – are regularly found in the human GIT (the former is more frequently found and in greater proportions). They ferment sugars to produce mainly lactate. Some species (including *Ent. faecalis*) can hydrolyse proteins, including collagen, casein, insulin, haemoglobin, fibrinogen and gelatin, and a number of peptides. They can also hydrolyse lipids and hyaluronic acid. They have complex nutritional needs and require a number of vitamins and amino acids. Enterococci are very hardy organisms and can tolerate a wide variety of growth conditions, including temperatures of 10°–45°C, as well as hypotonic, hypertonic, acidic, or alkaline environments. They are also resistant to bile salts, dessication, detergents, and many antimicrobial agents. Optimum growth occurs between 35°C and 37°C.

Ent. faecalis and, to a lesser extent, *Ent. faecium* can cause infections in humans, particularly in hospitalised individuals, and many of these are very difficult to treat because of the resistance of many strains to a wide variety of antibiotics. *Ent. faecalis* is responsible for 85–90% of enterococcal infections, and these mainly involve the urinary tract, wounds, and the bilary tract. It can also invade the bloodstream and cause meningitis and endocarditis. It produces a cytolysin, which is able to disrupt a variety of membranes, including those belonging to bacterial cells, erythrocytes, and other mammalian cells. It also produces proteases (a gelatinase and a serine protease), which can hydrolyse a number of proteins, including collagen and fibrinogen, thereby damaging host tissues. It can also degrade hyaluronic acid. A polysaccharide capsule protects the organism against phagocytosis. Most strains of *Ent. faecalis* and a few of *Ent. faecium* produce substantial amounts of extracellular superoxide which may cause damage to host cells. *Ent. faecalis* has a number of adhesins, including aggregation substance which not only promotes aggregation of the organism, but also mediates adhesion to epithelial cells; Ace (i.e., adhesin of collagen from enterococci) – an adhesin which enables binding of the organism to collagen and laminin, and a surface protein (Esp) which is involved in adhesion of the organism to epithelial cells of the urinary tract.

Recently, the genome of a vancomycin-resistant strain of *Ent. faecalis* has been sequenced. The chromosome of this organism, together with its three plasmids, has a total of 3,337 protein-encoding genes. A remarkable feature of the organism's genome is that more than one-quarter of it consists of mobile and/or exogenously acquired DNA – this is the highest proportion of mobile elements detected in any bacterium sequenced so far. These include thirty-eight insertion elements, seven regions probably derived from integrated phages, multiple conjugative and composite transposons, a putative pathogenicity island, and integrated plasmid genes. The genome of *Ent. faecalis*, therefore, is highly malleable and is likely to have undergone multiple rearrangement events. Many putative sugar-uptake systems are present, as are pathways for the metabolism of more than fifteen different sugars, thereby demonstrating its adaptation to the utilisation of carbohydrates in the GIT. It has a variety of cation homeostasis mechanisms which probably contribute to its ability to survive extreme pHs, dessication, and high salt and metal concentrations. A remarkable number (134) of putative surface proteins which may function as adhesins are present, and forty-seven of these may be involved in adhesion of the organism to choline- or integrin-containing substrata.

7.4.1.6 *Helicobacter pylori*

Hel. pylori is a microaerophilic, Gram-negative, motile, non-sporing, curved rod. The G+C content of its DNA is 36–38 mol%. It grows best in an atmosphere with reduced oxygen content (5–10%) and an elevated concentration of carbon dioxide (5–12%) and hydrogen (5–10%). Its optimum temperature for growth is 37°C, and its optimum pH is 7.0, although it will grow over the range of 6.0–8.0. It is unable to grow at a pH lower than 4.0 or higher than 8.2. It is catalase-, oxidase-, and urease-positive – properties which, along with its characteristic morphology, enable its presumptive

Table 7.5. | Virulence factors of *Helicobacter pylori*

Virulence factor	Activity
BabA (outer membrane protein)	adhesin, binds to Lewis b blood group antigen on gastric epithelial cells
Lewis x and Lewis y antigens on LPS	adhesins, receptors unknown
HpaA (a haemagglutinin)	adhesin, binds to sialic acid-containing components of epithelial cells
CagA	induces cytoskeletal changes following its translocation into epithelial cells; induces inflammation
CagE	induces IL-8 release from epithelial cells
VacA (vacuolating cytotoxin)	induces vacuolation and apoptosis in epithelial cells
OipA	induces IL-8 secretion by epithelial cells
NAP (neutrophil-activating protein)	stimulates phagocyte chemotaxis and release of reactive oxygen species which damage epithelial cells
LPS	contains Lewis x and Lewis y antigens (also expressed on gastric epithelium); involved in molecular mimicry, thereby evading host immune response
Type IV secretion system	transport of CagA into epithelial cells
urease	cytoplasmic and surface-associated; liberates NH_3 from urea, thereby raising local pH enabling survival
cecropins	antibacterial peptides that may kill competing organisms
phospholipase A and C	degrades epithelial cell membrane

Notes: IL = interleukin; LPS = lipopolysaccharide.

identification. During adverse conditions, the organism can revert to a coccoid form – this can occur as a result of nutritional deprivation, increased oxygen tension, alkaline pH, increased temperature, exposure to antibiotics, and aging. These coccoid structures remain viable for up to 4 weeks and can then revert to the normal bacillary form. It is possible that the coccoid form may enable survival of the organism in the environment and facilitate its transmission in water or via fomites.

The main habitat of *Hel. pylori* is the mucus layer overlying the gastric epithelium, as well as the epithelial surface itself. It reaches these sites by burrowing into and through the mucus gel powered by its monopolar flagella, and it has been suggested that its curved morphology and ability to move in a helical manner facilitate gel penetration. Within these habitats, it creates a microenvironment with a neutral pH by converting the urea present in gastric juice to ammonia by means of its surface-associated urease. *Hel. pylori* expresses urease at a level higher than that of any other known microbe, and this enzyme accounts for approximately 20% of its total cell protein. Ammonia produced by the organism is also thought to provide a protective "shell", enabling the organism to move through the low pH gastric fluid until it reaches the mucus layer. The organism possesses a number of adhesins that mediate its binding to the gastric epithelium, and these include BabA and LPS (Table 7.5). It produces a range of virulence factors, and these are listed in Table 7.5. Many of these are located on the organism's *cag* (i.e., cytotoxin-associated gene) pathogenicity island, which contains a total of thirty-one genes.

DNA fingerprinting of strains isolated from different individuals has shown that each strain is unique, unless the individuals are linked epidemiologically. This population structure is similar to that found in organisms such as *N. meningitidis* and is charactersitic of organisms that are naturally competent for genetic transformation. Consequently, not all strains of *Hel. pylori* possess all of the virulence factors listed in Table 7.5, and it is possible to use the possession of some of these virulence factors as markers of the disease-inducing ability of a particular strain. Hence, strains that are *cagA*$^+$ or have the type s1 *vacA* signal sequence are associated with an increased risk of severe gastritis, peptic ulcers, and gastric cancer (Section 7.5.1). Strains with the *iceA1* gene which is upregulated on contact of the organism with gastric epithelial cells or those expressing the Lewis b antigen, are associated with an increased risk of peptic ulcers.

7.4.1.7 *Enterobacteriaceae*

This is a large family comprising approximately 30 genera and more than 150 species, many of which are normal inhabitants of the human GIT. They are non-sporing, facultatively anaerobic Gram-negative bacilli that ferment glucose to produce acid and reduce nitrate to nitrite – most species are motile. They contain an antigenic polysaccharide known as the "enterobacterial common antigen". The G+C content of their DNA is 38–60 mol%. They grow well between 25° and 37°C. The genera most frequently present in the human GIT and in the greatest proportions are *Escherichia, Proteus, Citrobacter*, and *Enterobacter*. Space limitations preclude a description of all of the many species encountered in the human GIT and only *E. coli*, an important human pathogen, is described further.

E. coli is usually motile and ferments glucose and many other sugars to produce mainly acetic, lactic, and succinic acids, together with carbon dioxide and hydrogen. The classification system used for distinguishing between different strains of the organism is based on the type of O antigen (i.e., lipopolysaccharide [LPS]), K antigen (i.e., capsule), and H antigen (i.e., flagellum) possessed by the strain. Currently, more than 170 O antigens, 80 K antigens, and 56 H antigens have been recognised. A sub-set of clones of *E. coli* resident in the colon are particularly able to cause UTIs, and these are described as being uropathogenic *E. coli* (UPEC). In such organisms, the genes encoding virulence factors involved in the pathogenesis of UTIs are usually grouped together on regions of the chromosome known as pathogenicity islands. The expression of such genes is usually co-regulated, often in response to environmental factors. Although the success of UPEC as pathogens of the urinary tract is attributable to a number of such virulence factors, the most important of these are probably the adhesins, which enable colonisation of the uroepithelium despite the mechanical flushing action of urinary flow. A variety of adhesins and adhesive structures have been detected in UPEC strains, and these are summarised in Table 7.6. However, a particular strain may possess several or none of these fimbriae. Of the various fimbriae detected, the type I pili are the most prevalent among UPEC strains and carry the FimH adhesin, which mediates binding to host cells and matrix molecules via mannose-containing glycoproteins. The main receptor on bladder epithelial cells is the membrane glycoprotein uroplakin 1a. Interestingly, the FimH adhesin also mediates auto-aggregation and biofilm formation,

Table 7.6. Adhesins of uropathogenic strains of *E. coli*

Adhesin	Adhesive structure	Host receptor	Target cells/structures
FimH	type I pili	mannosylated glycoproteins (e.g., uroplakin 1a and CD48); Tamm-Horsfall protein; collagen; laminin; fibronectin	uroepithelial cells, PMNs, macrophages, erythrocytes, extracellular matrix, other bacteria
PapG	P pili	globotriasylceramide residues	kidney epithelial cells, erythrocytes
SfaS, SfaA	S/F1C pili	sialic-acid residues, plasminogen β-GalNac-1, 4-Gal	uroepithelial cells, endothelial cells, erythrocytes
Dr adhesin Family	fimbriae and cell surface	DAF (CD55), CD66e, type IV collagen	uroepithelial cells, PMNs, erythrocytes

Note: PMNs = polymorphonuclear leukocytes.

and can function as an invasin, thereby enabling invasion of the bladder epithelium. The adhesin of P pili is papG, which recognises globotriasylceramide residues in kidney cells and red blood cells, and is an important virulence factor in strains causing pyelonephritis. S/F1C pili have two major adhesins: SfaS and SfaA. SfaS recognises sialic-acid residues on kidney epithelial cells and endothelial cells, while SfaA mediates binding to glycolipids in endothelial cells and also to plasminogen. Strains with these pili are associated with cystitis, pyelonephritis, sepsis, and meningitis. The Dr family of adhesins include the Dr adhesin, which is located on fimbriae and a number of others that are not. These adhesins recognise certain amino-acid sequences in decay accelerator factor (CD55), which is expressed by a variety of tissues and by erythrocytes. They also mediate binding to cells expressing carcinoembryonic antigen (CD66e), and some can recognise type IV collagen. UPEC strains with Dr family adhesins appear well-adapted to persistent colonisation of the bladder and are associated with recurrent UTIs. Dr adhesins can also function as invasins, enabling entry into a number of cell types, and this may be associated with the ability of strains expressing these invasins to persist in tissues. Fimbriation in *E. coli* undergoes phase variation, which is likely to be important in undermining host defence mechanisms.

As well as adhesins and invasins, other virulence factors produced by UPEC include a capsule, LPS, three haemolysins (α-, β-, and γ-), cyotoxic necrotising factor 1 (cnf1), secreted autotransporter toxin (sat), and the siderophores aerobactin and enterochelin. The capsule is antiphagocytic and also protects the organism from lysis by human serum, while the LPS is able to stimulate the release of inflammatory cytokines. Cnf1 induces apoptosis in bladder epithelial cells and decreases the phagocytic ability of polymorphonuclear leukocytes (PMNs). Sat is a serine protease and is cytopathic to uroepithelial cells. The haemolysins are important in liberating iron-containing compounds from erythrocytes and may also behave as general cytotoxins. The α-haemolysin also decreases neutrophil chemotaxis and phagocytosis. The siderophores are important in iron acquisition in the low-iron environment within the urinary tract. UTIs caused by UPEC are described in Section 7.5.7.

E. coli is also an important cause of meningitis in neonates. A variety of entero-pathogenic strains of *E. coli* also exist but because they are not members of the indigenous microbiota of the GIT, they are not described further.

7.4.1.8 *Ruminococcus* spp.

This genus consists of non-sporing, non-motile, anaerobic Gram-positive cocci. The G+C content of their DNA is 37–48 mol%. Thirteen species are recognised and all use carbohydrates as an energy source and produce acetate – some also produce other fatty acids, such as formate, lactate, and succinate. They utilise NH_4^+ as a nitrogen source. Species frequently isolated from the human GIT include *Rum. obeum, Rum. torques, Rum. flavefaciens, Rum. gnavus*, and *Rum. bromii*. Many ruminococci produce sialidases and/or glycosidases and/or proteases, and consequently are able to partially degrade mucins and other glycoproteins. Although few bacterial species are able to produce the full complement of enzymes necessary to degrade mucins entirely, two such species are *Rum. torques* and *Rum. gnavus*. Ruminococci rarely cause infections in humans.

7.4.1.9 Methanogenic bacteria

These are organisms that produce large quantities of methane as a byproduct of their energy-generating reactions. Two main genera are found in the human colon: *Methanobrevibacter* and *Methanosphaera*. Both of these belong to the domain *Archea* and are only distantly related to the domain *Bacteria*.

Methanobrevibacter spp. are non-motile, anaerobic, Gram-positive short rods. The most common species isolated from the human colon is *Methanobrevibacter smithii*, which has a G+C content of 29–31 mol%, and grows best at a pH of 7.0 and at a temperature of 37°–39°C. It is nutritionally fastidious and requires acetate, amino acids, and B vitamins for growth. The organism obtains its energy from the oxidation of hydrogen using carbon dioxide as the electron acceptor – the latter being reduced to methane.

Methanosphaera spp. are anaerobic, Gram-positive cocci that occur singly, in tetrads, or in clusters. The species most frequently isolated from the human colon is *Methanosphaera stadtmaniae*, which has a G+C content of 26 mol%, and grows best at a temperature of 37°C and at a pH of 6.5–6.9. It is nutritionally fastidious and requires acetate, several amino acids, thiamin, and biotin for growth. Although hydrogen is used as an energy source, the electron acceptor is methanol, which is reduced to methane.

Hydrogen is produced by colonic bacteria during the fermentation of carbohydrates, and its utilisation by methanogens is an important means by which the gas is disposed of in this environment.

7.4.1.10 *Desulphovibrio* spp.

This genus consists of motile, anaerobic, Gram-negative curved rods with a G+C content of 46–61 mol%. They grow best at a temperature of 30°–38°C. Members of the genus utilise sulphate as an energy source and hydrogen, lactate, and ethanol as electron donors. This results in the generation of hydrogen sulphide. The main source of sulphate in the colon is mucin, from which it is liberated by the sulphatases produced by organisms such as *Bacteroides* spp. *Desulphovibrio* spp. constitute the major sulphate-reducing organisms in the human colon, and frequently encountered species include *Des. desulphuricans, Des. fairfieldensis*, and *Des. vulgaris*.

Table 7.7. | Presence of bacterial groups in faeces of breast-fed and formula-fed infants during the first week of life

	Breast-fed		Formula-fed	
Bacterial group	First isolated (days)	Highest count (days)	First isolated (days)	Highest count (days)
Bifidobacterium spp.	2.0 ± 1.3	6.0 ± 1.2	1.0 ± 0	5.2 ± 1.5
Bacteroides spp.	1.8 ± 1.0	4.9 ± 1.7	1.0 ± 1.0	5.0 ± 1.0
Enterobacteria	1.1 ± 0.3	3.3 ± 1.7	1.0 ± 1.0	3.2 ± 1.6
Streptococcus spp.	1.4 ± 0.8	3.3 ± 1.0	1.5 ± 0.7	5.3 ± 0.6

Notes: Data (mean ± standard deviation) are based on 12 independent studies. Reproduced with permission from: Development of intestinal microbiota. Conway, P. In: *Gastrointestinal microbiology*, Vol. 2. Mackie, R.I., White, B.A., and Isaacson, R.E. (eds.). Copyright © 1996, Kluwer Academic/Plenum Publishers. First isolated = the day on which a member of the group was first isolated. Highest count = the day on which the highest count of that group was noted.

7.4.2 Acquisition of the intestinal microbiota

A variety of microbes can be cultured from the stomach of neonates within 10 minutes of delivery, and the species found are mainly typical members of the cervical and vaginal microbiota. Organisms from the mother's GIT – such as *Bacteroides* spp. and *E. coli* – may also be present. In a study of thirty-five neonates, a mean of 3.4 organisms were cultured from gastric aspirates, with 40% of these being obligate anaerobes. The most frequently isolated species were *Bacteroides* spp., *Staph. epidermidis*, coryneforms, viridans streptococci, *Propionibacterium* spp., *Lactobacillus* spp., *Bacillus* spp., *Micrococcus* spp., *Peptococcus* spp., and *E. coli*. These organisms are acquired by ingestion during delivery, and many are merely transients and do not become established as members of the intestinal microbiota. Other important sources of potential colonisers of the neonate's GIT include the skin and saliva of the mother (and other individuals) and the environment. Although babies born by caesarean section are also colonised by maternally derived microbes, they are at a greater risk of colonisation by environmental organisms and by organisms from hospital staff and equipment. This represents a potential hazard to the neonate because microbes from hospital staff and equipment are frequently resistant to one or more antibiotics.

During the first few days following birth, the neonate comes into contact with a wide variety of microbes from different sites on its mother and other individuals, and from the different environments to which it may be exposed. Consequently, a diverse range of species can be isolated from its GIT, especially during the first 24 hours. However, within the first week of life, a distinct pattern of colonisation emerges, and this is profoundly affected by whether the neonate has been fed with breast milk or a milk-based formula (Table 7.7). In both breast- and formula-fed infants, the GIT is initially colonised by streptococci and enterobacteria, and these create the anaerobic conditions and low redox potential necessary for the establishment of the anaerobic *Bacteroides* spp. and *Bifidobacterium* spp. The redox potential of the meconium is approximately +175 mV, but within 1–2 days, the faeces becomes more reduced (Eh = −113 mV). Only after weaning does the redox potential reach values similar to those found in adults. Table 7.8 lists the major groups of bacteria that have been found in the faeces of neonates. However, Table 7.8 obscures differences between breast- and

Table 7.8. Bacterial composition of faeces in neonates (less than 1 week old)

Bacterial group	Viable count ($\log_{10}$ cfu/g faeces)
Bacteroides spp.	≥ 9
Bifidobacterium spp.	≥ 9
enterobacteria	6–8
enterococci	6–8
Propionibacterium spp.	6–8
Streptococcus spp.	6–8
Eubacterium spp.	6–8
Fusobacterium spp.	6–8
Peptostreptococcus spp.	6–8
Veillonella spp.	6–8
Bacillus spp.	≤ 5
corynebacteria	≤ 5
Lactobacillus spp.	≤ 5
Micrococcus spp.	≤ 5
Staphylococcus spp.	≤ 5
Clostridium spp.	≤ 5
Ruminococcus spp.	≤ 5

Note: Data are summarised from the results of 15 studies.

formula-fed neonates, and investigations have revealed differences between these two groups (Table 7.9). After 1 week, *Bifidobacterium* spp. dominate the faecal microbiota of breast-fed infants, with much lower proportions of other genera. In contrast, the faecal microbiota of formula-fed infants is more complex, with high proportions of *Bifidobacterium* spp., *Bacteroides* spp., enterobacteria, and *Streptococcus* spp., with no particular group being dominant. Another notable difference is that formula-fed infants have much higher counts of *Clostridium* spp. than breast-fed infants. In older infants, a number of differences between breast- and formula-fed groups are also apparent. Hence, the counts of clostridia and bacteroides are much higher in the faeces of formula-fed than breast-fed infants, whereas the counts of bifidobacteria are similar. Another notable difference is that *Bacteroides* spp. and *Streptococcus* spp. comprise a much greater proportion of the faecal microbiota of formula-fed than breast-fed infants. That such

Table 7.9. Microbial composition of faeces from breast-fed and formula-fed infants

Bacterial group	Viable count ($\log_{10}$ cfu/g; mean $\pm$ standard deviation)			
	1 week		1–19 weeks	
	BF	FF	BF	FF
Bifidobacterium spp.	9.0 ± 1.1	7.8 ± 1.8	9.8 ± 2.9	9.7 ± 0.9
Bacteroides spp.	7.3 ± 1.6	7.4 ± 1.0	7.5 ± 1.8	9.0 ± 0.8
Clostridium spp.	3.5 ± 3.3	5.1 ± 1.5	4.9 ± 3.0	6.6 ± 0.8
Enterobacteria	7.5 ± 2.0	8.3 ± 1.1	8.1 ± 0.8	8.7 ± 0.9
Lactobacillus spp.	ND	ND	7.2 ± 0.7	7.1 ± 0.7
Streptococcus spp.	7.1 ± 0.7	7.7 ± 0.7	7.3 ± 0.9	8.5 ± 0.9

Notes: Data are obtained from eight (1 week) and seventeen (1–19 weeks) independent studies. BF = breast-fed; FF = formula-fed; ND = not determined.

differences exist between the faecal microbiota of these two groups is not surprising when the composition of breast and formula milk are compared. Although manufacturers have endeavoured to make the composition of formula milk as close to that of human milk as possible, profound differences between the two exist. Hence, unlike formula milk, breast milk contains a number of hormones that are important in promoting gut maturation, as well as a range of effector molecules of the innate and acquired immune systems – these include antibodies, lysozyme, lactoperoxidase, lactoferrin, and cytokines (IL-6, IL-10, tumour necrosis factor-α). The principal antibody in human milk is secretory IgA, and it has been estimated that a breast-fed infant receives between 0.5 and 1.0 g of this immunoglobulin per day. Another important difference is the relative buffering capacity of the two feeds. Breast milk has a poor buffering capacity, compared with formula milk, and this leads to marked differences in the pH of the colon of breast- and formula-fed infants – 5.1 and 6.5, respectively. This low pH promotes the growth of *Bifidobacterium* spp. and lactobacilli, but is inhibitory to many other bacteria. Furthermore, a number of peptides capable of stimulating the growth of several *Bifidobacterium* spp. have recently been isolated from human milk. Another factor that could contribute to the dominance of bifidobacteria in the faeces of breast-fed infants is the presence in human milk of glycoproteins, glycolipids, fucose, neuraminic acid, lactose, *N*-acetylglucosamine, and a variety of oligosaccharides based on lactose. These compounds are not assimilated to a great extent in the infant small intestine, and their presence in the colon could stimulate the growth of bifidobacteria, which can use these compounds as carbon and energy sources.

The breast milk of healthy mothers may contain as many as 10^6 viable bacteria/ml – these are derived predominantly from the nipples and surrounding skin. The organisms most frequently isolated from breast milk are coagulase-negative staphylococci (CNS), but streptococci, corynebacteria, lactobacilli, micrococci, propionibacteria, and bifidobacteria are also often present. It is reasonable to assume that this constant "inoculation" of the GIT would have an effect on the GIT microbiota, especially as acid secretion by the stomach of neonates is not fully developed for the first 2 to 3 weeks and does not protect the lower GIT from the ingress of microbes. Indeed, some studies have demonstrated that the faeces of breast-fed neonates have consistently higher counts of staphylococci than those of formula-fed neonates. While formula milk may also be contaminated with microbes, the organisms present are likely to include environmental species derived from the manufacturer's machinery, from domestic equipment used during feed preparation, and from the water used to reconstitute the feed. The latter can be a serious problem in developing countries where access to safe water supplies is limited. The species most commonly isolated from the faeces of infants are listed in Table 7.10.

The next critical stage in the development of the intestinal microbiota and the acquisition of a microbiota characteristic of that of adults occurs during weaning. The introduction of non-milk food to the diet, together with the development of intestinal function in the immature gut, results in profound changes in the intestinal microbiota. At the onset of weaning, the infant is exposed for the first time to many different complex carbohydrates, and starch, for example, will be incompletely digested because of the inability of the infant to chew and because of low levels of salivary and pancreatic amylases. Once weaning begins, the microbiota of breast-fed infants becomes similar to that of formula-fed infants. The results of a study of the faecal microbiota of seven

Table 7.10. Species most frequently isolated from the faeces of infants

Bif. breve, Bif. infantis, Bif. longum, Bif. bifidum, Bif. adolescentis
L. acidophilus, L. salivarius, L. fermentum, L. brevis, L. plantarum
B. distasonis, B. vulgatus, B. fragilis, B. ovatus, B. uniformis, B. thetaiotaomicron
Cl. perfringens, Cl. difficile, Cl. butyricum, Cl. tertium, Cl. paraputrificum
Eub. rectale, Eg. lenta (Eub. lentum), Col. aerofaciens (Eub. aerofaciens)
V. parvula
Pep. productus, Pep. anaerobius
Staph. epidermidis, Staph. aureus
Ent. faecalis, Ent. faecium
E. coli
Enter. cloacae
K. pneumoniae
Pr. mirabilis
Cit. freundii
Ps. aeruginosa

breast-fed and seven formula-fed infants during their first year of life are shown in Figure 7.10. The average age of the infants when weaning commenced was 17 weeks, and the figure shows that this corresponds to a dramatic increase in the numbers of bacteroides and clostridia in the breast-fed infants. By the second year of life, the faecal microbiota of breast- and formula-fed infants begins to resemble that of adults.

While molecular-based studies of the intestinal microbiota of neonates have confirmed many of the findings of culture-based studies, some differences in the results obtained using these different methodologies have emerged. Hence, in one recent study using PCR-DGGE, it was found that for several months one of the predominant organisms in the faecal microbiota of neonates was a *Ruminococcus* sp. Furthermore, the

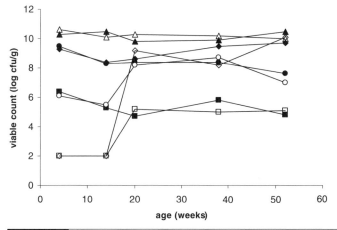

Figure 7.10 Concentrations of various bacterial groups in the faeces of breast- and formula-fed infants during the first year of life. □ = *Clostridium* spp., breast-fed; ■ = *Clostridium* spp., formula-fed; △ = *Bifidobacterium* spp., breast-fed; ▲ = *Bifidobacterium* spp., formula-fed; ◇ = *Bacteroides* spp., breast-fed; ◆ = *Bacteroides* spp., formula-fed; ○ = enterobacteria, breast-fed; ● = enterobacteria, breast-fed. Reproduced with permission from: Development of intestinal microbiota. Conway, P. In: *Gastrointestinal microbiology*. Vol. 2. Mackie, R.I., White, B.A., and Isaacson, R.E (eds.). Copyright © 1996, Kluwer Academic/Plenum Publishers.

Table 7.11. Microbes isolated from the oesophagus of 10 adults

Organism	Frequency of isolation (%)	Number of viable bacteria present (cfu)
Streptococcus spp.	40	100
Staphylocccus spp.	20	1,000
Corynebacterium spp.	10	1,000
Lactobacillus spp.	10	1,000
Peptostreptococcus spp.	10	100
Enterococcus spp.	0	0
Klebsiella spp.	0	0
Veillonella spp.	0	0

Note: Microbes were cultivated from only four of these individuals. Frequency of isolation = proportion of study population from which the organism was isolated.

sequences of 56% of the 16S rDNA bands obtained did not correspond to those of known species.

7.4.3 Community composition at different sites

7.4.3.1 Oesophagus

The lumen of the oesophagus differs from that of other regions of the GIT in that it contains material for only short periods of time (i.e., it is a passageway rather than a receptacle). The only site of colonisation for microbes, therefore, is the oesophageal wall. Few investigations of the microbiota of the oesophagus have been carried out due to the difficulties in obtaining samples from this region, and the organisms isolated are usually considered to be transients, resulting from the swallowing of microbe-laden oropharyngeal and nasopharyngeal fluids. The studies that have been carried out have found streptococci to be the predominant cultivable organisms present. In a typical study of ten adults, from whom samples were obtained during endoscopy, microbes were cultured from only four individuals. Streptococci were present in all four of the samples exhibiting microbial growth (Table 7.11).

In a culture-independent study, DNA was extracted from oesophageal biopsies obtained from 4 adults and the bacterial 16S rRNA genes present were amplified by PCR and the resulting products cloned and sequenced. A total of 833 unique sequences belonging to 95 taxa from 41 genera were identified with *Streptococcus* spp., *Prevotella* spp. and *Veillonella* spp. being the most prevalent. Most (82%) of the taxa identified corresponded to species that could be cultivated in the laboratory. Fourteen taxa were detected in all four samples and these accounted for 64% of the clones analysed. These taxa included *Strep. mitis*, *Strep. thermophilus*, *Strep. parasanguis*, *V. atypica*, *V. dispar*, *Roth. mucilaginosus*, *Megasphaera micronuciformis*, *Granulicatella adiacens*, and *Prev. pallens*. The microbiota of the oesophagus was, in general, found to be similar to that of the oropharynx and it is likely that many of the organisms present were transients from oropharyngeal secretions. However, the absence of spirochaetes and many uncultivable organisms found in the oral cavity suggest that the oesophageal microbiota is distinctly different from that of the oral cavity. Microscopic examination of the biopsies revealed the presence of bacteria closely associated with the mucosal surface at a density of approximately 10^4 bacteria per mm^2.

7.4.3.2 Stomach

The number of viable microbes that can be recovered from stomach contents is in the order of 10^3 cfu/ml, although counts 100 to 1,000 times greater than this can be reached after a meal because of the transient increase in pH caused by the buffering action of food (see Figure 7.8). This sparse population is a consequence of the low pH of the stomach and the relatively rapid transit of material through this organ. Microbes frequently isolated from gastric juice (i.e., from the gastric lumen) are mainly acid-tolerant species of streptococci and lactobacilli, although these are considered to be transients from the oral and/or nasal cavity that have managed to survive the low gastric pH. Organisms that have been cultivated from stomach contents include viridans streptococci (*Strep. sanguis, Strep. salivarius*), lactobacilli, *Staph. aureus, Staph. epidermidis, Neisseria* spp., *Haemophilus* spp., *Bacteroides* spp., *Micrococcus* spp., *Bifidobacterium* spp., coryneforms, and *Veillonella* spp.

The microbiota associated with the gastric mucosa is similar to that found in the gastric lumen, except that an additional organism – *Hel. pylori* – may be present in up to 50% of the population. Unlike most of the organisms previously mentioned, *Hel. pylori* is not acidophilic or acid-tolerant, but grows optimally at a pH of 7.0. Because *Hel. pylori* can grow and reproduce in the stomach and can persist for long periods of time in an individual, it has been suggested that only this organism should be regarded as a member of the indigenous gastric microbiota. *Hel. pylori* is mainly found in the antrum of the stomach either within the overlying mucus layer or attached to the gastric epithelium. Within these habitats, it secretes urease which liberates ammonia from the urea present in gastric juices (at a concentration of between 1 and 3 mM), thereby increasing the pH locally and creating a microenvironment suitable for its growth and proliferation. The antrum is the preferred site of colonisation because little acid production occurs in this region – its main functions are the secretion of mucus, pepsinogen, and gastrin and, to some extent, absorption of nutrients. Acidification of the mucus layer and the epithelial surface of the antrum can only occur by diffusion of acid from the lumenal contents. Presumably, the ammonia-generating ability of *Hel. pylori* is sufficient to counteract this and provide a microenvironment with a pH suitable for its growth (i.e., between 6.0 and 8.0). In contrast, the organism appears to be unable to produce sufficient ammonia to neutralise the acid produced in the corpus (i.e., the main acid-secreting region of the stomach). *Hel. pylori* cannot survive at a pH lower than 4.0 and, consequently, is vulnerable within the gastric lumen, although it may be able to survive by transformation to a coccoid form (Section 7.4.1.6).

Initial colonisation by *Hel. pylori* occurs during childhood when the gastric pH may be lower than in adults, the immune system may not be fully developed, and as a consequence of children frequently sampling their environment through their mouths. Once acquired, the organism generally persists in the individual for life. It has been suggested that prior to the twentieth century, all humans were colonised by *Hel. pylori*, but that improved living conditions have resulted in a decrease in the prevalence of the organism – this is particularly evident in developed countries, in which prevalence rates continue to fall. Currently, approximately half of the world's population is colonised by *Hel. pylori*, but there are marked regional variations in colonisation rates (Table 7.12) which also vary with age, ethnicity, and socioeconomic status. In general, colonisation rates are higher in developing countries than in developed countries and are declining

| Table 7.12. | Prevalence of gastric colonisation by *Helicobacter pylori* in various countries and population groups |

Country/population group	Proportion of individuals colonised by *Hel. pylori* (%)
USA	30–40
Alaska	
overall	75
children ($\leq$4 years)	40
children ($\leq$10 years)	70
Southern China	
rural areas	39
urban areas	52
Southern Estonia	
adults	87
children ($\leq$15)	56
Novosibirsk (25–34 year olds)	91
Holland	
children (6–8 years)	9
children (12–15 years)	11

in the latter. The frequency of colonisation increases with age among children and adolescents, levels off during middle age, and then often decreases in the elderly.

7.4.3.3 Small intestine

A combination of factors render the upper regions of the small intestine (duodenum and jejunum) inhospitable for microbial colonisation (Section 7.3.3). These include a low pH, the rapid peristalsis leading to a short transit time for the lumenal contents, the flushing action of chyme and intestinal secretions, and the presence of a range of antimicrobial compounds (e.g., bile, proteolytic enzymes, antimicrobial peptides, and proteins). The rapid peristalsis is particularly important in this respect because any procedure that reduces peristalsis has been found to result in microbial overgrowth in the small intestine.

Most studies of samples obtained from the lumen or mucosa of the duodenum have either failed to detect viable microbes or have found them to be present only in low concentrations. Most of the organisms detected are acid-tolerant species (mainly *Lactobacillus* spp. and *Streptococcus* spp.) derived from the oral and nasal cavities. However, *Bacteroides* spp., enterobacteria, *Bifidobacterium* spp., *Veillonella* spp., *Staphylococcus* spp., and yeasts may also be isolated in smaller numbers. The time of sampling in relation to food intake has a profound effect on the recovery of viable microbes. Hence, in a study involving serial sampling of the duodenum before and after a meal, bacteria were detected in three of five individuals at a time corresponding to the entry of chyme from the stomach. Within 1 hour, the viable counts then returned to baseline values of less than 10^3 cfu/ml.

The jejunum also appears to have a sparse microbiota, although viable organisms tend to be more frequently recovered, and in greater numbers, from this region than from the duodenum. In a study of five individuals, microbes were recovered from most samples of the mucosa and lumen of the jejunum. The viable count of aerobes ranged from 10 to 5,000 cfu (except one individual who had no aerobes), while that of anaerobes ranged from 200 to 30,000 cfu. The organisms found in the lumen contents were similar to those reported from the duodenum and included streptococci, lactobacilli, staphylococci, *Veillonella* spp., and yeasts with an occasional fusiform. Microscopy of the mucosal samples revealed the presence of rods and cocci in the mucus layer, and streptococci and lactobacilli were cultured from the samples. In another study, microbes were cultivated from the jejunum of thirteen of eighteen healthy adults. The dominant organisms were enterococci, followed by lactobacilli, yeasts, and unidentified fungi. The main species of lactobacilli were *L. fermentum, L. reuteri, L. gasseri,* and *L. salivarius.*

In contrast to the results obtained from populations in developed countries, the jejunum of individuals from developing countries appears to have a more substantial and varied microbiota. Hence, in a study of ten healthy adults in India, microbes were recovered in relatively high numbers from both the lumen (eight individuals) and mucosa (all individuals) of the jejunum. Viable counts ranged from 10^2 to 10^8 cfu/ml of lumen contents and from 10^3 to 10^6 cfu/g of mucosa. Organisms isolated included enterobacteria, enterococci, streptococci, *Neisseria* spp., lactobacilli, *Bifidobacterium* spp., *Fusobacterium* spp., *Bacteroides* spp., *Veillonella* spp., *Ps. aeruginosa*, Gram-positive anaerobic cocci (GPAC), CNS, and yeasts. Interestingly, the mucosa of the small intestine of Asians differs considerably from that of individuals in developed countries in being relatively flat and having leaf-like villi. Whether these differences contribute to the establishment of a qualitatively and quantitatively different microbiota has not been established.

In the distal region of the small intestine (the ileum), peristalsis is slower than in the upper regions, and intestinal juice dilutes the antimicrobial effects of pancreatic enzymes and bile. Furthermore, significant absorption of bile acids occurs in the jejunum, resulting in depleted concentrations of these antimicrobial compounds in the ileum. The pH is also neutral or slightly alkaline. These factors enable the establishment of an ileal microbiota derived from organisms arriving in material from the jejunum and by the reflux of caecal contents through the ileocaecal sphincter, which separates the ileum from the large intestine. Most of the information available on the ileal microbiota has been obtained from patients with an ileostomy (i.e., individuals who have had their colon removed and an opening made from the ileum to the abdomen to serve as an anus). The lumenal contents of the ileum contain between 10^6 and 10^8 cfu/ml, and the microbiota is dominated by facultative bacteria (mainly streptococci and coliforms), which are between 20 and 50 times more numerous than obligate anaerobes (mainly *Veillonella* spp., *Clostridium* spp., and *Bacteroides* spp.). The main organisms isolated are listed in Table 7.13. The results of a typical study showing the relative proportions of the various groups of bacteria present in the lumen of the ileum are shown in Table 7.14.

In contrast to the situation in the lumen, the microbiota associated with the mucosal surface of the ileum appears to be dominated by obligate anaerobes. In a study of forty adults, biopsies were taken, and the bacteria adherent to the mucosa of the ileum were determined. The ratio of obligate to facultative anaerobes was 60:1. Approximately 10% of the obligate anaerobes were *Bacteroides* spp., while enterobacteria comprised 67% of the facultative anaerobes. Each individual had at least three different *Bacteroides* spp. Other anaerobes isolated included *Col. aerofaciens* (formerly *Eubact. aerofaciens*) – the second most frequent isolate – as well as clostridia, bifidobacteria, peptostreptococci, eubacteria, and propionibacteria.

Table 7.13. | Microbiota of lumenal contents of the ileum

Organisms present in large numbers	Organisms present in smaller numbers
viridans streptococci	*Eubacterium* spp. (and former members of the genus)
Enterococcus spp.	*Bifidobacterium* spp.
Ent. faecalis	*Fusobacterium* spp.
Ent. faecium	*Peptococcus* spp.
	Staphylococcus spp.
Enterobacteria	*Peptostreptococcus* spp. (and former members
E. coli	of the genus)
Klebsiella spp.	
Proteus spp.	
Bacteroides spp.	
B. fragilis group	
B. asaccharolyticus	
Veillonella spp.	
V. parvula	
Clostridium spp.	
Cl. bifermentans	
Cl. perfringens	
Lactobacillus spp.	

Notes: From: Factors controlling the microflora of the healthy upper gastrointestinal tract. Hill, M.J. In: *Human microbial ecology.* Hill, M.J. and Marsh, P.D. (eds.). Copyright © 1989, by CRC Press. Reprinted with permission of CRC Press (via the Copyright Clearance Center).

The composition of the ileal microbiota is profoundly affected by diet, demonstrating that food ingested by the host, rather than host secretions, constitutes the main source of nutrients for microbes in this region. Hence, a diet rich in fat has been shown to increase the proportions of obligate anaerobes (mainly *Bacteroides* spp. and *Clostridium* spp.), while one rich in protein increases the proportion of facultative anaerobes – particularly coliforms, enterococci, and streptococci. In a study of two groups of ileostomy patients, one group was fed on a diet rich in sucrose and low in dietary fibre, while the other was low in sucrose and rich in unrefined cereal. The ileostomy fluid was dominated by facultative anaerobes (mainly streptococci) and, while there

Table 7.14. | Bacterial composition of the ileal contents of five adults 5–6 hours after a meal rich in sucrose and low in dietary fibre (i.e., characteristic of a Western diet)

Bacterial group	Viable count (cfu/g)
streptococci	3.3×10^6
enterococci	1.3×10^6
enterobacteria	5.2×10^4
Veillonella spp.	1.7×10^4
Clostridium spp.	7.2×10^3
lactobacilli	3.8×10^3
Bacteroides spp.	1.5×10^3
facultative anaerobes	5.6×10^6
total anaerobes	1.0×10^5

was no difference between the groups in terms of the relative proportions of the various groups of organisms, those on a low-sucrose/high-fibre diet produced a fluid with a 10-fold higher concentration of viable bacteria than those on a high-sucrose/low-fibre diet. Easily digestible carbohydrates (e.g., sucrose and polysaccharides present in the high-sucrose/low-fibre diet) are rapidly broken down in the oral cavity, stomach, and duodenum, and the sugars produced are then rapidly absorbed in the upper regions of the small intestine – hence, reducing the quantity of nutrients available to intestinal microbes. In contrast, the polysaccharides present in the high-fibre diet are less easily digested, are only slowly broken down, and persist for longer periods within the GIT. Interestingly, the concentrations of glucose and oligosaccharides in the ileostomy fluid were higher in those on a low-sucrose/high-fibre diet, and these higher concentrations may reflect the greater availability of carbon and energy sources for the growth of intestinal microbes and, hence, account for the greater numbers of bacteria present.

The microbiota of the ileum is intermediate between that of the upper small intestine (sparse, mainly Gram-positive species, many facultative anaerobes) and that of the large intestine (huge numbers of microbes, mainly Gram-negative species, dominated by obligate anaerobes).

7.4.3.4 Large intestine

The total mass of microbes in the colon is approximately 90 g, making this the most densely populated region of the body. Although many of the microbial inhabitants of the colon are associated with the mucosal surface, most are within the lumen, and each gram of faeces contains approximately 10^8 facultative anaerobes and 10^{11} obligate anaerobes (i.e., approximately 20 times the number of human beings on earth). The microbiota of the colon is extrememly complex, and more than 400 species from more than 190 genera have been isolated. However, many of these are present in small numbers, and it has been estimated that between thirty and forty species belonging to six genera account for 99% of the cultivable microbiota. The number of different species that can be cultivated from the colon of an average adult appears to be between 13 and 30. It is important to bear in mind that most of what we know about the microbiota of the large intestine is derived from the results of microscopy- and culture-based studies. Unfortunately, estimates of the proportion of the colonic microbiota that can be cultured in the laboratory have been as low as 15–58%, depending on the culture technique used. The results of studies using molecular methodologies will add significantly to our knowledge of the composition of the microbiota of this region and, if backed by appropriate studies of the physiology and ecology of these organisms, will also lead to a greater understanding of these complex communities. Another problem with studies of the microbiota of the large intestine is that because of the difficulties associated with obtaining representative samples (especially from the upper regions), most studies have used faeces. It has been suggested that, because faeces are obtained from the terminal region of the colon, they may not be representative of the contents of the upper regions. However, studies have shown that the microbial composition of faeces is similar to that found in several regions of the colon (although the absolute numbers may differ), apart from the caecum which appears to have a different microbiota (described later in this section). Another perceived problem is that because faeces consists of material from the lumen of the colon, analysis of such material will

Table 7.15. Frequency of isolation of various organisms from the faeces of 62 adults

Organism	Frequency of isolation (%)
Bacteroides spp.	100
Clostridium spp.	100
Streptococcus spp.	100
anaerobic cocci	98
Gram-negative facultative anaerobes	98
Eubacterium spp.[*]	95
Bifidobacterium spp.	79
Lactobacillus spp.	73
Ruminococcus spp.	45
Peptococcus spp.	37
Peptostreptococcus spp.[†]	35

[*] Includes *Eubacterium*-like organisms (e.g., *Eg. lenta* and *Col. aerofaciens*).
[†] Including former members of this genus.
Frequency of isolation = proportion of study population from which the organism was isolated.

provide no information regarding the microbial populations associated with the mucosa, although faeces may, of course, contain organisms initially attached to the mucosa.

The organisms most frequently isolated from faeces include *Bacteroides* spp., *Clostridium* spp., *Streptococcus* spp., anaerobic cocci, Gram-negative facultative anaerobes, and *Eubacterium* spp. (including former members of this genus) (Table 7.15). With regard to the relative proportions of the various organisms present, the microbiota of the colon is dominated by obligate anaerobes, which are approximately 1,000-fold more abundant than facultative anaerobes. The main anaerobic genera are *Bacteroides, Eubacterium* (including *Eubacterium*-like organisms), and *Bifidobacterium*, while *Streptococcus* spp. and enterobacteria comprise the main facultative organisms. Species belonging to the genus *Bacteroides* are the most numerous organisms and constitute 20–30% of the cultivable microbiota – *B. thetaiotaomicron* and *B. vulgatus* are generally the predominant species present. *Eubacterium* spp. (including *Eubacterium*-like organisms) and *Bifidobacterium* spp. are the next most abundant organisms. Together, these three genera often account for approximately 90% of the cultivable microbiota of the colon. The predominant genera present in human faeces are shown in Table 7.16. Although the relative proportions of these genera vary considerably between individuals, the composition of faeces from an individual (at the genus level) is stable. The particular species present in the faeces of an individual, however, may vary markedly from day to day. Other genera that are regularly present in human faeces, but in smaller proportions than those listed in Table 7.16, include *Veillonella* spp., *Enterococcus* spp., *Staphylococcus* spp., *Acidaminococcus* spp., *Candida* spp., and other yeasts. The genus *Acidaminococcus* consists of non-motile, non-sporing, anaerobic Gram-negative cocci which belong to the family *Veillonellaceae*. The G+C content of their DNA is 56.6 mol%, and they are similar to *Veillonella* spp. (Section 8.4.1.3), except that they degrade amino acids to acetate and butyrate. The most frequently encountered species within the various genera are shown in Table 7.17.

Table 7.16. Predominant genera isolated from human faeces

Organism	Abundance in faeces ($\log_{10}$ per g dry wt of faeces)	
	Mean	Range
Bacteroides spp.	11.3	9.2–13.5
Eubacterium spp.[*]	10.7	5.0–13.3
Bifidobacterium spp.	10.2	4.9–13.4
Ruminococcus spp.	10.2	4.6–12.8
Peptostreptococcus spp.[†]	10.1	3.8–12.6
Peptococcus spp.	10.0	5.1–12.9
Clostridium spp.	9.8	3.3–13.1
Lactobacillus spp.	9.6	3.6–12.5
Propionibacterium spp.	9.4	4.3–12.0
Actinomyces spp.	9.2	5.7–11.1
Streptococcus spp.	8.9	3.9–12.9
Methanobrevibacter spp.	8.8	7.0–10.5
Escherichia spp.	8.6	3.9–12.3
Desulphovibrio spp.	8.4	5.2–10.9
Fusobacterium spp.	8.4	5.1–11.0

[*] Includes *Eubacterium*-like organisms (e.g., *Eg. lenta* and *Col. aerofaciens*).
[†] Includes former members of this genus.

Table 7.17. Most frequently encountered species of common genera present in human faeces

Genus/group	Most frequently encountered species
Bacteroides	*B. thetaiotaomicron, B. vulgatus, B. distasonis, B. eggerthii, B. fragilis, B. ovatus*
Eubacterium and *Eubacterium*-like organisms	*Col. aerofaciens, Eg. lenta, Eub. contortum, Eub. cylindroids, Eub. rectale, Eub. biforme, Eub. ventriosum*
Bifidobacterium	*Bif. adolescentis, Bif. infantis, Bif. catenulatum, Bif. pseudocatenulatum, Bif. breve, Bif. longum*
Clostridium	*Cl. ramosum, Cl. bifermentans, Cl. butyricum, Cl. perfringens, Cl. difficile, Cl. indolis, Cl. septicum, Cl. sporogenes*
Enterococcus	*Ent. faecalis, Ent. faecium*
Fusobacterium	*F. prausnitzii, F. mortiferum, F. necrophorum, F. nucleatum, F. varium, F. russii*
Gram-positive anaerobic cocci	*Pep. productus, Micromonas micros (Pep. micros), Anaerococcus prevotii (Pep. prevotii), Schleiferella asaccharolytica, (Pep. asaccharolyticus), Finegoldia magna (Pep. magnus)*
Ruminococcus	*Rum. albus, Rum. obeum, Rum. torques, Rum. flavefaciens, Rum. gnavus, Rum. bromii*
Enterobacteria	*E. coli, Enter. aerogenes, Pr. mirabilis*
Lactobacillus	*L. acidophilus, L. brevis, L. casei, L. salivarius, L. plantarum, L. gasseri, L. ruminis, L. crispatus*
Actinomyces	*A. naeslundii, A. odontolyticus*
Propionibacterium	*P. acnes, P. avidum*
Streptococcus	*Strep. salivarius, Strep. bovis, Strep. equinus*

| Table 7.18. | Influence of diet on the composition of the microbiota of the colon |

Type of diet	Effect on microbiota
high meat diet	increased ratio of obligate: facultative anaerobes
fibre supplements	increase in total viable count of *Clostridium* spp.
addition of gum arabic	increase in proportion of organisms able to degrade this polymer
vegetarian diet	decreased prevalence of anaerobic Gram-positive cocci and *Fusobacterium* spp.; increased prevalence of *Actinomyces* spp.
inulin supplementation	increase in *Bifidobacterium* spp.

That diet can affect the composition of the colonic microbiota is well established in infants, where characteristic differences are observed between those who are breast- and formula-fed (Section 7.4.2). Although the effect of diet on the colonic microbiota of adults has been studied extensively, the complexity of the microbiota and the large variations in its composition between individuals have made the results of such studies difficult to interpret. Nevertheless, it has long been known that individuals in the United Kingdom and the United States have a higher ratio of anaerobes to facultative anaerobes than individuals in Japan, India, or Uganda, and this has been attributed to dietary influences. The most likely constituents to evade digestion in the upper GIT and, consequently to affect the colonic microbiota are complex carbohydrates, and the effects of these "prebiotics" are discussed in Section 10.2. The results of a number of studies on the effect of diet on the colonic microbiota are summarised in Table 7.18. Although modification of the colonic microbiota due to diet has been difficult to demonstrate, many studies have shown that the diet has a profound impact on the metabolic activities of the microbiota of the colon.

The information on the colonic microbiota provided in this section so far has been derived from culture-based studies. Increasingly, however, a variety of molecular techniques are being used to explore the microbiota of this region and have produced some interesting findings. Unfortunately, just as variations in the methodology used in culture-based studies (e.g., media used, incubation conditions, length of incubation) has often made inter-study comparisons difficult, the results of studies using molecular techniques, likewise, are often difficult to compare. Hence, differences in the efficiency of nucleic-acid extraction, amplification conditions, and the primers used (in PCR-based approaches), and differences in probe specificity and hybridisation conditions (in oligonucleotide probe-based studies) often make comparisons difficult. Nevertheless, some interesting features of the colonic microbiota are emerging from these molecular-based analyses. Firstly, it would appear that large proportions (up to 75% in some studies) of the microbes present in the colon are unknown species. Secondly, the presence of high proportions of bifidobacteria in the microbiota suggested by culture-based studies is not supported by the results of molecular-based studies. Both of these features are evident in the results of a detailed analysis of the faeces of an adult using microscopic, cultural, and molecular techniques. The numbers of organisms present using the different techniques were as follows: microscopy, 10.6×10^{11} cells/g (dry weight); anaerobic cultivation, 2.2×10^{11} cfu/g (dry weight); and

Table 7.19.	Examples of species belonging to three of the main phylogenetic groups of organisms comprising the colonic microbiota
Phylogenetic group	Examples of organisms within the group
Bacteroides	*Bacteroides* spp. (at least 13 species), including *B. distasonis*, *B. fragilis*, *B. thetaiotaomicron*; *Prevotella* spp. (at least 8 species); *Porphyromonas* spp. (at least 5 species); *Rikenella microfusus*; *Cytophaga fermentans*
Cl. coccoides-Eubt. rectale	*Clostridium* spp. (at least 12 species), including *Cl. coccoides*; *Eubacterium* spp. (at least 9 species), including *Eub. rectale*; *Ruminococcus* spp. (at least 5 species); *Butyrivibrio* spp.; *Strep. hansenii*; *Coprococcus eutactus*
Cl. leptum	*Clostridium* spp. (at least 3 species), including *Cl. leptum*; *Eubacterium* spp.; *Ruminococcus* spp. (at least four species); *F. prausnitzii*.

hybridisation with an oligonucleotide probe recognising the *Bacteria* domain, 7.1×10^{11} cells/g (dry weight). Hence, organisms capable of growing anaerobically on a blood-containing medium accounted for only 21% of those observed microscopically and 32% of those detected using an oligonucleotide probe. DNA was then extracted from the sample, the bacterial 16S rRNA genes present were amplified by PCR, and the resulting products were cloned. A total of eighty-two taxa were detected, of which only twenty corresponded to known cultivated species. All of the species identified were ones that have previously been reported to be present in faeces and included *Bacteroides* spp., *Fusobacterium* spp., *Eubacterium* spp., and *Streptococcus* spp. However, some well-known residents of the colon, including *Bifidobacterium* spp., *Col. aerofaciens*, and *Rum. productus*, were not detected – this may be attributable to the fact that only one faeces sample was analysed and/or to technical problems such as DNA denturation. Of the clones obtained, 95% could be classified into three major monophyletic groups: *Bacteroides* group (31%), *Cl. coccoides-Eub. rectale* group (44%), and the *Cl. leptum* group (20%). Because a number of molecular-based studies have reported similar results, some of the species present in each of these groups are listed in Table 7.19. In fact, the results of many molecular-based studies suggest that most members of the colonic microbiota fall within four phylogenetic groups: the three shown in Table 7.19 plus the genus *Bifidobacterium*. Other important groups include the *Lactobacillus/Enterococcus/Streptococcus* group and the enterobacteria. Fluorescent *in situ* hybridisation (FISH) analysis using group-specific 16S rRNA-targeted oligonucleotide probes on samples from a larger group of individuals also showed that the *Bacteroides* and *Cl. coccoides-Eub. rectale* groups accounted for large proportions of the faecal microbiota: 20% and 29%, respectively. Again, *Bifidobacterium* spp. accounted for only a small proportion (3%) of the microbiota. A panel of probes has been used to quantify the relative proportion of rRNAs of various microbial groups in faecal samples from ten healthy adults using quantitative dot-blot hybridisations. *Archaea* and *Eukarya* represented only very small proportions of the total rRNA – 0.28% and 2.02%, respectively. Of the remaining rRNA, 92.4% could be attributed

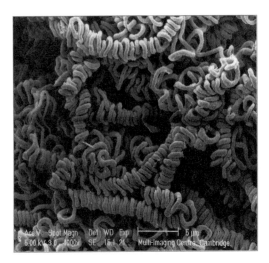

Figure 7.11 Scanning electron micrograph of large spiral bacterial forms associated with the colonic epithelium. Reproduced with permission from: Bacterial growth on mucosal surfaces and biofilms in the large bowel. Macfarlane, S. and Macfarlane, G.T. In: *Medical implications of biofilms*. Wilson, M. and Devine, D. (eds.). Copyright © 2003, Cambridge University Press.

to six bacterial groups, as follows: *Bacteroides* (35.5 ± 4.5%), enterics (2.7 ± 0.7%), *Lactobacillus* (<1.0%); *Cl. leptum* (30.4 ± 6.5%); *Cl. coccoides* (19.7 ± 2.2%), and *Bifidobacterium* (4.1 ± 0.7%).

While the results of molecular-based approaches support many of the findings derived from culture-based studies (e.g., the dominance of obligate anaerobes over facultative anaerobes, the presence of high proportions of *Bacteroides, Clostridium*, and *Eubacterium* species), they also suggest that cultural studies have possibly overestimated the proportions of bifidobacteria in the colonic microbiota. Furthermore, they have revealed that the identity of many, if not most, of the species present remains to be established.

As well as the microbiota varying in composition along the GIT, it must be remembered that there is also a cross-sectional stratification at any point along the tract. Hence, as well as the lumen, a number of other habitats are available for microbial colonisation, including the mucus layer, the epithelial surface, and the intestinal crypts. In a number of animals other than humans, the mucus layer, mucosa, crypts, and goblet cells have all been shown to be colonised by distinct groups of microbes. In rats, cats, dogs, pigs, and monkeys, the major groups of mucosa-associated organisms have a spiral morphology. In humans, the situation is not so clear. While there have been many reports of the presence of mucosa-associated microbes, most of these have found that the organisms are typical of those present in the lumen and do not represent a distinct population. This is surprising in view of the different environment at the mucosal surface. Hence, the oxygen concentration is considerably higher than in the lumen, and the range of nutrients is likely to differ due to the abundance of mucins and host-cell secretory and excretory products. Furthermore, the concentration of host antimicrobials is likely to be much greater at the mucosal surface. These factors would be expected to enable the establishment of a microbial community different from that existing in the lumen. The failure to find such communities may be attributable to sampling problems and the difficulty of obtaining samples uncontaminated by lumenal contents. Microscopic examination of the colonic mucosa has revealed the presence of organisms with an unusual morphology (Figure 7.11). Although *Bacteroides* spp. are most frequently mentioned as being mucosa-associated, there are reports of the isolation

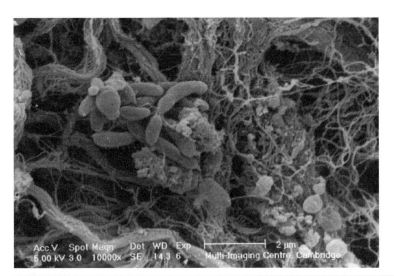

Figure 7.12 Scanning electron micrograph showing microbes attached to the mucosa of the colon. Photomicrograph kindly supplied by Dr. S. Macfarlane, MRC Microbiology and Gut Biology Group, University of Dundee.

from the colonic mucosa of members of all of the major bacterial genera found in the lumen (Figure 7.12). There is also little agreement with regard even to the relative proportions of obligate and facultative anaerobes, although almost all studies have reported more of the former than the latter. Unfortunately, few studies have reported the density of colonisation of the mucosa – many having recorded the number of microbes per gram of tissue (this ranges from 10^6 to 10^9/g), which is not very informative. However, at least one study has shown that the mean number of cultivable microbes per square millimeter of mucosa ranged from 7×10^3 to 1.5×10^6 cfu, with approximately two-thirds of these being *Bacteroides* spp. Because of technical difficulties and ethical considerations, it is also often impossible to distinguish between bacterial populations adhering to the epithelium and those present within the mucus layer itself. A number of investigators have concluded that all "mucosa-associated" microbes are, in fact, present within the mucus layer overlying the epithelium, whereas others assert that bacteria do actually adhere to the epithelial cells. In a recent study of forty healthy individuals, biopsies from the ileum and ascending and sigmoid colon were examined for the presence of microbes adhering to the epithelium using a combination of quantitative PCR, FISH, culture, and electron microscopy. Before analysis, the surfaces of the biopsies were washed to remove the mucus layer. Bacteria were generally absent from the biopsies or, when found, were present in only low numbers. In contrast, patients with inflammatory bowel disease investigated in this study generally had high numbers of bacteria adhering to the epithelium, and these were mainly *Bacteroides* spp.

As described in Section 7.3.4, the environment in the caecum differs substantially from that of other regions of the large intestine in having an acidic pH, fluid contents, and a relatively high concentration of easily fermentable compounds. The microbiota of the caecum has been compared with that of faeces using both culture-based and molecular methodologies. Using traditional culture techniques, the concentration of all bacterial groups (other than facultative anaerobes) was significantly lower in the

Table 7.20. Analysis of the microbial communities present in the caecum and faeces of eight adults using a traditional culture-based approach

Microbial group	Counts ($\log_{10}$ cfu/ml $\pm$ SEM) in	
	Caecum	Faeces
total anaerobes	8.0 $\pm$ 0.22	10.4 $\pm$ 0.12
facultative anaerobes	7.4 $\pm$ 0.23	7.8 $\pm$ 0.29
Bifidobacterium	6.7 $\pm$ 0.45	8.9 $\pm$ 0.22
Bacteroides	7.4 $\pm$ 0.33	8.9 $\pm$ 0.42

caecum than in faeces (Table 7.20). Furthermore, facultative anaerobes comprised a much greater proportion (25%) of the total counts in the caecal contents than in the faeces, where they comprised only 0.25% of the total anaerobic count. The molecular analysis was based on hybridisation assays using rRNA-targeted probes recognising six main groups of organisms (Table 7.21) and total RNA extracted from the samples. Sixty-eight percent and 53.6%, respectively of the RNA extracted from the caecal and faecal samples could be attributed to organisms from these six groups. The results obtained supported the culture-based findings in that (1) the caecum had much greater proportions of facultative anaerobes (i.e., *E. coli* plus the *Lactobacillus/Enterococcus* group) than did faeces – the proportions being 49.6% and 6.6%, respectively; and (2) the proportions of *Bifidobacterium* spp. were similar in both communities. The faeces appeared to contain much greater proportions of *Bacteroides* spp. and members of the *Cl. coccoides-Eub. rectale* and *Cl. leptum* groups.

The microbiota of the colon is extremely complex and is a stable, climax community which is maintained in this state because of the stable environment provided by the host (which derives many benefits from the microbiota – as described in Chapter 9), together with a plethora of positive and negative interactions between its constituent members.

7.4.4 Microbial interactions in the gastrointestinal tract

It is not surprising that, within the extremely complex microbiotas of the GIT – particularly in the terminal ileum, caecum, and colon – a wide range of microbial interactions have been found to occur. Some of these enable microbial consortia to degrade complex

Table 7.21. Analysis of the microbial communities present in the caecum and faeces of eight adults using a culture-independent approach

Microbial group	% RNA (mean $\pm$ SEM)	
	Caecum	Faeces
Bacteroides spp.	1.2 $\pm$ 0.22	8.0 $\pm$ 0.32
Bifidobacterium spp.	5.8 $\pm$ 0.37	3.2 $\pm$ 0.55
Cl. coccoides-Eub. rectale group	10.0 $\pm$ 0.55	22.8 $\pm$ 2.2
Cl. leptum group	1.4 $\pm$ 0.11	13.0 $\pm$ 0.78
E. coli	26.8 $\pm$ 7.4	0.02 $\pm$ 0.02
Lactobacillus/Enterococcus	22.8 + 2.1	6.6 $\pm$ 0.23
Total	68.0	53.6

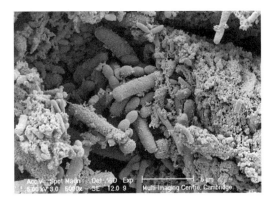

Figure 7.13 Scanning electron micrograph showing microbes attached to food particles in the human colon. Photomicrograph kindly supplied by Dr. S. Macfarlane, MRC Microbiology and Gut Biology Group, University of Dundee.

substrates (e.g., glycoproteins and plant polymers), while others result in a particular species being able to dominate a habitat.

7.4.4.1 Positive interactions

At birth, the GIT is an aerobic region, but colonisation by aerobes and facultative anaerobes results in a reduction in the oxygen content and the redox potential which enables the survival of microaerophiles and obligate anaerobes. Even in the GIT of adults, oxygen is present at mucosal surfaces due to diffusion from underlying tissues, yet obligate anaerobes (e.g., *Bacteroides* spp., *Clostridium* spp., *Eubacterium* spp.) are able to colonise these surfaces because of local utilisation of oxygen by co-colonising facultative anaerobes. Because the diet of the host is so varied, the GIT regularly receives a diverse range of compounds, which serve as nutrients for its microbial inhabitants. Some of these are relatively simple compounds (e.g., monosaccharides, amino acids, fatty acids) that can be used directly as sources of energy, carbon, nitrogen, etc., by individual species. Macromolecules, however, can present a problem to many organisms unable to secrete hydrolases, and these must rely on the activities of macromolecule-degrading organisms to provide assimilable degradation products. Examples of organisms able to degrade dietary polysaccharides to simple compounds are given in Table 7.3. Some macromolecules are so complex (e.g., mucins) that very few individual species are able to degrade the molecule entirely. Microbial consortia are able to achieve this in a sequential manner to the benefit of all their members. Some microbial communities form biofilms on particles composed of insoluble dietary polysaccharides and proteins which exist in the lumenal contents of the colon. The levels of enzymes involved in the degradation of insoluble plant cell wall components have been found to be higher in bacteria associated with particulate matter, which suggests that these adherent organisms are important in the digestion of complex carbohydrates. It has been estimated that approximately 5% of lumenal microbes are associated with food particles (Figure 7.13).

Many residents of the GIT (e.g., *Bifidobacterium* spp., *Enterococcus* spp. *Streptococcus* spp., *Lactobacillus* spp.) produce lactate as a metabolic end-product, and this can be utilised by a wide range of microbes, including desulphovibrios, bacteroides, veillonella, propionibacteria, clostridia, and enterobacteria. Other important metabolic end-products include acetate, ethanol, propionate, and butyrate, and these can also be utilised by other organisms, including sulphate-reducing bacteria. Hydrogen is produced in large quantities by many species during fermentation and is utilised by methanogenic

Table 7.22. Positive interactions occurring between members of the intestinal microbiota

Process	Benefit to other organisms
oxygen utilisation by aerobes/facultative anaerobes	creates environment suitable for growth of microaerophiles and anaerobes
degradation of polysaccharides	provides monosaccharides
degradation of proteins	provides amino acids
degradation of mucins and other glycoproteins	provides sugars, amino acids, and sulphate
excretion of metabolic end-products (e.g., lactate, ethanol, hydrogen, ammonia)	serve as nutrient sources for other organisms
quorum sensing	production of bacteriocins and virulence factors

and sulphate-reducing bacteria to produce methane and hydrogen sulphide, respectively. The main methanogens in the human colon are *Methanobrevibacter smithii* and *Methanosphaera stadtmaniae*, both of which have an obligate requirement for hydrogen. The other main hydrogen utilisers are the sulphate-reducing bacteria and in the human colon; there is an inverse relationship between the presence of these organisms and methanogens. The two groups of organisms compete for hydrogen, and this is an example of a negative interaction known as "nutritional competition".

Many colonic species are unable to ferment carbohydrates and are dependent on amino acids as carbon and energy sources – these include *Peptococcus* spp., *Eubacterium* spp., *Acidaminococcus* spp., and some *Fusobacterium* spp. Furthermore, many saccharolytic species utilise amino acids because a nitrogen source. The chyme entering the colon has a low content of free amino acids as most of these will have been absorbed during passage through the small intestine. The amino acids needed by colonic organisms, therefore, have to be supplied by the proteolytic activities of organisms, such as *Bacteroides* spp., *Propionibacterium* spp., *Clostridium* spp., and *Fusobacterium* spp. Amino-acid fermentation also requires electron acceptors (e.g., other amino acids, keto acids, fatty acids, hydrogen) produced by the metabolic activities of other organisms. The products of these reactions include ammonia, amines, SCFAs, BCFAs, phenols, hydroxy fatty acids, and α-keto acids. Ammonia is an important nitrogen source for many colonic organisms.

The possible role of quorum sensing in microbial communities of the GIT is receiving considerable attention, and some examples of activities regulated by the population density of intestinal species (*E. coli*, *Enterococcus* spp., *Lactobacillus* spp., *Clostridium* spp.) were given in Section 1.1.3 (see Table 1.4). Examples of positive interactions between intestinal bacteria are shown in Table 7.22.

7.4.4.2 Negative interactions

There are four main ways in which an organism can exert a detrimental effect on other members of the intestinal microbiota: (1) by consuming an essential nutrient, (2) by occupying a site for adhesion, (3) by creating an environment unsuitable for the growth of other species, and (4) by producing antimicrobial substances.

Competition for nutrients between species would be expected to exert an effect on the relative proportions of microbes in GIT communities. However, there is little direct evidence of this occurring. One example of the existence of such competition is provided by the inverse relationship between sulphate-reducing bacteria (SRB) and methanogens in the colon – individuals tend to have either one group of organisms or the other. Both of these groups of organisms require hydrogen for growth, but SRB have a greater substrate affinity for hydrogen, and generate greater energy and growth yields. SRB, therefore, will always outcompete methanogens for hydrogen, resulting in the displacement of the latter from the colonic microbiota. Why then do some individuals have methanogens rather than SRB? This is because SRB are dependent on the presence of sulphate in the colon; thus, the aforementioned nutrient competition only occurs in the presence of adequate levels of sulphate. SRB cannot colonise the colon of individuals with low sulphate levels so that the nutrient competition described will not take place, resulting in high levels of methanogens in the colonic microbiota. In continuous culture models of the colon, competition for a carbohydrate has been shown to be responsible for suppressing growth of both *E. coli* and *Cl. difficile*.

The ability of one species to adhere to a receptor site (whether on an artificial substratum or a host cell) and prevent adhesion of other strains has been well documented in the laboratory. It is apparent also that some species can actually displace already adherent organisms from adhesion sites. However, whether this occurs *in vivo* and so affects the composition of the microbial community at a site within the GIT remains to be established.

Acid production by many species can result in a decreased pH locally and inhibit the growth of other organisms. This is considered to be an important means by which lactic acid producing bacteria (e.g., lactobacilli, bifidobacteria, and streptococci) control the growth of various facultative and obligate anaerobes, including a range of intestinal pathogens.

Many intestinal organisms produce microbiostatic or microbicidal compounds, such as organic acids, H_2S, hydrogen peroxide, and bacteriocins. Although many studies have shown that bacteriocins can inhibit or kill intestinal microbes and pathogens *in vitro*, there is little evidence that these compounds influence the microbiotas of the GIT. However, many laboratory studies have demonstrated that organic acids, H_2S, and unconjugated bile acids (produced from the conjugated forms by intestinal bacteria) can all affect the composition of microbial communities *in vitro* and are likely to influence the composition of the colonic microbiota. The ability of specific microbes to influence the various microbiotas of the GIT (and of those at other anatomical sites) is discussed in greater detail in Sections 10.1 and 10.4.6.

7.4.5 Dissemination of organisms from the gastrointestinal tract

Microbes are disseminated from the GIT in faeces, which can be transferred either directly or indirectly to other individuals and also throughout the external environment. As faeces contains such enormous numbers of microbes (approximately 10^{10} per gram), large numbers of viable bacteria will be present even in small quantities of faecal matter. Faecal material may be present on skin regions adjacent to the anus, as well as on clothing and on the hands – the amounts present will depend on the level of personal hygiene of the individual. Inadequate hand washing after defaecation can lead

Table 7.23.	The ability of anaerobes and microaerophiles to survive in air	

Organism	Survival time (hours)
Pep. anaerobius	0.3–0.75
Eub. lentum (Eg. lenta)	0.75–1.0
F. nucleatum	0.75–1.0
B. vulgatus	4–8
P. acnes	4–8
B. fragilis	3–48
Bif. adolescentis	24–48
Cl. perfringens (vegetative cells)	>72
L. plantarum	>72

Note: This table shows the survival times of a variety of such organisms when a suspension of each is aerated.

to the transfer of faecal microbes to other parts of the body, to other individuals, and to food and water. The transfer of intestinal microbes to food constitutes an excellent means of dissemination because the organisms may be able to proliferate in this nutritious environment and the food may be subsequently distributed to many individuals. Once they have left the protection of the GIT, the ability of intestinal microbes to survive will depend on the particular species and on the nature of the new environment. Some organisms – such as enterococci – are extremely hardy and can survive for long periods of time on inanimate surfaces, even in direct sunlight. Clostridia, of course, can form spores if the new environment is not conducive to their growth and can then be dispersed to environments more suitable for their growth and reproduction. Non-sporing anaerobes (e.g., *Bacteroides* spp., *Fusobacterium* spp., and *Peptostreptococcus* spp.) and microaerophiles are less able to survive in an aerobic environment (Table 7.23). The ability of a number of intestinal (and other) species to survive in a variety of external environments (soil, river water, and drinking water) is shown in Table 7.24, from which it can be seen that non-sporing organisms – such as *E. coli* and other enterobacteria – can survive for surprisingly long periods (several months) in soil. Such prolonged survival times increase the possibility of transmission to a new host, offering a more conducive environment for growth and reproduction.

7.4.6 Effect of antibiotics and other interventions on the microbiotas of the gastrointestinal tract

7.4.6.1 Antibiotics

An enormous number of studies have been carried out on the effect of the administration of antibiotics on the microbiotas of the GIT – most of which have focussed on the colonic microbiota. In general, antibiotics tend to simplify the intestinal microbiota rather than eliminate it, although the effects produced vary with the class of antibiotic used and also between antibiotics from the same class. Such variation would be expected due to differences between antibiotics with respect to their pharmacokinetics, extent of inactivation within the GIT, and *in vitro* activities. Antibiotics such as ampicillin, clindamycin, and cefoperazone have major effects on the microbiota of the colon, causing dramatic reductions in the numbers of *Bacteroides* spp., anaerobic cocci,

| Table 7.24. | Survival of intestinal (and other) bacteria in soil, river water, and drinking water |

| | Survival in | | |
| | Garden soil at 20°C | Drinking water at 22°C | River water at 20°C |
Organism	Survival time (days)	Survival time (days)	Decimal reduction time (days)
Citrobacter sp.	75	–	–
Enter. cloacae	180	>48	–
E. coli	120	>34	8
Hafnia sp.	120	–	–
Klebsiella sp.	15	–	–
Enter. aerogenes	–	>48	5
Ent. faecalis	–	–	13
Ent. faecium	–	–	>20
Sal. typhimurium	–	–	7
Strep. bovis	–	–	1
Cit. freundii	–	>34	–
Hafnia alvei	–	>48	–
K. pneumoniae	–	>34	–
Alcaligenes faecalis	–	>48	–
Ps. aeruginosa	–	>48	–
Serratia marcescens	–	>34	–

Note: The decimal reduction time is the time taken for the original bacterial population to decrease to one-tenth of its initial value.

bifidobacteria, and lactobacilli. In contrast, antibiotics such as sulphonamides, penicillin G, metronidazole, chloramphenicol, and cefotaxime appear to have only minor effects on the microbiota. The effects of some commonly used antibiotics on the colonic microbiota are summarised in Table 7.25. The severe disruption of the colonic microbiota due to the administration of antibiotics, such as ampicillin and clindamycin, can result in the overgrowth of organisms such as *Cl. difficile* as a consequence of the elimination of microbes that normally control its growth. Furthermore, in those individuals

| Table 7.25. | Effects of antibiotics on the colonic microbiota |

Antibiotic	Main effects
amoxycillin and ampicillin	decrease in numbers of obligate and facultative anaerobes; increase in proportion of amoxycillin-resistant enterobacteria
cephalosporins	reduced proportions of enterobacteria; increased proportion of enterococci; increased isolation of *Cl. difficile*
quinolones	decreased proportions of enterobacteria; little effect on proportions of obligate anaerobes or enterococci
erythromycin	decreased proportions of obligate anaerobes; no effect on enterococci or Gram-positive bacilli
clindamycin	decreased proportions of obligate anaerobes

not already colonised by the organism (approximately 97% of adults), the antibiotic-induced changes to the colonic microbiota reduces its colonisation resistance, thus enabling infection with *Cl. difficile*, which is particularly prevalent in hospitals. The result is antibiotic-associated diarrhoea or the life-threatening condition pseudomembranous colitis.

7.4.6.2 Artificial nutrition

The use of a chemically defined diet by healthy volunteers has been shown to have a dramatic effect on the faecal microbiota, with decreased numbers of enterococci and increased numbers of enterobacteria. In hospitalised individuals, the use of total enteral nutrition (TEN) has also been shown to affect the faecal microbiota, with the ratio of obligate to facultative anaerobes decreasing from 878:1 in controls to 0.42:1 in the TEN patients. There were marked increases in the counts of facultative and aerobic species, such as *Enterococcus* spp., *Citrobacter* spp., and *Acinetobacter* spp., as well as decreases in the counts of many *Eubacterium* spp. and *Bacteroides* spp. The species diversity of TEN patients was significantly lower than that found in individuals on a normal diet, and this was due entirely to a reduction in the number of anaerobic species present. Patients fed by total parenteral nutrition (TPN) had an even lower species diversity, and many of the usual inhabitants of the colon were absent, including *Eubacterium* spp. and many *Bacteroides* spp. *Clostridium* spp. dominated the faecal microbiota of TPN patients. The dramatic change in the faecal microbiota of TPN patients is a consequence of the absence of dietary substrates in their GIT. The diet of the TEN patients was free of fibre, lactose, and gluten, and diarrhoea is commonly encountered in such patients – these factors may account for their abnormal faecal microbiota.

7.4.6.3 Probiotics and prebiotics

The effects of the administration of probiotics and prebiotics on the microbiotas of the GIT are described in Sections 10.1 and 10.2.

7.5 | Diseases caused by members of the intestinal microbiota

7.5.1 Diseases due to *Helicobacter pylori*

There is considerable evidence implicating *Hel. pylori* as the causative agent of a number of diseases, including acute gastritis, chronic atrophic gastritis, gastric ulcers, duodenal ulcers, gastric adenocarcinoma, and MALT lymphomas. Acute gastritis may occur in individuals within 2 weeks of colonisation by the organism – this is characterised by abdominal pain, nausea, and vomiting. However, in most individuals, colonisation occurs in a symptomless fashion and results in chronic gastritis, which goes unnoticed in most individuals – fewer than 15% of those colonised by the organism develop any symptoms of disease. *Hel. pylori* has a number of virulence factors which can damage host cells (Section 7.4.1.6), and this can lead to breaches in the mucosa exposing the underlying muscularis mucosae, which is then damaged by gastric acid resulting in a lesion known as an ulcer. Ulcers may occur in the stomach (mainly at the border between the antrum and corpus) or in the duodenum, and it has been estimated that 60–80% of gastric ulcers and 70–95% of duodenal ulcers are attributable to *Hel.*

pylori infection. *Hel. pylori* usually colonises the antrum of the stomach, resulting in inflammation that is confined mainly to this region. In some individuals, however, colonisation is more widespread and results in inflammation of the whole of the stomach (i.e., pangastritis). Pangastritis, over a period of many years, induces atrophy of the mucosa and reduces the number of functional parietal cells in the corpus, resulting in decreased acid secretion – this condition is known as chronic atrophic gastritis. Chronic atrophic gastritis may eventually lead to gastric adenocarcinoma, although this is a rare event affecting only 1 in 40,000 of those infected with the organism. Carcinoma of the stomach is the second most common form of cancer worldwide, and gastric adenocarcinomas account for 95% of gastric carcinomas. Most of the remaining gastric carcinomas are MALT lymphomas, and these also are caused by *Hel. pylori*. The association between *Hel. pylori* and carcinomas of the GIT is so strong that, in 1994, the World Health Organisation declared the organism to be a Group I (i.e., definite) human carcinogen. The organism has also been implicated in a number of other conditions, including liver cancer, acne rosacea, atherosclerosis, and iron-deficiency anaemia.

Hel. pylori is found only in humans, other primates, and some mammals (e.g., cats, sheep), and has been detected in saliva, dental plaque, faeces, and vomit. The most likely mode of transmission is person-to-person, and this may occur via the oral-oral or faecal-oral routes. Evidence in support of this is provided by the greater prevalence of the organism in institutionalised individuals and in those living in crowded conditions, as well as intra-familial case clustering. The organism may also be able to survive in water sources as the coccoid form, and could be transmitted via inadequately treated water supplies.

Treatment of infections due to the organism involves the use of a combination of two antibiotics (usually amoxycillin and either clarithromycin or metronidazole), together with a proton pump inhibitor (PPI), such as omeprazole, for up to 14 days. The PPI causes a rise in gastric pH, which induces the organism to enter the log phase of growth, rendering it more susceptible to the antibiotics.

7.5.2 Irritable bowel syndrome

Irritable bowel syndrome (IBS) is a poorly understood condition which affects between 8% and 22% of the population. It is a disease that usually starts in early adult life and is characterised by abdominal pain, excessive flatulence, and variable bowel habit. Nearly half of patients with the condition report the onset of symptoms following a course of antibiotics, abdominal or pelvic surgery, or an episode of gastroenteritis. This suggests that IBS may be attributable, at least in part, to some alteration in the bowel microbiota. Studies have shown that the faecal microbiota of patients with IBS contains significantly higher numbers of facultative anaerobes, but lower numbers of lactobacilli and bifidobacteria. It has been suggested that this alteration in the gut microbiota results in abnormal colonic fermentation, which in turn generates excessive quantities of intestinal gases, such as hydrogen. Patients with IBS, in fact, have been shown to have greater quantities of intestinal gas than healthy controls. One approach to treatment of the syndrome would be to restore the gut microbiota to one that is compatible with health, and it may be possible to achieve this by using probiotics or prebiotics – there is currently considerable interest in this approach.

7.5.3 Inflammatory bowel disease

Crohn's disease (CD) and ulcerative colitis (UC) are chronic inflammatory conditions which affect, respectively, the whole of the GIT (i.e., mouth to anus) and the colon together with the rectum. Collectively, these diseases are referred to as inflammatory bowel disease (IBD). They mainly affect individuals in developed countries, where the incidence is 3–5 per 100,000 for CD and 10 per 100,000 for UC. The aetiology of IBD has not been established, but four main theories have been proposed: (1) it is a reaction to a specific infecting agent – organisms implicated include *Mycobacterium paratuberculosis*, measles virus, *Saccharomyces cerevisiae*, *E. coli, Lis. monocytogenes*, and *Helicobacter hepaticus*; (2) it is the result of alterations in the function and/or composition of the GIT microbiota; e.g., defective SCFA metabolism or increased levels of sulphate-reducing bacteria; (3) it is a consequence of some mucosal defect which results in its exposure to the GIT microbiota; and (4) it is an aberrant host response due to loss of tolerance to the indigenous microbiota. Studies using gnotobiotic and cytokine-knockout animals confirm that the normal GIT microbiota is essential for the development of IBD, but its exact role remains to be established.

7.5.4 Diseases due to *Clostridium* spp.

Cl. perfringens is responsible for gas gangrene, a rapidly progressive, potentially fatal infection that involves the breakdown of muscle tissue. The disease may follow the contamination of wounds by the organism or its spores, provided the conditions are suitable for its growth and proliferation – appropriate conditions include a low redox potential, a compromised blood supply at the site, and the presence of peptides and amino acids. The organism is capable of producing a range of exotoxins, and five different toxin types are recognised (A–E) on the basis of which exotoxins are produced. Type A is the one most frequently associated with gas gangrene in humans, and it produces α-toxin (phospholipase C), which is responsible for most of the tissue damage accompanying the disease. This exotoxin hydrolyses phosphatidylcholine and sphingomyelin in the membranes of a variety of host cells resulting in their lysis. Destruction of muscular tissue is accompanied by gas production (resulting in the formation of gas bubbles under the skin), a foul-smelling discharge, fever, toxaemia, shock, and death. Treatment involves surgical removal of the affected tissue, the administration of penicillin and, sometimes, hyperbaric oxygen.

Type A strains of *Cl. perfringens* are also responsible for food poisoning associated with undercooked meat or meat products. Spores surviving the initial cooking process germinate and are ingested with the food. In the small intestine, the vegetative cells undergo sporulation, and this is accompanied by the production of an enterotoxin. The enterotoxin binds to receptors on intestinal epithelial cells and damages their membranes, which alters their permeability and results in diarrhoea 7–15 hours after consumption of the contaminated food. The disease is usually mild, and patients recover after 2–3 days.

Cl. difficile is present in the colon of only approximately 3% of adults, and it is considered by many to not be a member of the indigenous microbiota. However, it can be detected in 49–66% of formula-fed infants and in 6–20% of breast-fed infants, and could be considered a constituent of the colonic microbiota of infants. The organism is responsible for pseudo-membranous colitis, and this is discussed in greater detail in Section 9.1.3.1.

7.5.5 Intra-abdominal infections

Large numbers of colonic microbes can gain access to the peritoneum as a consequence of penetrating trauma, certain surgical procedures, and diseases such as appendicitis, diverticulitis, or cancer. This can result in infections, such as peritonitis, subphrenic, and intra-abdominal abscesses and abscesses of the liver, spleen, and pancreas. Many of these can progress to a septicaemia, and the morbidity and mortality associated with these infections is high. Intra-abdominal infections are invariably polymicrobial and involve obligate anaerobes (mainly *Bacteroides* spp., GPAC, *Clostridium* spp., *Prevotella* spp.) and facultative anaerobes (mainly enterobacteria and *Enterococcus* spp.). The peritoneal cavity is an aerobic environment, and many members of the colonic microbiota are killed soon after they gain entry. However, growth of facultative anaerobes quickly reduces the oxygen content and redox potential of the cavity, allowing growth of obligate anaerobes. *B. fragilis* is one of the most frequently isolated anaerobes from such infections and, as well as being oxygen-tolerant, has a number of virulence factors enabling it to circumvent host defence systems. One of the most important of these is its capsular polysaccharide, which protects the organism against phagocytosis, stimulates the release of a range of pro-inflammatory cytokines from a variety of host cells, and induces abscess formation. It also produces a number of extracellular enzymes that can cause tissue damage; these include hyaluronidase, heparinase, chondroitin sulfatase, phosphorylase, DNase, and proteases.

7.5.6 Diseases due to *Enterobacteriaceae*

Some of the enterobacteria normally resident in the GIT of humans are able to cause disease in healthy individuals. For example, *E. coli* is the main organism responsible for UTIs in females (Section 7.5.7) and is an important cause of meningitis in neonates. Many of these organisms are also opportunistic pathogens that are capable of causing a variety of infections in compromised hosts (e.g., hospitalised patients), including UTIs, pneumonia, septicaemia, peritonitis, meningitis, and wound infections. They are responsible for more than 50% of nosocomial infections, with *E. coli* being the most frequent causative agent. Other organisms frequently involved in such infections include *Klebsiella* spp., *Enterobacter* spp., *Serratia* spp., *Citrobacter* spp., and species belonging to the *Proteus-Providencia-Morganella* group.

7.5.7 Urinary tract infections

The main causative agents of UTIs are inhabitants of the large intestine – these include *E. coli*, *Staph. saprophyticus*, *Enterococcus* spp., *Proteus* spp., *Klebsiella* spp., and *Enterobacter* spp.

7.5.7.1 Urinary tract infections in females

Strictly speaking, the term UTI can refer to an infection of any region of the urinary tract (Section 5.1.1) and would include infections of the urethra (urethritis), bladder

Table 7.26.	Groups with a high risk of contracting urinary tract infections

infants
pregnant women
the elderly
patients with spinal cord injuries
patients with catheters
patients with diabetes
patients with multiple sclerosis
patients with acquired immunodeficiency syndrome
patients with underlying urological abnormalities

(cystitis), and kidney (pyelonephritis). However, urethritis is most commonly due to exogenous pathogens that are transmitted by sexual intercourse and are beyond the scope of this book. Also, the most common infections of the urinary tract are cystitis and pyelonephritis. For these reasons, the term UTI is usually understood to mean an infection of the bladder or the kidney – these being referred to as a lower UTI (or cystitis) and an upper UTI (or pyelonephritis), respectively. The vast majority (95%) of UTIs affect only the bladder and are diagnosed on the basis of the presence of 10^5 viable bacteria per millilitre of urine (i.e., a "significant bacteriuria"). There are three possible routes by which bacteria can reach the kidney and bladder – via the bloodstream (haematogenous route), via the lymphatic system, or via the urethra (ascending route). The latter is by far the most common route of infection, whereas the lymphatic pathway remains speculative. Infection via the haematogenous route does occur, but usually as a consequence of a bacteraemia involving an organism not generally considered to be a typical uropathogen or member of the urethral microbiota (e.g., *Staph. aureus*, *Salmonella* spp., and *Mycob. tuberculosis*).

UTIs are among the most prevalent infectious diseases affecting humans, and it has been estimated that, in the United States, there are more than 8 million episodes per year. The disease is more common in women than in men, and between 40% and 50% of women will have at least one episode of UTI during their lifetime. Recurrent infections occur in 27–48% of healthy women who have no detectable abnormality of their urinary tract. Although all women are susceptible to UTIs, certain groups are at greater risk, and these are listed in Table 7.26. The aetiology of UTIs depends on a number of factors, and it is convenient to classify these infections as either "uncomplicated" or "complicated". An uncomplicated UTI is one involving a pre-menopausal, sexually active, non-pregnant woman with no genito-urinary abnormality who has not recently had genito-urinary tract instrumentation or received antibiotics. A complicated UTI is one in which the woman has a condition that increases the risk of treatment failure; this includes an abnormality of the urinary tract that interferes with urine flow, the presence of a foreign body (e.g., a catheter), infection with a multi-drug-resistant organism, immunosuppression, a systemic disease (e.g., diabetes), or pregnancy.

The majority of uncomplicated UTIs (80%) are caused by *E. coli*, with *Staph. saprophyticus* being responsible for between 10% and 15% of cases (Table 7.27). Less frequent causative agents include *Klebsiella* spp., *Enterobacter* spp., *Proteus* spp., and enterococci. The principal source of the organisms responsible for uncomplicated UTIs is the intestinal tract, although *Staph. saprophyticus*, as well as being a resident of the rectum,

| Table 7.27. | Aetiological agents of urinary tract infections in different female patient groups |

Patient group	Main aetiological agents
those with uncomplicated UTIs	*E. coli, Staph. saprophyticus, Klebsiella* spp.
paediatric hospitalised patients	*E. coli, Candida* spp., *Enterococcus* spp., Gram-negative non-fermenters, *Enterobacter* spp., *Pseudomonas* spp.
patients with diabetes	*E. coli, Klebsiella* spp., *Strep. agalactiae, Enterococcus* spp.
patients with spinal cord injuries	*E. coli, Pr. mirabilis, Pseudomonas* spp.
catheterised patients	*E. coli, Pr. mirabilis, Candida* spp., enterococci, *Ps. aeruginosa, Klebsiella* spp., *Enterobacter* spp., *Staph. aureus*
patients infected with HIV	*Enterococcus* spp., *E. coli*
pregnant patients	*E. coli, Proteus* spp., *Klebsiella* spp., *Strep. agalactiae*, coagulase-negative staphylococci
elderly patients	*E. coli, Proteus* spp., *Klebsiella* spp., *Serratia* spp., *Pseudomonas* spp., *Ent. faecalis, Staph. aureus, Staph. epidermidis*

Notes: HIV = human immunodeficiency virus; UTIs = urinary tract infections.

is also a member of the skin microbiota. Unlike UTIs due to *E. coli*, the incidence of *Staph. saprophyticus* UTIs shows a seasonal variation, with infections peaking during late summer/early autumn. This seasonal variation is similar to that seen with sexually transmitted diseases, and this observation – together with the fact that sexually active men and women are affected preferentially – implies that staphylococcal UTIs are sexually transmitted. Apart from recent sexual intercourse, other risk factors for UTIs due to *Staph. saprophyticus* include prior outdoor swimming, the use of condoms with vaginal spermicides, and working in the meat-processing industry (the organism is a frequent contaminant of raw meat, particularly beef and pork).

A number of studies have investigated the genetic similarity between *E. coli* strains isolated from the urine and rectum of patients with UTIs. Restriction fragment-length polymorphism analysis of such strains has demonstrated that the strain causing the UTI is also invariably present in the patient's rectum. While the *E. coli* strains responsible for UTIs are members of the intestinal microbiota, the organism gains entry to the bladder after phases of colonisation of the perineum, vagina, periurethra, and urethra. Colonisation of the vagina by *E. coli* invariably precedes cystitis, and factors affecting the ability of the organism to colonise the vagina also affect the risk of contracting a UTI. In women who suffer from recurrent UTIs, the vagina is more often colonised by *E. coli* and harbours greater numbers of the organism than individuals who have not had a UTI. Furthermore, *E. coli* can adhere in greater numbers to the vaginal epithelial cells of women with recurrent UTIs. The main factors associated with an increased risk of UTI are listed in Table 7.28. The ability of lactobacilli to control vaginal colonisation by *E. coli* and other uropathogens and, hence, the risk of cystitis is due to a number of factors. Firstly, lactobacilli are able to block the adhesion of uropathogens to epithelial cells. Cell-wall fragments from lactobacilli, in particular, a lipoteichoic acid-peptidoglycan

Table 7.28. | Risk factors for urinary tract infections in females

Factor	Effect
decrease in the proportion of lactobacilli in the vagina	increased colonisation by uropathogens; increased risk of UTI
sexual intercourse	introduces bacteria into the bladder; increased risk of UTI
use of spermicides and diaphragm	decrease in the proportion of lactobacilli in the vagina; increased risk of UTI
use of antibiotics	altered genital microbiota and increased colonisation by uropathogens; increased risk of UTI
onset of menopause	altered genital microbiota and increased colonisation by uropathogens; increased risk of UTI

Note: UTI = urinary tract infection.

complex – are also effective at blocking adhesion. Secondly, they are able to inhibit the growth of a range of uropathogens by a number of mechanisms – production of hydrogen peroxide, generation of a low pH by secreting acidic metabolic end-products, and the production of bacteriocins. Thirdly, lactobacilli can co-aggregate with *E. coli* and other uropathogens, thereby preventing their binding to epithelia and mucus, thus facilitating their removal. Finally, they produce biosurfactants that are able to prevent adhesion of uropathogens to epithelia.

Judging by the incidence of cystitis, however, uropathogens do often reach the bladder where they can initiate a disease process. The ability of *E. coli* to cause cystitis is an attribute of only certain strains of the organism, which are known as uropathogenic *E. coli* (UPEC). The first stage in the disease process is colonisation of the bladder epithelium, and UPEC possess a number of specialised adhesive structures for this purpose, the most

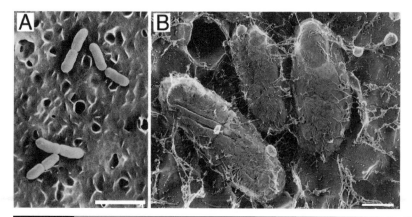

Figure 7.14 Scanning electron micrographs showing type I pilus-mediated attachment of uropathogenic *E. coli* to the bladder epithelium of mice. (A) Adherent cells of *E. coli*. Bar = 3 μm. (B) At a higher magnification, type I pili are evident. The pili can be seen linking bacteria to the bladder epithelium and also linking adjacent bacteria – these pili are also known to be involved in biofilm formation by the organism. Bar = 0.5 μm. Reproduced with permission from: Mulvey, M.A., Schilling, J.D., Martinez, J.J., and Hultgren, S.J. *Proceedings of the National Academy of Sciences USA* 2000;97:8829–8835. Copyright © 2000, National Academy of Sciences.

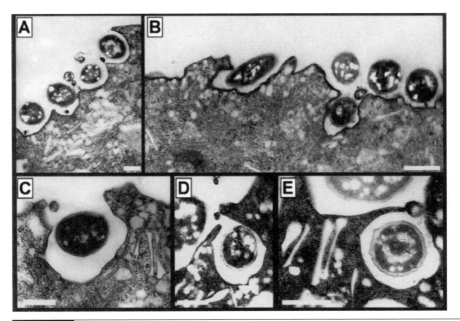

Figure 7.15 Transmission electron micrographs showing the internalisation of type 1-piliated uropathogenic *E. coli* (UPEC) by mouse bladder epithelium. Bars = 1.0 μm (A and B) and 0.5 μm (C–E). Reproduced with permission from: Mulvey, M.A., Schilling, J.D., Martinez, J.J., and Hultgren, S.J. *Proceedings of the National Academy of Sciences USA* 2000;97;8829–8835. Copyright © 2000, National Academy of Sciences.

important of which from the point of view of cystitis is the type 1 pili (Figure 7.14). Each pilus consists of a long thick cylinder joined to a short, thinner fibrillum, at the end of which is the adhesin FimH, which mediates binding to the epithelium. The host-cell receptors for FimH include a number of mannose-containing glycoproteins, but the main one is uroplakin 1a, which is a membrane glycoprotein expressed on all cells in the outermost layer of the bladder mucosa. Attachment to an epithelial cell induces a series of cytoskeletal rearrangements in the latter, which result in the envelopment of the bacterium and, ultimately, its uptake into the cell (Figure 7.15). This is a survival mechanism protecting the organism against removal by urine flow and also provides it with a far more nutritious environment than that supplied by urine. However, the host cell responds by undergoing apoptosis so that it will be shed, carrying with it any attached or internalised bacteria (Figure 7.16). Before this occurs, the organism must replicate, exit the cell, and adhere to other cells if it is to persist in the bladder. There is evidence that UPEC does multiply rapidly within the invaded cell, resulting in the formation of large inclusions. They then re-emerge from the invaded cell in an elongated filamentous form and adhere to neighbouring or underlying epithelial cells. Alternatively, bacteria may be able to escape from the exfoliated cell before this is expelled from the bladder (Figure 7.17). As well as initiating apoptosis, invasion of a host cell also induces (in response to the bacterium's LPS) the release of inflammatory cytokines (mainly IL-6 and IL-8), which results in the recruitment of PMNs and the disposal of the bacteria by phagocytosis. In most cases, the host response is effective at killing all of the UPEC; but, in some cases, a few bacteria persist in the bladder in a quiescent state and can manage to avoid detection by host surveillance mechanisms – these constitute a reservoir for

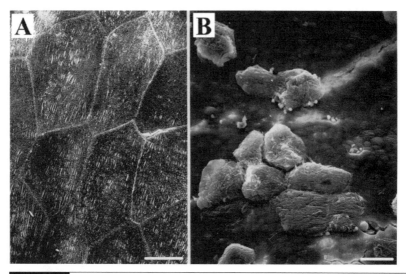

Figure 7.16 Scanning electron micrographs showing (A) typical hexagonal cells of the outer layer of the bladder epithelium of a mouse and (B) the appearance of the epithelium 6 hours after infection with type 1-piliated *E. coli*. Many of the infected, outermost cells have exfoliated, thereby exposing the smaller, underlying epithelial cells. Some of the remaining outermost, infected cells have lost their hexagonal shape. Bar = 50 μm. Reprinted with the permisssion of Blackwell Publishing from: Adhesion and entry of uropathogenic *Escherichia coli*. Mulvey, M.A. *Cellular Microbiology* 2002;4:257–271.

subsequent infections. As many as 25% of women suffer another UTI within 6 months of the initial infection; and, in many cases, the UPEC strain involved in both infections is the same. This recurrence could be due to the re-emergence of the organism from its quiescent state in the bladder or else it could be the result of re-colonisation of the bladder with the same strain from its primary habitat: the GIT.

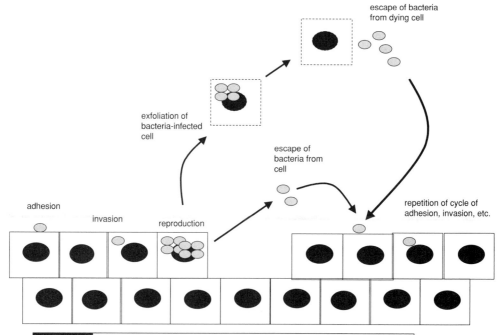

Figure 7.17 Persistence of uropathogenic *E. coli* in the bladder. See text for explanation.

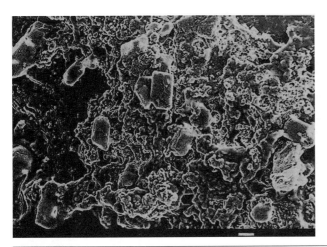

Figure 7.18 Scanning electron micrograph showing encrustation on a urethral catheter (bar = 10 μm). Crystals of struvite (coffin-shaped) are visible, together with smaller crystals of hydroxyapatite and brushite. Reprinted with permission from: Biofilm complications of urinary tract infections. Gorman, S.P. and Jones, D.S. In: *Medical implications of biofilms.* Wilson, M. and Devine, D. (eds.). Copyright © 2003, Cambridge University Press.

UTIs due to urease-producing organisms, such as *Proteus* spp., can result in an additional complication – the production of urinary stones. These stones consist of ammonium urate, struvite (magnesium ammonium phosphate), and/or carbonate apatite. Although there are other causes of urinary-stone formation, approximately 15% of stones are produced as a result of infection. Urease-producing organisms hydrolyse urea to ammonia and carbon dioxide, which results in a rise in pH until the urine becomes alkaline. The ammonium and bicarbonate ions produced then react with magnesium and or calcium ions to produce complex insoluble carbonate and phosphate compounds. The resulting crystals often nucleate on the bacterial surface and then may grow rapidly to produce visible stones within 4–6 weeks. Stones may form in the kidney or other parts of the urinary tract, or encrustations may accumulate on urinary catheters (Figure 7.18). It has also been suggested that nanobacteria may be responsible for stone formation, but this is a very controversial area and the issue remains unresolved.

The presence of an indwelling catheter is a major risk factor for UTIs, and between 10% and 20% of catheterised individuals develop a UTI. Both the lumen and the outer surface of the catheter rapidly become colonised by one or more members of the urethral microbiota, or by members of the colonic microbiota, such as *E. coli* and *Pr. mirabilis*. A substantial biofilm (up to 200 μm thick) often forms on the lumenal surface, and most infections appear to be due to organisms ascending into the bladder from this biofilm (Figure 7.19). The biofilm usually consists of a number of organisms, the most frequently isolated species being *Staph. epidermidis, Ent. faecalis, E. coli, Pr. mirabilis, Providencia stuartii,* and *Ps. aeruginosa*. There is considerable interest in developing means of preventing biofilm formation on the catheter and, hence, reducing the risk of a UTI, and this is described further in Section 10.3.4.

Acute uncomplicated pyelonephritis is associated with the same risk factors and has a similar bacteriology, with *E. coli* being the aetiological agent in approximately 85% of cases. However, it is far less common than cystitis and is accompanied by back, side,

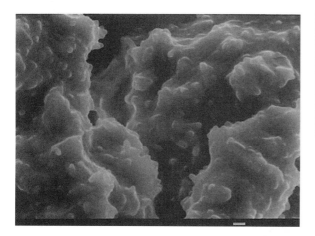

Figure 7.19 Biofilm on a urinary catheter (bar = 1.0 μm). Gram-negative bacilli (approximately 1–2 μm in length) are embedded within a confluent glycocalyx. Reprinted with permission from: Biofilm complications of urinary tract infections. Gorman, S.P. and Jones, D.S. In: *Medical implications of biofilms*. Wilson, M. and Devine, D. (eds.). Copyright © 2003, Cambridge University Press.

and groin pain; nausea; vomiting; fever; malaise; and blood in the urine. In about 30% of cases, pyelonephritis is complicated by bacteraemia.

7.5.7.2 Urinary tract infections in males

UTIs are much rarer in young males than in young females, and the prevalence in the 30- to 65-year age group (0.1%) is almost 100-fold lower than in women of comparable age. This is largely the result of three factors. Firstly, the much greater length of the male urethra makes it more difficult for bacteria to gain access to the bladder (Section 5.2.1). Secondly, in the female, the external urethral orifice is close to the main sources of uropathogenic bacteria (the anus and perineum), whereas in males it is not (Section 5.1.1). Thirdly, in females, the orifice is within the moist mucosa of the vaginal vestibule, which has a greater microbial density than the drier epithelium of the glans penis of males. Interestingly, uncircumcised males are more likely to suffer from a UTI than circumcised males; presumably, this is related to the finding that, in uncircumcised individuals, the glans penis not only has a greater population density of microbes, but also is more likely to contain facultatively anaerobic Gram-negative rods. When a UTI does occur in a young male, it is usually due to a structural or functional abnormality or is the consequence of some form of instrumentation. Such infections resemble the "complicated" UTIs of females rather than the "uncomplicated" form of UTI. The frequency of UTIs increases with age and disability, so that approximately half of male (and female) inmates of nursing homes have a UTI. The increase in the prevalence of UTI with age is mainly a consequence of prostatic enlargement, which causes partial obstruction of the urethra, thereby preventing complete emptying of the bladder. The continuous presence of urine in the bladder provides an opportunity for bacteria to grow and to reach levels that can overwhelm host defences.

Most UTIs (approximately 75%) in men are caused by Gram-negative facultative anaerobes, the rest being due to members of the urethral, GIT, and cutaneous microbiotas (e.g., *Ent. faecalis, Staph. epidermidis*, and *Staph. aureus*). Unlike the situation in females, a wider range of Gram-negative bacilli are involved in UTIs, and only 25–50% of infections are due to *E. coli*, with the remainder being caused by *Proteus* spp., *Providencia* spp., *Klebsiella* spp., *Enterobacter* spp., *Pseudomonas* spp., and *Citrobacter* spp. *Staph. saprophyticus*, a frequent cause of UTIs in females, is a rare uropathogen in males.

Table 7.29.	Carcinogens and mutagens produced by members of the colonic microbiota

Carcinogen/mutagen	Source
nitrosamines	reaction between nitrite and amines produced by the microbial decarboxylation of amino acids
deoxycholic acid, lithocholic acid, ursodeoxycholic acid	microbial action on primary bile acids
fecapentaenes	synthesised by bacteria from ether phospholipids
ethionine	microbial production from methionine
tryptophan metabolites	microbial metabolism of tryptophan

7.5.8 Colorectal cancer

Colorectal cancer (CRC) is the fourth most common cause of cancer-related mortality in the world and affects 6% of individuals by the age of 75. The incidence of the disease is much greater in developed than developing countries. Diet plays a significant role in the disease, but there is uncertainty regarding which dietary components are associated with an increased risk of CRC. However, there is evidence to suggest that diets rich in vegetables protect against CRC and that high-fibre diets may also be protective. Consumption of meat or high intakes of saturated fats appear to be associated with an increased risk of CRC. It has long been suggested that microbes inhabiting the colon may play a role in the initiation of CRC by converting dietary constituents to carcinogenic compounds (Table 7.29). Colonic microbes are also able to transform the bile acids produced by the host into potential carcinogens. Bile acids are produced by the liver from cholesterol and are then conjugated to either glycine or taurine. These conjugated forms are then secreted into the small intestine via the gallbladder, where they are involved in the emulsification, digestion, and absorption of dietary fats. Approximately 95% of the conjugated bile salts are reabsorbed by the small intestine (particularly by the ileum) and colon, and are then returned to the small intestine – this cycle is often referred to as the enterohepatic circulation. The unabsorbed bile salts enter the colon, where some undergo transformation by members of the gut microbiota, resulting in the production of a variety of metabolites. One of the main transformations accomplished by colonic microbes is deconjugation, resulting in the liberation of bile acids. This is achieved by many species, including *Bacteroides* spp., *Bifidobacterium* spp., *Fusobacterium* spp., *Clostridium* spp., and *Lactobacillus* spp. The free bile acids have greater antimicrobial activities than the conjugated forms, particularly against Gram-positive organisms. Following deconjugation, the primary bile acids cholic and chenodeoxycholic acids may be converted into deoxycholic and lithocholic acids. These reactions are achieved mainly by *Clostridium* spp. and *Eubacterium* spp. Unfortunately, both of these acids, particularly deoxycholic acid, are cancer promoters and co-carcinogens.

Although intestinal microbes have been shown to be capable of producing mutagens and/or carcinogens from a range of compounds present in the colon (Table 7.29), the involvement of these in the induction of CRC remains to be established. It has also been suggested that phenol produced by bacterial metabolism of tyrosine may be involved in the initiation of leukaemia. Tyrosine-metabolising organisms in the colon include *B. fragilis*, *Schleiferella asaccharolytica*, *E. coli*, and *Proteus* spp. On the other hand, it is thought that the production of butyrate by intestinal microbes may provide some protection against CRC.

7.5.9 Systemic infections resulting from bacterial translocation

In healthy individuals, bacteria are continually crossing the intestinal mucosa and are then transported in lymph to extra-intestinal sites, including the mesenteric lymph nodes, liver, kidney, spleen, and bloodstream. This phenomenon is known as bacterial translocation and occurs at a very low rate and involves very small numbers of microbes in healthy individuals – most of the organisms being killed by the normal host defence mechanisms. Bacteria vary with respect to their ability to tranlocate, with facultative anaerobes – such as *E. coli*, *K. pneumoniae*, and *Pr. mirabilis* – being the most effective, while obligate anaerobes are poor translocators. A number of factors are known to increase the rate of translocation, including (1) physical or chemical damage to the intestinal epithelium, thereby increasing its permeability; (2) disruption of the intestinal climax community (e.g., by antibiotics), leading to overgrowth by facultative anaerobes; and (3) a defect in any of the host defence systems. In such cases, the possibility of a systemic infection resulting from bacterial translocation is dramatically increased, and it has been shown that those at high risk of a systemic infection from translocating intestinal organisms are hospitalised individuals who are immunosuppressed, undergoing surgery, or trauma patients. The organisms most often associated with systemic infections in such patients are facultatively anaerobic Gram-negative rods (mainly *E. coli*, *K. pneumoniae*, *Enterobacter* spp., and *Pr. mirabilis*), *Enterococcus* spp., *Streptococcus* spp., and *Can. albicans*.

7.5.10 Contaminated small bowel syndrome

In comparison with the large intestine, the number of microbes in the small intestine is very small. This is due to a combination of factors, including the low pH of the region, the rapid peristalsis which reduces transit time, and local host defences. Impairment of any of these results in an increase in the microbial population, and this is known as "contaminated small bowel syndrome" or "bacterial overgrowth syndrome". The major causes include anatomical abnormalities of the gut, gut obstruction, disordered gut motility, reduced gastric acidity, and infection of the biliary duct. The main symptoms associated with the syndrome include steatorrhoea (increased faecal fat) and vitamin B_{12} deficiency. The increased microbial load in the small intestine results in rapid deconjugation of bile salts, creating water-insoluble free bile acids which are not reabsorbed and recycled. This eventually depletes the levels of bile salts in the small intestine so that micelle formation does not take place. Consequently, fat is not absorbed and is excreted. The small intestine is an important site of vitamin B_{12} absorption. This vitamin binds to intrinsic factor produced by the stomach, and the resulting complex is then absorbed. However, the vitamin is also needed by many bacterial species, and large numbers of bacteria will rapidly reduce its concentration, thus depriving the host of this important dietary constituent. Vitamin B_{12} deficiency ultimately leads to megaloblastic anaemia. Other manifestations of the syndrome include carbohydrate malabsorption, hypoproteinaemia, and diarrhoea. Treatment involves the surgical correction of any abnormalities, the administration of antibiotics, and vitamin supplementation.

7.6 | Further Reading

Books

Fuller, R. and Perdigon, G. (eds.). (2003). *Gut flora, nutrition, immunity, and health*. Oxford: Blackwell Publishing.

Gibson, G.R. and Macfarlane, G.T. (eds.). (1995). *Human colonic bacteria: role in nutrition, physiology, and pathology*. Boca Raton: CRC Press.

Gibson, G.R. and Roberfroid, M.B. (eds.). (1999). *Colonic microbiota, nutrition, and health*. Dordrecht: Kluwer Academic Publishers.

Hentges, D.J. (ed.). (1983). *Human intestinal microflora in health and disease*. New York: Academic Press.

Hill, M.J. (ed.). (1986). *Microbial metabolism in the digestive tract*. Boca Raton: CRC Press.

Mackie, R.I., White, B.A. and Isaacson, R.E. (eds.). (1997). *Gastrointestinal microbiology; Volume 2. Gastrointestinal microbes and host interactions*. New York: Chapman and Hall.

Rowland, I.R. (ed.). (1997). *Role of the gut flora in toxicity and cancer*. London: Academic Press.

Reviews and Papers

Adlercreutz, H. (1998). Evolution, nutrition, intestinal microflora, and prevention of cancer: a hypothesis. *Proceedings of the Society for Experimental Biology and Medicine* **217**, 241–246.

Andrieux, C., Membre, J.M., Cayuela, C., and Antoine, J.M. (2002). Metabolic characteristics of the faecal microflora in humans from three age groups. *Scandinavian Journal of Gastroenterology* **37**, 792–798.

Axon, A. (2002). Review article: gastric cancer and *Helicobacter pylori*. *Alimentary Pharmacology and Therapeutics* **16** (Suppl 4), 83–88.

Belley, A., Keller, K., Gottke, M., Chadee, K., and Goettke, M. (1999). Intestinal mucins in colonization and host defence against pathogens. *American Journal of Tropical Medicine and Hygiene* **60** (Suppl 4), 10–15.

Bennet, R., Eriksson, M., and Nord, C.E. (2002). The fecal microflora of 1–3-month-old infants during treatment with eight oral antibiotics. *Infection* **30**, 158–160.

Blaser, M.J. (1999). Hypothesis: the changing relationships of *Helicobacter pylori* and humans: implications for health and disease. *Journal of Infectious Diseases* **179**, 1523–1530.

Blaser, M.J. and Atherton, J.C. (2004). *Helicobacter pylori* persistence: biology and disease. *Journal of Clinical Investigation* **113**, 321–333.

Blaut, M., Collins, M.D., Welling, G.W., Dore, J., van Loo, J., and de Vos, W. (2002). Molecular biological methods for studying the gut microbiota: the EU human gut flora project. *British Journal of Nutrition* **87** (Suppl 2), S203–S211.

Blum, S., Alvarez, S., Haller, D., Perez, P., and Schiffrin, E.J. (1999). Intestinal microflora and the interaction with immunocompetent cells. *Antonie Van Leeuwenhoek* **76**, 199–205.

Comstock, L.E. and Coyne, M.J. (2003). *Bacteroides thetaiotaomicron*: a dynamic, niche-adapted human symbiont. *Bioessays* **25**, 926–929.

Cummings, J.H. (1998). Dietary carbohydrates and the colonic microflora. *Current Opinion in Clinical Nutrition and Metabolic Care* **1**, 409–414.

Cummings, J.H., Macfarlane, G.T., and Macfarlane, S. (2003). Intestinal bacteria and ulcerative colitis. *Current Issues in Intestinal Microbiology* **4**, 9–20.

Dai, D. and Walker, W.A. (1999). Protective nutrients and bacterial colonisation in the immature human gut. *Advances in Pediatrics* **46**, 353–382.

Donskey, C.J., Hujer, A.M., Das, S.M., Pultz, N.J., Bonomo, R.A., and Rice, L.B. (2003). Use of denaturing gradient gel electrophoresis for analysis of the stool microbiota of hospitalized patients. *Journal of Microbiological Methods* **54**, 249–256.

Edwards, C.A. and Parrett, A.M. (2002). Intestinal flora during the first months of life: new perspectives. *British Journal of Nutrition* **88** (Suppl 1), S11–S18.

Falk, P.G., Hooper, L.V., Midtvedt, T., and Gordon, J.I. (1998). Creating and maintaining the gastrointestinal ecosystem: what we know and need to know from gnotobiology. *Microbiology and Molecular Biology Reviews* **62**, 1157–1170.

Fallingborg, J. (1999). Intraluminal pH of the human gastrointestinal tract. *Danish Medical Bulletin* **46**, 183–196.

Fanaro, S., Chierici, R., Guerrini, P., and Vigi, V. (2003). Intestinal microflora in early infancy: composition and development. *American Journal of Clinical Dermatology* **4**, 641–654.

Farrell, R.J. and LaMont, J.T. (2002). Microbial factors in inflammatory bowel disease. *Gastroenterology Clinics of North America* **31**, 41–62.

Fihn, S.D. (2003). Acute uncomplicated urinary tract infection in women. *New England Journal of Medicine* **349**, 259–266.

Ghose, C., Perez-Perez, G.I., Dominguez-Bello, M.G., Pride, D.T., Bravi, C.M., and Blaser, M.J. (2002). East Asian genotypes of *Helicobacter pylori* strains in Amerindians provide evidence for its ancient human carriage. *Proceedings of the National Academy of Sciences of the USA* **99**, 15107–15111.

Gilmore, M.S. and Ferretti, J.F. (2003). The thin line between gut commensal and pathogen. *Science* **299**, 1999–2002.

Go, M.G. (2002). Natural history and epidemiology of *Helicobacter pylori* infection. *Alimentary Pharmacology and Therapeutics* **16** (Suppl 1), 3–15.

Goldman, A.S. (2000). Modulation of the gastrointestinal tract of infants by human milk. Interfaces and interactions. An evolutionary perspective. *Journal of Nutrition* **130** (2S Suppl), 426S–431S.

Gregg, C.R. (2002). Enteric bacterial flora and bacterial overgrowth syndrome. *Seminars in Gastrointestinal Disease* **13**, 200–209.

Guarner, F. and Malagelada, J.R. (2003). Gut flora in health and disease. *Lancet* **361**, 512–519.

Gunn, J.S. (2000). Mechanisms of bacterial resistance and response to bile. *Microbes and Infection* **2**, 907–913.

Hart, A.L., Stagg, A.J., Frame, M., Graffner, H., Glise, H., Falk, P., and Kamm, M.A. (2002). The role of the gut flora in health and disease and its modification as therapy. *Alimentary Pharmacology and Therapeutics* **16**, 1383–1393.

Hayashi, H., Sakamoto, M., and Benno, Y. (2002). Phylogenetic analysis of the human gut microbiota using 16S rDNA clone libraries and strictly anaerobic culture-based methods. *Microbiology and Immunology* **46**, 535–548.

Hebuterne, X. (2003). Gut changes attributed to ageing: effects on intestinal microflora. *Current Opinion in Clinical Nutrition and Metabolic Care* **6**, 49–54.

Hecht, G. (1999). Innate mechanisms of epithelial host defence: spotlight on intestine. *American Journal of Physiology* **277**, C351–C358.

Heller, F. and Duchmann, R. (2003). Intestinal flora and mucosal immune responses. *International Journal of Medical Microbiology* **293**, 77–86.

Hill, M.J. (1998). Composition and control of ileal contents. *European Journal of Cancer Prevention* **7** (Suppl 2), S75–S78.

Hooton, T.M. (2000). Pathogenesis of urinary tract infections; an update. *Journal of Antimicrobial Chemotherapy* **46** (Suppl S1), 1–7.

Hopkins, M.J., Sharp, R., and Macfarlane, G.T. (2002). Variation in human intestinal microbiota with age. *Digestive and Liver Disease* **34** (Suppl 2), S12–S18.

Hoy, C.M. (2001). The role of infection in necrotising enterocolitis. *Reviews in Medical Microbiology* **12**, 121–129.

Huijsdens, X.W., Linskens, R.K., Mak, M., Meuwissen, S.G.M., Vandenbroucke-Grauls, C.M.J.E., and Savelkoul, P.H.M. (2002). Quantification of bacteria adherent to gastrointestinal mucosa by real-time PCR. *Journal of Clinical Microbiology* **40**, 4423–4427.

Justice, S.S., Hung, C., Theriot, J.A., Fletcher, D.A., Anderson, G.G., Footer, M.J., and Hultgren, S.J. (2004). Differentiation and developmental pathways of uropathogenic *Escherichia coli* in urinary tract pathogenesis. *Proceedings of the National Academy of Science USA* **101**, 1333–1338.

Kalsi, J., Arya, M., Wilson, P., and Mundy, A. (2003). Hospital-acquired urinary tract infection. *International Journal of Clinical Practice* **57**, 388–391.

Lacy, B.E. and Rosemore, J. (2001). *Helicobacter pylori*: ulcers and more: the beginning of an era. *Journal of Nutrition* **131**, 2789S–2793S.

Lievin, V., Peiffer, I., Hudault, S., Rochat, F., Brassart, D., Neeser, J.R., and Servin, A.L. (2000). Bifidobacterium strains from resident infant human gastrointestinal microflora exert antimicrobial activity. *Gut* **47**, 646–652.

Linskens, R.K., Huijsdens, X.W., Savelkoul, P.H., Vandenbroucke-Grauls, C.M., and Meuwissen, S.G. (2001). The bacterial flora in inflammatory bowel disease: current insights in pathogenesis and the influence of antibiotics and probiotics. *Scandinavian Journal of Gastroenterology Supplement* **234**, 29–40.

Lu, L. and Walker, W.A. (2001). Pathologic and physiologic interactions of bacteria with the gastrointestinal epithelium. *American Journal of Clinical Nutrition* **73**, 1124S–1130S.

Macfarlane, G.T. and Macfarlane, S. (1997). Human colonic microbiota: ecology, physiology, and metabolic potential of intestinal bacteria. *Scandinavian Journal of Gastroenterology Supplement* **222**, 3–9.

MacFie, J., O'Boyle, C., Mitchell, C.J., Buckley, P.M., Johnstone, D., and Sudworth, P. (1999). Gut origin of sepsis: a prospective study investigating associations between bacterial translocation, gastric microflora, and septic morbidity. *Gut* **45**, 223–228.

Mackie, R.I., Sghir, A., and Gaskins, H.R. (1999). Developmental microbial ecology of the neonatal gastrointestinal tract. *American Journal of Clinical Nutrition* **69**, 1035S–1045S.

Madden, J.A. and Hunter, J.O. (2002). A review of the role of the gut microflora in irritable bowel syndrome and the effects of probiotics. *British Journal of Nutrition* **88** (Suppl 1), S67–S72.

Marshall, J.C. (1999). Gastrointestinal flora and its alterations in critical illness. *Current Opinion in Clinical Nutrition and Metabolic Care* **2**, 405–411.

Marteau, P., Pochart, P., Doré, J., Be'Ra-Maillet, C., Bernalier, A., and Corthier, G. (2001). Comparative study of bacterial groups within the human caecal and faecal microbiota. *Applied and Environmental Microbiology* **67**, 4939–4942.

Matsuki, T., Watanabe, K., Tanaka, R., Fukuda, M., and Oyaizu, H. (1999). Distribution of bifidobacterial species in human intestinal microflora examined with 16S rRNA-gene-targeted species-specific primers. *Applied and Environmental Microbiology* **65**, 4506–4512.

Matute, A.J., Schurink, C.A., Krijnen, R.M., Florijn, A., Rozenberg-Arska, M., and Hoepelman, I.M. (2002). Double-blind, placebo-controlled study comparing the effect of azithromycin with clarithromycin on oropharyngeal and bowel microflora in volunteers. *European Journal of Clinical Microbiology and Infectious Diseases* **21**, 427–431.

McCartney, A.L. (2002). Application of molecular biological methods for studying probiotics and the gut flora. *British Journal of Nutrition* **88** (Suppl 1), S29–S37.

McCracken, V.J. and Lorenz, R.G. (2001). The gastrointestinal ecosystem: a precarious alliance among epithelium, immunity, and microbiota. *Cellular Microbiology* **3**, 1–11.

McLaughlin, S.P., and Carson, C.C. (2004). Urinary tract infections in women. *Medical Clinics of North America* **88**, 417–429.

Mitchell, H. and Megraud, F. (2002). Epidemiology and diagnosis of *Helicobacter pylori* infection. *Helicobacter* **7** (Suppl 1), 8–16.

Monstein, H.-J., Tiveljung, A., Kraft, C.H., Borch, K., and Jonasson, J. (2000). Profiling of bacterial flora in gastric biopsies from patients with *Helicobacter pylori*-associated gastritis and histologically normal control individuals by temperature gradient get electrophoresis and 16S rDNA sequence analysis. *Journal of Medical Microbiology* **49**, 817–822.

Montecuccoa, C., Papinib, E., de Bernarda, M., and Zorattia, M. (1999). Molecular and cellular activities of *Helicobacter pylori* pathogenic factors. *FEBS Letters* **452**, 16–21.

Mountzouris, K.C., McCartney, A.L., and Gibson, G.R. (2002). Intestinal microflora of human infants and current trends for its nutritional modulation. *British Journal of Nutrition* **87**, 405–420.

Naumann, M., and Crabtree, J.E. (2004). *Helicobacter pylori*-induced epithelial cell signalling in gastric carcinogenesis. *Trends in Microbiology* **12**, 29–36.

Neal, E.N. (1999). Host defence mechanisms in urinary tract infections. *Urologic Clinics of North America* **26**, 677–686.

O'Sullivan, D.J. (2000). Methods for analysis of the intestinal microflora. *Current Issues in Intestinal Microbiology* **1**, 39–50.

Peek, R.M., Jr. (2001). The biological impact of *Helicobacter pylori* colonization. *Seminars in Gastrointestinal Disease* **12**, 151–166.

Pei, Z., Bini, E.J., Yang, L., Zhou, M., Francois, F., and Blaser, M.J. (2004). Bacterial biota in the human distal esophagus. *Proceedings of the National Academy of Science USA.* **101**, 4250–4205.

Poxton, I.R., Brown, R., Sawyer, A.F., and Ferguson, A. (1997). The mucosal anaerobic Gram-negative bacteria of the human colon. *Clinical Infectious Diseases* **25** (Suppl 2), S111–S113.

Probert, H.M. and Gibson, G.R. (2002). Bacterial biofilms in the human gastrointestinal tract. *Current Issues in Intestinal Microbiology* **3**, 23–27.

Pryde, S.E., Duncan, S.H., Hold, G.L., Stewart, C.S., and Flint, H.J. (2002). The microbiology of butyrate formation in the human colon. *FEMS Microbiology Letters* **217**, 133–139.

Rath, H.C. (2003). The role of endogenous bacterial flora: bystander or the necessary prerequisite? *European Journal of Gastroenterology and Hepatology* **15**, 615–620.

Reuter, G. (2001). The *Lactobacillus* and *Bifidobacterium* microflora of the human intestine: composition and succession. *Current Issues in Intestinal Microbiology* **2**, 43–53.

Saito, N., Konishi, K., Sato, F., Kato, M., Takeda, H., Sugiyama, T., and Asaka, M. (2003). Plural transformation-processes from spiral to coccoid *Helicobacter pylori* and its viability. *Journal of Infection* **46**, 49–55.

Sanderson, I.R. (1999). The physicochemical environment of the neonatal intestine. *American Journal of Clinical Nutrition* **69** (Suppl), 1028S–1034S.

Savage, D.C. (1977). Microbial ecology of the gastrointestinal tract. *Annual Review of Medicine* **31**, 107–133.

Schell, M.A., Karmirantzou, M., Snel, B., Vilanova, D., Berger, B., Pessi, G., Zwahlen, M., Desiere, F., Bork, P., Delley, M., Pridmore, D., and Arigoni, F. (2002). The genome sequence of *Bifidobacterium longum* reflects its adaptation to the human gastrointestinal tract. *Proceedings of the National Academy of Sciences USA* **99**, 14422–14427.

Schmidt, H. and Martindale, R. (2001). The gastrointestinal tract in critical illness. *Current Opinion in Clinical Nutrition and Metabolic Care* **4**, 547–551.

Schneider, S.M., Le Gall, P., Girard-Pipau, F., Piche, T., Pompei, A., Nano, J.L., Hebuterne, X., and Rampal, P. (2000). Total artificial nutrition is associated with major changes in the faecal flora. *European Journal of Nutrition* **39**, 248–255.

Segal, I.I. (2000). Colonic microbiota, nutrition, and health. *Gut* **47**, 741C–741C.

Sghir, A., Gramet, G., Suau, A., Rochet, V., Pochart, P., and Dore, J. (2000). Quantification of bacterial groups within human faecal flora by oligonucleotide probe hybridization. *Applied and Environmental Microbiology* **66**, 2263–2266.

Shanahan, F. (2001). Inflammatory bowel disease: immunodiagnostics, immunotherapeutics, and ecotherapeutics. *Gastroenterology* **20**, 622–635.

Silva, S.H., Vieira, E.C., Dias, R.S., and Nicoli, J.R. (2001). Antagonism against *Vibrio cholerae* by diffusible substances produced by bacterial components of the human faecal microbiota. *Journal of Medical Microbiology* **50**, 161–164.

Sussman, M. and Gally, D.L. (1999). The biology of cystitis: host and bacterial factors. *Annual Review of Medicine* **50**, 149–158.

Swidsinski, A., Ladhoff, A., Pernthaler, A., Swidsinski, S., Loening-Baucke, V., Ortner, M., Weber, J., Hoffmann, U., Schreiber, S., Dietel, M., and Lochs, H. (2002). Mucosal flora in inflammatory bowel disease. *Gastroenterology* **122**, 44–54.

Swift, S., Vaughan, E.E., and de Vos, W.M. (2000). Quorum sensing within the gut ecosystem. *Microbial Ecology in Health and Disease* **12** (Suppl 2), 81–92.

Tannock, G.W. (1999). Analysis of the intestinal microflora: a renaissance. *Antonie Van Leeuwenhoek* **76**, 265–278.

Tannock, G.W. (1999). The bowel microflora: an important source of urinary tract pathogens. *World Journal of Urology* **17**, 339–344.

Tannock, G.W. (2000). The intestinal microflora: potentially fertile ground for microbial physiologists. *Advances in Microbial Physiology* **42**, 25–46.

Tannock, G.W. (2002). The bifidobacterial and Lactobacillus microflora of humans. *Clinical Reviews in Allergy and Immunology* **22**, 231–253.

Vaira, D., Holton, J., Ricci, C., Basset, C., Gatta, L., Perna, F., Tampieri, A., and Miglioli, M. (2002). *Helicobacter pylori* infection from pathogenesis to treatment – a critical reappraisal. *Alimentary Pharmacology and Therapeutics* **16** (Suppl 4), 105–113.

Vaughan, E.E., Schut, F., Heilig, H.G.H.J., Zoetendal, E.G., de Vos, W.M., and Akkermans, A.D.L. (2000). A molecular view of the intestinal ecosystem. *Current Issues in Intestinal Microbiology* **1**, 1–12.

Whelan, K., Judd, P.A., Preedy, V.R., and Taylor, M.A. (2004). Enteral feeding: the effect on faecal output, the faecal microflora, and SCFA concentrations. *Proceedings of the Nutrition Society* **63**, 105–113.

Wold, A.E. and Adlerberth, I. (2000). Breast feeding and the intestinal microflora of the infant–implications for protection against infectious diseases. *Advances in Experimental Medicine and Biology* **478**, 77–93.

Xu, J., Chiang, H.C., Bjursell, M.K., and Gordon, J.I. (2004). Message from a human gut symbiont: sensitivity is a prerequisite for sharing. *Trends in Microbiology* **12** 21–28.

Yamamoto, S., Tsukamoto, T., Terai, A., Kurazono, H., Takeda, Y., and Yoshida, O. (1997). Genetic evidence supporting the faecal-perineal-urethral hypothesis in cystitis caused by *Escherichia coli*. *Journal of Urology* **157**, 1127–1129.

Zoetendal, E.G., Collier, C.T., Koike, S., Mackie, R.I., and Gaskins, H.R. (2004). Molecular ecological analysis of the gastrointestinal microbiota: a review. *Journal of Nutrition* **134**, 465–472.

8

The oral cavity and its indigenous microbiota

The oral cavity is not simply the entrance to the gastrointestinal tract (GIT), but consists of a complex system of tissues and organs with a variety of functions which together are involved in selecting food that is suitable for intake and processing the food into a form that is suitable for passage into the rest of the GIT. Another major function of the oral cavity is speech production. With regard to its feeding function, the oral cavity contains several sensory systems which are involved in perceiving the taste, smell, touch, and temperature of the food. This sensory information is analysed in the central nervous system and used to determine the acceptability of the food. If it is regarded as acceptable, then saliva is secreted, chewing is initiated, and eventually swallowing takes place.

8.1 | Anatomy and physiology of the oral cavity

The oral cavity is formed from the cheeks, the hard and soft palates, and the tongue (Figure 8.1). It contains accessory digestive structures, the teeth, and is connected to the pharynx by an opening known as the fauces. The total surface area of the oral cavity is approximately 200 cm^2. The surfaces of the teeth comprise 20% of this, with the remainder being attributable to the oral mucosa. The oral mucosa, like other mucosal surfaces, consists of two layers – an epithelium and an underlying layer of connective tissue, the lamina propria. The oral epithelium varies in structure, depending on its function, and three basic types are recognised: masticatory, lining, and specialised (Table 8.1). Thirty percent of the surface area of the oral cavity is comprised of keratinised mucosa, while 50% is non-keratinised mucosa. The epithelium of masticatory mucosa is keratinised and moderately thick, and covers those regions of the oral cavity that are subjected to abrasion and shear stress during chewing. In contrast, the epithelium of lining mucosa is not keratinised and is thicker – the epithelium of the cheek, for example, may be 500 μm thick. These features render the lining mucosa more flexible and able to withstand the stretching that occurs in the regions in which it is found. The tongue contains regions of keratinised and non-keratinised epithelia and is highly extensible.

The cheeks comprise the lateral walls of the oral cavity and, at the entrance to the oral cavity, they terminate in fleshy folds known as the lips (labia) which are covered on the outside by skin. The hard and soft palates comprise the roof of the mouth and these consist of bone and muscle, respectively. The hard palate separates the oral and nasal cavities, while the soft palate separates the oropharynx and nasopharynx.

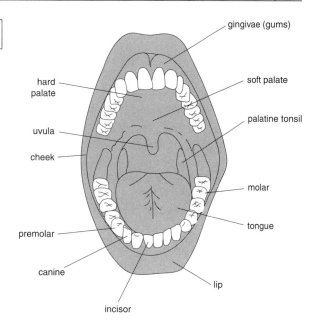

Figure 8.1 The major anatomical features of the oral cavity.

gingivae (gums)

hard palate

soft palate

palatine tonsil

uvula

cheek

molar

tongue

premolar

canine

lip

incisor

Hanging from the soft palate is a muscular process known as the uvula. The floor of the oral cavity is formed from the tongue, which is a muscular, accessory digestive structure. Tongue movements are important in mastication and are involved in forming food into a rounded mass known as a bolus and propelling this to the back of the mouth for swallowing. The upper surfaces and sides of the tongue are covered in projections known as papillae, between which are crypts that can be several millimetres deep. Over the anterior two-thirds of the tongue (often known as the "body"), these projections are conical, whitish in colour, contain no taste buds, and are covered by a keratinised epithelium – they are known as filiform papillae. Scattered among these, but more numerous at the tip of the tongue, are reddish, mushroom-shaped projections, which are covered in a non-keratinised epithelium – these are known as fungiform papillae, and they usually contain taste buds. On the posterior surface are ten to twelve circumvallate papillae, which contain taste buds and have non-keratinised walls, but a keratinised upper surface. The epithelium between the papillae is non-keratinised, making it flexible and extensible.

Table 8.1. The three types of oral mucosa and the main features of the epithelial layer associated with each

Mucosal type	Characteristic features of epithelial layer	Anatomical location
masticatory	keratinised, stratified, squamous	hard palate; gingiva
lining	non-keratinised, stratified, squamous	soft palate; floor of mouth; cheek; underside of tongue; inside of lips
specialised	keratinised and non-keratinised, stratified, squamous	tongue

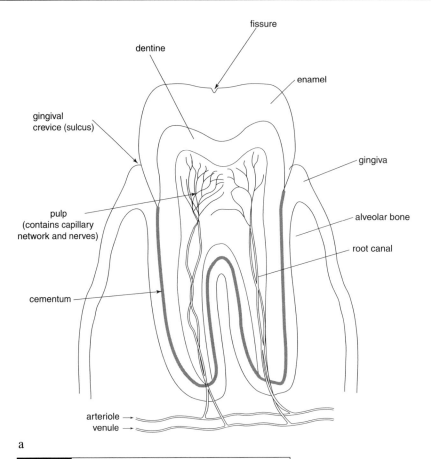

Figure 8.2 (a) Gross structure of a tooth and adjacent tissues.

The teeth are mineralised structures which protrude into the oral cavity from sockets within alveolar bone (Figure 8.2). The upper regions of alveolar bone are covered by the gingivae (gums), which extend a short distance into the socket, resulting in a shallow depression surrounding the tooth. This depression is approximately 1.8 mm deep and is known as the gingival crevice or sulcus. From the gingival crevice, a serum-like ex-udate (gingival crevicular fluid − GCF) continually enters the oral cavity. The gingivae are covered by a stratified, squamous, keratinised epithelium which merges with the non-keratinised "sulcular" epithelium in the gingival crevice. Within the socket, the embedded region of the tooth (the root) is attached to the alveolar bone by a lining of dense, fibrous tissue known as the periodontal ligament. The tooth itself is composed mainly of a bone-like material known as dentine, within which is a pulp cavity contain-ing blood vessels, lymphatics, and nerves. These vessels and nerves are connected to the rest of the circulatory, lymphatic, and nervous systems via a tube (root canal) that runs from the pulp cavity through each root of the tooth. Each tooth may have one or more roots. The dentine is covered by other mineralised tissues − enamel in the region of the tooth protruding into the oral cavity (the crown) and cementum in the case of the root. Dentine consists mainly of a mineral, hydroxyapatite $[Ca_{10}(PO_4)_6(OH)_2]$, with smaller quantities of organic matter (21%) and water (10%). The main organic com-ponent is type I collagen, which is present as a cross-linked network of fibres along which the hydroxyapatite crystals form. Other proteins present are type V collagen,

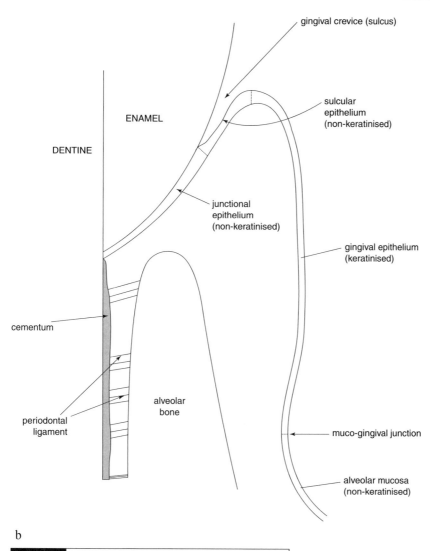

gingival crevice (sulcus)

ENAMEL

DENTINE

sulcular
epithelium
(non-keratinised)

junctional
epithelium
(non-keratinised)

gingival epithelium
(keratinised)

cementum

periodontal
ligament

alveolar
bone

muco-gingival junction

alveolar mucosa
(non-keratinised)

b

Figure 8.2 (b) Structure of the supporting tissues of the tooth.

phosphoproteins, glycoproteins, and proteoglycans. Enamel is harder than dentine and is, in fact, the hardest substance in the body. It forms a layer of varying thickness (approximately 2 μm thick on the crown of the tooth) and protects the tooth from the wear associated with chewing and also from acids present in the diet. Its hardness is a consequence of its high mineral content (96–97%) and low content of water (2–3%) and organic matter (1%). The mineral content consists mainly of hydroxyapatite and fluorapatite [$Ca_{10}(PO_4)_6F_2$], with the latter rendering enamel more resistant to acid attack. Because saliva is supersaturated with calcium and phosphate ions, it functions as a remineralising agent, replacing any of these ions lost from the enamel surface as a result of acid dissolution, etc. Unlike dentine, the main organic constituents of enamel are members of two families of proteins: amelogenins (90%) and enamelins (10%). Cementum consists of hydroxyapatite (60%), organic material (23%), and water (17%). Most of the organic material is type I collagen, with smaller amounts of type III collagen. Its main function is to provide attachment points for the periodontal ligament which anchors the tooth in the alveolar bone. Teeth begin to appear in the oral cavity approximately 6 months after birth, and by the age of 3 years, most individuals have a complete set of what are known as "deciduous" or "milk" teeth (i.e., the primary dentition). These teeth are gradually lost between the ages of 6 and 12, and are replaced by a set of up to thirty-two teeth known as the "permanent dentition". The permanent dentition comprises (1) eight chisel-shaped incisors (for cutting into food) and four pointed ("cuspid") canines (for tearing and shredding) – all of which are single-rooted; (2) eight bicuspid premolars (for crushing and grinding) – single-rooted apart from the upper first premolars, which sometimes have two roots; and (3) up to 12 four-cusped molars (for crushing and grinding) – upper and lower molars have three and two roots, respectively.

Saliva is a dilute, aqueous fluid with a number of functions, including lubrication, digestion, temperature regulation, and host defence. It also acts as a buffer, thus preventing extreme pH fluctuations and, because of its content of calcium and phosphate ions, is involved in the remineralisation of teeth. Although some saliva is produced by the buccal glands present in the oral mucosa (collectively known as "minor salivary glands"), most (approximately 90%) is produced by the salivary glands (parotid, submandibular, and sublingual), which are located outside of the oral cavity. The parotid glands are located in front of the ears, and each has a duct with an opening next to the upper second molar tooth. The submandibular glands are beneath the base of the posterior region of the tongue, and their ducts open behind the central incisors. The sublingual glands are located in front of the submandibular glands and open into the oral cavity below the tongue. The composition of the secretions produced by each of these types of glands varies, with the submandibular and sublingual glands producing saliva with a higher mucus content than that produced by the parotid glands. The rate of secretion of saliva and its composition are inter-related and are highly variable. Factors influencing the secretion rate and composition of saliva include the time of day, the presence of food (or a non-food item) in the mouth, the smell or site of food, as well as a variety of other stimuli. It is usually convenient to recognise two types of saliva – that produced in the absence of any stimulus ("resting saliva") and that produced in response to some stimulus ("stimulated saliva"). In resting saliva, approximately 20% is derived from the parotid glands, 65% from the submandibular glands, 7% from the sublingual glands, and the rest comes from the minor salivary glands. The resting rate of salivary flow is approximately 0.3 ml/minute, and this increases to 2.5–5.0 ml/minute during stimulation. Because little saliva is produced during sleep, the average daily production of saliva is likely to be between 700 and 800 ml.

Saliva contains a variety of electrolytes, including sodium, potassium, calcium, magnesium, bicarbonate, and phosphates. High-molecular-mass components include immunoglobulins, proteins, enzymes, and mucins. Nitrogenous compounds – such as urea, uric acid, amino acids, and ammonia – are also present. Bicarbonates, phosphates, urea, amino acids, and proteins contribute to the buffering capacity of saliva, thus preventing dramatic fluctuations in pH. Calcium, phosphates, and proteins modulate tooth demineralisation and remineralisation. Immunoglobulins, enzymes, other proteins, and mucins have a variety of functions, including an involvement in host defence, as well as microbial co-aggregation and adhesion. The composition of saliva is described in Section 8.3.2.

Food processing in the oral cavity is accomplished by a combination of mechanical and enzymatic activities, with the teeth and tongue being primarily responsible for the former while saliva is involved in both. The first stage in the process is mastication, which breaks apart the food into smaller particles and mixes these with saliva. This produces a soft, flexible mass which is more suitable for swallowing. In addition, the increased surface area of the food particles facilitates their solubilisation and their degradation by digestive enzymes both in the oral cavity and in other regions of the GIT. Amylase and lipase present in saliva probably contribute to the digestion of starch and lipids in the oral cavity, although in view of the short residence time of food in the oral cavity, it is unlikely that much digestion takes place. However, lipase is active at the low pH of the stomach and salivary amylase survives for 15–30 minutes before being inactivated by the low gastric pH. Following mastication, the food is formed into a bolus and swallowed.

8.2 | Antimicrobial defence mechanisms of the oral cavity

Because of its main function as a food processor, considerable mechanical and hydrodynamic forces are produced within the oral cavity. Hence, biting, chewing, and tongue movements – as well as salivary flow – will dislodge microbes from surfaces within the oral cavity and so discourage colonisation. This is augmented by the presence in saliva of mucins and various proteins (statherins, histadine-rich proteins, and proline-rich proteins), which bind to microbes and/or encourage their co-aggregation so that they remain in suspension in saliva and, along with microbes dislodged by mechanical and hydrodynamic forces, are eventually swallowed. Swallowing occurs approximately 2,500 times each day, and this delivers the dislodged microbes to the stomach, where most are killed by the low pH. This mechanical clearance of microbes from the oral cavity is not restricted to periods of eating and drinking, but also occurs during talking which involves continuous tongue movements. Enormous numbers of microbes are removed from the oral cavity in this way. Hence, based on a daily rate of production of 750 ml and a content of 10^8 bacteria/ml, the total number of bacteria swallowed per day is approximately 8×10^{10}.

Many of the surfaces available for colonisation within the oral cavity are mucosal surfaces, and the presence of a mucus gel coating and the shedding of superficial cells are important means of reducing microbial colonisation. Replacement of the outermost cells of the oral mucosa occurs every 24–48 hours. Uniquely, however, the oral cavity also has non-shedding surfaces – the teeth – available for microbial colonisation. Most of the microbes colonising the oral cavity are, in fact, found on tooth surfaces, and this illustrates the effectiveness of epithelial-cell shedding as a defence mechanism.

Saliva contains a range of antimicrobial compounds, including lysozyme (11 mg/ 100 ml), lactoperoxidase, secretory leukocyte protease inhibitor, lactoferrin, transferrin, and a number of antimicrobial peptides produced by a variety of oral tissues (Table 8.2). The antimicrobial activities of these compounds were described previously (Sections 2.2, 3.2, and 4.2). However, in addition to its antibacterial activities, lysozyme can also agglutinate bacteria, thus facilitating their removal from the oral cavity. The oral mucosa, like other epithelia, produces a number of antimicrobial peptides, each of which has a characteristic antimicrobial spectrum. Studies have shown that human β-defensin (HBD)-1 mRNA is constitutively expressed in human gingivae, buccal

Table 8.2. Antimicrobial defence mechanisms operating in the oral cavity

Defence mechanism	Function/effect
mechanical forces due to chewing and talking	microbes dislodged from oral surfaces and swallowed
hydrodynamic forces due to salivary flow	microbes dislodged from oral surfaces and swallowed
mucus layer on mucosal surfaces	reduces microbial colonisation of mucosal surfaces
shedding of epithelial cells	reduces microbial colonisation of mucosal surfaces
lysozyme	antibacterial; agglutinates bacteria
lactoferrin, transferrin	deprive microbes of iron
apo-lactoferrin	antimicrobial
lactoperoxidase	antimicrobial
secretory leukocyte protease inhibitor	antimicrobial
adrenomedulin	antimicrobial
defensins, LL-37, histatins	kill or inhibit the growth of microbes; neutralise pro-inflammatory activities of bacterial components
sIgA	prevents microbial adhesion; agglutinates microbes
IgG and IgM	prevent microbial adhesion, activate complement, function as opsonins
complement	microbial lysis and opsonisation
phagocytes	kill microbes

Note: sIgA = secretory IgA.

mucosa, tongue, submandibular glands, parotid glands, and minor salivary glands. The peptide has also been detected in saliva. HBD-2 is produced by the gingivae, tongue, buccal mucosa, labial mucosa, and salivary glands. Its expression is up-regulated in response to lipopolysaccharide (LPS) (5-fold) or interleukin (IL)-1β (16-fold). It is present in saliva at a concentration of approximately 150 ng/ml. The antimicrobial peptide cathelicidin, LL-37 (derived from the pro-peptide precursor hCAP-18), is constitutively expressed at both RNA and protein levels in the buccal mucosa and the tongue, and has been detected in the saliva of healthy adults. It has a wide antimicrobial spectrum and is active against *E. coli, Staph. aureus, Strep. pneumoniae, Staph. epidermidis, Ent. faecalis, Ps. aeruginosa, Lis. monocytogenes, Sal. typhimurium, Act. actinomycetemcomitans, F. nucleatum,* and *Capnocytophaga* spp., but not against *Strep. sanguis* or *Por. gingivalis.* Furthermore, it is able to neutralise the biological activities of LPS and has been shown to protect pigs against endotoxic shock. Although it is active at isotonic and hypotonic NaCl concentrations, its antimicrobial activity is reduced at high NaCl concentrations and by serum. Expression of HBD-3 mRNA has been demonstrated in the gingivae, tongue, buccal mucosa, labial mucosa, and salivary glands. Expression is up-regulated by tumour necrosis factor (TNF) and bacteria. The peptide is active against *Staph. aureus, Strep. pyogenes, Ps. aeruginosa, Ent. faecium, E. coli,* and *Can. albicans* at low microgram/millilitre concentrations. Furthermore, it is active at low and physiological salt concentrations. Its mode of action appears to differ from that of other antimicrobial peptides, in that it does not seem to target the cytoplasmic membrane but interferes with peptidoglycan synthesis. The morphological changes accompanying killing of *Staph. aureus* are similar to those brought about by penicillin. Oral epithelial cells *in vitro* have also been shown to produce adrenomedulin constitutively, and the production of the peptide is increased in response to *Por. gingivalis, Strep. mutans*, and *Eikenella corrodens*, but not to *Can. albicans*. The peptide has been shown to inhibit the growth of *Por. gingivalis, Act. actinomycetemcomitans, Eik. corrodens, A. naeslundii*, and *Strep. mutans*. It has been detected in saliva at a concentration of 32 pM. Histatins comprise a group of twelve small (3–4 kDa) histidine-rich proteins, which are able to kill *Can. albicans* and other yeasts. They also have some inhibitory effect against Gram-positive species and *Por. gingivalis*, can inhibit coaggregation between *Por. gingivalis* and *Strep. mitis*, and suppress the induction of cytokines by *Por. gingivalis* outer-membrane proteins. A number of α-defensins produced by neutrophils (human neutrophil peptides 1–4) are also found in the oral cavity. These peptides have broad-spectrum antimicrobial activity and reach high concentrations in the gingival crevice where neutrophils accumulate.

With regard to the acquired immune defence system, intra-epithelial lymphocytes (IELs) and Langerhans cells are present in the oral mucosa. IELs can recognise a limited number of antigens, but are mainly involved in repair processes while Langerhans cells are important antigen-presenting cells. The continuous flow of GCF (Section 8.3.2) into the oral cavity brings with it lymphocytes, monocytes, and polymorphonuclear leukocytes (PMNs), and thus these are present at relatively high levels in the gingival crevice but, because of dilution, at lower concentrations in saliva. The predominant immunoglobulin in the oral cavity is secretory IgA (sIgA), which is present in saliva at a concentration of approximately 19 mg/100 ml. sIgA prevents binding of microbes to oral surfaces and also functions as an agglutinin, resulting in the formation of microbial aggregates which are removed from the oral cavity during swallowing. IgG and IgM enter the oral cavity in GCF and, because of dilution by saliva, are present at low concentrations at most oral sites. Nevertheless, IgG in particular is an important effector molecule of the acquired immune defence system in the gingival crevice and within gingival tissues, where it is involved in opsonisation, complement activation, neutralisation of microbial enzymes and toxins, and prevention of microbial adhesion. However, while opsonisation and complement activation may be able to achieve killing of planktonic microbes, they are undoubtedly less effective against microbes in biofilms. Hence, once a biofilm has formed in the gingival crevice or on supra-gingival surfaces, access of complement components, antibodies, and PMNs to the constituent organisms will be severely impaired by the biofilm matrix. One consequence of this is the discharge of lysosomal granules of PMNs with their range of enzymes and free radicals, which can damage host tissues directly, as well as indirectly by exacerbating the inflammatory response.

8.3 | Environmental determinants at the various sites within the oral cavity

8.3.1 Mechanical determinants

Nowhere else in the body are microbes subjected to such extreme mechanical forces as those that operate in the oral cavity. The forces generated during routine chewing of food, such as meat, are 70–150 N, and the maximum biting force is between

500 and 700 N. As well as these forces, oral microbes also have to contend with the forces generated by movement of the tongue during mastication and talking. It has been estimated that the force exerted by the tongue when resting against the teeth is $0.2\,N/cm^2$ and is as high as $5.0\,N/cm^2$ during swallowing. The continuous flow of saliva and, to a much lesser extent, GCF also create hydrodynamic shear forces. In the case of saliva, these are of the order of 0.1–$1.0 \times 10^{-5}N/cm^2$. Furthermore, detachment forces (approximately 2×10^{-7} N) are also generated during the formation and disruption of air/fluid interfaces, which continually occur on oral surfaces as a consequence of salivary flow, chewing, and talking. To put the magnitude of these forces into perspective, atomic force microscopy has shown that bacterial adhesion forces are of the order of 4–7×10^{-9} N. The ability to adhere to some surface in the oral cavity in the face of such powerful removal forces is, therefore, of prime importance for any potential microbial coloniser. However, there are sites within the oral cavity that are protected to a greater or lesser extent from these mechanical and hydrodynamic shear forces. Such sites include the gaps between adjacent teeth (known as "approximal" regions), the naturally occuring fissures on the occlusal (i.e., biting) surfaces of pre-molar and molar teeth, the gaps between the papillae on the surface of the tongue, and the gingival crevice. Such regions tend to have much higher population densities, as well as a more varied microbiota, than exposed sites such as the smooth buccal (i.e., facing the cheek) and lingual (i.e., facing the inside) surfaces of teeth.

8.3.2 Nutritional determinants

Although the main source of nutrients for many microbial inhabitants of the oral cavity is saliva, there are a number of other important nutrient sources. For example, microbes adhering to mucosal surfaces may obtain their nutrients directly from the products secreted and excreted by the cells to which they are attached. Furthermore, microbes colonising the gingival crevice are likely to obtain many of their nutrients from GCF rather than from saliva. In addition, as at any body site, the excreted and secreted products of the microbes colonising the oral cavity can also serve as nutrients. Saliva contains not only the fluids produced by the salivary glands, but also GCF, substances secreted and excreted by the oral mucosa, and constituents of the host's diet. Oral microbes can undoubtedly obtain nutrients (particularly easily fermentable carbohydrates) from the host's diet, and this can have a profound effect on the composition of the oral microbiota – which is described further in Section 8.5.1. Nevertheless, the relative importance of diet-derived nutrients and those available from host secretions is difficult to determine. Endogenous nutrients supplied by the host can certainly support the development of complex oral microbial communities, as shown by laboratory studies and investigations involving humans and other animals fed via nasogastric tubes.

As described in Section 8.1, the amount of saliva produced and its composition depend on a number of factors, particularly on whether it is produced under resting or stimulatory conditions. Because stimulated saliva contributes between 80% and 90% of the saliva produced each day, its composition reflects that usually present in the oral cavity. Table 8.3 shows the concentration of the main constituents of stimulated saliva, but does not include all of the substances that may be present. The main constituents of saliva are proteins and glycoproteins. Salivary mucins comprise approximately 25%

Substance	Concentration (mg/100 ml)
Table 8.3. The main constituents of human saliva	
total solids	0.5% (w/v)
protein	140–640
amylase	38
peroxidase	0.3
lysozyme	11
glycoprotein sugars	110–300
free carbohydrate	2
amino acids	3.4–4.8
IgA	19
IgG	1.4
IgM	0.2
lipid	2–3
glucose	1.0
urea	13
uric acid	3
citrate	0.2–2.0
cAMP	50
potassium	80
chloride	100
bicarbonate	200
phosphate	12
sodium	60
calcium	6
thiocyanate	1–3
ammonia	3

of the total protein content, while amylase, IgA, and lysozyme are also major constituents. The main salivary mucins are MG1 (MUC5B) and MG2 (MUC7). MG1 is highly glycosylated and exists as a polymer with a molecular mass greater than 1 MDa and is able to form gels. The gel produced covers and protects the mucosal surfaces of the oral cavity from dessication, mechanical damage, and microbes. MG1 also binds tightly to enamel and forms complexes with other salivary proteins and, therefore, is a major constituent of the acquired enamel pellicle (Section 8.4.3.1), which protects the enamel surface from acid dissolution. MG2 exists as a less heavily glycosylated monomer, with a molecular mass of 200–250 KDa. It is soluble and although it also binds to enamel, it is easily displaced and forms aggregates with bacteria, thereby aiding their clearance from the oral cavity. A wide range of enzymes other than amylase and lysozyme has been detected in saliva, and these include peroxidase, lipase, acid phosphatase, esterases, aldolase, succinic dehydrogenase, β-glucuronidase, and carbonic anhydrase. Low concentrations of eighteen amino acids have been detected in saliva, as well as low concentrations of vitamin C and some B vitamins. A wide range of ions is present with bicarbonate, chloride, and potassium predominating. Calcium and phosphate ions are present at supersaturation concentrations, and this is important with regard to maintaining the integrity of tooth enamel.

The oral cavity can only contain approximately 1.1 ml of saliva before the swallowing reflux is initiated. Swallowing removes approximately 25% of this, leaving

Table 8.4.	The main constituents of gingival crevicular fluid

Constituent	Concentration
total protein	68–92 g/L
albumin	40 g/L
IgG	12.3 g/L
IgA	1.6 g/L
IgM	1.2 g/L
complement components	1.6 g/L
transferrin	3 g/L
lactoferrin	0.6 g/L
α_1-antitrypsin	3 g/L
α_2-macroglobulin	2.7 g/L
haptoglobin	2.1 g/L
carbohydrate	1.3 g/L
glucose	0.1 g/L
vitamin K	NA
urea	NA
sodium	105–222 mM
calcium	5.2–23.8 mM
magnesium	0.8 ± 0.5 mM

Note: NA = data not available.

approximately 0.8 ml of saliva. Given that the total surface area of the oral cavity is approximately 200 cm^2, this means that all of the oral surfaces are covered with a film of saliva, with a depth of approximately 100 μm. Such a thin coating enables rapid transfer of materials present in saliva (nutrients, gases, ions, etc.) to microbes adhering to oral surfaces. Therefore, although saliva is continually being removed from the oral cavity, the oral surfaces over which it flows (i.e., mucosal and tooth surfaces) are continually being supplied with a wide range of nutrients.

In addition to the mucins secreted by the salivary glands, the oral mucosa itself also produces a number of mucins – MUC1 and MUC4 – which remain attached to the epithelial cell membrane where, like MG1, they have a protective function. These mucins may also serve as nutrients for oral microbes able to bring about their degradation.

GCF is a transudate that flows continuously from the gingival crevice at low rates (approximately 8 μl/tooth/hour) in healthy individuals. However, the flow rate increases markedly in individuals with gingivitis (i.e., most adults), where it is approximately 14 μl/tooth/hour. In those with periodontitis, it is in the region of 44 μl/tooth/hour. This means that the amount of GCF flowing into the oral cavity per day is between 6 and 34 ml, depending on the oral health of the individual. Because most of the adult population has some degree of gingival inflammation, GCF will constitute a regular and abundant source of nutrients (see Table 8.4) to microbes inhabiting the gingival crevice. GCF is a nutrient-rich fluid, the main constituents of which are host cells, a range of proteins, carbohydrates, and a number of inorganic ions (Table 8.4). But the amounts of each are markedly affected by the flow rate, which depends on the degree of inflammation of the adjacent tissues. GCF contains desquamated cells from the junctional and sulcular epithelia, as well as leukocytes and, in fact, constitutes the main source of leukocytes in the oral cavity. Most (95–97%) of the leukocytes are PMNs, with 1–2% lymphocytes and 2–3% monocytes. The protein content of GCF is similar to that of serum, and reported values range from 6.8 to 9.2 g/100 ml. Proteins detected in GCF include IgA, IgG, IgM, albumin, transferrin, complement components, fibrinogen, α_1-antitrypsin, α_2-macroglobulin, ceruloplasmin, and β-lipoprotein. Unlike saliva, the main immunoglobulin present is IgG, rather than IgA. Free amino acids detected in GCF include alanine and lysine. Glucose, hexosamine, and hexuronic acids have all been detected in GCF, with the concentration of glucose being approximately 0.1 g/l. Lipids, mainly phospholipids and neutral lipids, are also present. Vitamin K and haemin are found in GCF, and these are essential growth factors for a number of oral bacteria. It also contains a number of ions, including potassium (6.3–69 mM), calcium (5.2–23.8 mM), and magnesium (0.8 ± 0.5 mM). The fluid is usually alkaline due to the presence of urea, and estimates of its pH range between 7.5 and 8.0.

Although microbes colonising the gingival crevice are supplied with an abundance of a range of nutrients from GCF, the dilution of this fluid by saliva means that organisms at other oral sites are less well provided for. The presence of only low concentrations of free carbohydrates and amino acids in saliva means that, unless such compounds are available from other microbes or from the host's diet, oral microbes must obtain most of their carbon and energy sources from host proteins and glycoproteins. The urea, ammonia, and amino acids present in saliva and GCF – together with amino acids liberated from proteins and glycoproteins – constitute the main nitrogen sources for oral microbes. As emphasised in Section 1.5.3, the complete degradation of complex glycoproteins, such as mucins, requires a number of hydrolytic enzymes (sialidases, proteases, glycosidases) and is usually achieved by microbial consortia rather than by an individual species. Nevertheless, partial degradation of mucins can supply sufficient carbohydrates to support the growth of a number of oral bacteria (e.g., *Strep. oralis, Strep. mitis* biovar 1, and *Strep. sanguis*). Collectively, oral microbes produce a wide range of proteases, peptidases, sialidases, and glycosidases, which enable them to hydrolyse a variety of proteins and glycoproteins (Table 8.5). A number of studies have shown that saliva can support the growth of some, but not all, oral bacteria *in vitro*. However, growth is often far more substantial and more rapid when the saliva is inoculated with several species simultaneously, thereby demonstrating that microbial consortia are usually needed to fully utilise the nutrients locked up in salivary glycoproteins.

8.3.3 Physicochemical determinants

The temperature of the oral cavity is usually between 35° and 36°C, and is suitable for the growth of a wide range of microbes. However, the temperature of subgingival regions can be as high as 39°C in individuals with periodontitis. Dramatic fluctuations in temperature occur when cold or hot food or drinks are being consumed, but these short-term changes are unlikely to have a significant effect on microbial growth.

Saliva is the main determinant of pH in the oral cavity (except for the gingival crevice), although microbial activity can have a dramatic, localised effect on the pH at individual oral sites. The pH of saliva varies with its rate of secretion and, in general, resting saliva is more acidic than stimulated saliva, with the ranges encountered being 6.5–6.9 and 7.0–7.5, respectively. Because, as previously mentioned, individuals produce stimulated saliva most of the time, the pH of saliva will usually be approximately neutral,

Table 8.5. | Oral species able to contribute to the degradation of glycoproteins

Organism	Sialidase	Glycosidase	Protease/peptidase
Tannerella forsythensis	+	+	+
B. capillosus	+	NA	NA
Prev. denticola	+	+	+
Prev. loescheii	+	+	+
Prev. buccalis	+	+	NA
Prev. buccae	+	+	NA
Prev. disiens	+	−	+
H. aphrophilus	NA	+	NA
Strep. oralis	+	+	+
Strep. intermedius	+	+	NA
Strep. mitis	+	+	+
Strep. sanguis	−	+	+
Strep. mitior	NA	+	+
Strep. mutans	−	NA	+
Strep. sobrinus	−	NA	+
Strep. anginosus	−	NA	+
Strep. pneumoniae	+	+	+
A. naeslundii genospecies 1	+	+	−
A. naeslundii genospecies 2	+	+	+
Por. gingivalis	+	+	+
Por. endodontalis	−	−	+
Por. asaccharolytica	NA	+	+
Prev. intermedia	−	+	+
Prev. nigrescens	NA	NA	+
Prev. oralis	+	+	+
T. denticola	+	NA	+
T. vincentii	NA	NA	+
Act. actinomycetemcomitans	NA	NA	+
Cap. sputigena	+	+	+
Cap. gingivalis	+	+	+
Cap. ochracea	+	+	+
Sel. sputigena	NA	+	NA
Eub. nodatum	NA	NA	+
P. acnes	+	NA	+
Bifidobacterium spp.	+	+	+

Notes: + = some strains produce the enzyme; − = most strains do not produce the enzyme; NA = data not available.

which is suitable for the growth of many types of microbes. Because of the presence of proteins, amino acids, phosphates, and bicarbonates, saliva has good buffering capacity. The buffering action is particularly effective over the pH range of 6.0–7.5 due mainly to the bicarbonate/carbonic acid system, which has been estimated to contribute between 64% and 85% of the total buffering capacity of saliva. At lower pHs, only a weak buffering action is found, and this is due to the presence of proteins and amino acids. The buffering action of saliva is unable to prevent dramatic falls in pH due to acid production by microbes when the host's diet is rich in easily fermentable carbohydrates, such as sucrose. The frequent consumption of sucrose can select for colonisation by acidophilic and/or aciduric species which, because of continuous acid production, can give rise to dental caries (Section 8.5.1). The fact that the bacteria on the tooth surface

Table 8.6. | Partial pressure of oxygen at various oral sites and the effect of oral hygiene abstention on these values 5 days later

Site	Partial pressure of oxygen (mm Hg)	
	Day 0	Day 5
plaque on labial and lingual tooth surfaces	61	23
gingival margin plaque	12	7
approximal plaque	18	6
labial and buccal mucosa	10–15	9
tongue (anterior)	8	8
whole, pooled saliva	65	48
saliva from parotid gland	40^a–110^b	ND
saliva from submandibular gland	35^a–65^b	ND

Notes: ^{a} = stimulated; ^{b} = unstimulated; ND = not determined.

are growing as biofilms (known as dental plaques; see Sections 1.1.2 and 8.4.3.1), which can partially exclude the buffering components of saliva and also impede the removal of bacterially produced acids, exacerbates this problem.

The main determinant of the pH of the gingival crevice is GCF, which has a pH of between 7.5 and 8.0. In individuals with periodontitis, the pH of subgingival dental plaque ranges from 7.5 to 8.5.

Although all regions of the oral cavity are exposed to a plentiful supply of air, a number of anatomical and microbial factors render many regions of the oral cavity suitable for the growth of microaerophiles and obligate anaerobes. In fact, surprisingly few members of the oral microbiota are obligate aerobes. Anatomical regions to which there is restricted access of oxygen-rich saliva and which, therefore, provide habitats with a low oxygen content and/or high redox potential include the following: (1) gaps between adjacent teeth, (2) the regions between tongue papillae, and (3) the gingival crevice. The growth of obligate aerobes and/or facultative anaerobes at these sites can rapidly exhaust the available oxygen and produce reducing compounds, thereby creating conditions suitable for colonisation by capnophiles, microaerophiles, and obligate anaerobes. Another important microbial factor leading to the provision of a range of micro-environments with different oxygen and redox conditions is the biofilm mode of growth. Within a biofilm, gradients of oxygen and redox potential will form as a result of microbial consumption of oxygen and inadequate replenishment by saliva. Hence, within a biofilm, environments suitable for the growth of aerobes, facultative anaerobes, capnophiles, microaerophiles, and obligate anaerobes are likely to be present at different locations (Figure 1.5). Gradients are also produced within the biofilm with respect to the concentration of nutrients supplied by saliva and GCF. The concentration of endogenous nutrients (i.e., metabolic end-products of organisms within the biofilm), as well as the pH (due to microbial acid production) will vary at different locations within the biofilm (Section 1.1.2).

Surprisingly few measurements have been made of the oxygen content or redox potential of oral sites. Available data are shown in Tables 8.6 and 8.7. These values will be a consequence of the anatomical features of the site and the metabolic activities of microbes at the site. From Table 8.6, it can be seen that fresh unstimulated saliva secreted from the salivary glands has a high oxygen content, whereas the saliva

Table 8.7.	Redox potential at various sites within the oral cavity	
Site	Redox potential (mV)	
saliva	+273	
tongue – centre	+139	
gingival mucosa	+123	
gingival crevice	+74	
gingiva – between teeth	−73	

that accumulates in the oral cavity is much lower. This reduction is probably a consequence of rapid oxygen consumption by respiring bacteria. Regions with the lowest oxygen concentrations are the tongue and the gingival crevice, and it is within these regions that obligate anaerobes are particularly abundant. Failure to practice oral hygiene has a dramatic effect on the oxygen content of the plaques accumulating on various surfaces; this is presumably due to oxygen consumption by the constituent bacteria. Table 8.7 shows that sites with the lowest redox potentials are those at the gingiva–tooth interface, particularly between adjacent teeth. As described in Sections 8.4.3.1 and 8.4.3.2, these regions usually have high proportions of obligate anaerobes. Apart from saliva, all of the sites listed have a redox potential (Eh) lower than the +200 mV found in aerobically metabolising host tissues. Once plaque is allowed to accumulate, oxygen consumption and the production of reducing compounds can lead to a dramatic reduction in Eh. This is illustrated in Figure 8.6 from which it can be seen that within 7 days, the Eh of plaque can be as low as −150 mV. Even greater reducing conditions can be reached in the subgingival plaques present in the periodontal pockets of individuals suffering from periodontitis (Section 8.5.2), where the Eh has been reported to reach −300 mV. The main features of the environment of each of the main oral sites are summarised in Table 8.8.

8.4 | The indigenous microbiota of the oral cavity

More than 200 microbial species have been cultured from the oral cavity of humans and between 400 and 500 additional taxa (or phylotypes) have been detected by 16S

Table 8.8.	Main features of the environment of key habitats within the oral cavity				
	Mucosa	Tongue	Fissures	Approximal	Gingival crevice
nutrient source	saliva, diet	saliva, diet	saliva, diet	saliva, diet, GCF	GCF
oxygen level	high	low	high	low	low
Eh	intermediate	intermediate (low in crypts)	high	low	low
pH	neutral	neutral	neutral/acidic	neutral/acidic	alkaline
mechanical abrasion	high	high (not in crypts)	low	low	low

Notes: Eh = redox potential; GCF = gingival crevicular fluid.

rRNA gene analysis of oral samples. Of these possible colonisers of the oral cavity (a total of more than 700), a more limited number are present in the oral cavity of a healthy individual at any one time. Hence, it is usual to culture between twenty and thirty organisms from an individual plaque sample so that the total number of species present is likely to be between fifty and seventy-five. The proportion of the microbiota that has been cultivated (approximately 40%) is higher for the oral cavity than for many other body sites. This reflects the great interest that has been shown in the oral microbiota because of its role in two of the most common infections of humans – caries and periodontal diseases – and is also a consequence of the ease of access to samples for analysis. Hence, unlike many other body sites, samples can be obtained very easily and without discomfort or embarrassment to the individual. Like many of the other anatomical sites or organ systems described in this book, the oral cavity contains a variety of habitats, each with its characteristic microbiota. However, one unique feature of the oral cavity which is not encountered elsewhere in the body is the presence of non-shedding surfaces (i.e., the enamel and cementum of the teeth) that are exposed to the external environment. These enable the formation of biofilms and, although biofilms are found on other body surfaces, they are generally not as substantial or as microbially diverse as those present on the surfaces of teeth. Biofilm formation is a very important and characteristic feature of the organisms inhabiting the oral cavity, and the structure of dental plaques is described later (Section 8.4.3.1).

8.4.1 Main characteristics of key members of the oral microbiota

The predominant genera detected in the oral cavity include *Streptococcus, Actinomyces, Veillonella, Fusobacterium, Porphyromonas, Prevotella, Treponema, Neisseria, Haemophilus, Eubacterium, Lactobacillus, Bifidobacterium, Capnocytophaga, Eikenella, Leptotrichia, Peptostreptococcus, Staphylococcus,* and *Propionibacterium*. Most of these genera have been described in previous chapters, and this section briefly outlines the main characteristics of those that have not.

8.4.1.1 Oral streptococci and other Gram-positive cocci

The genus *Streptococcus* has been described in Section 4.4.1.3, and this section deals only with the oral streptococci. These organisms can be divided into four main groups: (1) mitis group – found mainly in dental plaque – includes *Strep. mitis, Strep. gordonii, Strep. oralis, Strep. sanguis, Strep. parasanguis, Strep. crista, Strep. infantis,* and *Strep. peroris*; (2) mutans group – also found mainly in dental plaque and are associated with dental caries – includes *Strep. mutans* and *Strep. sobrinus*; (3) salivarius group – found mainly on mucosal surfaces – includes *Strep. salivarius* and *Strep. vestibularis*; and (4) anginosus group – found mainly in the gingival crevice – includes *Strep. anginosus, Strep. constellatus,* and *Strep. intermedius*. Oral streptococci are among the dominant members of the oral microbiota and are important primary colonisers of mucosal and tooth surfaces. Their adhesins have been extensively studied, and some are listed in Tables 8.9 and 8.10. Many are able to synthesise polysaccharides, which constitute the principal components of the matrix of dental plaque and can also serve as carbohydrate sources for other oral species once they have been degraded by exopolysaccharidases. They are acidogenic, and many species are acidophilic – for example, *Strep. mutans* exhibits vigorous growth

Table 8.9. | Oral bacterial adhesins and their receptors

AEP receptor(s)	Adhesin(s)	Approximate size (kDa)	Species
parotid salivary agglutinin, salivary glycoproteins, proline-rich proteins, collagen	Antigen I/II family [Ag I/II, AgB, PI (SpaP), PAc, Sr, SpaA, PAg, SspA, SspB, SoaA]	160–175	S. mutans, S. sobrinus, S. gordonii, S. oralis, S. intermedius
salivary components in pellicle	Lral family [FimA, SsaB]	35	S. parasanguis, S. sanguis
salivary components in pellicle	FapI	200	S. parasanguis
α-amylase	amylase-binding proteins	82, 65, 20, 15, 12,	S. gordonii, S. mitis, S. crista
73-kDa submandibular salivary protein	antigen complex	80, 62, 52	S. gordonii
salivary glycoprotein presenting N-acetylneuraminic acid	surface lectins	96, 70, 65	S. oralis, S. mitis
proline-rich proteins, statherin	Type I fimbriae-associated protein	–	A. naeslundii

Notes: AEP = acquired enamel pellicle. Reprinted from: Dental plaque formation. Rosan, B. and Lamont, R.J. In: *Microbes and infection.* Vol. 2, pp. 1599–1607. Copyright © 2000, with permission from Elsevier.

at a pH as low as 4.5. As well as causing caries (mutans group – Section 8.5.1), some oral streptococci (mitis group) are frequently associated with endocarditis (Section 8.5.3) and abscess formation (anginosus group – Section 8.5.3).

Organisms with complex nutritional requirements (e.g., a requirement for vitamin B_6), previously known as nutritionally variant streptococci, have been assigned to a new genus *Abiotrophia*. The genus includes *Ab. elegans, Ab. defectiva*, and *Ab. adjacens*. They have been shown to be responsible for some cases of endocarditis and bacteraemia. *Gemella* spp. are sometimes misidentified as viridans streptococci or *Abiotrophia* spp. They are facultatively anaerobic, non-motile, catalase-negative Gram-positive cocci occurring in pairs or short chains.

8.4.1.2 *Actinomyces* spp.

The genus *Actinomyces* consists of non-motile, non-sporing, Gram-positive rods which may be short, filamentous, pleomorphic, or branching. The G+C content of their DNA is 55–71 mol%. Twelve species are recognised, of which seven are found in the human oral cavity. These include *A. naeslundii* genospecies 1, *A. naeslundii* genospecies 2 (includes strains formerly classified as *A. viscosus*), *A. israelii, A. gerencseriae, A. georgiae, A. odontolyticus*, and *A. meyeri*. Most species are facultative anaerobes, and their growth is stimulated by high concentrations of carbon dioxide; some species are obligate anerobes. They can grow over the temperature range of 20°–45°C, but growth is optimal at 35°–37°C. They ferment carbohydrates, producing succinic, lactic, formic, and acetic acids as

Table 8.10. Adhesins and receptors involved in co-adhesion and co-aggregation of oral bacteria

Species	Adhesin	Approximate size (KDa)	Receptor(s)	Partner species
Strep. gordonii, Strep. mitis, Strep. oralis	antigen I/II family (SspA SspB)	171, 160	bacterial surface proteins, yeast mannoproteins	Por. gingivalis, Strep. mutans, A. naeslundii, Can. albicans, Strep. oralis
Strep. gordonii	CshA, CshB	259, ca. 245	bacterial surface proteins, yeast mannoproteins	A. naeslundii, Can. albicans, Strep. oralis
Strep. gordonii	LraI family (ScaA)	35		A. naeslundii
Strep. gordonii	co-aggregation-mediating adhesin	100	carbohydrate containing lactose or lactose-like moieties	Streptococcus spp.
Strep. salivarius	fibrillar antigen B (VBP)	320		V. parvula
A. naeslundii	type 2 fimbriae-associated protein	95	cell wall polysaccharide containing Galβ1→3 GalNAc and GalNAcβ1→3Gal glycosidic linkages	Streptococcus spp.
Por. gingivalis	fimbrillin	43	surface proteins	Strep. gordonii, Strep. oralis, A. naeslundii
Por. gingivalis	outer membrane protein	35	surface protein	Strep. gordonii
Por. gingivalis	outer membrane protein	40		A. naeslundii
Prev. loescheii	PlaA	75	cell wall polysaccharide containing Galβ1→ 3GalNAc and GalNAcβ1→3Gal glycosidic linkages	Streptococcus spp.
Prev. loescheii	fimbria-associated protein	45		A. israelii
F. nucleatum	outer membrane proteins	42, 30	galactose-containing carbohydrate	Por. gingivalis
F. nucleatum	corn cob receptor	39		Strep. cristatus
V. atypica	outer membrane protein	45	carbohydrate containing lactose or lactose-like moieties	Streptococcus spp.
Cap. gingivalis	outer membrane protein	140	cell wall carbohydrate	A. israelii
Cap. ochracea	outer membrane protein	155	cell wall carbohydrate	Strep. oralis
Treponema medium	outer membrane protein	37	fimbriae	Por. gingivalis

Note: Reprinted from: Dental plaque formation. Rosan, B. and Lamont, R.J. *Microbes and infection.* Vol. 2, pp. 1599–1607. Copyright © 2000, with permission from Elsevier.

end-products, and they are aciduric. They are generally not proteolytic, but do produce sialidases and glycosidases and can, therefore, contribute to mucin degradation.

Actinomyces spp. are able to adhere to both tooth and mucosal surfaces, mainly by means of fimbriae. Two types of fimbriae have been identified (types 1 and 2); the first of these mediates binding to proline-rich proteins present in the acquired enamel pellicle and is involved in dental plaque formation. Type 2 fimbriae bind to glycoproteins present on epithelial cells and enable adhesion of the organism to the oral mucosa. They also recognise carbohydrate receptors on the surfaces of many viridans streptococci and other oral bacteria, and are responsible for co-aggregation with such species. Many *Actinomyces* spp. also produce a polysaccharide which contributes to the plaque matrix.

A. israelii is responsible for actinomycosis (Section 8.5.7), while other *Actinomyces* spp. are associated with caries (particularly root-surface caries), gingivitis, and periodontitis.

8.4.1.3 *Veillonella* spp.

These are non-motile, anaerobic, Gram-negative cocci which occur in pairs, clusters, or chains. The G+C content of their DNA is 40.3–44.4 mol%. Seven species are recognised, of which *V. parvula, V. atypica*, and *V. dispar* are residents of the oral cavity. They are able to grow over the temperature range of 24°–40°C, but growth is optimal at 30°–37°C. They cannot use carbohydrates as an energy source, but use lactate, pyruvate, fumarate, malate, and some purines for this purpose and generate propionate, acetate, and hydrogen as metabolic end-products. *Veillonella* spp. are aciduric and can grow well at a pH as low as 4.5. They are able to co-aggregate with a variety of oral bacteria (including species belonging to the genera *Streptococcus, Fusobacterium, Eubacterium, Actinomyces, Neisseria, Rothia*, and *Propionibacterium*) and are, therefore, important in plaque formation and in colonisation of oral mucosal surfaces. They are rarely responsible for human infections.

8.4.1.4 Anaerobic Gram-negative bacilli

The genus *Fusobacterium* consists of non-sporing, non-motile, anaerobic, Gram-negative rods which are often pleomorphic and/or filamentous. Many species are spindle-shaped with tapering ends (i.e., they have a "fusiform" morphology). Twelve species are recognised and the G+C content of most of these is 26–34 mol%. The species most commonly isolated from the oral cavity are *F. nucleatum, F. periodonticum, F. sulci*, and *F. alocis*. Their optimum temperature for growth is 35°–37°C, and they grow best at a pH of 7.0. They are generally unable to ferment sugars and obtain their energy by the fermentation of peptides and amino acids to butyric and acetic acids. However, they have only weak proteolytic activity and are dependent on amino acids and peptides provided by the host or produced by the action of proteases of other oral bacteria. *Fusobacterium* spp. can co-aggregate with most oral bacterial species (see Figure 8.7) and are considered to be important in plaque formation, where they act as "bridging" organisms between the primary colonisers of the tooth surface and later colonisers.

Porphyromonas spp. are non-motile, non-sporing, anaerobic, Gram-negative rods or coccobacilli. They produce porphyrin pigments, and colonies on blood agar become black after several days incubation due to the production of protohaem. The G+C content of their DNA is 46–54 mol%. They are proteolytic and use peptides and amino acids as an energy source – their growth is unaffected by the presence of carbohydrates. Butyrate, acetate, and propionate are the main metabolic end-products. They require haemin and vitamin K for growth, and their optimum growth temperature is 37°C. The main oral species are *Por. gingivalis, Por. endodontalis*, and *Por. asaccharolytica. Por. gingivalis* is one of the main aetiological agents of periodontitis (Section 8.5.2.3) and has a variety of virulence factors, including a number of potent proteases, nucleases, sialidases, and glyosaminoglycan-degrading enzymes. Its capsule protects it from phagocytosis and interferes with chemotaxis of PMNs. It has fimbriae which, along with its haemagglutinins, mediate adhesion to host cells and other bacteria. It also has a number of cytokine-inducing components and produces cytotoxic substances, including ammonia, H_2S, methylmercaptan, butyrate, and indole. It is able to invade epithelial cells and can inhibit bone formation and induce bone destruction. *Por. endodontalis* is one of several organisms associated with endodontic infections (Section 8.5.6).

Prevotella spp. are morphologically similar to *Porphyromonas* spp. but, unlike the latter, are saccharolytic. The G+C content of their DNA is 40–60 mol%. Their optimum growth temperature is 37°C, but they can grow over the range of 25°–42°C. They ferment carbohydrates, producing mainly acetate and succinate, and some species are proteolytic. The oral species of the genus can be broadly classified into two main groups: pigmented and non-pigmented. The former include *Prev. intermedia, Prev. nigrescens, Prev. melaninogenica, Prev. loeschii, Prev. corporis*, and *Prev. denticola*. Non-pigmented species include *Prev. buccae, Prev. buccalis, Prev. oralis*, and *Prev. oulora*. Some of these species (particularly *Prev. intermedia* and *Prev. nigrescens*) have been implicated in the pathogenesis of periodontitis, while others have been isolated from infections at other body sites.

Tannerella forsythensis (formerly known as *Bacteroides forsythus*) is believed to be an important aetiological agent of periodontitis. This is a fusiform bacillus which requires N-acetyl muramic acid for growth, and its DNA has a G+C content of 46 mol%. It uses peptides and amino acids as energy sources and produces mainly acetic, butyric, isovaleric, and propionic acids as end-products. It produces a sialidase, a number of glycosidases, as well as proteases and can, therefore, degrade mucins. One of the proteases produced, a trypsin-like enzyme, is considered to be a major virulence factor, and a cell surface protein has been identified which can induce apoptosis in host cells.

Leptotrichia buccalis is a non-sporing, non-motile, Gram-negative fusiform. The G+C content of its DNA is 25 mol%. Although it is an anaerobe when first isolated, it becomes aerotolerant on subculture and can eventually grow in a carbon dioxide-enriched atmosphere. It uses carbohydrates as an energy source and produces lactic acid as the main end-product of metabolism.

Selenomonas spp. are motile, anaerobic, Gram-negative curved rods which are bile sensitive. They are fermentative and produce mainly acetate and propionate from glucose.

Campylobacter spp. are motile (apart from *Camp. gracilis*), anaerobic, Gram-negative thin rods which require formate and fumarate for growth. They are asaccharolytic, oxidase-positive, and produce mainly succinate as a metabolic end-product.

8.4.1.5 Spirochaetes

Spirochaetes are large, spiral-shaped organisms which are motile despite not having external flagella. Those found in the oral cavity of humans belong to the genus *Treponema*. This genus consists of helical, tightly coiled, motile, anaerobic, Gram-negative bacteria which are 5–20 μm long and have a diameter of 0.1–0.4 μm. At least ten oral species have been classified, but a large number of additional taxa (approximately seventy) have been detected using molecular techniques. They are nutritionally fastidious and very difficult to grow in the laboratory. They prefer a slightly alkaline pH (approximately 7.5) for growth and grow best at a temperature of 37°C. The main oral species include *T. denticola*, *T. vincentii*, and *T. socranskii*. *T. denticola* is considered an important periodontopathogen, and a number of virulence factors have been identified, including adhesins mediating adhesion to host cells and extracellular matrix molecules, chymotrypsin-like and trypsin-like proteinases, peptidases, a haemolysin, and a pore-forming cytotoxic protein. It can invade epithelial cells and excretes ammonia and H_2S, which are toxic to host cells. The organism also produces glycine and pyruvate from the glutathione present in the subgingival region and can, therefore, support the growth of a number of other oral bacteria colonising this region.

8.4.1.6 Facultatively anaerobic Gram-negative bacilli

Actinobacillus actinomycetemcomitans is a non-motile, facultatively anaerobic, Gram-negative coccobacillus whose growth is enhanced by the presence of carbon dioxide. The G+C content of its DNA is 42.7 mol%. It grows over the temperature range of 25°–42°C, but grows best at 37°C. Its optimum pH for growth is between 7.0 and 8.0. It is nutritionally fastidious, and growth is stimulated by certain steroid hormones, including estrogen, progesterone, and testosterone. It is fermentative and can utilise a number of carbohydrates for energy generation, including glucose, fructose, and mannose. The organism is an important aetiological agent of localised aggressive periodontitis (Section 8.5.2.3) and is also responsible for diseases at extra-oral sites, including endocarditis, septicaemia, and meningitis. It produces a number of virulence factors, including (1) a leukotoxin that induces apoptosis in PMNs and macrophages; (2) a cytolethal distending toxin that inhibits host-cell cycle progression; (3) an immunosuppressive protein that inhibits lymphokine production by host cells; (4) an inhibitor of neutrophil chemotaxis; (5) a number of bone resorbing factors, including chaperonin 60, an outer membrane protein, and a capsular polysaccharide; and (6) a collagenase. It is also able to invade epithelial cells. The organism produces a bacteriocin, actinobacillin, which is active against some streptococci and *Actinomyces* spp.

Eikenella corrodens is the sole member of the genus *Eikenella*, and its DNA has a G+C content of 56–58 mol%. It is a non-sporing, facultatively anaerobic, Gram-negative coccobacillus whose growth is enhanced by the presence of carbon dioxide. Although it does not have flagella, it exhibits twitching motility. It can grow over the temperature range of 27°–43°C, but grows best at 35°–37°C. Its optimum pH for growth is 7.0–8.0. The organism characteristically produces pitting on agar surfaces. As well as being implicated in the pathogenesis of periodontal diseases, the organism also causes endocarditis, meningitis, brain abscesses, osteomyelitis, septic arthritis, pneumonia, postsurgical infections, and soft-tissue infections.

Capnocytophaga spp. are facultatively anaerobic, Gram-negative slender rods that exhibit gliding motility. Growth is enhanced by carbon dioxide, and the optimum temperature for growth is 35°–37°C. The G+C content of their DNA is 37.6–39.1 mol%. They all ferment carbohydrates to produce acetic and succinic acids. Species found in the oral cavity include *Cap. ochracea*, *Cap. sputigena*, *Cap. gingivalis*, *Cap. granulosa*, and *Cap. haemolytica*. Virulence factors produced by the organisms include an epitheliotoxin, proteases able to degrade collagen, elastin and immunoglobulins, peptidases, acid and

Table 8.11.	Development of the oral microbiota with age
Age range	**Organisms frequently isolated**
0–2 months	*Strep. salivarius, Strep. oralis, Strep. mitis* biovar 1, *A. odontolyticus, Neisseria* spp., *Staphylococcus* spp., *Veillonella* spp., *Prev. melaninogenica* group
2–6 months	as above plus: *F. nucleatum, Por. catoniae,* non-pigmented *Prevotella* spp., and *Leptotrichia* spp.
6–12 months	as above plus: *Strep. sanguis, Capnocytophaga* spp., *Eik. corrodens,* and *Fusobacterium* spp. other than *F. nucleatum, Actinomyces* spp. other than *A. odontolyticus*
12 months – adolescence	as above plus: *Strep. mutans, Selenomonas* spp., *Prev. nigrescens, Prev. intermedia, Prev. pallens, Por. gingivalis, Act. actinomycetemcomitans, Clostridium* spp., and *Peptostreptococcus* spp.
adolescence	as above plus: increased proportions of black-pigmented anaerobes and spirochaetes; increased proportions of *Prev. intermedia* in females

alkaline phosphatases, phospholipase A_2, sialidase, and glycosidases. In addition to being associated with periodontal diseases, *Capnocytophaga* spp. have also been isolated from cases of septicaemia, endocarditis, and abscesses at a variety of body sites.

8.4.2 Acquisition of the oral microbiota

Immediately after birth, small numbers of a variety of bacteria may be detected in the oral cavity of most babies, and these are derived from a number of sources, including the mother's birth canal, the environment, and the skin and oral cavity of the mother and birth attendants. After approximately 8 hours, the number of bacteria in the oral cavity of neonates increases markedly, and the organisms that can be isolated include streptococci, lactobacilli, staphylococci, *Neisseria* spp., coliforms, *Ps. aeruginosa, Veillonella* spp., and enterococci. However, many of these are present only transiently, and during the first few days, the types of organisms isolated fluctuate markedly, but invariably include viridans streptococci, which are the first persistent colonisers of the oral cavity and dominate the oral microbiota. The most frequently encountered viridans streptococci are *Strep. oralis, Strep. mitis* biovar 1, and *Strep. salivarius* (Table 8.11). Genetic analysis of the strains isolated from neonates has shown that many are identical to strains present in the oral cavity of the mother or other family members. Staphylococci also persist in the oral cavity of neonates, but are isolated in lower numbers than the streptococci. Over the next few weeks, the diversity of the oral microbiota increases and by the age of 1 month, most infants are colonised by more than one streptococcal species. Local alterations in the environment due to microbial activity (e.g., oxygen depletion, pH changes, production of metabolic end-products), together with the provision of new sites for colonisation (i.e., the bacteria themselves), promotes autogenic succession and within 2 months, *Neisseria* spp., staphylococci, facultative or microaerophilic *Actinomyces* spp. (e.g., *A. odontolyticus*), and obligate anaerobes (*Veillonella* spp. and members of the *Prev. melaninogenica* group) are frequently isolated and constitute appreciable proportions of the oral microbiota. Between 2 months and 6 months (prior to tooth eruption), colonisation by *F. nucleatum*, *Por. catoniae*, non-pigmented *Prevotella* spp., and *Leptotrichia* spp. takes place and the frequency of isolation of *Actinomyces* spp. increases.

Up until the age of approximately 6 months, the only surfaces available for microbial colonisation in the oral cavity of neonates are mucosal surfaces. However, the eruption of teeth offers a number of new habitats for colonisation, including the different regions of the tooth surface, the gingival sulcus, and the gap between adjacent teeth. The presence of the non-shedding tooth surfaces enables the formation of biofilms, each of which provides a wide range of microhabitats (Sections 1.1.2 and 8.4.3.1). Furthermore, the gingival crevice provides a very different habitat from those existing previously because it is bathed in GCF – a fluid with a composition very different from that of saliva (Section 8.3.2). Not surprisingly, therefore, the range of organisms that can be isolated from the oral cavity markedly increases following the emergence of teeth. One of the most dramatic changes due to tooth eruption (an example of allogenic succession) is the establishment of *Strep. sanguis* in the oral cavity. This organism is adept at colonising enamel rather than mucosal surfaces. Other organisms making their first appearance include *Capnocytophaga* spp., *Eikenella corrodens*, and *Fusobacterium* spp. other than *F. nucleatum*. The frequency of isolation of *Por. catoniae*, *F. nucleatum*, *Actinomcyes* spp., and non-pigmented *Prevotella* spp. also increase following tooth eruption. Furthermore, while only *A. odontolyticus* is usually isolated prior to tooth eruption, a number of additional species can be detected subsequent to this event so that by the age of 1 year, *A. naeslundii*, *A. viscosus*, *A. gerencseriae*, and *A. graevenitzii* are also often frequently isolated. Somewhat later (after approximately 1 year), additional organisms found in the oral cavity include *Selenomonas* spp., *Prev. nigrescens*, *Prev. pallens*, *Clostridium* spp., *Peptostreptococcus* spp., *Strep. mutans*, as well as the periodontopathogens *Prev. intermedia*, *Por. gingivalis*, and *Act. actinomycetemcomitans*. With regard to colonisation by the important cariogenic organism, *Strep. mutans*, a "window of infectivity" has been identified as occurring between the ages of 19 and 31 months. Although the reasons why colonisation peaks during this period have not been established, eruption of the primary molars also occurs at this time (between 16 and 29 months), and these teeth have fissured occlusal surfaces that are more easily colonised by mutans streptococci than smooth surfaces.

Hormonal changes in females occurring at puberty have been shown to influence the acquisition of certain species. Hence, the proportions of *Prev. intermedia* in plaque show a positive correlation with the levels of oestradiol and progesterone, which can substitute for vitamin K – an essential growth factor for the organism. The frequency of isolation and proportions of black-pigmented anaerobes and spirochaetes in plaque from the gingival crevice are also much greater in adolescents than in younger children, although the reasons for these increases have not yet been established.

8.4.3 Community composition at different sites

The oral cavity has two main types of surfaces for microbial colonisation: shedding (mucosa) and non-shedding (teeth). Because of anatomical variations, each of these types of surfaces provides a range of habitats, each with a characteristic microbiota. Although saliva contains large numbers of bacteria (approximately 10^8/ml), it is not considered to have an indigenous microbiota because its high flow rate and low content of free carbohydrates do not enable significant multiplication to occur *in vivo*. The organisms found in saliva are those shed by or dislodged from oral surfaces. The type of organisms found, and their relative proportions, are similar to those of the dorsal surface of the tongue.

8.4.3.1 Supragingival plaque

The term supragingival plaque refers to those biofilms that form on the tooth surface above the level of the gingival margin. A number of different types of supragingival plaque are recognised, depending on their location on the tooth surface – each of these has a characteristic composition dictated by the environmental conditions operating at the particular site. The main types of supragingival plaque recognised are smooth surface plaque (found on the lingual and buccal surfaces), approximal plaque (found between adjacent teeth), and fissure plaque (found in the fissures on the biting surfaces of premolars and molars). Plaque formed at the gingival margin, although technically "supragingival", has a composition distinct from that of other supragingival plaques because of the very different environmental selecting factors operating in this region – it is, therefore, described separately (Section 8.4.3.2). In this section, the formation of plaque on the supragingival tooth surface is outlined in general terms and then the composition of the microbial communities at different anatomical sites are described.

The first stage in the formation of supragingival plaque is adhesion of bacteria to the enamel surface. However, bacteria rarely come into contact with enamel itself because, like all oral surfaces, the tooth is covered with a complex layer of adsorbed molecules known as the "acquired enamel pellicle" (AEP), or "acquired pellicle". Proteins and, to a lesser extent, lipids and glycolipids, are rapidly adsorbed by an enamel surface within seconds of cleaning and continue to accumulate for approximately 2 hours to form a tenacious layer 0.1–1 μm thick. Investigations of the composition of the AEP have been hampered by its complexity and the small quantities of material available. Nevertheless, there is general agreement that the AEP does not contain all of the substances present in saliva, indicating that adsorption is selective and that the adsorbed molecules form distinct globular and fibrillar structures rather than an amorphous mass. Proline-rich proteins, histatins, and statherin are consistently detected in the AEP, and studies have shown that histatins tend to accumulate near the outer surface, while the other two groups of proteins are distributed throughout the layer (Figure 8.3). Recently, two-dimensional gel electophoresis and matrix-assisted laser desorption/ionisation time-of-flight mass spectrometry have been used to investigate the proteinaceous constituents of AEP and have shown it to contain histatins, lysozyme, statherin, cytokeratins, and calgranulin B. Other constituents identified include MG1, MG2, amylase, IgA, IgG, cystatin SA-1, parotid saliva agglutinin, serum albumin, and carbonic anhydrase. Bacterial components and products, including glucosyltransferases and glucans, have also been detected in the AEP. As well as acting as a substratum for bacterial adhesion, the AEP has an important role in maintaining the integrity of enamel by protecting it against demineralisation and enhancing its remineralisation, thus hindering its dissolution by acids.

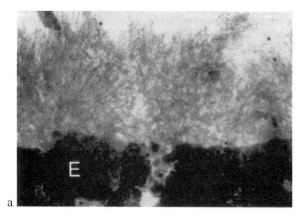

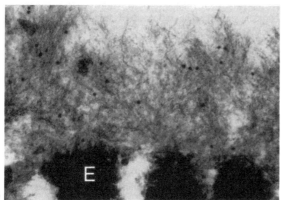

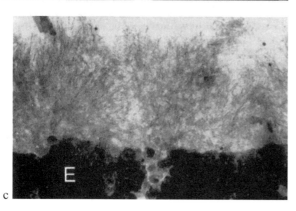

Figure 8.3 Transmission electron micrographs of the pellicles formed on bovine enamel (E) discs from whole human saliva (a–c). In (a), the primary antibody was goat anti-proline-rich protein 1 (PRP 1), and the secondary antibody was gold-labeled rabbit anti-goat antibody. In (b), the primary antibody was rabbit anti-PRP 3, and the secondary antibody was gold-labeled goat anti-rabbit antibody. Because both primary antibodies react with the acidic PRPs, the gold particles may indicate the presence of any of the anionic PRPs. PRPs appeared to be widely distributed throughout the pellicle. In (c), the primary antibody was rabbit anti-histatin 3, and the secondary antibody was gold-labeled goat anti-rabbit antibody. Because the primary antibody against histatin 3 can also react with histatin 1 and histatin 5, the gold particles may indicate the presence of any of the major histatins. The histatins tended to localise near the outer, surface region. Magnification in (a–c) is the same. Reproduced with the permission of Blackwell Publishing from: Electron-microscopic demonstration of proline-rich proteins, statherin, and histatins in acquired enamel pellicles *in vitro*. Schupbach, P., Oppenheim, F.G., Lendenmann, U., Lamkin, M.S., Yao, Y., and Guggenheim, B. *European Journal of Oral Science* 2001;109:60–68.

The first organisms to adhere to the AEP will be those with adhesins able to bind to receptors displayed by the constituent molecules of the pellicle. A wide range of receptors for bacterial adhesins has been identified in the AEP (Table 8.9). Bacteria, mainly streptococci, cover between 12% and 32% of the enamel surface within a few hours resulting in a population density of between 2.5 and 6.3×10^5 cells/mm^2. Those of the adherent organisms that can grow and reproduce under the environmental conditions operating at the site will become established and constitute the pioneer community. The supragingival surface of the tooth is an aerobic environment with a relatively high redox potential and a neutral pH, and the main pioneer species include viridans streptococci (mainly *Strep. sanguis*, *Strep. oralis*, and *Strep. mitis* biovar 1), *Neisseria*

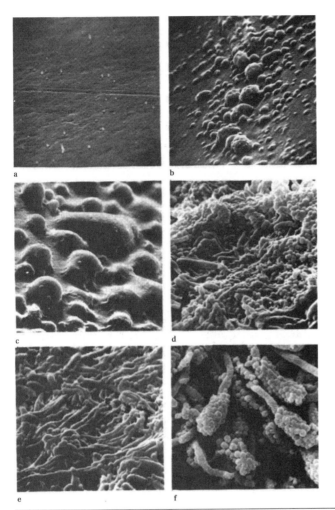

Figure 8.4 Electron micrographs showing the development of dental plaque. (a) A recently (<1 hour) cleaned tooth surface coated with the acquired enamel pellicle (AEP). (b) Attachment of individual bacteria (1–4 hours). (c) Formation of microcolonies (4–24 hours). (d) Increasing species diversity due to microbial succession (>24 hours). (e) Mature plaque (>24 hr). (f) An example of a frequently encountered structure in mature plaque – a "corn-cob". Reprinted with permission of Cambridge University Press from: Novel microscopic methods to study the structure and metabolism of oral biofilms. Bradshaw, D.J. and Marsh, P.D. In: *Medical implications of biofilms*, 173–188. Wilson, M. and Devine, D. (eds.). Copyright © 2003.

spp., *Haemophilus* spp., and *Actinomyces* spp. (mainly *A. naeslundii* and *A. odontolyticus*). Microcolonies of Gram-positive cocci (presumably streptococci) can be detected on the tooth surface within 2–4 hours and these then grow rapidly with doubling times of between 1 and 3 hours (Figure 8.4). Once the cell density reaches 2.5–4.0×10^6 cells/mm^2, there is a dramatic increase in the growth rate suggestive of cell density-dependent growth regulation (i.e., quorum sensing). Growth of the attached bacteria proceeds exponentially for the first few days, then decreases. The growth of the pioneer community alters the environment in a number of ways as a result of (1) oxygen utilisation, (2) the production of reducing compounds, (3) the excretion of metabolic end-products, and (4) the degradation of host macromolecules. The environment, therefore, becomes less aerobic, has a lower redox potential, contains a more diverse range of nutrients,

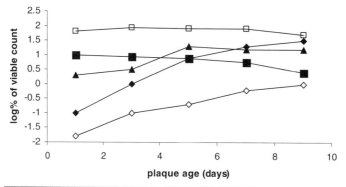

Figure 8.5 Establishment of anaerobic genera (*Veillonella* and *Fusobacterium*) in dental plaque as a consequence of environmental changes brought about by initial colonisers of the tooth surface (i.e., autogenic succession). □ = streptococci; ■ = *Neisseria* spp.; ▲ = *Veillonella* spp.; ♦ = *Actinomyces* spp.; ◇ = *Fusobacterium* spp. Reprinted from: Microbial population shifts in developing human dental plaque. Ritz, H.L. *Archives of Oral Biology* 1967;12:1561–1568. Copyright © 1967, with permission from Elsevier.

and provides additional substrata – the bacteria themselves and the polymers they produce – for microbial adhesion. This enables colonisation by organisms which are unable to either adhere to, or survive on, the original tooth surface – such organisms are often termed "secondary colonisers". This autogenic succession results in a more diverse community, as can be seen in Figures 8.5 and 8.7. The aerobic *Neisseria* spp. and facultative streptococci which initially dominate the community comprise decreasing proportions of the microbiota with time, while the proportions of anaerobic *Fusobacterium* spp. and *Veillonella* spp. increase (Figure 8.5). The changes that occur in the redox potential of plaque as it develops on the tooth surface have been followed and are shown in Figure 8.6. Microscopic examination of the plaque at different time points reveals that, during the first 2 days, it is dominated by Gram-positive cocci and rods (presumably streptococci and *Actinomyces* spp., respectively), whereas by day 3, Gram-negative cocci (presumably *Veillonella* spp.) appear. By day 7, Gram-negative rods and filaments (presumably *Fusobacterium* spp.) are also present. Comparison of Figures 8.5 and 8.6 shows that the increasing proportions of *Veillonella* spp. and *Fusobacterium* spp. correlate with the decreasing Eh of the developing plaque.

Colonisation of the growing and diversifying plaque by additional bacterial species present in saliva then occurs, and this is mediated by two principal mechanisms: (1)

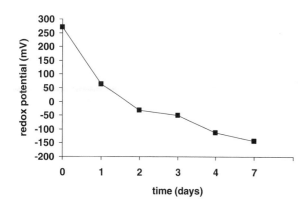

Figure 8.6 Changes in redox potential as plaque develops on a tooth surface.

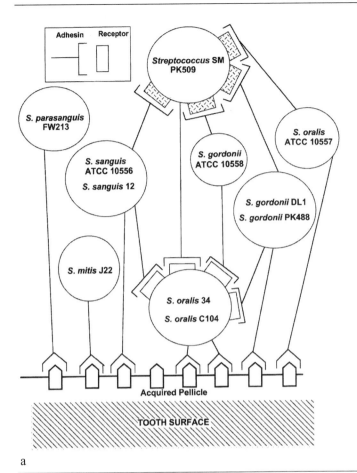

a

Figure 8.7 Adhesive interactions between bacteria in dental plaque. Adhesins on early colonisers bind to receptors on the acquired pellicle which coats the tooth surface (a), then secondary colonisers adhere to the early colonisers (b). Interactions occurring in mature plaque are depicted in (c). Several kinds of co-aggregations are depicted as complementary sets of symbols of different shapes. One set is depicted in the box at the top. Proposed adhesins (symbols with a stem) represent surface components that are sensitive to heat and proteases; their complementary receptors (symbols without a stem) are not usually affected by heat or proteases, as they are often a carbohydrate. Identical symbols represent components that are functionally similar, but may not be structurally identical. Reproduced with kind permission of Dr. Paul Kolenbrander, National Institute of Dental and Craniofacial Research, Bethesda, MD. (*Continued*)

co-adhesion – this involves the adhesion of a planktonic cell in saliva to a microbe already present in the biofilm and (2) co-aggregation. Co-aggregation involves the formation of a microbial aggregate from planktonic bacteria present in saliva – the resulting aggregate then adheres to a microbe already present in the biofilm. The adhesin–receptor interactions enabling both coadhesion and coaggregation have been determined for many oral microbes, and examples are shown in Table 8.10. The secondary coloniser *F. nucleatum* is particularly important in both co-adhesion and co-aggregation because it has been shown to bind to virtually all oral bacterial species (Figure 8.7c). Co-aggregation is not only important for the physical development and spatial organisation of plaque, but is also thought to facilitate nutritional and physiological interactions between organisms in these communities. Hence, obligate anaerobes such as *V. dispar* and *Por. gingivalis* can survive in aerobic environments only when they

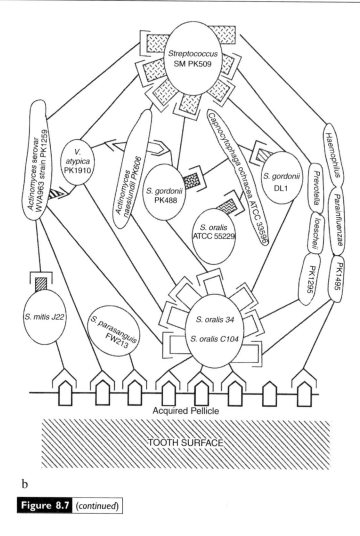

b

Figure 8.7 (continued)

co-aggregate with *F. nucleatum* and aerobic or facultative species, such as *N. subflava* and streptococci. Glucose utilisation by *Actinomyces* spp. and streptococci is also enhanced by co-aggregation of the species. Apart from co-adhesion and co-aggregation, the production of exopolymers by plaque bacteria, as previously mentioned, provides additional substrata for bacterial colonisation. Furthermore, bacterial degradation of host macromolecules in the AEP can reveal previously unavailable receptors (known as "cryptitopes") for bacterial adhesins.

So far, only factors contributing to an increase in the size and complexity of plaque have been described (i.e., bacterial proliferation and deposition of organisms from saliva). It must be remembered, however, that as the biofilm grows in size, it will become increasingly susceptible to mechanical and hydrodynamic shear forces, which will tend to decrease its size and, possibly, its diversity due to detachment of constituent organisms. Several studies have shown that, in the absence of tooth brushing and other oral hygiene measures, plaque growth reaches a maximum after 3–4 days. Although there is little further increase in plaque mass, changes in its composition may continue to take place, leading to increased proportions of Gram-negative and anaerobic species.

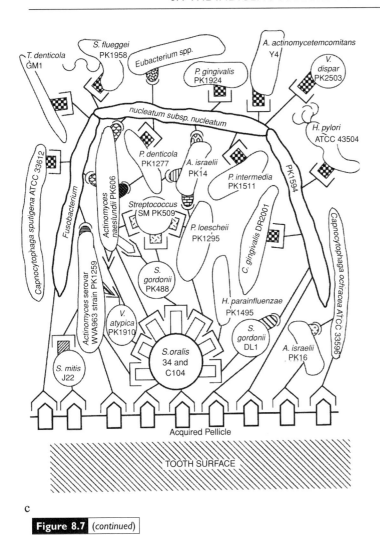

c

Figure 8.7 (continued)

Detachment due to mechanical abrasion constitutes a greater problem for biofilms on exposed tooth surfaces and can limit their size and diversity. In contrast, biofilms growing in anatomically protected regions will be largely unaffected until they protrude beyond the limits of their sheltered site. Detachment of bacteria and biofilm sections from the "parent" biofilm provides a means by which biofilm formation can be initiated at other sites once reattachment has taken place.

The secondary colonisers in the plaque biofilm will, of course, also alter the environment enabling colonisation by yet other organisms. Microbial succession eventually results in the formation of a stable, climax community with a high species diversity. The members of this climax community exist in a state of dynamic equilibrium because of complex spatial, physiological, and nutritional interactions, both beneficial (Figure 8.8) and antagonistic (Table 8.12), and their relative proportions will remain constant provided that there is no change in any of the environmental determinants at the site. Any changes that do occur may result in a shift in the relative proportions of the constituent members, with the possible elimination of some and recruitment of others,

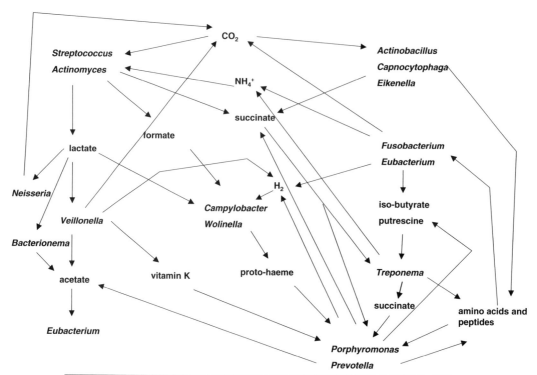

Figure 8.8 Some potential nutritional interactions that may occur among plaque bacteria.

Table 8.12. | Antagonistic interactions among oral bacteria

Organism	Mechanism	Target organism(s)
Strep. mutans	bacteriocin (mutacin)	other strains of Strep. mutans, other streptococci, Actinomyces spp.
streptococci, Actinomyces spp., Lactobacillus spp.	generation of low pH	wide range of non-aciduric spp. especially Gram-negatives
streptococci (many species)	hydrogen peroxide	Act. actinomycetemcomitans, Cap. sputigena, F. nucleatum, Camp. rectus, Bacteroides spp.
C. matruchotii	bacteriocins	Actinomyces spp., L. casei, L. acidophilus, L. fermentum, F. nucleatum, Strep. salivarius
Act. actinomycetemcomitans	bacteriocin (actinobacillicin)	other strains of the organism, Strep. sanguis, A. viscosus, Strep. uberis
Por. gingivalis	haematin	streptococci, Actinomyces spp., C. matruchotii, P. acnes
Strep. sanguis	sanguicin	Por. gingivalis
Strep. mutans	lactic acid	T. denticola

tooth surface

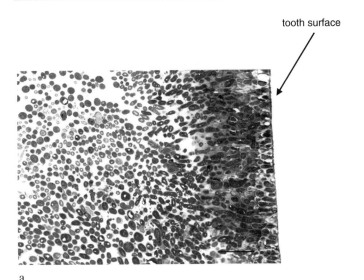

a

Tooth surface

b

> **Figure 8.9** (a) and (b) Transmission electron micrographs of supragingival plaque from a healthy volunteer. In (b), a microcolony (circled) is apparent within the plaque. Bar = 3 μm. Images kindly supplied by Mrs. Nicola Mordan, Eastman Dental Institute, University College London.

until a new climax community is established. Studies of the composition of dental plaques over several years have shown that the relative proportions of the predominant organisms remain remarkably stable despite the fact that allochthonous species are continually being introduced into the oral cavity and autochthonous microbes are repeatedly being removed by oral-hygiene procedures.

The main constituents of dental plaque are microbes, their exopolymers, and host macromolecules. The nature of the host macromolecules has already been described, and the types of bacteria present are detailed later. Oral bacteria produce a variety of exopolymers which function as adhesins, provide protection against host defences, and help to maintain the integrity of the plaque. They can also act as receptors for the adhesins of other bacteria and be utilised as carbon and energy sources. Both homopolysaccharides and heteropolysaccharides are produced, but the main exopolymers present in plaque are the glucans and fructans synthesised by streptococci. These polymers are produced from dietary sucrose by the enzymes glucosyltransferase and fructosyltransferase, respectively. Both water-soluble and water-insoluble polymers are present in plaque, with the latter being the most important from the point of view of maintaining plaque structure, and these consist mainly of α1–3-linked glucose residues. Although streptococci are probably the most important producers of the plaque matrix, other oral bacteria are able to synthesise exopolymers, which could also contribute to the structure of the biofilm. These include *Actinomyces* spp., *Eubacterium* spp., *Rothia dentocariosa* (a nonmotile, non-sporing, Gram-positive pleomorphic rod with a fermentative metabolism), *Neisseria* spp. and *Lactobacillus* spp.

Studies of the structure of dental plaque have been carried out using a variety of techniques. Electron microscopy of plaques generally reveals tightly packed bacteria near the tooth surface with a more open structure within the outermost layers (Figure 8.9). Characteristic arrangements of bacteria are often discernible, such as microcolonies and "corn-cobs" (Figure 8.4) consisting of cocci (*Strep. crista*) surrounding a central filamentous organism (*F. nucleatum* or *C. matruchotii*). However, the specimen preparation essential for electron microscopy (dehydration, fixation, and staining) is likely to alter the native plaque structure because the biofilm matrix is composed mainly of water – often as much as 97%. Greater use is now being made of confocal laser scanning microscopy (CLSM) to examine dental plaques, and this enables the biofilms to be viewed in their hydrated state, thereby preserving their native structure. Such studies have revealed that some dental plaques have a more open structure than that implied by electron microscopy, and this is in keeping with the structures found in other biofilms (Figure 8.10a and b). Hence, water channels permeating "stacks" of bacterial microcolonies enclosed in an exopolymeric matrix have been observed. A microcolony forms at the particular location within a stack that has the appropriate combination of environmental factors (due to gradients in nutrient concentration, pH, oxygen content, and the concentration of metabolic products as previously mentioned) suitable for the survival and growth of that organism. The water channels may function as a primitive circulatory system, bringing fresh supplies of nutrients and oxygen while removing metabolic waste products. As yet, few studies of the structure of dental plaques by CLSM have been carried out, and the technique has not yet been applied to investigate plaques from the various sites within the oral cavity which could well display a variety of structures.

The aforementioned description of biofilm formation on tooth surfaces will, in general terms, apply to biofilm formation on any non-shedding oral surface, including enamel, dentine, cementum, as well as oral appliances and prostheses, such as dentures, implants, and artificial crowns. However, the composition of the climax community will be affected by the nature of the environmental determinants operating at the particular site. Many studies have been carried out to determine the composition of dental plaques at a particular site, but the results are affected by a number of factors (in addition to the usual variables mentioned in Section 1.1.1), including (1) the age of the plaque, (2) whether or not the subjects have been using oral hygiene measures (i.e., tooth brushing, flossing) during the study, and (3) the diet of the subjects involved.

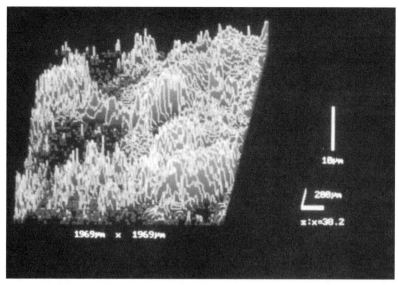

a

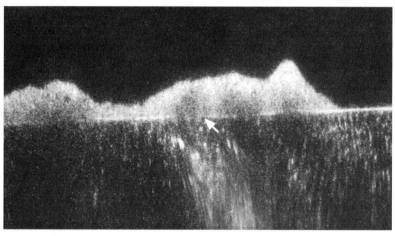

b

Figure 8.10 (a) Confocal laser scanning micrograph showing the topography of 3-day-old dental plaque formed on the enamel surface in a human volunteer. (b) Confocal laser scanning micrograph showing a z-section of 2-day-old plaque formed on enamel in a human volunteer. The arrow indicates the border between the plaque and the enamel surface. The plaque ranges in depth from 0 μm (at the right of the image) to a maximum of 32 μm. Reprinted from: A pilot study of confocal laser scanning microscopy for the assessment of undisturbed dental plaque vitality and topography. Netuschil, L., Reich E., Unteregger, G., Sculean, A., and Brecx, M. *Archives of Oral Biology* 1998;43:277–285. Copyright © 1998, with permission from Elsevier.

Consequently, a wide range of values has been reported for the composition of plaque from a particular site.

As can be seen from Table 8.13, the predominant cultivable bacteria in supragingival plaques on smooth surfaces in approximal regions and in fissures are streptococci and *Actinomyces* spp., which together comprise more than 50% of the cultivable microbiota. These organisms are consistently present at all three sites in the majority of individuals. *Veillonella* spp., *Neisseria* spp., and lactobacilli are often (but not always) present

Table 8.13. The predominant cultivable microbiota of the three main types of supragingival plaque

Organism	Smooth surface Median (%)	Approximal Median (%)	Approximal Range (%)	Fissure Median (%)	Fissure Range (%)
streptococci	54.8	15.4	0.4–70	44.9	7.9–86.3
mutans streptococci	0.3	0.11	0–23.0	24.7	0–86.3
Actinomyces spp.	27.2	39.1	4–81	18.2	0–45.9
Veillonella spp.	17.2	9.8	0–59	3.3	0–44.4
Neisseria spp.	0.3	0.1	0–44	ND	
lactobacilli	0	0	0–1.9	0	0–28.6

Note: ND = not determined.

and may also comprise appreciable proportions of the microbiotas. Smooth surface plaques tend to be thinner and have a less diverse microbiota than other supragingival regions because they are less protected and, thus, are more exposed to mechanical abrasion. This limits their growth and, consequently, the range of microhabitats that can develop within the plaque due to autogenic succession. Approximal plaques tend to be dominated by *Actinomyces* spp. rather than streptococci. Their anatomical location means that they are largely protected from mechanical abrasion, and they have a lower oxygen content because of poor penetration by saliva. Gram-negative anaerobes (mainly *Prevotella* spp.) are frequently present, although in low proportions. Occlusal fissures often contain impacted food, and this abundance of nutrients may contribute to the diversity of the microbiota at these sites. Mutans streptococci are found more often and in higher proportions in plaques at these sites, and this correlates with occlusal fissures being the most likely sites of dental caries (Section 8.5.1). Staphylococci are also frequently present, although in low proportions. Although, as can be seen from Table 8.13, the predominant cultivable organisms in supragingival plaques include streptococci, *Actinomyces* spp., and *Veillonella* spp. – a wide variety of other organisms are often isolated; these are listed in Table 8.14.

Only a very limited number of culture-independent studies of the supragingival plaque microbiota have been carried out. 16S rRNA genes have been amplified from the DNA extracted from the supragingival plaque of one healthy child and then cloned, and fifty of the clones obtained were sequenced. Fifty percent of the clones corresponded to *Strep. sanguis*, 14% to *Ab. defectiva*, 12% to *N. mucosa*, and 6% to a *Leptotrichia* sp. A total of eleven species or taxa were detected, including one that was novel. The presence or absence of twenty-three oral species in the supragingival plaques of children has been determined by amplification of 16S rRNA genes in DNA extracted from the samples followed by hybridisation with oligonucleotide probes. *A. naeslundii* genospecies 2 and *A. gerencseriae* were the dominant species and together comprised more than 60% of the microbiota. Seven streptococcal species and *Veillonella* spp. accounted for most of the rest of the microbiota.

The composition of supragingival plaque in healthy adults has been studied by probing DNA extracted from samples with whole genomic DNA probes specific for forty bacterial taxa. Plaque samples were taken from a number of supragingival sites in each individual and pooled. The most frequently detected species in the twenty-two

Table 8.14.	Bacteria that may be isolated from supragingival plaque

Genus	Examples
Actinomyces	A. naeslundii genospecies 1, A. naeslundii genospecies 2, A. odontolyticus, A. israelii, A. gerencseriae, A. meyeri
Streptococcus	Strep. gordonii, Strep. sanguis, Strep. intermedius, Strep. oralis, Strep. mitis, Strep. mutans, Strep. anginosus, Strep. salivarius
Abiotrophia	Ab. defectiva, Ab. adiacens
Lactobacillus	L. casei, L. plantarum, L. fermentum, L. acidophilus
Enterococcus	Ent. faecalis
Leptotrichia	Lep. buccalis
Propionibacterium	P. acnes
Haemophilus	H. parainfluenzae, H. segnis, H. paraphrophilus, H. haemolyticus
Neisseria	N. subflava, N. sicca, N. mucosa
Veillonella	V. parvula, V. dispar, V. atypica
Eubacterium	Eub. nodatum, Eub. brachy
Rothia	Roth. dentocariosa
Porphyromonas	Por. gingivalis, Por. endodontalis
Selenomonas	Selenomonas sputigena
Bifidobacterium	Bif. dentium
Capnocytophaga	Cap. ochracea
Fusobacterium	F. nucleatum, F. vincentii, F. polymorphum
Prevotella	Prev. intermedia, Prev. nigrescens
Actinobacillus	Act. actinomycetemcomitans
Peptostreptococcus	Pep. micros
Gemella	Gem. morbillorum
Arachnia	Arachnia propionica

individuals were *A. naeslundii* genospecies 2, *A. israelii*, *A. naeslundii* genospecies 1, *A. gerencseriae, A. odontolyticus, Lep. buccalis, Gem. morbillorum, F. nucleatum, Strep. sanguis, Strep. oralis, Eub. nodatum*, and *Micromonas micros* (formerly *Pep. micros*) – all of which were present in at least 50% of individuals.

The diet of the host can have a profound effect on the composition of the plaque communities. Easily fermentable carbohydrates, such as sucrose, are metabolised by many members of the plaque microbiota to organic acids (mainly lactic and acetic acids), and this lowered pH selects for aciduric species, such as mutans streptococci and lactobacilli. Such species are also acidogenic, and exposure of the enamel surface to such low pHs for long periods of time can induce demineralisation of enamel and result in dental caries. On the other hand, consumption of milk and cheese can protect against dental caries by a number of mechanisms. Proteins present in milk and cheese can exert a buffering action and can become incorporated into the salivary pellicle, where they enhance remineralisation by sequestering calcium phosphate. Their presence in the salivary pellicle reduces adhesion of mutans streptococci, and their degradation by plaque bacteria can result in the release of pH-raising amines and ammonia.

8.4.3.2 Gingival crevice

For a number of reasons, the gingival crevice is a very different habitat from those existing on the supragingival surfaces of the tooth. Hence, although saliva will be

		Proportion (%) of cultivable microbiota	
Organism	Isolation frequency (%)	Range	Mean
Gram-positive facultatively anaerobic cocci (mainly streptococci)	100	2.4–73.2	39.6
Gram-positive facultatively anaerobic rods (mainly *Actinomyces* spp.)	100	9.8–62.5	35.1
Gram-negative anaerobic rods	100	7.5–20.4	12.7
Gram-positive anaerobic rods	86	0–36.6	9.5
Gram-negative anaerobic cocci (predominantly *Veillonella* spp.)	57	0–4.9	2.0
Gram-positive anaerobic cocci (predominantly *Peptostreptococcus* spp. and former members of the genus)	14	0–5.6	0.8
Gram-negative aerobic/facultatively anaerobic cocci (predominantly *Neisseria* spp.)	14	0–1.8	0.3

Table 8.15. | Predominant cultivable microbiota of the gingival crevice

present in this region, the predominant source of nutrients is likely to be GCF with its high protein content. Furthermore, because of its anatomical location, it is less exposed to the mechanical abrasion that occurs in supragingival regions. Finally, penetration of oxygen-rich saliva into the region is hindered by the outward flow of GCF, so that the oxygen content and redox potential will be lower than those found on the more exposed tooth surfaces. As a consequence of these differences, the microbiota of the gingival crevice differs substantially from those of other supragingival plaques. Essentially, the microbiota, although dominated by streptococci and *Actinomyces* spp., has a greater species diversity with higher proportions of obligate anaerobes, including spirochaetes which are very rarely found in supragingival plaques. Most oral spirochaetes have not yet been grown in the laboratory and often do not appear in assessments of the total cultivable microbiota of the gingival crevice. Their presence is usually determined by phase-contrast or dark-ground microscopy. The results of a study of the cultivable microbiota of the healthy gingival crevice are shown in Table 8.15. Some of the organisms that have been isolated from the healthy gingival crevice are listed in Table 8.16.

Despite the extensive number of species isolated by culture, culture-independent studies have shown that only approximately half of the gingival microbiota are known species. For example, in a 16S rRNA gene analysis of plaque obtained from the gingival crevice of five healthy individuals, 268 clones were examined but only 58% of the sequences corresponded to those of known species. A total of seventy-two taxa were detected. 16S rRNA genes were amplified from the DNA extracted from the gingival crevice plaque of one healthy adult and then cloned, and 264 of the clones obtained were sequenced. The sequences of 52.5% of the clones did not correspond to any of those in public databases. The composition of the gingival microbiota has also been studied by probing DNA extracted from samples with whole genomic DNA probes specific for

Table 8.16.	Bacterial species detected in the gingival crevice

Type of organism	Examples
spirochaetes	*T. denticola, T. pectinovorum, T. socranskii, T. vincentii*
Actinomyces spp.	*A. meyeri, A. naeslundii, A. odontolyticus, A. gerencseriae, A. israelii*
Eubacterium spp.	*Eub. saburreum, Eub. nodatum*
Mogibacterium spp.	*Mogibacterium timidum* (formerly *Eub. timidum*)
Rothia spp.	*Roth. dentocariosa*
streptococci	*Strep. anginosus, Strep. gordonii, Strep. intermedius, Strep. oralis, Strep. salivarius, Strep. sanguis*
Haemophilus spp.	*H. aphrophilus, H. segnis, H. haemolyticus*
Neisseria spp.	*N. mucosa, N. elongata*
Veillonella spp.	*V. parvula*
Porphyromonas spp.	*Por. gingivalis, Por. endodontalis*
Selenomonas spp.	*Sel. sputigena*
Capnocytophaga spp.	*Cap. ochracea, Cap. sputigena, Cap. gingivalis, Cap. haemolytica.*
Fusobacterium spp.	*F. nucleatum, F. vincentii, F. polymorphum, F. periodonticum*
Prevotella spp.	*Prev. intermedia, Prev. nigrescens, Prev. melaninogenica*
Actinobacillus spp.	*Act. actinomycetemcomitans*
Campylobacter spp.	*Camp. concisus, Camp. gracilis, Camp. showae, Camp. rectus*
Peptostreptococcus spp.	*Pep. micros*
Eikenella spp.	*Eik. corrodens*
Leptotrichia spp.	*Lep. buccalis*

forty bacterial taxa. The most frequently detected species in twenty-two periodontally healthy individuals were *A. israelii, Lep. buccalis, A. gerencseriae, A. naeslundii* genospecies 1 and 2, *F. polymorphum*, and *F. nucleatum*, all of which were present in approximately 50% of individuals.

8.4.3.3 Tongue

Unlike other oral mucosal surfaces, the epithelium of the tongue is highly convoluted owing to the presence of numerous papillae. Access of oxygenated saliva to the crypts between the papillae is therefore limited, resulting in oxygen-depleted habitats suitable for the growth of microaerophiles and obligate anaerobes. The crypts also provide some protection to the resident microbial communities from mechanical abrasion, and it is possible that substantial biofilms are formed in these regions. The papillate structure of the tongue also results in the retention of appreciable quantities of food, desquamated epithelial cells, and other debris which furnish resident microbes with an abundant supply of nutrients. Consequently, the population density on the tongue is higher than that of other oral mucosal surfaces, and the microbiota is more diverse. Estimates of the microbial population density on the tongue range from 10^7 to 10^9 cfu/cm^2, with the greater densities being found towards the back of the tongue. Microscopy has revealed that the number of bacteria per epithelial cell is approximately 100.

Culture-based studies of the tongue microbiota have revealed a diverse microbiota with streptococci, *Veillonella* spp., *Actinomyces* spp., and Gram-negative anaerobic rods being frequently isolated and comprising appreciable proportions of the microbiota.

Table 8.17. | Cultivable microbiota of the tongue

Organism	Frequency of isolation (%)	Viable count cfu/cm^2 ($\times 10^6$)
streptococci	100	17.4
Porphyromonas/Prevotella spp.	94	4.3
Veillonella spp.	88	3.9
Actinomyces spp.	82	2.4
Fusobacterium spp.	65	0.7
facultatively anaerobic branching rods (mainly Rothia spp. and Corynebacterium spp.)	65	0.7
Neisseria spp.	59	0.4
Stomatococcus spp.	59	0.4
Selenomonas spp.	41	0.2
Leptotrichia spp.	41	0.2
unidentified Gram-negative anaerobes	41	0.3
Peptostreptococcus spp.	29	0.3
Propionibacterium spp.	24	0.1
Gram-negative facultative rods	24	0.1
Haemophilus spp.	18	0.1
Capnocytophaga spp.	12	0.1
Lactobacillus spp.	12	0.1
Eubacterium spp.	12	0.1

Note: Frequency of isolation = proportion of study population from which the organism was isolated.

The results of a culture-based analysis of the tongue microbiota of a group of seventeen healthy adults are shown in Table 8.17. The predominant streptococcal species were *Strep. salivarius* and *Strep. mitis* biovar 2. In a culture-independent study of the tongue microbiota of ten adults (five with and five without halitosis), 16S rRNA gene sequences were amplified from the DNA extracted from the tongue samples, and then cloned and sequenced. The diversity of the tongue microbiota can be appreciated from the fact that ninety-two different taxa were detected – only approximately 40% of the sequences corresponded to those of known species. Thirty-two percent of the taxa appear to be unique to the tongue because they have not been found among more than 6,000 sequences obtained from other oral sites. Between sixteen and twenty-two taxa were detected in the microbiota of each of the five adults without halitosis, but the ten most frequently detected taxa comprised 69–85% of those present (Table 8.18). Streptococci were detected in all individuals with *Strep. salivarius* and *Strep. parasanguis*, comprising the predominant organisms in most cases. *Rothia mucilaginosa* and *Strep. infantis* were also present in all individuals, but comprised lower proportions of the microbiota.

In another culture-independent study, the microbiotas of the dorsal, lateral, and ventral surfaces of the tongue in 225 healthy adults were analysed. DNA was extracted from samples from each site and probed with whole genomic DNA probes specific for forty bacterial taxa. Table 8.19 shows the ten most prevalent members of the microbiotas of these sites; the other thirty species were all detected at each of the three sites, although they always comprised less than 5% of the microbiota of each site. These findings again illustrate the diversity of the microbiotas present on the surfaces of the tongue. The

| Table 8.18. | The ten most frequently detected species on the dorsum of the tongue of five healthy adults as determined by 16S rRNA gene sequencing |

| | % of clones | | Frequency of detection (%) |
	Range	Mean	
Strep. salivarius	12–41	23.0	100
Strep. parasanguis	7–32	18.4	100
Strep. infantis	1–9	4.6	100
Roth. mucilaginosa	3–10	5.6	100
Streptococcus strain HalT4-E3	0–24	5.4	60
Ab. adiacens	0–21	9.6	80
N. flavescens	0–7	2.2	40
Strep. pneumoniae	0–1	0.2	20
V. parvula/V. dispar	0–11	4.6	80
Atopobium parvulum	0–3	0.6	20
number of species detected	16–22	18.6	–
% of total detected	69–85	73.2	–

Note: The percentage of the total number of taxa detected that these ten species comprise is also shown. Frequency of detection = proportion of the study population in which the organism was detected.

dorsal surface of the tongue was dominated by Gram-negative anaerobes, which differs from the findings of culture-based studies and those of the culture-independent study described previously (Table 8.18). Interestingly, in the latter study, the predominant organisms were *Strep. salivarius* and *Strep. parasanguis* – no probes for these organisms were used in the study from which the data shown in Table 8.19 were derived. Of interest was the consistent finding on the tongue surface of a number of organisms associated with periodontal diseases – *Por. gingivalis, Prev. intermedia, Act. actinomycetemcomitans*, and *T. denticola* (Section 8.5.2). This suggests that the tongue could act as a reservoir for these

| Table 8.19. | Most prevalent members of the microbiotas of the dorsal, lateral, and ventral surfaces of the tongue in 225 healthy adults based on the use of whole genomic DNA probes recognising forty different oral species |

| Organism | Tongue surface | | |
	Dorsal	Lateral	Ventral
V. parvula	11	8	2
Prev. melaninogenica	11	5	3
Cap. gingivalis	5	7	10
Strep. mitis	3	8	12
Eik. corrodens	6	6	3
Gem. morbillorum	5	4	9
N. mucosa	5	5	3
Strep. oralis	2	4	5
Eub. saburreum	5	3	4
F. periodonticum	5	3	2

Note: Each number represents the percentage of the total DNA probe count and does not necessarily correspond to the proportion of the organism in the microbiota of the site because of the likely presence of organisms other than those recognised by the probes used.

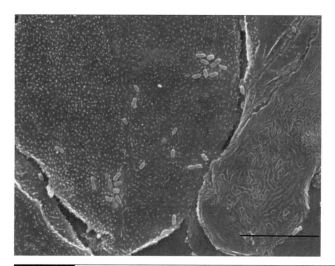

Figure 8.11 Scanning electron micrograph of an epithelial cell from the cheek. A limited number of bacteria are present, and these consist mainly of pairs of oval-shaped cocci – probably streptococci. Bar = 10 μm. Image kindly supplied by Mrs. Nicola Mordan, Eastman Dental Institute, University College London.

organisms from which recolonisation of the subgingival region could occur following successful periodontal therapy.

8.4.3.4 Other mucosal surfaces

Surprisingly few studies have investigated the microbiotas of the various mucosal sites within the oral cavity, and many of these have involved young children and have been described in Section 8.4.2. The surfaces of the oral mucosa are predominantly aerobic regions, with the major sources of nutrients being saliva and the mucus gel covering the epithelium. The ability to adhere to the epithelium or the mucus gel will be an important requirement for species colonising these regions. Microscopy has revealed that mucosal surfaces other than the tongue have a relatively sparse microbiota, compared with those found on the tooth surface. Hence, the average number of bacteria per epithelial cell is usually between five and twenty-five (Figure 8.11). Presumably, desquamation and mechanical abrasion are major factors in reducing colonisation of these surfaces.

The microbiota of the buccal mucosa is dominated by facultatively anaerobic and capnophillic species. Viridans streptococci (mainly *Strep. oralis, Strep. mitis* biovar 1, *Strep. sanguis, Strep. salivarius*, and *Strep. vestibularis*) are the dominant organisms, with *Neisseria* spp. and *Haemophilus* spp. also invariably present in high numbers. The main *Haemophilus* spp. found are *H. parainfluenzae, H. segnis*, and *H. aphrophilus*. Organisms also often present, but in lower proportions include staphylococci, *Veillonella* spp., lactobacilli, *Actinomyces* spp., and *Propionibacterium* spp. A morphologically unique member of the buccal microbitoa is *Simonsiella muelleri*, which is a motile organism consisting of eight to twelve flat, wide cells joined together to form a filament. Thin filaments protrude at right angles from the ventral surface, and the organism uses these to glide over the mucosal surface.

The microbiota of the hard palate is again dominated by viridans streptococci, with high proportions of *Actinomyces* spp., but lower proportions of *Neisseria* spp. and

Organism	Floor of mouth	Buccal mucosa	Hard palate	Lip	Attached gingivae
Strep. mitis	13	13	9	17	15
Cap. gingivalis	8	8	11	8	6
Strep. oralis	5	5	4	6	6
Gem. morbillorum	6	5	9	6	5
Eik. corrodens	4	4	4	4	5
V. parvula	4	6	4	2	5

Table 8.20. Microbiotas of different mucosal sites in the oral cavity based on the use of whole genomic DNA probes recognising forty different oral species

Note: Each number represents the percentage of the total DNA probe count and does not necessarily correspond to the proportion of the organism in the microbiota of the site because of the likely presence of organisms other than those recognised by the probes used.

Haemophilus spp. *Corynebacterium* spp., *Prevotella* spp., and *Veillonella* spp. are also usually isolated, but in much lower proportions.

Recently, a culture-independent approach has been used to characterise the microbiotas of various mucosal sites in 225 healthy adults – the floor of the mouth, buccal mucosa, hard palate, lip, and the attached gingivae. DNA was extracted from samples from each site and probed with whole genomic DNA probes specific for forty bacterial taxa (Table 8.20). The results agree in some respects with those obtained from culture-based studies in that streptococci (mainly *Strep. mitis* and *Strep. oralis*) predominate at all of the mucosal sites. Another frequently detected species was *Gemella morbillorum*. This is a Gram-positive coccus which is difficult to identify and has many physiological and biochemical properties similar to those of viridans streptococci – it is possible that culture-based studies have misidentified this organism as a viridans streptococcus. Unfortunately, the probes used included only one for a *Neisseria* species – *N. mucosa* – and no probes were included for any *Haemophilus* spp., which are frequently detected in culture-based studies. Surprisingly, a number of anaerobic species were detected at all sites, although each of these tended to comprise a low ($\leq$5%) proportion of the microbiota of each site – these included *V. parvula*, *Lep. buccalis*, *Eub. saburreum*, *F. periodonticum*, and *Prev. intermedia*. Presumably, their survival at these sites would be dependent on oxygen utilisation by aerobic and facultative anaerobes (e.g., streptococci, *Capnocytophaga* spp., *Eik. corrodens*, and others which were not screened for in the study).

8.4.4 Dissemination of organisms from the mouth

Mechanical abrasion continually removes microbes from oral surfaces, and these organisms accumulate in saliva until this is swallowed. Saliva contains large numbers (approximately 10^8 cfu/ml) of a variety of microbes, and these are expelled from the mouth during talking, coughing, and sneezing, as described in Section 4.4.5. Saliva can also be transferred between individuals directly (e.g., by kissing) and indirectly via fingers and objects after these have been in the mouth. Genetic typing of oral bacteria has enabled the transmission of specific strains between family members to be demonstrated. Another means of dissemination of oral bacteria is via the GIT. A number of oral bacteria – including streptococci (*Strep. sanguis*, *Strep. salivarius*, *Strep. oralis*,

Strep. intermedius), *V. parvula*, *Actinomyces* spp., and *Lactobacillus* spp. – can survive passage through the stomach and the rest of the GIT, and are excreted in faeces.

8.4.5 Effect of antibiotics and other interventions on the oral microbiota

8.4.5.1 Antibiotics

An important consideration in the use of antibiotics to treat plaque-related diseases is the well-documented recalcitrance of biofilms to antimicrobial agents. Bacteria in biofilms display reduced susceptibility to antibiotics, and this has been attributed to one or more of a variety of factors, including (1) binding of the antimicrobial agent to the extracellular matrix of the biofilm, thereby limiting its penetration; (2) inactivation of the antimicrobial agent by enzymes trapped in the biofilm matrix; (3) the reduced growth rate of bacteria in biofilms renders them less susceptible to the antimicrobial agent; (4) the altered microenvironment within the biofilms (e.g., pH, oxygen content) can reduce the activity of the agent; and (5) altered gene expression by organisms within the biofilm can result in a phenotype with reduced susceptibility to the antimicrobial agent. The effect of antimicrobial chemotherapy on the oral microbiota depends, of course, on the particular antibiotic. Antibiotics such as ciprofloxacin, for example – which are active mainly against aerobic/facultative Gram-negative bacteria, have only moderate activity against facultative Gram-positive species, and poor activity against anaerobes – have little effect on the oral microbiota. In contrast, the systemic or topical administration of a number of other antibiotics has been found to have a substantial effect (Table 8.21). Amoxycillin and tetracycline are particularly disruptive and, as well as altering the relative proportions of organisms comprising the oral microbiota, they facilitate colonisation of the oral cavity by *Can. albicans* and enterobacteria.

A number of antibiotics are currently used in the treatment of periodontal diseases – particularly amoxycillin, metronidazole, and tetracyclines – and attention has focussed mainly on the effects of these antibiotics on the subgingival microbiota rather than on other sites within the oral cavity. Of obvious interest is the possibility of resistance development in individuals treated with these three antibiotics. The effects of oral administration of amoxycillin, metronidazole, and doxycycline for 14 days to periodontitis patients has been investigated with regard to the prevalence of bacteria resistant to these antibiotics. In all cases, the proportion of bacteria resistant to the administered antibiotic increased rapidly during chemotherapy, but then declined slowly once the antibiotic had been withdrawn. After approximately 90 days (76 days after the withdrawal of the antibiotic), the proportions of resistant organisms in saliva and subgingival plaque were similar to those found prior to antibiotic administration. Similarly, local application of tetracyclines to periodontal pockets for long periods of time (up to 9 months) has been shown to result in only transient increases in the proportion of resistant organisms – the proportion of resistant organisms quickly returns to pre-treatment levels. However, although individual studies appear to show that antibiotic administration results in only transient increases in antibiotic-resistant organisms, there is increasing evidence that the susceptibility of oral bacteria to many antibiotics is decreasing. Hence, a comparison of the antibiotic susceptibilities of strains isolated from subgingival plaque during 1991–1995 with those isolated during 1980–1985 revealed a substantial increase in antibiotic resistance. The proportions of strains

Table 8.21. Effects of antibiotics on the oral microbiota

Antibiotic	Antimicrobial spectrum	Main effects on oral microbiota
cefadroxil	cephalosporin; good activity against Gram +ves, moderate activity against Gram −ves	decrease in streptococci, *Actinomyces* spp., *Veilllonella* spp., *Bacteroides* spp., *Fusobacterium* spp., *Leptotrichia* spp.; facultative anaerobes; increase in *Neisseria* spp.
phenoxymethylpenicillin	penicillin; active against Gram +ve aerobes and anaerobes	decrease in streptococci, *Veillonella* spp., *Bacteroides* spp., *Fusobacterium* spp. *Leptotrichia* spp., obligate anaerobes; increase in *Neisseria* spp.
ciprofloxacin	4-quinolone; active against facultative Gram −ves; moderate activity against Gram +ves; poor activity against anaerobes	decrease in *Neisseria* spp.
mecillinam	an amidinopenicillin; active mainly against facultative Gram −ves	no effect
tetracyclines	broad-spectrum; active against many Gram +ves and Gram −ves	decrease in total viable count; decrease in number of different facultative and anaerobic species; increase in frequency of isolation of enterobacteria and *Can. albicans*
metronidazole	active against obligate anaerobes	reduction in proportions (sometimes elimination) of species belonging to several anaerobic genera including *Fusobacterium*, *Veillonella*, *Leptotrichia*, *Actinomyces*, and unspeciated Gram −ve rods; no effect on streptococci, lactobacilli, and Gram +ve anaerobic cocci
amoxycillin	broad spectrum	decrease in proportions of *Fusobacterium* spp., *Leptotrichia* spp., and anaerobic Gram-positive cocci; increase in frequency of isolation of enterobacteria and *Can. albicans*

Note: These effects were generally assessed by cultivation of microbes present in one or more of the following samples: saliva, the oral mucosa, and dental plaques.

resistant to tetracycline, penicillin, amoxycillin, erthromycin, and clindamycin were found to have increased by between 108% and 238% during this period.

8.4.5.2 Mechanical oral-hygiene measures

The mechanical removal of plaque is the most important means of preventing and treating plaque-related diseases. The most widely practiced of such procedures is tooth-brushing which constitutes the only procedure used by vast numbers of individuals on a daily basis in order to prevent infectious diseases (i.e., caries and periodontal diseases). Toothbrushing can achieve considerable reductions in the amount of dental plaque on the supragingival surfaces of the teeth, but its effectiveness correlates with the brushing time, and this is not always optimal. It has been reported that individuals who regularly brush their teeth have 40–60% of their tooth surfaces covered with plaque. Although toothbrushing achieves considerable reductions in the number of viable bacteria on the tooth surface, even after careful toothbrushing, there are still approximately 10^6 bacteria/mm^2 adhering to the enamel. As well as a general reduction in the number of viable organisms, toothbrushing also appears to alter the relative proportions of organisms remaining – the proportions of obligate anaerobes are decreased while those of facultative anaerobes increase.

Toothbrushing is designed primarily for the removal of supragingival plaque and, even though bristles may penetrate subgingivally to a depth of between 0.9 and 1.5 mm, it has no effect on the proportions of periodontopathogenic species in subgingival plaque. Toothbrushing is also unable to remove approximal plaque, which is unfortunate as the gaps between the teeth are important sites of caries formation. Plaque removal from this region requires floss or toothpicks, which are used by only approximately 10% of the population. Such inter-dental cleaning can remove 30–40% of approximal plaque. A wide range of antimicrobial agents have been incorporated into toothpastes in order to improve the levels of oral hygiene achievable by toothbrushing. Such agents include chlorhexidine, triclosan, cetylpyridinium chloride, phenolic compounds, sanguinarine, fluoride, tin compounds, zinc compounds, and enzymes. The most effective of these appears to be chlorhexidine.

Removal of subgingival plaque (subgingival debridement) is accomplished by dentists and dental hygienists using a variety of devices, ranging from simple, sharp-ended, hand instruments to power-driven ultrasonic scalers. Complete removal of all subgingival plaque is rarely achieved because access to and visibility of the root surface is limited. A thorough debridement of the root surface can remove approximately 99% of viable bacteria, but substantial numbers can remain (approximately 10^5 cfu per root) in dentinal tubules, surface indentations, or adjacent soft tissues. Recolonisation of the root surface then occurs and viable counts can reach pre-treatment values within 3 to 7 days. However, the composition of the newly formed subgingival microbiota is often different from that existing prior to debridement with substantially decreased proportions of periodontopathogenic species. Nevertheless, recolonisation of the subgingival plaque by periodontopathogens can occur from reservoirs of these organisms, such as the tongue and other mucosal surfaces.

8.4.5.3 Prosthetic devices

Bacterial biofilms are able to form on any non-shedding surface and, as well as teeth, a number of prostheses and devices may be present in the oral cavity which are able to support biofilm growth. These include dentures, implants, bridges, crowns, and the wires and bands used in orthodontic procedures. Of these, the most widely used are dentures, which are worn by approximately 20% of individuals over the age of 65 years in the United Kingdom. While biofilms (denture plaques) form on the surface of the denture in contact with the palate, as well as on the exposed surface, most attention has been focussed on the former because it is considered to be important in the aetiology of denture stomatitis (Section 8.5.4). The composition of denture plaque is similar to that found on enamel surfaces – particularly fissure plaque. The predominant organisms are viridans streptococci (approximately one-half of the cultivable microbiota) and *Actinomcyes* spp. (approximately one-third of the cultivable microbiota). *Veillonella* spp. are also frequently present in denture plaque and generally comprise the next most abundant group of cultivable organisms. Unusual for the oral cavity, *Staph. aureus* is often present but in low proportions.

Implants are increasingly being used in dentistry and, following their installation in the oral cavity, they rapidly become colonised by microbes. Titanium is one of the most frequently used implant materials and is very resistant to attack by microbial products, including acids. Corrosion, therefore, is not generally regarded as a problem. However, as happens with natural teeth, a gap is formed between the surface of the implant and the adjacent tissues – this is known as the peri-implant sulcus (cf. gingival sulcus). The microbes colonising this region are similar to those found in the gingival sulcus – mainly streptococci and Gram-positive rods with small numbers of Gram-negative rods. In a small proportion of cases, inflammation of the tissues adjacent to the implant occurs (peri-implantitis), and this can eventually lead to alveolar bone destruction and loss of the implant. The microbiota associated with these failing implants is similar to that associated with periodontitis. However, it has not been established whether this is a cause or a consequence of the peri-implantitis.

8.4.5.4 Immunosuppressive chemotherapy

Infections are a major cause of morbidity and mortality among cancer patients during therapeutic immunosuppression. Because of the variety of therapeutic regimes used by the cancer patients and differences in the type of microbiological investigations undertaken (particularly the sampling site), comparisons between studies are difficult. Nevertheless, it is possible to discern some common trends. Many studies have reported an increase in the proportion of facultatively anaerobic Gram-negative rods in supragingival plaque and at other oral sites – most commonly, these are enterobacteria (particularly *Klebsiella* spp. and *Enterobacter* spp.) or *Pseudomonas* spp. Enteric Gram-negative rods have been reported to comprise as much as 17% and 14% of the microbiota of supragingival and subgingival plaques, respectively. Between 2- and 3-fold increases in the proportions of *F. nucleatum* have been observed in the supragingival and subgingival plaques of adults undergoing chemotherapy, and this organism can cause septicaemia in these patients. Virtually all studies have demonstrated increases in the proportion of staphylococci in supragingival and subgingival plaques during immunosuppressive chemotherapy. Increased colonisation of the oral cavity by yeasts is also a very frequent occurrence. They have been detected in the saliva of 80% of leukaemia patients and may be among the predominant members of the oral microbiota in these patients. The most frequently encountered species are *Can. albicans* and *Can. tropicalis*.

Many of the organisms whose proportions are increased in the oral cavity of patients undergoing therapeutic immunosuppression are responsible for infections (often septicaemia) in these patients. Such organisms include enterobacteria, *Pseudomonas* spp., staphylococci, yeasts, and *F. nucleatum*. Surprisingly, there have been no reports of an increase in the proportion of *Capnocytophaga* spp., which are frequent causes of a range of infections in immunosuppressed individuals.

8.5 | Diseases caused by members of the oral microbiota

8.5.1 Dental caries

Dental caries is a disease characterised by the localised destruction of the tissues of the tooth by acids produced by oral microbes. The acids bring about demineralisation of enamel to produce a cavity ("enamel caries") and, if untreated, further demineralisation can take place until eventually the dentine is involved ("dentinal caries") and the pulp becomes inflamed and necrotic. The various sites on the tooth surface exhibit differing susceptibilities to caries with the susceptibility decreasing in the order: fissures on occlusal surfaces > approximal surfaces of molars and pre-molars > approximal surfaces of maxillary incisors > approximal surfaces of mandibular incisors. Dental caries is one of the most prevalent diseases of humans. In the United States, for example, dental caries affects 59% of children between the ages of 5 and 17. The disease has a multi-factorial aetiology, with the main elements involved being (1) plaque bacteria, (2) the host's diet, (3) the composition and flow rate of saliva, and (4) the structure and composition of enamel. The main organisms involved are members of the indigenous oral microbiota and include *Strep. mutans*, *Strep. sobrinus*, *Lactobacillus* spp., and *A. naeslundii*. *Strep. mutans* and *Strep. sobrinus* (collectively known as "mutans streptococci") are considered to be the most important species in the initiation of enamel caries, *A. naeslundii* is frequently involved in root-surface caries, and *Lactobacillus* spp. are important in advanced caries lesions. All of these cariogenic organisms are able to produce acids (mainly lactic and acetic acids) from carbohydrates present in the host's diet (i.e., they are acidogenic) and are able to survive and grow at the low pHs produced (i.e., they are acidophilic and/or aciduric).

If an individual consumes large quantities of fermentable carbohydrates (e.g., sucrose) at frequent intervals, then the buffering capacity of saliva is overcome, and the plaque pH is not only lowered, but also remains low for long periods of time. Such an environment favours the emergence of a plaque community dominated by cariogenic species, and this community eventually produces sufficient acid to initiate enamel demineralisation. In a classic series of experiments in 1944, R.M. Stephan measured the change in pH of plaque on the tooth surface following a glucose rinse and found that a characteristic pattern emerged (known as a Stephan curve – Figure 8.12). Immediately after the rinse, there is a dramatic fall in pH due to acid production by plaque bacteria. The pH then remains steady for a short period and then gradually returns to the initial value due to a combination of factors, including exhaustion of the carbohydrate supply, clearance of acid and glucose by saliva, production of basic compounds by plaque bacteria, acid utilisation by plaque bacteria (e.g., *Veillonella* spp.), and the inhibition of bacterial metabolism by the low pH. Interestingly, in individuals with caries, the initial pH is lower than in those who are caries-free (presumably

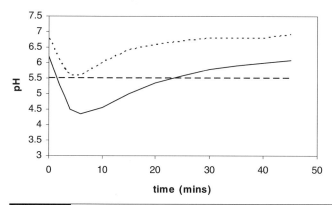

Figure 8.12 Effect of a glucose rinse (taken at time = 0 minute) on the pH of dental plaque. The initial plaque pH is higher in individuals who do not have caries (....) than in those who do have the disease (—). There is a rapid fall in pH immediately after the glucose rinse, and then the pH gradually returns to its initial value. In those with caries, the plaque pH often falls below the critical pH for enamel dissolution (pH 5.5) and remains there for a relatively long period of time. In those who do not have caries, the pH may or may not fall below 5.5.

because of the presence of significant numbers of acidogenic and aciduric organisms such as mutans streptococci), and the pH resulting from the sugar rinse decreases to below 5.5. – the point at which enamel demineralisation occurs. In these individuals, the pH remains below this critical value for considerable periods of time. Studies have shown that, in cariogenic communities, *Strep. mutans* comprises approximately 10% of the total cultivable microbiota. However, other organisms present in plaque can reduce the cariogenicity of the community. Hence, *Veillonella* spp. utilise lactic acid (the main acid responsible for enamel demineralisation) as a carbon and energy source and convert it to the much weaker propionic and acetic acids which are less damaging to enamel. The amount of acid required to induce demineralisation in a particular individual will be influenced by the composition and structure of the enamel – particularly its fluoride content. Demineralisation of enamel generally occurs when the pH falls below 5.5 but, provided that calcium and phosphate are present, remineralisation can occur when the pH rises above this value. The consumption of fermentable carbohydrates in many individuals, therefore, can result in alternating periods of demineralisation and remineralisation. In individuals who frequently consume carbohydrates, there will be corresponding frequent periods of demineralisation. The whole process has been elegantly summarised in the "ecological plaque hypothesis" outlined in Figure 8.13. Caries can be avoided by scrupulous oral hygiene (i.e., regular toothbrushing and flossing), coupled with a reduction in the amount of dietary sucrose. Fluoride (from the water supply or in toothpaste) is an effective anti-caries agent for a number of reasons. Firstly, it becomes incorporated into enamel to form fluorapatite, which is more resistant to acid dissolution than hydroxyapatite. Secondly, it reduces acid production by plaque bacteria by inhibiting sugar transport and reducing glycolysis.

8.5.2 Periodontal diseases

These are inflammatory diseases affecting the tooth-supporting tissues. They can be broadly classified into two major groups: gingivitis and periodontitis. Gingivitis is a

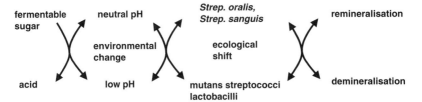

Figure 8.13 The ecological plaque hypothesis and the aetiology of caries. The frequent intake of easily fermentable sugars (particularly sucrose) in the diet results in the production of acids by plaque bacteria, which creates an environment suitable for acidophilic species such as mutans streptococci and lactobacilli. The plaque community becomes dominated by these cariogenic species which produce sufficient acid to initiate demineralisation of enamel. Reprinted with permission from: Microbial ecology of dental plaque and its significance in health and disease. Marsh, P.D. *Advances in Dental Research* 1994;8:263–271.

term describing inflammatory conditions when only the gingivae are involved, whereas in periodontitis, other tissues – including the periodontal ligament and alveolar bone – are affected.

8.5.2.1 Chronic gingivitis

Inflammation of the gingival tissues can result from the administration of a number of drugs, trauma, the presence of a foreign body, and the presence of bacteria at the gingival margin. In this section, only the latter is described. Chronic gingivitis is a very common disease affecting most of the adult population and can persist in an individual for many years. It is characterised by swollen red gingivae and is often accompanied by bleeding and halitosis, but the underlying tissues are not affected. The condition is a consequence of poor oral hygiene which results in an increase in the quantity of plaque in the gingival crevice – the plaque mass can be 10–20 times greater than that present in the healthy gingival crevice. As would be expected, this increase in plaque mass is accompanied by changes in its composition due to community succession, and the microbiota (initially dominated by streptococci and other Gram-positive species) has increased proportions of Gram-negative species and obligate anaerobes – obligate anaerobes comprise approximately 45% of the cultivable microbiota. These changes arise mainly because of the increased flow of GCF, which is a consequence of the inflammation induced by the increased plaque mass. Although large numbers of PMNs are present in the GCF, they are unable to penetrate the biofilm matrix and, consequently, have little effect on the bacteria present. Detailed studies of the cultivable microbiota of the gingival crevice in individuals with and without gingivitis have implicated a wide range of bacteria in the aetiology of the disease on the basis of their increased proportions in individuals with gingivitis. Such organisms include *Actinomyces* spp., *Strep. oralis*, *Strep. anginosus*, *Sel. sputigena*, *Por. endodontalis*, *Por. gingivalis*, *F. nucleatum*, *Prev. loeschii*, *Prev. intermedia*, *Eub. nodatum*, *Eub. brachy*, *Camp. concisus*, and *Eik. corrodens*.

Treatment of the disease involves the introduction of good oral-hygiene procedures.

8.5.2.2 Acute necrotising ulcerative gingivitis

This condition is also known as Vincent's infection and is characterised by gingivitis; bleeding; and the presence of a pseudomembrane consisting of bacteria, leukocytes, fibrin, and necrotic tissue. It is a consequence of poor oral hygiene and malnutrition and is more frequently encountered in developing than developed countries. It is a polymicrobial infection involving mainly *Treponema* spp. and *F. nucleatum*, although some studies have also implicated *Prev. intermedia* in addition. The disease is accompanied by tissue invasion and is treated by administering metronidazole and introducing oral-hygiene procedures.

8.5.2.3 Periodontitis

The term "periodontitis" is applied to those diseases in which gingival inflammation is accompanied by damage to the supporting tissues of the tooth (i.e., periodontal ligament and alveolar bone). In the United Kingdom, it has been estimated that approximately 69% of adults suffer from periodontitis. The main forms of the disease are "localised aggressive periodontitis" (LAP – previously known as "localised juvenile periodontitis"), adult periodontitis, and refractory periodontitis. Adult periodontitis is the most common of these and accounts for approximately 95% of cases of periodontitis.

Periodontitis is a consequence of the accumulation of plaque at the gingival margin and its subsequent growth along the root surface, which induces migration of the junctional epithelium down the root resulting in the formation of a gap known as a periodontal pocket. The composition of subgingival plaque within the periodontal pocket is very different from that of the plaque present in the gingival crevice of individuals with healthy gingivae. Culture-based studies have revealed that the plaque is dominated by Gram-negative anaerobes which comprise approximately 75% of the microbiota, while the proportion of streptococci is reduced considerably. Spirochaetes and motile organisms also comprise much greater proportions (determined by phase-contrast or dark-field microscopy) of the microbiota than they do either in health or in gingivitis. The subgingival environment is very different from that found supragingivally, in that the main source of nutrients is GCF, which has a high protein content compared with that of saliva; the oxygen content and Eh are much lower (because of limited ingress by oxygen-rich saliva), and the pH is higher (7.4–7.8) due to metabolites produced by protein and amino-acid metabolism by the resident microbiota. Failure to remove plaque regularly from the gingival crevice (due to poor oral hygiene) results in an increase in plaque mass, which induces an inflammatory response. This results in an increased flow of GCF, which is not only rich in proteins (thereby providing nutrients for proteolytic species), but also contains substances (e.g., iron and haem-containing compounds) which are usually present at only growth-limiting concentrations. Oxygen utilisation and the production of reducing substances by the increased microbial load lead to anaerobic conditions and a low Eh. According to the ecological plaque hypothesis (Figure 8.14), this results in an ecological shift to a microbiota dominated by proteolytic "periodontopathogenic" species (mainly Gram-negative anaerobes) that can damage host tissues directly (e.g., proteolytic enzymes, toxins, tissue matrix-degrading enzymes) and/or induce a damaging host response due to the over-production of pro-inflammatory cytokines. As previously mentioned, the PMNs in GCF are ineffective against biofilm bacteria, and the discharge of their lysosomal contents due to "frustrated phagocytosis" serves only to exacerbate the tissue destruction taking place.

Adult periodontitis is a common disease and affects between 70% and 80% of adults. As well as inflammation, the disease is accompanied by the formation of a periodontal pocket and destruction of the alveolar bone; bleeding is also common. If untreated,

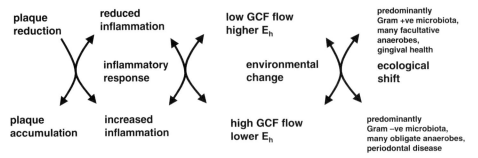

Figure 8.14 Ecological plaque hypothesis in relation to periodontal diseases – gingivitis and periodontitis. Accumulation of plaque causes inflammation of adjacent tissues (gingivitis) and other environmental changes that favour the growth of Gram-negative anaerobes and proteolytic species, including periodontopathogens. The increased proportions of such species results in destruction of periodontal tissues (i.e., periodontitis). E_h = redox-potential; GCF = gingival crevicular fluid; Gram +ve = Gram-positive; Gram −ve = Gram-negative. Reprinted with permission from: Microbial ecology of dental plaque and its significance in health and disease. Marsh, P.D. *Advances in Dental Research* 1994;8:263–271.

tooth loss occurs as a result of extensive bone destruction – the disease is the main cause of tooth loss in those over 25 years of age. Tissue damage is due to the direct effects of certain bacterial products (e.g., enzymes and toxins) and the host's response to these and other bacterial products and constituents. Innumerable culture-dependent and culture-independent studies have established that several members of the indigenous oral microbiota able to colonise the subgingival region are the aetiological agents of the disease. These include *Por. gingivalis, T. denticola, Tannerella forsythensis, Prev. intermedia, Prev. loeschii, F. nucleatum, Eubacterium* spp., *Act. actinomycetemcomitans, Capnocytophaga* spp., *Eik. corrodens*, and *Pep. micros*. These organisms can be isolated from a variety of oral sites (particularly the gingival crevice and tongue) in low numbers, but can become dominant members of subgingival plaque given the appropriate environmental conditions as previously described (Figure 8.14).

Localised aggressive periodontitis is a disease that starts in adolescence and affects only certain teeth (the incisors and pre-molars) and causes rapid destruction of alveolar bone leading to tooth loss. In the United States, the mean prevalence of LAP among adolescents of all racial origins is 0.53%, which makes it a reasonably common disease in this age group. However, adolescents of African-American descent have a 15-fold higher incidence of disease than Caucasian Americans. A number of studies have implicated *Act. actinomycetemcomitans*, a member of the indigenous oral microbiota, as the main aetiological agent of the disease. Interestingly, the subgingival plaque of affected teeth is usually quite sparse, but *Act. actinomycetemcomitans* may comprise 70% of the cultivable organisms present. Other organisms implicated in the aetiology of the disease include *Eik. corrodens, Capnocytophaga* spp., *Prev. intermedia*, and *Camp. rectus*.

Refractory periodontitis refers to the disease status of those periodontitis patients who do not respond to conventional therapy (i.e., mechanical removal of subgingival plaque, surgery, antimicrobial chemotherapy). It has been estimated that between 10% and 20% of periodontitis cases fall into this group. A limited number of studies have investigated the composition of subgingival plaque in these cases and have shown that, in addition to the periadontopathogens detected in cases of adult periodontitis, there are often significantly higher proportions of a number of other species, including *Strep. constellatus, Micromonas micros, Cap. ochracea*, and *Cap. sputigena*.

Periodontitis can be treated by removing subgingival plaque, and this is often supplemented by the local or systemic use of antimicrobial agents, such as metronidazole, tetracyclines, or amoxycillin.

8.5.3 Bacterial endocarditis and other extra-oral infections

A number of studies have shown that oral bacteria, their components, and their secreted products (including toxins) can gain access to the bloodstream. This is not surprising in view of the huge numbers of bacteria in the oral cavity; the presence of a thin, highly vascularised tissue (the gingivae) with a large surface area (approximately 10 cm^2); the frequent generation of large compressive forces during chewing; the presence of some degree of gingival inflammation in the majority of the population; and the abrasive nature of daily oral-hygiene measures. Studies have shown that chewing, toothbrushing, and flossing all generate transient bacteraemias and that these are of greater magnitude and duration in individuals with poor oral health. Dental procedures such as tooth extraction, root-surface debridement, endodontic treatment, and periodontal surgery also cause bacteraemias. In most cases, the bacteraemia is short-lived and involves only a small number of organisms. Following the extraction of a single tooth, for example, the concentration of bacteria in the bloodstream is usually between 1 and 10 per ml, and this lasts between 15 and 30 minutes. These organisms are rapidly eliminated in individuals with fully functioning defence systems, but any defect in these may allow them to survive. The organisms may then attach to and colonise tissues at any of a number of sites and initiate pathology. Colonisation will be facilitated by any abnormality (e.g., damaged heart tissue due to rheumatic heart disease or the presence of a prosthetic device). Disease may also result from the influx of bacterial components, rather than viable organisms, into the bloodstream. A variety of bacterial constituents (LPSs, peptidoglycans, lipoteichoic acids, outer membrane proteins, etc.) are potent inducers of pro-inflammatory cytokines, and the massive release of these mediators can result in pathology. LPSs are likely to be particularly important in this respect because very large quantitites have been detected on the root surfaces of periodontally diseased teeth.

One of the most well-documented diseases resulting from a bacteraemia involving oral bacteria is bacterial endocarditis. This life-threatening condition has an incidence of between 1 and 5 cases/100,000 population/year and is associated with a high degree of morbidity and mortality. Despite antimicrobial chemotherapy, the mortality rate remains at 15–30%. It generally occurs more frequently in the elderly, but young intravenous drug users also constitute a high-risk group. It is rare in children and affects mainly those with a congenital cardiac abnormality. The disease arises as a result of colonisation of the endocardium by bacteria which induce platelet aggregation and fibrin deposition and the formation of a vegetation (thrombus). Ultimately, cardiac function is compromised due to tissue damage, either as a result of direct bacterial action or indirectly via immune-mediated mechanisms. In approximately 50% of cases, the organism responsible is a viridans streptococcus, usually *Strep. sanguis, Strep. oralis, Strep. mitis, Strep. salivarius, Strep gordonii*, or *Strep. mutans*. Congenital heart abnormalities, rheumatic heart disease, a history of endocarditis, and the presence of a prosthesis are all risk factors which facilitate colonisation by streptococci. As well as being responsible for endocarditis, evidence is accumulating that implicates oral bacteria, particularly periodontopathogenic species, in the aetiology of atherosclerosis and myocardial infarction.

Once they have gained access to the bloodstream, oral bacteria can induce diseases at a variety of sites and have been reported to be responsible for brain, liver, and lung abscesses; pneumonia; sinusitis; orbital cellulitis; osteomyelitis; and infections of prosthetic devices. The organisms responsible are often obligate anaerobes, although streptococci are also frequently involved – particularly *Strep. intermedius*, *Strep. anginosus*, and *Strep. constellatus*. Pneumonia can also arise from aspiration of saliva into the lungs, and this occurs mainly in the elderly, those with chronic lung disease, patients on mechanical ventilation or intubation, immunosuppressed individuals, and patients suffering from diabetes mellitus or congestive heart failure. The organisms usually involved include *Por. gingivalis*, *Act. actinomycetemcomitans*, *A. israelii*, *Capnocytophaga* spp., *Eik. corrodens*, *Prev. intermedia*, and *Fusobacterium* spp.

Finally, there is increasing evidence that periodontitis during pregnancy increases the risk of pre-term, low-birth weight infants. This may be due to the induction of labour by cytokines (IL-1, IL-6, and TNFα) and prostaglandins, which are produced by the host in response to LPS and other oral bacterial components entering the bloodstream.

8.5.4 Denture stomatitis

This is an inflammatory condition of that part of the oral mucosa in contact with the fitting surface of a denture. Estimates of its prevalence vary enormously – from 9% to 97% of denture wearers. Although for many years *Can. albicans* was considered to be the aetiological agent of denture stomatitis, few studies have actually confirmed any association of the yeast with the condition. Even when it is present in the denture plaque of patients with the disease, the organism rarely comprises more than 1% of the cultivable microbiota. The denture plaque of patients with the disease is thicker and has a higher viable count than that of individuals without inflammation. This observation, coupled with the finding that simple cleaning of the dentures without antifungal chemotherapy results in resolution of the inflammation, suggests that the condition is simply the response of the mucosa to the presence of an increased microbial load.

8.5.5 Halitosis

Halitosis, or oral malodour, is a distressing condition which affects approximately 50% of adults and is due to the presence in the breath of volatile compounds (VCs) with an unpleasant odour. In more than 80% of cases, the VCs are produced by bacteria resident in the oral cavity, with the remaining cases being attributable to VCs produced by the host and secreted in saliva and/or excreted by the lungs. Most of the VCs (60–70%) originating from the oral cavity are produced by microbes present on the dorso-posterior surface of the tongue, with the rest being due to bacteria on the teeth or in the gingival crevice. The most important VCs contributing to halitosis are volatile sulphur compounds (VSCs), such as hydrogen sulphide and methyl mercaptan, amines (e.g., cadaverine, putrescine, and trimethylamine), indole, and volatile acids (e.g., butyric and isovaleric acids). While a variety of oral microbes are able to produce one or more of these compounds, many Gram-negative anaerobes are able to produce a whole range of odiferous compounds, including VSCs, amines, indole, and acids. The most effective substrates for the production of VCs by oral anaerobes include cysteine, methionine, ornithine, lysine, and tryptophan. Such compounds are present in saliva and GCF or can be produced by the microbial degradation of host and dietary proteins. Because of its large surface area and the presence of considerable numbers of crypts, the tongue can retain appreciable quantities of food, desquamated epithelial cells, and other debris – which provide a plentiful supply of substrates for odour-producing microbes. On the basis of their ability to produce odiferous compounds and their presence on the tongue, organisms thought to be important in the aetiology of halitosis include *Por. gingivalis, Prev. intermedia, T. denticola, Tannerella forsythensis, Por. endodontalis*, and *Eubacterium* spp. Culture-dependent approaches have also demonstrated increased proportions of some of these organisms on the tongues of patients with halitosis, compared with healthy individuals. However, the results of a culture-independent investigation failed to find an association with halitosis of many of the Gram-negative organisms in the afore-mentioned list. In this study, which involved five healthy individuals and five patients with halitosis, DNA was extracted from tongue samples, and the 16S rRNA genes were amplified, cloned, and sequenced. Comparison of the tongue microbiotas of the patients and controls revealed an association between a number of Gram-positive and Gram-negative anaerobes (as well as unidentified taxa) and halitosis; these included *Atopobium parvulum* (a nonmotile, non-sporing, Gram-positive anaerobic rod producing mainly lactic acid as a metabolic end-product), *F. sulci, F. periodonticum*, and *Solobacterium moorei*. However, this is the first reported study of this kind, and the number of patients involved was limited.

A variety of treatments for the condition has been suggested. These involve (1) reducing the number of bacteria on the tongue (by scraping), (2) reducing nutrient availability on the tongue (by scraping), (3) reducing the total numbers of bacteria in the oral cavity (by broad-spectrum antiseptics), and (4) conversion of odiferous compounds to odour-free compounds (by oxidising agents). Each of these has met with varying degrees of success.

8.5.6 Endodontic infections

These are infections involving the root canal and associated tissues. Infections of the pulp within the root canal are invariably polymicrobial, with anaerobic species predominating. The species involved are usually members of the following genera: *Bacteroides, Porphyromonas, Prevotella, Fusobacterium, Treponema, Peptostreptococcus, Eubacterium*, and *Campylobacter*. Facultative or microaerophilic streptococci are also frequently present. An infection of the root canal may also lead to the formation of an acute periradicular abscess, also known as a periapical, dental, or dento-alveolar abscess.

8.5.7 Actinomycosis

This is a chronic infection characterised by abscess formation and tissue fibrosis. Reports of its incidence vary widely and, in the United Kingdom, approximately one person per million is affected each year. Although between 60% and 65% of cases involve the face or neck, the disease can also affect other body sites, including the lungs, the ileocaecal region of the gut, and the central nervous system. Most infections (more than 90%) are caused by *A. israelii* which, in a few cases, is accompanied by other organisms, such as *Act. actinomycetemcomitans, Porphyromonas* spp., *Prevotella* spp., and *Propionibacterium* spp. *A. israelii* is often present in the lesion as filamentous aggregates, which are known as "sulphur granules" because of their yellow colour.

8.6 │ Further Reading

Books

Busscher, H.J. and Evans, L.V. (eds.). (1999). *Oral biofilms and plaque control.* Amsterdam: Harwood Academic Publishers.

Kuramitsu, H.K. and Ellen, R.P. (eds.). (2000). *Oral bacterial ecology: the molecular basis.* Wymondham: Horizon Scientific Press.

Marsh, P. and Martin, M.V. (1999). *Oral microbiology.* Oxford: Wright.

Newman, H.N. and Wilson, M. (1999). *Dental plaque revisited: oral biofilms in health and disease.* Cardiff: BioLine.

Reviews and Papers

Auschill, T.M., Artweiler, N.B., Netuschil, L., Brecx, M., Reich, E., and Sculean, A. (2001). Spatial distribution of vital and dead microorganisms in dental biofilms. *Archives of Oral Biology* **46**, 471–476.

Bowden, G.H.W. (2000). The microbial ecology of dental caries. *Microbial Ecology in Health and Disease* **12**, 138–148.

Burne, R.A. (1998). Oral streptococci . . . products of their environment. *Journal of Dental Research* **77**, 445–452.

Burton, R. and Lamont, R.J. (2000). Dental-plaque formation. *Microbes and Infection* **2**, 1599–1607.

Drucker, D.B. and Natsiou, I. (2000). Microbial ecology of the dental root canal. *Microbial Ecology in Health and Disease* **12**, 160–169.

Duncan, M.J. (2003). Genomics of oral bacteria. *Critical Reviews in Oral Biology and Medicine* **14**, 175–187.

Dunsche, A., Acil, Y., Dommisch, H., Siebert, R., Schroder, J.M., and Jepsen, S. (2002). The novel human beta-defensin-3 is widely expressed in oral tissues. *European Journal of Oral Sciences* **110**, 121–124.

Feres, M., Haffajee, A.D., Allard, K., Som, S., Goodson, S., and Socransky, S.S. (2002) Antibiotic resistance of subgingival species during and after antibiotic therapy. *Journal of Clinical Periodontology* **29**, 724–735.

Gendron, R., Grenier, D., and Maheu-Robert, L.-F. (2000). The oral cavity as a reservoir of bacterial pathogens for focal infections. *Microbes and Infection* **2**, 897–906.

Hegde, S. and Munshi, A.K. (1998). Influence of the maternal vaginal microbiota on the oral microbiota of the newborn. *Journal of Clinical Pediatric Dentistry* **22**, 317–321.

Henderson, B., Nair, S.P., Ward, J.M., and Wilson, M. (2003). Molecular pathogenicity of the oral opportunistic pathogen *Actinobacillus actinomycetemcomitans. Annual Review of Microbiology* **57**, 29–55.

Henderson, B. and Wilson, M. (1998). Commensal communism and the oral cavity. *Journal of Dental Research* **77**, 1674–1683.

Humphrey, S.P. and Williamson, R.T. (2001). A review of saliva: normal composition, flow, and function. *Journal of Prosthetic Dentistry* **85**, 162–169.

Kazor, C.E., Mitchell, P.M., Lee, A.M., Stokes, L.N., Loesche, W.J., Dewhirst, F.E., and Paster, B.J. (2003). Diversity of bacterial populations on the tongue dorsa of patients with halitosis and healthy patients. *Journal of Clinical Microbiology* **41**, 558–563.

Kleinberg, I. (2002). A mixed-bacteria ecological approach to understanding the role of the oral bacteria in dental caries causation: an alternative to *Streptococcus mutans* and the specific-plaque hypothesis. *Critical Reviews in Oral Biology and Medicine* **13**, 108–125.

Koga, T., Oho, T., Shimazaki, Y., and Nakano, Y., (2002). Immunization against dental caries. *Vaccine* **20**, 2027–2044.

Kolenbrander, P.E., Andersen, R.N., Blehert, D.S., Egland, P.G., Foster, J.S., and Palmer, R.J., Jr. (2002). Communication among oral bacteria. *Microbiology and Molecular Biology Reviews* **66**, 486–505.

Kononen, E. (2000). Development of oral bacterial flora in young children. *Annals of Medicine* **32**, 107–112.

Kononen, E., Kanervo, A., Takala, A., Asikainen, S., and Jousimies-Somer, H. (1999). Establishment of oral anaerobes during the first year of life. *Journal of Dental Research* **78**, 1634–1639.

Koolstra, J.H. (2002). Dynamics of the human masticatory system. *Critical Reviews in Oral Biology and Medicine* **13**, 366–376.

Kroes, I., Lepp, P.W., and Relman, D.A. (1999). Bacterial diversity within the human subgingival crevice. *Proceedings of the National Academy of Sciences USA* **96**, 14547–14552.

Lenander-Lumikari, M. and Loimaranta, V. (2000). Saliva and dental caries. *Advances in Dental Research* **14**, 40–47.

Li, X., Kolltveit, K.M., Tronstad, L., and Olsen, I. (2000). Systemic diseases caused by oral infection. *Clinical Microbiology Reviews* **13**, 547–558.

Liljemark, W.F. (2000). Microbial ecology of marginal gingivitis. *Microbial Ecology in Health and Disease* **12**, 149–159.

Lockhart, P.B. and Durack, D.T. (1999). Oral microflora as a cause of endocarditis and other distant site infections. *Infectious Disease Clinics of North America* **13**, 833–850.

Loesche, W.J. and Grossman, N.S. (2001). Periodontal disease as a specific, albeit chronic, infection: diagnosis and treatment. *Clinical Microbiological Reviews* **14**, 727–752.

Loesche, W.J. and Kazor, C. (2002). Microbiology and treatment of halitosis. *Periodontology 2000* **28**, 256–279.

Mager, D.L., Ximenez-Fyvie, L.-A., Haffajee, A.D., and Socransky, S.S. (2003). Distribution of selected bacterial species on intra-oral surfaces. *Journal of Clinical Periodontology* **30**, 644–654.

Marcotte, H. and Lavoie, M.C. (1998). Oral microbial ecology and the role of salivary immunoglobulin A. *Microbiology and Molecular Biology Reviews* **62**, 71–109.

Marsh, P.D. (2000). Role of the oral microflora in health. *Microbial Ecology in Health and Disease* **12**, 130–137.

Marsh, P.D. (2003). Are dental diseases examples of ecological catastrophes? *Microbiology*, **149**, 279–294.

Moreillon, P., and Que, Y.A. (2004). Infective endocarditis. *Lancet* **363**, 139–149.

Murakami, M., Ohtake, T., Dorschner, R.A., and Gallo, R.L. (2002). Cathelicidin antimicrobial peptides are expressed in salivary glands and saliva. *Journal of Dental Research* **81**, 845–850.

Nakano, Y., Yoshimura, M., and Koga, T. (2002). Correlation between oral malodor and periodontal bacteria. *Microbes and Infection* **4**, 679–683.

Okuda, K., Kato, T., and Ishihara, K. (2004). Involvement of periodontopathic biofilm in vascular diseases. *Oral Diseases* **10**, 5–12.

Petersilka, G.J., Ehmke, B., and Flemmig, T.F. (2002). Antimicrobial effects of mechanical debridement. *Periodontology 2000* **28**, 56–71.

Rudney, J.D. (2000). Saliva and dental plaque. *Advances in Dental Research* **14**, 29–39.

Sakamoto, M., Takeuchi, Y., Umeda, M., Ishikawa, I., and Benno, Y. (2003). Application of terminal RFLP analysis to characterise oral bacterial flora in saliva of healthy subjects and patients with periodontitis. *Journal of Medical Microbiology* **52**, 79–89.

Sarkonen, N., Kononen, E., Summanen, P., Kanervo, A., Takala, A., and Jousimies-Somer, H. (2000). Oral colonisation with *Actinomyces* spp. in infants by 2 years of age. *Journal of Dental Research* **79**, 864–867.

Sixou, J.L., de Medeiros-Batista, O., and Bonnaure-Mallet, M. (1996). Modifications of the microflora of the oral cavity arising during immunosuppressive chemotherapy. *European Journal of Cancer. Part B, Oral Oncology* **32B**, 306–310.

Slots, J. and Ting, M. (2002). Systemic antibiotics in the treatment of periodontal disease. *Periodontology 2000* **28**, 106–176.

Slots, J. and Kamma, J.J. (2001). General health risk of periodontal disease. *International Dental Journal* **51**, 417–427.

Socransky, S.S., Smith, C., and Haffajee, A.D. (2002). Subgingival microbial profiles in refractory periodontal disease. *Journal of Clinical Periodontology* **29**, 260–268.

Tanner, A.C., Milgrom, P.M., Kent, R., Jr., Mokeem, S.A., Page, R.C., Riedy, C.A., Weinstein, P., and Bruss, J. (2002). The microbiota of young children from tooth and tongue samples. *Journal of Dental Research* **81**, 53–57.

Wade, W. (2002). Unculturable bacteria – the uncharacterized organisms that cause oral infections. *Journal of the Royal Society of Medicine* **95**, 81–83.

Ximenez-Fyvie, L.A., Haffajee, A.D., and Socransky, S.S. (2000). Comparison of the microbiota of supra- and subgingival plaque in health and periodontitis. *Journal of Clinical Periodontology* **27**, 648–657.

Zalewska, A., Zwierz, K., Zolkowski, K., and Gindzieñski, A. (2000). Structure and biosynthesis of human salivary mucins. *Acta Biochimica Polonica* **47**, 1067–1079.

Role of the indigenous microbiota in maintaining human health

It has been recognised for many years that the indigenous microbiota has an important role to play in maintaining the health of an individual by preventing colonisation by exogenous pathogens. However, this "barrier" function is far from being the only contribution that the microbiota makes to the well-being of its host (Table 9.1). It is well established that resident microbes contribute significantly to human development and nutrition and are an effective means of neutralising potentially toxic dietary components. Most (approximately 10^{14}) of the microbial inhabitants of a human being are located in the colon, which also harbours one of the most diverse of the resident microbial communities. The activities of this microbiota, therefore, are likely to have the greatest impact on human development and health, and have been the most extensively investigated. Studies of the role played by microbial communities inhabiting other sites in maintaining human health have largely been devoted to their ability to prevent colonisation by exogenous pathogens. Ethical considerations limit the type of experiments that can be carried out using humans; therefore, much of our knowledge of the role of the indigenous microbiota on mammalian development and health have come from studies involving animals.

9.1 | Colonisation resistance

9.1.1 Exclusion of exogenous microbes

Co-evolution has ensured that each exposed surface of a human being is colonised by microbes exquisitely adapted to that particular environment. Hence, the microbes colonising a particular region can grow and reproduce under the conditions (physical, chemical, and biological) that exist there and can remain associated with that particular region by adhering to some substratum within it. Previous chapters have shown that the microbiotas of all but the most extreme environments (e.g., the stomach and duodenum) available for colonisation are very complex. The climax community that develops at each site will consist of microbes able to adhere to the existing substrata and utilise the available nutrients and is in a state of dynamic equilibrium as a result of the many interactions occurring between its constituent members. Any exogenous microbe attempting to colonise such a site will, therefore, be faced with a very difficult task, and the microbiota of that site is said to exhibit "colonisation resistance" as a consequence of its members having occupied all of the available physical, physiological, and metabolic niches. Studies have shown that the dose of exogenous organisms

Table 9.1.	Role of the indigenous microbiota in mammalian health

exclusion of exogenous pathogens
development of immune functions
tissue and organ differentiation and development
development of nutritional capabilities
provision of nutrients
provision of vitamins
detoxification of harmful dietary constituents
prevention of bowel cancer

required to colonise germ-free animals can be between 1,000- and 100,000-fold lower than that required to colonise those with an indigenous microbiota. The exclusion of an exogenous organism from a particular site is achieved by a combination of factors, including those attributable to the host and the indigenous microbiota of that site i.e., the microbiota's "colonisation resistance". The term is also sometimes used to refer to the ability of a microbiota to control the number of potentially pathogenic members that may be present in that community. Hence, in the colon of healthy individuals, potentially pathogenic, Gram-negative facultative anaerobes, enterococci, and yeasts generally comprise only a small proportion of the microbiota.

The anatomical and physiological characteristics of a particular site in the host can deter colonisation by exogenous organisms in a number of ways, although it must be pointed out that these characteristics will often be modified by the indigenous microbiota of the site. Firstly, the site will have only a limited range of substrata available for adhesion so that only organisms with appropriate adhesins will be able to adhere. This contributes to the phenomenon known as "tissue tropism" (i.e., the predilection of a particular organism for a certain body site). However, the indigenous microbiota itself will provide additional substrata for microbial adhesion. In the mouth, for example, the tooth surface is colonised by streptococci, which provide adhesion sites for organisms such as *Fusobacterium* spp., which rarely adhere to the tooth surface itself. The environment provided by the host (e.g., pH, moisture content) will limit the range of organisms able to survive and grow at any given site. Examples of such restrictions to microbial colonisation imposed by the host are given in Table 9.2. However, again, the activities of indigenous microbes can alter the environmental conditions operating at a site. For example, the large intestine of a neonate is an aerobic environment with a high redox potential, but microbial colonisation rapidly converts it to an anaerobic environment with a very low redox potential. The range of host molecules (cell constituents, secretions, etc.) available at a particular site are limited and will restrict colonisation to those organisms able to utilise these as nutrients. Once again, compounds produced by the indigenous microbiota, as well as those released when they die, extend the range of potential nutrients at a site.

Both the innate and acquired immune responses are also thought to be involved in controlling microbial colonisation, and the means by which this is achieved is described in Section 1.5.4. Individuals who are immunocompromised in some way often experience abnormal patterns of colonisation (i.e., one or more body sites are colonised by species not usually found at such sites).

Table 9.2. | Host factors involved in restricting microbial colonisation

Host factor	Example
availability of receptors for bacterial adhesins	many examples; contributes to tissue tropism
high salt content	restricts range of microbes found on skin
low moisture content	restricts range of microbes found on skin
low pH	restricts range of microbes colonising the stomach, duodenum, vagina, and skin
lipids as major available nutrients	restricts range of microbes found in hair follicles and on certain skin regions
high shear forces	restrict colonisation of surfaces in the oral cavity, urinary tract, and small intestine
peristalsis	restricts colonisation in upper regions of the GIT
bile acids	restricts colonisation in lower regions of the GIT
innate and acquired immune response	influences colonisation at most sites

Note: GIT = gastrointestinal tract.

9.1.2 Mechanisms involved in colonisation resistance

Colonisation resistance by the indigenous microbiota of a site involves a number of mechanisms, including (1) occupation of adhesion sites, (2) alteration of the physico-chemical environment, (3) production of antagonistic substances, and (4) utilisation of all available nutrients within a site. Unfortunately, while many examples of colonisation resistance have been documented, the mechanism involved has often not been determined and is likely to involve a combination of several of the aforementioned mechanisms.

9.1.2.1 Occupation of adhesion sites

Adhesion to some substratum within a site is an important first step in colonisation, although this does not apply to the microbiotas of the lumen of some of the lower regions of the gastrointestinal tract (GIT). If all of the available adhesion sites are occupied by members of the indigenous microbiota, then exogenous organisms will have difficulty colonising the region. This phenomenon has been termed "competitive exclusion" and is considered to be an important means by which lactobacilli dominate the vaginal microbiota (Section 6.4.3.1.1). Studies have demonstrated that competitive exclusion is one of the main means by which lactobacilli prevent colonisation of the vagina by organisms such as *G. vaginalis*, *Can. albicans*, and uropathogens such as *E. coli*. Hence, the presence of *L. acidophilus* has been shown to reduce the numbers of *Can. albicans* and *G. vaginalis* that adhere to vaginal epithelial cells. This lactobacillus and *G. vaginalis* appear to bind to the same receptors on the surfaces of vaginal epithelial cells, but the former can displace the latter from these cells, suggesting that it has a greater affinity for the receptors. Lactobacilli also co-aggregate with *G. vaginalis*, *Can. albicans*, *E. coli*, and *Strep. agalactiae* – thereby preventing them from attaching to the vaginal mucosa. Lactobacilli can also prevent colonisation of the urethra by uropathogenic organisms. In an *in vitro* study, pre-incubation of uroepithelial cells with lactobacilli was found to completely inhibit adhesion of *E. coli* and *K. pneumoniae* and reduce adhesion of *Ps. aeruginosa* by 90%. That this phenomenon can also occur in the GIT has been demonstrated in a number of *in vitro* and animal studies. *L. acidophilus* has been shown to inhibit the adhesion of enteropathogenic *E. coli* to intestinal epithelial cells *in vitro*. Inoculation of mice and pigs with a non-enterotoxin-producing strain of *E. coli* with the K88 adhesin was found to inhibit subsequent colonisation by a K88-expressing enterotoxin-producing strain but not by strains without the K88 adhesin. Also, using a mouse model, it has been shown that *Cl. indolis*, *Cl. cocleatum*, and a *Eubacterium* sp. can prevent colonisation by *Cl. difficile* by blocking mucosal receptors for the organism.

Corynebacterium spp. have been shown to prevent colonisation of the anterior nares by *Staph. aureus*, and this has been attributed to the greater binding affinity of *Corynebacterium* spp. for nasal mucin.

9.1.2.2 Alteration of the physico-chemical environment

The maintenance of a low pH in the vagina due to lactic-acid production by lactobacilli is able to prevent colonisation by exogenous neutrophilic and alkaliphilic organisms. Exogenous organisms thought to be excluded in this way include *Chlamydia* spp., *Trichomonas* spp., *Mycoplasma* spp., *Mobiluncus* spp., *N. gonorrhoeae*, and human immunodeficiency virus (HIV; see Section 6.4.3.1.1). The low pH of the colon of breast-fed infants resulting from acid production by bifidobacteria restricts colonisation of this region to acidophilic organisms (Section 7.4.2). Although the pH of the contents of the colon of adults is almost neutral, acid production by many species can result in localised regions with a low pH and is able to inhibit the growth of other organisms. This is thought to be an important means by which lactic-acid-producing bacteria (e.g., lactobacilli, bifidobacteria, and streptococci) inhibit the growth of exogenous enteric pathogens, such as *Salmonella* spp., enteropathogenic *E. coli*, and *Staph. aureus*. Acid production by propionibacteria, staphylococci, and coryneforms also contributes to the low pH of the skin surface, and thereby hinders colonisation by a range of organisms.

Oxygen utilisation and the creation of a low redox potential by facultative organisms can render a site unsuitable for colonisation by obligate aerobes. This occurs in various regions of the GIT and in dental plaque.

9.1.2.3 Production of antagonistic substances

Many indigenous microbes are known to produce substances that are antagonistic to other organisms, including exogenous species. The range of such compounds is wide and includes bacteriocins, volatile fatty acids, hydrogen peroxide, and enzymes. These have been described throughout the text, and some examples are listed in Table 9.3. A limited number of *in vivo* studies have shown that the production of such substances by a particular species may be responsible for the exclusion of other organisms. For example, a bacteriocin-producing strain of *Staph. albus* was able to protect mice against infection with *Cl. septicum*. Guinea pigs injected with a bacteriocin-producing strain of an enterococcus were protected against infections by *Cl. perfringens*. Furthermore, A bacteriocin-producing strain of *Staph. aureus* has been shown to prevent skin infection by *Strep. pyogenes* in hamsters. The presence of lactobacilli in the vagina is important in excluding pathogens such as *N. gonorrhoeae* and HIV, although the exact means by which this is achieved is not known. However, it is thought to involve the production of bacteriocins, hydrogen peroxide, and fatty acids. Similarly, the exclusion of *N. gonorrhoeae* from the urethra by the urethral microbiota (staphylococci, streptococci,

Table 9.3.	Production of antagonistic substances by the indigenous microbiota

Antagonistic substance	Produced by
short-chain fatty acids	many intestinal and oral microbes, lactobacilli in the vagina, cutaneous microbes
long-chain fatty acids	cutaneous microbes
hydrogen peroxide	lactobacilli, streptococci
bacteriocins	bacteria at all anatomical sites
hydrogen sulphide	gut bacteria
lysostaphin	*Staph. simulans* present on skin and in the urogenital tract
other bacterial enzymes	propionibacteria, *Abiotrophia* spp.

bifidobacteria, and lactobacilli) is thought to be due to the production of bacteriocins and fatty acids.

9.1.2.4 Utilisation of the available nutrients within a site

One of the ways in which climax communities prevent the intrusion of exogenous species and thereby contribute to their stability is by utilising all of the available nutrients in a web of cross-feeding interactions. Consequently, very few nutrients will be available to support the growth of any new organisms arriving at the site. Many examples of such interactions are given in Chapters 2–8.

To become established at a site with a pre-existing climax community, an exogenous organism will, therefore, have to compete with members of the indigenous microbiota for the nutrients that exist within the habitat. An organism that is successful in doing so may be able to displace a member of the climax community. This could have dramatic consequences for the community, as it would most likely disrupt those food webs in which the displaced member was involved. This could have a "knock-on" effect, leading to wholesale shifts in the composition of the microbiota. Evidence in support of such "nutrient competition" occurring in human beings *in vivo* is lacking. However, it has been shown that in the mouse gut, where gluconate is present in low concentrations, even small numbers of a strain of *E. coli* which is efficient at gluconate utilisation is able to colonise in the presence of established strains of *E. coli* and other organisms. Furthermore, the importance of nutrient competition in determining community composition has certainly been demonstrated in continuous culture studies *in vitro*. For example, successful competition for a number of nutrients (glucose, N-acetylglucosamine, and sialic acid) by members of the intestinal microbiota of mice can prevent the establishment of *Cl. difficile* in the community. Furthermore, the exclusion from intestinal communities of *Shigella flexneri* by coliforms can be overcome by addition of glucose, suggesting that competition for carbon and energy sources is the basis of the inhibitory activity. Biofilms of oral bacteria grown in a laboratory model of the oral cavity have been shown to be resistant to colonisation by *Strep. mutans*, and this is likely to be attributable to the inability of the organism to utilise the glycoprotein provided as the main carbon and energy sources for the community.

In an interesting series of experiments, the effect of the composition of the oral microbiota on its ability to prevent colonisation by the cariogenic organism, *Strep. mutans*, has been investigated. Rats with a simple oral microbiota – consisting of approximately 96% *Strep. bovis* with small proportions of a *Veillonella* sp. and an unidentified Gram-negative rod – were easily colonised by *Strep. mutans* when fed a sugar-containing diet. In fact, the *Strep. mutans* soon established itself as the dominant member of the microbiota. However, when the microbiota also contained *A. viscosus* and the community was allowed to equilibrate for 7 days before challenge with *Strep. mutans*, colonisation by this cariogenic organism was completely inhibited. The more complex *A. viscosus*-containing oral community is more similar to that found *in vivo* and demonstrates the ability of climax communities to exclude other species, possibly because of efficient utilisation of all of the available carbon and energy sources.

9.1.3 Disruption of colonisation resistance

9.1.3.1 Role of microbial factors

It is well known that using antibiotics can result in severe disruption of the indigenous microbiota, and numerous examples of this phenomenon are given in Chapters 2–8. One of the consequences of such disruption is that the colonisation resistance of the microbiota is reduced, and the individual becomes more susceptible to colonisation by exogenous organisms, including pathogenic species. Another frequent occurrence is an increase in proportions of potentially pathogenic members of the indigenous microbiota, which are normally present in only low proportions – this is often termed "superinfection".

In a study of the colonisation of human volunteers by a bovine strain of *E. coli*, daily oral administration of a therapeutic level (1 g) of tetracycline for 9 days enabled colonisation of the GIT by the organism when 1×10^6 cfu were given orally 4 days after the start of the course of the antibiotic. However, no such colonisation was observed in controls who did not receive tetracycline or in those who were administered only 50 mg/day. In a series of studies on healthy volunteers, the administration of amoxycillin, cefotaxime, clindamycin, or co-trimoxazole have all been shown to facilitate colonisation by *K. pneumoniae* and *Enterobacter cloacae*, with which the individuals were subsequently challenged. Furthermore, antibiotic administration was generally accompanied by increased proportions of enterocci, Gram-negative facultative bacilli, and yeasts in the faeces of the volunteers. In one such study, six volunteers ingested a suspension of 10^6 cfu of *K. pneumoniae* before and after being given co-trimoxazole. Prior to antibiotic administration, the challenge organism was not detectable in the faeces of any of the volunteers after 4 days. In contrast, after taking the co-trimoxazole, *K. pneumoniae* persisted in faeces for 7 to 18 days. The number of Gram-negative bacilli and yeasts in the faeces of four and five individuals, respectively, showed a statistically significant increase after administration of co-trimoxazole. Furthermore, the number of staphylococci and yeasts in the saliva of two and four individuals, respectively, showed a statistically significant increase after antibiotic administration. Two of the volunteers suffered from glossitis due to *Can. albicans* after receiving the antibiotic. Superinfection of the oral cavity, the vagina, or the GIT with *Can. albicans* is a frequent consequence of the administration of antibiotics. This organism is normally present in low numbers at all of these sites, but outgrowth is possible once controlling organisms have been reduced or eliminated. Other examples of superinfection with indigenous microbes following antimicrobial chemotherapy are given in Table 9.4.

Few investigations have determined the ability of the indigenous microbiota to exclude environmental organisms. However, in a study involving healthy adult

Table 9.4. Examples of superinfection following antibiotic administration

Antibiotic	Superinfecting organism
vancomycin	vancomycin-resistant enterococci
clindamycin	vancomycin-resistant enterococci, *Cl. difficile*
cephalosporins	vancomycin-resistant enterococci, *Cl. difficile*
metronidazole	vancomycin-resistant enterococci
third-generation cephalosporins (e.g., ceftazidime)	ESBL-KP
ampicillin	*Cl. difficile, Enterococcus* spp., *Enterobacter* spp., *Staph. aureus, E. coli*
ceftriaxone	*Enterococcus* spp., *Enterobacter* spp., *Staph. aureus, Cl. difficile, Candida* spp.
cefaclor	*Haemophilus* spp., *Staph. aureus*
amoxycillin	*Can. albicans, Enterobacter* spp., *Klebsiella* spp., *Staph. aureus*
ciprofloxacin	*Can. albicans, Klebsiella* spp.

Note: ESBL-KP = extended-spectrum β-lactamase producing *K. pneumoniae*.

volunteers, *Ps. aeruginosa* was detected in the faeces of 44% of those treated with β-lactam antibiotics, but not in the faeces of those who had not received antibiotics. The organism was not present in the faeces prior to antibiotic therapy and was presumably acquired from the environment.

One of the most disturbing consequences of the breakdown of colonisation resistance due to antimicrobial chemotherapy is infection by *Cl. difficile* resulting in the life-threatening disease, pseudomembranous colitis. While the resistance to colonisation by *Cl. difficile* exerted by the indigenous colonic microbiota is very effective, it is easily disrupted by certain antibiotics, particularly clindamycin, cephalosporins, and broad-spectrum penicillins to which the organism is resistant. *Cl. difficile* is the most frequent cause of antibiotic-associated diarrhoea (AAD), which can vary in severity ranging from mild abdominal discomfort to pseudomembranous colitis. Most of these infections involve hospitalised individuals, particularly the elderly, and it has been estimated that, in the United States, approximately 1.2% of hospitalised patients are affected. Although approximately 3% of healthy individuals harbour the organism in their colon, most infections are caused by strains acquired from exogenous sources. For example, DNA fingerprinting of more than 200 isolates recovered from patients with *Cl. difficile* diarrhoea and from the ward environment during a 6-month period revealed that only six distinct types were present. The pattern of acquisition of an infection due to the organism appears to involve colonisation of the patient by exogenous strains of the organism after entry into hospital, breakdown of colonisation resistance due to the administration of one of the antibiotics previously mentioned, and then growth of the organism in the colon to disease-inducing levels. The organism is acquired mainly from the contaminated hands of health-care workers or from the environment – surveys have shown that the organism is present in 20–70% of sampled sites in a hospital, depending on the level of cleaning and the extent of faecal soiling. The organism can survive in the environment as spores for several months, possibly years, and is resistant to many disinfectants. Infected individuals are usually treated with metronidazole, although probiotics and replacement therapy have sometimes proved to be effective (Sections 10.1.1 and 10.4.6).

Although *Cl. difficile* is an important cause of AAD, it has been estimated that it is responsible for only between 10% and 20% of cases. Some of the remaining cases arise from a breakdown in the colonisation resistance of the colonic microbiota, thus resulting in the overgrowth of potentially pathogenic members of the microbiota, including *Cl. perfringens, Staph. aureus, K. oxytoca*, and *Candida* spp. However, many cases of AAD are also attributable to antibiotic-induced disturbances of intestinal carbohydrate or bile acid metabolism, to allergic and toxic effects of antibiotics on the intestinal mucosa, or to their effects on gut motility.

9.1.3.2 Role of host factors

While the administration of certain antibiotics disrupts colonisation resistance primarily by their effect on the indigenous microbiota, it must be remembered that the phenomenon of colonisation resistance stems from host as well as microbial factors. The contribution made by the host to preventing colonisation by exogenous microbes can be appreciated by considering the results of studies involving individuals suffering from a variety of disorders, especially those which compromise the host's immunocompetence. Hence, cancer (and the chemotherapy and radiotherapy used in its treatment), diabetes, HIV infection, and a range of immunological disorders all result in the host being immunocompromised in some way. Because this immunocompromised state also often affects the composition of the indigenous microbiota, both the host and microbial contributions to colonisation resistance will be diminished. Such individuals are susceptible to infections by a wide range of both indigenous and exogenous microbes, and these are often referred to as "opportunistic infections" (Table 9.5). Immunocompromised individuals, however, are not the only ones at risk of opportunistic infections, as any impairment of the host's defences reduces its ability to exclude exogenous microbes. Hence, injuries such as burns and wounds which disrupt the integrity of the skin or mucosal surfaces will facilitate opportunistic infections, as will skin disorders such as psoriasis and eczema. Furthermore, the use of catheters and other medical devices may enable ingress of either exogenous microbes or the indigenous microbiota of the site to underlying tissues or may act as a nidus for infections. Other factors that predispose an individual to an opportunistic infection include a previous or current infection with an organism that impairs host defence mechanisms (e.g., viral infections such as influenza).

9.2 | Host development

There are several examples in the animal and plant kingdoms of indigenous microbes affecting the development of host tissues. Hence, *Vibrio fischeri* is responsible for inducing the development of the light organ in the bobtail squid *Euprymna scolopes*, and various soil bacteria induce the formation of root nodules in plants. Such interactions have been investigated extensively for many years, and we now have a good understanding of the mechanisms involved. In contrast, although it has been well established that many aspects of mammalian development are influenced by the indigenous microbiota, our understanding of the mechanisms involved is limited. Most information

Table 9.5. Organisms responsible for opportunistic infections in individuals suffering from anatomical, physiological, or immunological disorders or those who are undergoing some medical or surgical procedure or are receiving certain therapeutic agents

Disorder/intervention	Common infecting organisms
burns and wounds	*Staph. aureus, Staph. epidermidis*, enterococci, *E. coli, Bacteroides* spp., **Pseudomonas spp.**
eczema	*Staph. aureus, Strep. pyogenes, Can. albicans*, GPAC, *Bacteroides spp., Prevotella spp., Fusobacterium spp.*
psoriasis	*Staph. aureus, Enterococcus* spp., FGNB, *Bacteroides* spp., GPAC
atopic dermatitis	*Staph. aureus, Strep. pyogenes, Bacteroides* spp., GPAC, *Fusobacterium* spp.
intravenous catheter	*Staph. epidermidis, Staph. aureus*, FGNB, *Candida* spp.
urinary catheters	*E. coli*, **Pseudomonas spp.**, enterococci, *Proteus mirabilis, K. pneumoniae, Staph. epidermidis, Candida* spp.
HIV infection	*Candida* spp., **Pneumocystis carinii, Mycobacterium spp.**, *Strep. pneumoniae, H. influenzae*, FGNB
bone marrow transplantation	*Staph. epidermidis*, FGNB, *Strep. pneumoniae, Candida* spp., viridans streptococci, *H. influenzae*
cystic fibrosis	**Pseudomonas spp., Burkholderia cenocepacia**, *Staph. aureus, H. influenzae*
diabetes	*Candida* spp., *Staph. aureus*
leukaemia	*Candida* spp., **other fungi**, *Staph. epidermidis*, viridans streptococci, FGNB, **Pseudomonas spp.**
chemotherapy or radiotherapy of malignancies	*Staph. aureus*, FGNB, *Candida* spp., *H. influenzae, Strep.* pneumoniae, **Pseudomonas spp., Pneumocystic carinii**
medical implants	*Staph. epidermidis, Staph. aureus, E. coli, Propionibacterium* spp., streptococci
chronic granulomatous disease	staphylococci, FGNB
infection by influenza viruses	*Strep. pneumoniae, H. influenzae, Mor. catarrhalis, Staph. aureus*
treatment with corticosteroids	*Candida* spp., **Aspergillus spp., Cryptococcus spp., Nocardia spp.**
neutropaenia	FGNB
Hodgkin's disease	*Candida* spp., *Staphylococcus* spp., **Aspergillus spp., Cryptococcus spp., Nocardia spp., Histoplasma spp.**

Notes: Some of these organisms (in **boldface**) are important causative agents of such infections, but are not members of the indigenous microbiota of humans. FGNB = facultatively anaerobic Gram-negative bacilli; GPAC = Gram-positive anaerobic cocci; HIV = human immunodeficiency virus.

Table 9.6.	Attributes of germ-free animals, compared with their counterparts with an indigenous microbiota

decrease in mass of heart, lung, and liver
increase in size of caecum (may be eight times larger)
decreased water absorption by large intestine
increased redox potential of large intestine
decreased concentration of deconjugated bile salts
increased pH of stomach
altered surface epithelial mucins
decreased mass of small intestine
absence of immune cells in lamina propria
decrease in intestinal peristalsis
prolonged intestinal epithelial cell cycle time
shorter, more slender villi in intestinal tract
increased length of microvilli
decreased surface area of intestinal mucosal
decreased number of lymphocytes in lamina propria
decreased size of Peyer's patches, mesenteric lymph nodes, spleen, and thymus
decreased macrophage chemotaxis and phagocytic activity
decrease in plasma cells in lamina propria and Peyer's patches
decreased production of IgG and IgA
decreased number of TCR$\alpha\beta^+$ intra-epithelial lymphocytes

Note: TCR = T-cell receptor.

regarding the role of microbes in mammalian development has been obtained by comparative studies involving germ-free animals, animals with their indigenous microbiota, and those colonised with specific microbial species. The absence of an indigenous microbiota can have dramatic effects on the anatomy and physiology of an animal and examples of these are listed in Table 9.6 – these are often termed "germ-free animal characteristics". Those aspects of the host's anatomy, immunology, physiology, or biochemistry that are influenced by the indigenous microbiota have been termed "microbiota-associated characteristics" and, as is evident from Table 9.6, these are many and varied. Many of the abnormalities observed in germ-free animals can be reversed by inoculation with the indigenous microbiota or constituents of the microbiota. Hence, the characteristic caecal enlargement in germ-free rodents can be eliminated by monoinfection with *Micromonas micros*, while gut motility can be restored by reintroducing the normal gut microbiota.

One interaction between a member of the indigenous intestinal microbiota and the gut mucosa that has been investigated in detail at the molecular level is the ability of *B. thetaiotaomicron* (one of the most abundant members of the colonic microbiota of humans) to induce the expression of fuc-α1,2-Gal-terminating glycans, which it can then utilise as carbon and energy sources (Figure 9.1). The organism has a molecular sensor (FucR) which responds to fucose in the environment. When fucose is plentiful (indicative of ongoing hydrolysis of the glycan by the organism's extracellular enzymes), then the FucR allows expression of fucose-degrading enzymes, but turns off a signal telling the epithelial cells to synthesise more glycans. When fucose is scarce, FucR allows the production of the signal, which induces the epithelial cells to produce more glycans. The benefits of this interaction to the organism are obvious – a constant supply of a source of carbon and energy. However, this microbe-induced developmental

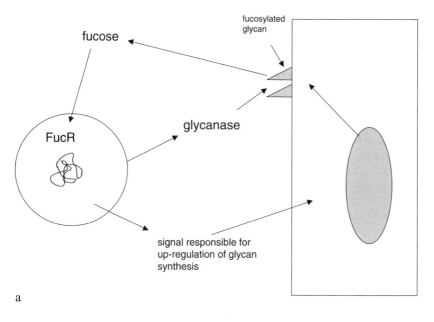

a

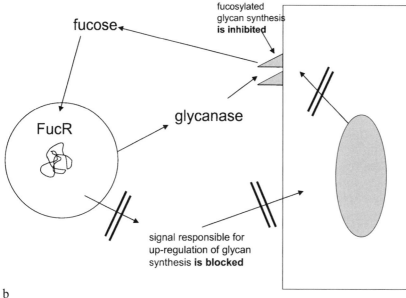

b

Figure 9.1 Induction of fuc-α1,2-Gal-terminating glycan synthesis in host intestinal epithelial cells by *Bacteroides thetaiotaomicron*. (a) When fucose is scarce, the bacterium produces a signal that stimulates the epithelial cell to produce the fucose-containing glycan which can then be hydrolysed to fucose by bacterial glycosidases. (b) When fucose is plentiful, the signal is not produced and no more glycan is synthesised.

change is also of benefit to the host. Hence, it encourages colonisation by an organism with broad carbohydrate- and protein-degrading activities, thus enabling the host to scavenge sources of carbon and energy (in the form of short-chain fatty acids – SCFAs – produced by the organism) from polymers that it is unable to digest itself. Secondly, it encourages colonisation by an organism of low virulence which will contribute towards "colonisation resistance" against exogenous pathogens.

Table 9.7.	Effect of colonisation by *B. thetaiotaomicron* on gene expression in the intestinal tract of mice

increased ileal expression of Na$^+$/glucose co-transporter – main mediator of glucose uptake

increased expression of pancreatic lipase-related protein – responsible for breakdown of triacylglycerols

increased expression of a cytosolic fatty-acid-binding protein – involved in intracellular trafficking of fatty acids

increased expression of an apolipoprotein – mediates export of triacylglycerols from enterocytes

increased expression of an epithelial copper transporter

Colonisation of germ-free mice with *B. thetaiotaomicron* also results in changes in the expression of several host genes involved in the processing and absorption of carbohydrates, lipids, and micronutrients (Table 9.7), and thereby contributes to the development of the host's nutritional capabilities. This may account for the finding that germ-free animals require much higher calorie intakes (approximately 30% more in rodents) than their conventional counterparts to sustain their body mass.

Germ-free adult mice have a greatly reduced capillary network in their intestinal villi compared with conventional mice, and it has been shown that in conventional mice, the development of the capillary network coincides with the establishment of a complex intestinal microbiota. Recently, it has been reported that inoculation of germ-free mice with *B. thetaiotaomicron* induces the formation of a normal capillary network, thereby greatly increasing the host's ability to absorb nutrients. The organism is thought to stimulate the production of angiogenic molecule(s) by Paneth cells, which are located in the crypts at the base of the villi. However, the identity of the bacterial signal responsible for stimulating the release of the angiogenic molecule(s) from the Paneth cells remains to be established.

B. thetaiotaomicron is known to have a number of other effects that are beneficial to the gut mucosa. Hence, inoculation of germ-free mice with the organism was found to result in increased expression of (1) decay-accelerating factor, which prevents damage resulting from the microbial activation of complement components; (2) a receptor for trefoil peptides which help in the repair of damaged epithelium; and (3) Sprr2a, which is a cross-linking protein belonging to the family of small proline-rich peptides that are involved in maintaining epithelial barrier functions – this was increased more than 200-fold. These remarkable studies involving *B. thetaiotaomicron* demonstrate the intimate relationship that exists between a mammal and at least one member of its intestinal microbiota. The bacterium "informs" its host of its needs for a key nutrient and is thereupon supplied with it, while the bacterium increases the host's ability to utilise and absorb a range of dietary constituents, as well as increasing its ability to maintain an effective epithelial barrier.

One of the major roles of the intestinal microbiota is in stimulating the growth and differentiation of intestinal epithelial cells. Germ-free rodents have fewer crypt cells than conventional animals, and the rate of production of such cells is reduced. Hence, in conventional rodents, the rate of enterocyte turnover is almost twice that found in germ-free animals. The microbially induced proliferation and differentiation of epithelial cells is mediated by the SCFAs produced by fermentation of carbohydrates and amino acids. Although all three major SCFAs (i.e., butyrate, acetate, and propionate) are able to induce this trophic effect, butyrate is the most potent in this respect and has been shown to alter the expression of a number of genes in epithelial cells *in vitro*. Butyrate can also inhibit DNA synthesis in, and proliferation of, neoplastic cells, and it has been suggested that this may account for the protective effect that dietary fibre exhibits against bowel cancer.

Gut microbes also appear to have a role in the maturation of the gut that occurs during weaning in mice. During this period, the levels of ileal epithelial lactase (which hydrolyses lactose present in milk) decrease, and this coincides with colonisation of the gut by *B. thetaiotaomicron*. Monoinfection of mice with this organism has been shown to result in decreased levels of lactase. It is known that adenosine deaminase and polyamines are involved in post-natal maturation of the gut in mice, and expression of adenosine deaminase and ornithine decarboxylase antizyme (a regulator of polyamine synthesis) are up-regulated following colonisation of the gut of germ-free mice by *B. thetaiotaomicron*.

It is well established that the indigenous microbiota plays a key role in the development of a competent immune system. Because the gut-associated lymphoid tissue (GALT) contains the largest collection of immunocompetent cells in the human body, most studies have involved the GIT and its microbiota. As shown in Table 9.1, the immune system of germ-free animals has a number of structural and functional abnormalities, including low densities of lymphoid cells in the gut mucosa, circulating concentrations of antibodies are low, specialised follicle structures are small, etc. However, exposure of the gut mucosa to the indigenous microbiota has a dramatic effect on GALT. Hence, the number of intra-epithelial lymphocytes expands greatly; germinal centres with antibody-producing cells appear in follicles and in the lamina propria; the levels of circulating antibodies increase; and increased quantities of IgA are secreted into the gut lumen. In general, the use of single species or simple mixtures of species belonging to the indigenous gut microbiota are not as effective as samples of the entire gut microbiota at inducing the development of normal immune responses in germ-free animals. Many studies have shown that the indigenous microbiota stimulates the secretory IgA system and B-lymphocyte function in general. In mice, for example, bacterial colonisation soon after birth induces the development of IgA-secreting cells in the small intestine. Furthermore, during weaning, with its concomitant increase in the diversity of the gut microbiota, the number of IgA-secreting plasma cells characteristically increases.

9.3 | Host nutrition

Until the 1970s, the main functions of the colon were considered to be the absorption of water and salts, and the disposal of the waste products of digestion. However, it is now recognised that the microbiota of the colon makes a significant contribution to the nutritional requirements of its host, and this microbial community – which has a mass similar to that of the spleen – may be considered to be an important digestive "organ". The colonic microbiota functions as an effective scavenger of dietary constituents that the stomach and small intestine are unable to digest (mainly complex carbohydrates), have failed to digest (carbohydrates, proteins, peptides), or have failed to absorb (amino acids and monosaccharides). Through the concerted

Table 9.8. Carbohydrates available to microbes in the colon

Carbohydrate	Source
resistant starch	grains, seeds, certain fruits, certain peas, and beans
cellulose, hemi-cellulose, pectin	plant cell walls
inulin, guar	plant storage polymers
karaya	plant exudates
fructo-oligosaccharides	onions, garlic, artichokes, and wheat
lactose	milk and milk products
sugar alcohols	artificial sweeteners
host-derived carbohydrate-containing molecules	mucins, hyaluronic acid, glycosphingolipids, and chondroitin sulphate

activities of its diverse constituents, the colonic microbiota degrades these materials to assimilable molecules that serve as nutrients for the host as well as for resident microbes. Although protein-degrading and amino-acid-fermenting organisms are present in the colon, saccharolytic bacteria predominate both numerically and in metabolic terms.

9.3.1 Short-chain fatty acids

Quantitatively, the most important dietary constituent to escape digestion in the upper regions of the GIT is starch, and the digestion of the various forms of this polysaccharide is a major function of the colon. Although a considerable quantity of starch is digested by the action of salivary and pancreatic enzymes, a large proportion (known as "resistant starch") reaches the colon. Starch may be resistant to digestion in the upper GIT for a number of reasons. Firstly, it may be present in forms which render it physically inaccessible to salivary and pancreatic amylases (e.g., when it is inside grains or seeds). Secondly, various crystalline forms of starch exist, including ones that are particularly resistant to pancreatic enzymes – such starches are present in uncooked potato, banana, and certain peas and beans. Thirdly, cycles of heating and cooling can generate enzyme-resistant forms of starch – this occurs with potato, maize, and rice. Finally, the formation of complexes with other dietary constituents can render the starch resistant to enzymatic hydrolysis. In individuals with a typical Western diet, approximately 10% of dietary starch reaches the colon, but the proportion can be much higher in countries where starchy foods constitute a larger proportion of the diet. The other main group of undigested polysaccharides reaching the colon are the non-starch polysaccharides which constitute what is known as dietary fibre (Table 9.8). These include plant cell wall components (e.g., cellulose, hemi-cellulose, pectins), storage polymers (e.g., inulin, guar), and exudates (e.g., karaya). Studies have shown that none of these polymers are broken down in the stomach or small intestine, yet approximately 50% of cellulose and 70–90% of hemi-celluloses are broken down by bacteria in the large intestine. The average daily intake of dietary fibre in individuals on a typical Western diet is between 10 and 25 g.

Apart from polysaccharides, a number of low-molecular mass carbohydrates also reach the colon, either because they cannot be digested by pancreatic enzymes or because they are incompletely absorbed by the small intestine. Examples of the former compounds include lactose, galacto-oligosaccharides, and fructo-oligosaccharides (e.g., raffinose, stachyose). With regard to the incomplete absorption of carbohydrates, it has been estimated that between 2% and 4% of ingested sucrose escapes digestion and absorption by the small intestine. Finally, host macromolecules – such as mucins, hyaluronic acid, glycosphingolipids, and chondroitin sulphate – are also important sources of carbohydrates for colonic microbes. Although a variety of colonic microbes can digest the complex carbohydrates reaching the colon (Table 7.3), the most effective species are those belonging to the genera *Bacteroides* and *Bifidobacterium*. Many species belonging to these genera are able to hydrolyse a variety of carbohydrates on their own, but the more complex polymers, particularly mucins, generally require microbial consortia for their complete degradation. Hence, the complete degradation of the oligosaccharide side chains of mucins requires the sequential action of a number of sulphatases and glycosidases, such as sialidase, α-glycosidases, β-D-galactosidase, and β-N-acetylglucosaminidase.

In addition to carbohydrates, the colon also receives proteins and peptides from several sources: the diet, exfoliated epithelial cells, and pancreatic enzymes. These are rapidly degraded by a variety of microbial proteases and peptidases, and these may be of significant nutritional value to the host. For example, it has been estimated that between 1% and 20% of plasma lysine and threonine in adults is derived from the intestinal microbiota. Furthermore, many colonic microbes can ferment these amino acids to generate a range of products, including SCFAs which are of great nutritional value to the host (see later). Other fermentation products of amino acids include branched-chain fatty acids, other organic acids, amines, phenols, ammonia, indoles, hydrogen, and carbon dioxide.

The main products of carbohydrate and amino-acid fermentation in the colon are the SCFAs, which include mainly acetic, propionic, and butyric acids. The yields from different substrates range from as little as 10 mg/g for oat, pea, and corn brans to as much as 600 mg/g for starch and guar. The total quantities produced per day have been estimated to be between 150 and 600 mmoles, of which at least 95% is absorbed and metabolised by the host and thereby provides up to 9% of the host's energy requirements. The concentration of SCFAs in the colon is between 100 and 180 mmoles/kg of gut contents with the relative proportions of acetate, propionate, and butyrate being approximately 3:1:1. The acids are rapidly absorbed by the colonic mucosa by diffusion and active transport. Colonocytes can utilise each of the three SCFAs as an energy source, with butyrate being the most important and acetate the least important in this respect. It has been estimated that the colonic epithelium derives up to 70% of its energy from these SCFAs. The acids are also used as precursors for the synthesis of mucosal lipids. Apart from its function as a major energy source and its involvement in lipid synthesis, butyrate has a number of effects on colonocytes. Hence, it is able to stimulate cell growth and proliferation, induce differentiation, alter gene expression, and induce apoptosis. Acetate and propionate are also able to stimulate the growth and proliferation of colonic epithelial cells. SCFAs, therefore, appear to play a key role in maintaining gut integrity. There is also some evidence to suggest that the butyrate produced by colonic microbes exerts a protective effect against large bowel cancer. A major function of the colon is the absorption of ions, particularly sodium, calcium, magnesium, and iron, and uptake of all of these ions has been shown to be increased by the presence of SCFAs.

Table 9.9. Vitamins known to be produced by microbes indigenous to humans

Vitamin	Producer organism
vitamin K	*Bacteroides* spp., *Eg. lenta, Propionibacterium* spp., *Veillonella* spp., staphylococci, enterococci, enterobacteria
folic acid	*Bif. bifidum, Bif. infantis, Bif. breve, Bif longum, Ent. faecalis, E. coli,* streptococci
vitamin B_{12}	*E. coli, Bifidobacterium* spp., *Klebsiella* spp., *Veillonella* spp., *Fusobacterium* spp., *Eubacterium* spp., *Clostridium* spp.
vitamin B_2	*E. coli, Cit. freundii, K. pneumoniae*
biotin	*E. coli, Bif. bifidum*
thiamine	*Bifidobacterium* spp.
nicotinic acid	*Bifidobacterium* spp.
pyridoxine	*Bifidobacterium* spp.

The SCFAs not used by the colonocytes pass into the portal vein, where their total concentrations range from 155 to 378 μmoles/litre, depending on the time elapsed since feeding. The molar ratios of acetate:propionate:butyrate are 71:21:8 (compared with 3:1:1 in the colon), demonstrating that considerable quantities of butyrate are utilised by colonocytes. Acetate is a valuable energy source for a variety of tissues, particularly for cardiac and skeletal muscles and for the brain. The fate of propionate in humans is uncertain.

The colonic microbiota, therefore, constitutes a means by which the host can achieve maximum recovery of the nutrients present in its diet without it having to elaborate the vast range of enzymes which would be needed to degrade a wide range of dietary constituents. Furthermore, the ability of the microbiota of the colon to utilise host-derived polysaccharides, glycoproteins, and glycolipids also enables the recycling of the nutrients tied up in such materials rather than having them lost to the environment as waste products. The host reciprocates by providing the microbial community with a warm, moist stable environment and a constant supply of nutrients. The colon and its microbiota, therefore, constitute an excellent example of a symbiosis.

9.3.2 Vitamins

A number of vitamins are present in the colon and are derived both from the diet and from the colonic microbiota – particularly *Bifidobacterium* spp., *Bacteroides* spp., *Clostridium* spp., and enterobacteria. Vitamins produced by colonic bacteria include biotin, vitamin K, nicotinic acid, folate, riboflavin, pyridoxine, vitamin B_{12}, and thiamine (Table 9.9).

Vitamin K is a coenzyme that is essential for the synthesis of several clotting factors, including prothrombin – a deficiency results in delayed clotting and excessive bleeding. A variety of microbes found in the small and large intestines synthesise menaquinones (vitamin K_2), including many *Bacteroides* spp., *Eg. lenta, Propionibacterium* spp., *Veillonella* spp., staphylococci, enterococci, and enterobacteria. It has been estimated that colonic bacteria provide as much as half of the host's requirements for this vitamin.

Folic acid is essential for the synthesis of DNA and RNA, and is involved in the production of red and white blood cells – a deficiency results in megaloblastic anaemia. It is synthesised by some members of the colonic microbiota, including *Bif. bifidum, Bif.*

infantis, Bif. breve, Bif longum, Ent. faecalis, E. coli, and some streptococci. Individuals who have received antibiotics may suffer from folate deficiency, and this implies that the colonic microbiota supplies a significant proportion of the host's requirement for this vitamin.

Vitamin B_{12} (cyanocobalamin) is essential for normal DNA synthesis and the production of red blood cells – a deficiency results in pernicious anaemia. The amount needed by an adult each day is approximately 1 μg. It is present in meat and animal products, but not in plants; thus, the diet of vegans must be supplemented with the vitamin. Absorption of the vitamin occurs in the small intestine after it has been conjugated to a molecule known as intrinsic factor, which is produced in the stomach. Although several members of the colonic microbiota synthesise the vitamin, it is not absorbed by the colon and is excreted in faeces. However, some organisms detected in the small intestine, such as *E. coli, Bifidobacterium* spp., *Klebsiella* spp., *Veillonella* spp., *Fusobacterium* spp., *Eubacterium* spp., and *Clostridium* spp., can synthesise the vitamin. This may constitute an additional source for the host.

Vitamin B_2 (riboflavin) is a component of certain co-enzymes involved in energy-yielding reactions and is synthesised by some members of the GIT microbiota (e.g., *E. coli, Cit. freundii,* and *K. pneumoniae*). However, the contribution of microbially synthesised riboflavin to the host's requirements has not been established.

Biotin is essential for the metabolism of fats and carbohydrates and is also involved in the synthesis of some amino acids. The content of this vitamin in faeces is independent of diet and can be much higher than the dietary intake – this suggests that colonic bacteria are a major source of the vitamin. As biotin can be absorbed by the colon, bacteria are likely to be an important supplier of the vitamin. It is synthesised by a number of species, including *E. coli* and *Bif. bifidum.*

Many bifidobacteria are also able to produce thiamine, nicotinic acid, and pyridoxine.

9.4 | Detoxification

Considerable attention has been directed at elucidating the involvement of the intestinal microbiota in cancer by virtue of its ability to produce carcinogenic compounds (Section 7.5.8). However, a number of studies have shown that the tremendous metabolic capabilities of the colonic microbiota can also achieve detoxification of potentially harmful dietary constituents.

Heterocyclic aromatic amines (HAAs) are pyrolysis products of amino acids found in cooked meat and fish products, and are suspected of a role in the aetiology of colon cancer in humans. A number of studies have shown that several microbial inhabitants of the human colon can reduce the mutagenicity of HAAs by binding to them and/or by altering their structure in some way. In general, Gram-positive species (e.g., *Lactobacillus* spp., *Clostridium* spp., and *Bifidobacterium* spp.) are more effective at reducing the mutagenicity of HAAs than Gram-negative species. Binding of HAAs to Gram-positive species appears to involve carbohydrate moieties on their cell walls. Live bacteria are generally more effective than heat-inactivated cells, implying that as well as binding the HAA, the bacteria either produce metabolites which contribute to their detoxification or transform the HAA to a less mutagenic form. Studies in humans have shown that the consumption of *L. casei* or *L. acidophilus* results in a greatly reduced urinary and faecal mutagenicity following the ingestion of meat.

Methylmercury is a highly toxic compound which is readily absorbed by the tissues of aquatic organisms and, because it is not easily eliminated, is transferred up the food chain and accumulates in predators. The populations most at risk are those who consume large quantities of fish. It is rapidly absorbed from the gut and accumulates in lipid-rich regions of the central nervous system, where it exerts its toxic effects. Detoxification occurs in the liver and a methylmercury–glutathione complex is secreted in bile. However, once this enters the gut, it is rapidly reabsorbed so that an enterohepatic circulation of mercury is established. Conversion of the complex to the mercuric ion and mercury, both of which are poorly absorbed, breaks this cycle and leads to excretion of the element in faeces. That the gut microbiota can achieve this has been demonstrated in studies using germ-free animals and animals treated with antibiotics to reduce their intestinal microbiota. The results of such studies have shown that mercury levels in the brain are increased by up to 45% in the germ-free or antibiotic-treated animals. Faecal suspensions from adult humans are able to demethylate methylmercury, whereas those from unweaned infants have little demethylating activity, suggesting that they are more susceptible to the neurotoxic effects of the compound.

Plant lignans – such as secoisolariciresinol and matairresinol, which are present in fibre-containing foods – are converted by intestinal facultative anaerobes to enterolactone and enterodiol. Soya beans have a high content of phyto-oestrogens (e.g., glycosides of genistein, daidzein, biochin A, and formonometin), and these are converted by the intestinal microbiota to a number of compounds, including the isoflavan equol. Enterolactone, enterodiol, and equol have structures similar to that of the synthetic oestregen diethylstilboestrol and exhibit both weakly oestrogenic and anti-oestrogenic properties. There is some evidence that such compounds protect against cancers of the breast, prostate, and colon.

Plant glycosides are non-toxic, low-molecular mass compounds widely distributed in fruits, vegetables, tea, and wine. In the colon, the sugar moieties of these compounds are removed by β-glycosidases, yielding aglycones. Many of these compounds are toxic, mutagenic, or carcinogenic. However, the flavonoids liberated from flavonoid glycosides in the colon also have protective effects against other mutagens and carcinogens. Such compounds include quercetin, rutin, myricetin, and morin.

9.5 | Further Reading

Books
Gibson, S.A.W. (ed.). (1994). *Human health: the contribution of microorganisms.* London: Springer-Verlag.

Reviews and Papers
Akpan, A. and Morgan, R. (2002). Oral candidiasis. *Postgraduate Medical Journal* **78**, 455–459.
Beaugerie, L. and Petit, J.C. (2004). Antibiotic-associated diarrhoea. *Best Practice and Research Clinical Gastroenterology* **18**, 337–352.
Cebra, J.J. (1999). Influences of microbiota on intestinal immune system development. *American Journal of Clinical Nutrition* **69**, 1046S–1051S.
Cummings, J.H. and Macfarlane, G.T. (1997). Role of intestinal bacteria in nutrient metabolism. *Journal of Parenteral and Enteral Nutrition* **21**, 357–365.

Elson, C.O., Cong, Y., Iqbal, N., and Weaver, C.T. (2001). Immuno-bacterial homeostasis in the gut: new insights into an old enigma. *Seminars in Immunology* **13**, 187–194.

Gorbach, S.L., Barza, M., Giuliano, M., and Jacobus, N.V. (1988). Colonisation resistance of the human intestinal microflora: testing the hypothesis in normal volunteers. *European Journal of Clinical Microbiology and Infectious Diseases* **7**, 98–102.

Hill, M.J. (1997). Intestinal flora and endogenous vitamin synthesis. *European Journal of Cancer Prevention* **6** (Suppl 1), S43–S45.

Hirayama, K. and Rafter, J. (1999). The role of lactic acid bacteria in colon cancer prevention: mechanistic considerations. *Antonie Van Leeuwenhoek* **76**, 391–394.

Hogenauer, C., Hammer, H.F., Krejs, G.J., and Reisinger, E.C. (1998). Mechanisms and management of antibiotic-associated diarrhea. *Clinical Infectious Diseases* **27**, 702–710.

Hooper, L.V., Bry, L., Falk, P.G., and Gordon, J.I. (1998). Host-microbial symbiosis in the mammalian intestine: exploring an internal ecosystem. *Bioessays* **20**, 336–343.

Hooper, L.V. and Gordon, J.I. (2001). Commensal host-bacterial relationships in the gut. *Science* **292**, 1115–1118.

Hooper, L.V. and Gordon, J.I. (2001). Glycans as legislators of host-microbial interactions: spanning the spectrum from symbiosis to pathogenicity. *Glycobiology* **11**, 1R–10R.

Hooper, L.V., Midtvedt, T., and Gordon, J.I. (2002). How host-microbial interactions shape the nutrient environment of the mammalian intestine. *Annual Review of Nutrition* **22**, 283–307.

Hooper, L.V. (2004). Bacterial contributions to mammalian gut development. *Trends in Microbiology* **12**, 129–134.

Hopkins, M.J. and Macfarlane, G.T. (2003). Nondigestible oligosaccharides enhance bacterial colonization resistance against *Clostridium difficile in vitro*. *Applied and Environmental Microbiology* **69**, 1920–1927.

Kaplan, J.E., Hanson, D., Dworkin, M.S., Frederick, T., Bertolli, J., Lindegren, M.L., Holmberg, S., and Jones, J.L. (2000). Epidemiology of human immunodeficiency virus-associated opportunistic infections in the United States in the era of highly active antiretroviral therapy. *Clinical Infectious Diseases* **30**, S5–S14.

Kelly, D. and Coutts, A.G. (2000). Early nutrition and the development of immune function in the neonate. *Proceedings of the Nutrition Society* **59**, 177–185.

Knasmuller, S., Steinkellner, H., Hirschl, A.M., Rabot, S., Nobis, E.C., and Kassie, F. (2001). Impact of bacteria in dairy products and of the intestinal microflora on the genotoxic and carcinogenic effects of heterocyclic aromatic amines. *Mutation Research* **480–481**, 129–138.

Lidbeck, A., Nord, C.E., Gustafsson, J.A., and Rafter, J. (1992). Lactobacilli, anticarcinogenic activities and human intestinal microflora. *European Journal of Cancer Prevention* **1**, 341–353.

McFall-Ngai, M.J. (2002). Unseen forces: the influence of bacteria on animal development. *Developmental Biology* **242**, 1–14.

McFarland, L.V. (2000). Normal flora: diversity and functions. *Microbial Ecology in Health and Disease* **12**, 193–207.

Metges, C.C. (2000). Contribution of microbial amino acids to amino-acid homeostasis of the host. *Journal of Nutrition* **130**, 1857S–1864S.

Neish, A.S. (2002). The gut microflora and intestinal epithelial cells: a continuing dialogue. *Microbes and Infection* **4**, 309–317.

Ouwehand, A., Isolauri, E., and Salminen, S. (2002). The role of the intestinal microflora for the development of the immune system in early childhood. *European Journal of Nutrition* **41** (Suppl 1), I32–I37.

Peterson, D.L. (2004). "Collateral damage" from cephalosporin or quinolone antibiotic therapy. *Clinical Infectious Diseases* **38** (Suppl 4), S341–S345.

Payne, S., Gibson, G., Wynne, A., Hudspith, B., Brostoff, J., and Tuohy, K. (2003). *In vitro* studies on colonization resistance of the human gut microbiota to *Candida albicans* and effects of tetracycline and *Lactobacillus plantarum* LPK. *Current Issues in Intestinal Microbiology* **4**, 1–8.

Pryde, S.E., Duncan, S.H., Hold, G.L., Stewart, C.S., and Flint, H.J. (2002). The microbiology of butyrate formation in the human colon *FEMS Microbiology Letters* **217**, 133–139.

Schierholz, J.M. and Beuth, J. (2001). Implant infections: a haven for opportunistic bacteria. *Journal of Hospital Infection* **49**, 87–93.

Schiffrin, E.J. and Blum, S. (2002). Interactions between the microbiota and the intestinal mucosa. *European Journal of Clinical Nutrition* **56** (Suppl 3), S60–S64.

Sepkowitz, K.A. (2002). Opportunistic infections in patients with and patients without acquired immunodeficiency syndrome. *Clinical Infectious Diseases* **34**, 98–107.

Stappenbeck, T.S., Hooper, L.V., and Gordon, J.I. (2002). Developmental regulation of intestinal angiogenesis by indigenous microbes via Paneth cells. *Proceedings of the National Academy of Sciences USA.* **99**, 15451–15455.

Stoddart, B. and Wilcox, M.H. (2002). *Clostridium difficile. Current Opinion in Infectious Diseases* **15**, 513–518.

Sweeney, N.J., Klemm, P., McCormick, B.A., Moller-Nielsen, E., Utley, M., Schembrm, M.A., Laux, D.C., and Cohen, P.S. (1996). The *Escherichia coli* K-12 *gntP* gene allows *E. coli* F-18 to occupy a distinct nutritional niche in the streptomycin-treated mouse large intestine. *Infection and Immunity* **64**, 3497–3503.

Wilcox, M.H. (2003). *Clostridium difficile* infection and pseudomembranous colitis. *Best Practice & Research Clinical Gastroenterology* **17**, 475–493.

Xu, J., Bjursell, M.K., Himrod, J., Deng, S., Carmichael, L.K., Chiang, H.C., Hooper, L.V., and Gordon, J.I. (2003). A genomic view of the human – *Bacteroides thetaiotaomicron* symbiosis. *Science* **299**, 2074–2076.

Xu, J. and Gordon, J.I. (2003). Honor thy symbionts. *Proceedings of the National Academy of Sciences USA* **100**, 10452–10459.

Young, V.B. and Schmidt, T.M. (2004). Antibiotic-associated diarrhea accompanied by large-scale alterations in the composition of the fecal microbiota. *Journal of Clinical Microbiology* **42**, 1203–1206.

Manipulation of the indigenous microbiota

Although the previous chapter has emphasised the benefits which we derive from the presence of our indigenous microbiota, it must be remembered that the microbiota of each body site also contains organisms that are able to cause disease. Methods of altering the microbiota of a site to increase the benefits it confers or to decrease its ability to cause disease are, therefore, of great interest and practical importance. An obvious way of altering the microbiota of a body site is to use antibiotics, and these chemotherapeutic agents have been used widely to prevent and treat infections due to indigenous microbes. The availability of effective and cheap antibiotics in the latter half of the twentieth century revolutionised the treatment of infectious diseases and, for developed countries at least, reduced their impact as major causes of death and disability. Indeed, the Nobel laureate immunologist, Macfarlane Burnett, stated in 1962 that "by the late twentieth century we can anticipate the virtual elimination of infectious disease as a significant factor in social life". However, the development of resistance to a range of antibiotics by some important pathogens, including many indigenous organisms (e.g., *Staph. aureus, Strep. pneumoniae, N. meningitidis, H. influenzae, E. coli*), has raised the possibility of a return to the pre-antibiotic "dark ages". Other adverse effects of antibiotic administration on the indigenous microbiota were also described in Chapters 2–9. There is, therefore, great interest in the development of other approaches to manipulating the indigenous microbiota that do not result in such side effects. The use of such alternative approaches would also reduce antibiotic use, thereby helping to slow the spread of antibiotic resistance. This chapter is concerned with those non-antibiotic approaches to disease prevention or treatment that involve modifying or manipulating the indigenous microbiota or its activities. It also considers the possible beneficial consequences of such manipulative procedures.

10.1 | Probiotics

Interest in the use of "harmless" bacteria to combat "dangerous" bacteria stretches back to the nineteenth century, when it was realised that some bacterial species were able to produce substances that inhibited or killed other species of bacteria. Early attempts at using microbes to treat infectious diseases included the application of staphylococci to the throats of diphtheria carriers to eliminate *Corynebacterium diphtheriae* and the spraying of "*Bacterium termo*" into patients' lungs to treat tuberculosis. The use of microbes in disease prevention or treatment was initially termed "bacterial interference", "bacteriotherapy", "competitive exclusion", or "replacement therapy".

Table 10.1.	Possible beneficial effects of probiotics

prevention of colonisation of a habitat by exogenous pathogens
displacement of a pathogen from a habitat
neutralisation/prevention of the detrimental activities of members of a microbial community
stimulation of the immune system
provision of metabolic activities lacking in the host

However, interest in this form of prophylaxis and therapy decreased rapidly with the arrival of the antibiotic era. Nevertheless, a belief with an even longer pedigree – that microbes and their products can contribute to the "well-being" of humans – continued to prosper. This idea is mentioned in the *Old Testament*, where it is stated in Genesis that "Abraham owed his longevity to the consumption of sour milk". Also, in 76 B.C., the Roman historian Plinius recommended the administration of fermented milk products for treating gastroenteritis. At the turn of the nineteenth century, Metchnikoff attributed the beneficial effects of fermented dairy products to changes in the microbial balance in the gut. From this emerged the concept of probiotics. The term "probiotic" was defined (in the late 1980s) as a "live microbial feed supplement which beneficially affects the host animal by improving its intestinal microbial balance". However, a survey of the literature reveals that the meaning of the term has broadened considerably and now appears to encompass the use of live microbes in the diet to provide some beneficial effect to the host without necessarily "improving its intestinal microbial balance" – it is in this sense that the term is used in this chapter. The probiotic field continues to focus mainly on the use of bacteria associated with dairy products (usually lactobacilli, bifidobacteria, and streptococci) to prevent or treat gastrointestinal infections or to confer some other "beneficial" effect. However, there is also some interest in the use of such organisms for the maintenance of vaginal health because the vaginal microbiota is dominated by lactic-acid bacteria and may be amenable to manipulation by the same probiotics used in the gastrointestinal tract (GIT). Because of its proximity to the anus, colonisation of the vagina by gut microbes is a frequent occurrence so that ingested organisms can and do reach the vagina. Infections at sites other than the GIT and vagina (and the urinary tract in females) are not likely to be influenced by the consumption of live microbes. If microbes are to be used to prevent or treat such infections, then they need to be applied directly to the site. The term "replacement therapy" has generally been retained for describing the use of microbes in this way. However, there is confusion over terminology, and some authors regard any microbe used to treat an infection as being a probiotic, despite the fact that it has not been ingested by the host. Most workers using microbes to prevent or treat infections other than those involving the GIT or vagina, however, appear to use the term "replacement therapy", and this convention is followed in this chapter.

There are several means by which live microbes might be able to confer some benefit on the host, and these are listed in Table 10.1. However, many of these benefits would require that the probiotic permanently colonised the host, and often there is little evidence that this actually happens – particularly in the case of microbes administered in order to confer some benefit on the host's GIT. In many cases, the

| Table 10.2. | Organisms frequently used as probiotics |

Lactobacillus spp.
   L. acidophilus
   L. johnsonii
   L. rhamnosus
   L. casei
   L. delbrueckii ss. *bulgaricus*
   L. reuteri
Bifidobacterium spp.
   Bif. adolescentis
   Bif. bifidum
   Bif. breve
   Bif. longum
   Bif. infantis
Strep. salivarius ss. *thermophilus*
Lactococcus lactis
Ent. faecium
Saccharomyces boulardii

only effect on the intestinal microbiota that can be detected is the transient presence of the probiotic itself – no major changes in the composition of the microbiota having taken place. Invariably, once intake of the probiotic has ceased, the organism can no longer be detected in the GIT or faeces. This is not really surprising, because the microbiota of any region represents a climax community which is, of course, intrinsically resistant to perturbation by exogenous or endogenous microbes (Section 1.2.1). It has been suggested by advocates of probiotics, therefore, that the administered organism does not necessarily have to affect the "microbial balance" within an ecosystem, but can exert a beneficial effect by other means (e.g., by affecting the immune system).

10.1.1 Probiotics and gastrointestinal health

Although the use of probiotics has been extended to sites other than the GIT, most interest and most studies are still centred on their use in the GIT. Ideally, a probiotic for intestinal use should (1) be of human origin, (2) be harmless, (3) be able to withstand manufacturing conditions, (4) survive passage through the intestinal tract, (5) colonise intestinal surfaces, (6) act against pathogens by several mechanisms, and (7) exert its beneficial effects rapidly. Although few microbes meet all of these requirements, a variety of species has been used in attempts to confer some improvement on gastrointestinal function or to prevent or treat gastrointestinal diseases in humans. These are listed in Table 10.2. The most widely investigated species are those belonging to the genera *Lactobacillus* and *Bifidobacterium*. Given the variety of organisms in the GIT of humans, it is important to question why attention has focussed on the use as probiotics of the organisms listed in Table 10.2, particularly lactobacilli and bifidobacteria. Although many of the organisms listed in the table have frequently been detected in human faeces, none of them are present in particularly high proportions. Hence, culture-based studies have shown that, in adults, bifidobacteria constitute approximately 10% of the colonic microbiota, while lactobacilli are usually present in lower proportions. Molecular-based studies, however, suggest that the proportions of bifidobacteria and lactobacilli are even lower – approximately 3% and <1.0%, respectively. If, as is the case, other organisms – particularly *Bacteroides* spp. and *Eubacterium* spp. – predominate in the faeces of healthy humans, would these not be more appropriate for use as probiotics? Furthermore, the choice of the actual species used also needs to be considered. More than thirty-two species of bifidobacteria have been recognised, but these are difficult to distinguish on the basis of phenotypic characteristics, and even differentiation on the basis of 16S rRNA gene sequences is difficult because of their high level of relatedness. The most frequently detected species in the faeces of adults are *Bif. longum*, *Bif. bifidum*, *Bif. catenulatum*, *Bif. pseudo-catenulatum*, and *Bif. adolescentis*. In contrast, two of the species often used as probiotics – *Bif. breve* and *Bif. infantis* – are not always present in adults, although they are usually the predominant species in the faeces of infants. In the case of lactobacilli, more than seventy species are recognised, of which the following are most frequently detected in human faeces: *L. acidophilus*, *L. brevis*, *L. casei*, *L. salivarius*, *L. plantarum*, *L. gasseri*, *L. ruminis*, and *L. crispatus*. However, longitudinal studies of the lactobacilli in human faeces have shown that, in approximately half of the individuals investigated, one or two species persist for long periods of time, whereas in the remaining individuals, the presence, identity, and numbers of lactobacilli fluctuate markedly. In the case of those with a stable population of lactobacilli, the species detected are *L. ruminis* and *L. salivarius* ss. *salivarius* – these are likely to be autochthonous species. In contrast, the species associated with individuals with a fluctuating population of lactobacilli belong mainly to the *L. casei* and *L. plantarum* groups, which are present in fermented foods and are probably transients ingested in such foods. It is possible, therefore, that the lactobacilli most often used as probiotics are not members of the indigenous microbiota of the GIT of humans. It is not surprising, therefore, that most studies of probiotic use have failed to demonstrate the persistence of the probiotic in human faeces once consumption has ceased.

In view of the relatively low prevalence of the organisms listed in Table 10.2 in human faeces, as well as the possibility that many of them are not indigenous to the human GIT, it is perhaps surprising that so much time and effort have been expended on studies involving their use as probiotics. This may have more to do with the following: (1) these organisms are already produced by the dairy industry, which obviously has facilities for their large-scale manufacture and preservation; (2) species belonging to these genera are very rarely implicated in infections of humans and, because they also have a long and safe history in the production of dairy products, they are categorised as "Generally regarded as safe" by the Food and Drug Administration of the United States.

Like many other members of the colonic microbiota, some lactobacilli and bifidobacteria are able to inhibit the growth of exogenous species and/or prevent their attachment to the intestinal epithelium. Despite the fact that they generally do not colonise the GIT, the slow transit time (approximately 65 hours from mouth to anus) of probiotic organisms may be sufficient to enable them to exert some effect within the gut. Furthermore, although the probiotic may not have a permanent effect on the composition of the intestinal microbiota, it may be able to exert a beneficial effect by other means (e.g., by affecting the immune system). Whole cells, as well as components of lactobacilli and bifidobacteria, have been shown to have a number of stimulatory effects on the host's immune system, including the ability to up-regulate the production

Probiotic	Reported benefit
L. rhamnosus GG	decreases course of erythromycin-induced diarrhoea
	decreases duration of diarrhoea in *Cl. difficile*-associated colitis
	protects against traveller's diarrhoea
	reduces duration of rotavirus-induced diarrhoea
	protects against rotavirus-induced diarrhoea
	reduces frequency of asthma in infants when given to their mothers during and after pregnancy
	reduces the risk of nosocomial-acquired diarrhoea in children
L. reuteri	reduces duration of rotavirus-induced diarrhoea
L. johnsonii	reduces colonisation of stomach by *Hel. pylori*
Bif. bifidum + *Strep. thermophilus*	protects against rotavirus-induced diarrhoea
Sac. boulardii	reduces relapse in patients with Crohn's disease
	reduces relapse in patients with recurrent *Cl. difficile*-associated diarrhoea
Bif. breve	reduces symptoms of irritable bowel disease
Bif. longum	decreases course of erythromycin-induced diarrhoea
Bif. lactis	protects against traveller's diarrhoea
	increases proportions of $CD4^+$, $CD25^+$, T lymphocytes, and NK cells in blood

Table 10.3. Clinical trials that have demonstrated a beneficial effect following the administration of a probiotic

Note: NK = natural killer.

of IgA, activate macrophages, and induce cytokine release from lymphocytes and dendritic cells. However, the biological significance of these findings remains uncertain. Furthermore, many other microbes colonising the human GIT, as well as exogenous species, are also able to exert such stimulatory effects.

The results of some clinical studies involving the administration of probiotics are briefly summarised below. A frequent side effect of antibiotic therapy is diarrhoea, due to disruption of the indigenous intestinal microbiota. Both *Bif. longum* and *L. rhamnosus* GG have been found to decrease the course of diarrhoea associated with the use of erythromycin. *L. rhamnosus* GG has also been shown to be beneficial in the treatment of antibiotic-induced colitis due to *Cl. difficile*. Another diarrhoeal disease, traveller's diarrhoea, is frequently due to infection with enterotoxigenic strains of *E. coli*, and *L. rhamnosus* GG is able to provide some protection against contracting the disease. Rotaviruses are the most frequent cause of diarrhoea in children, and many investigations have reported that the duration of rotavirus-induced diarrhoea in children is significantly reduced (by approximately 1 day) following the administration of *L. reuteri* or *L. rhamnosus* GG. Administration of *L. rhamnosus* or a mixture of *Bif. bifidum* and *Strep. thermophilus* has also been found to reduce the risk of contracting the infection. Other clinical trials showing the benefits of probiotics in treating or preventing pathological conditions are listed in Table 10.3.

Although a number of clinical trials have demonstrated that currently used probiotics have some ability to prevent and treat gastrointestinal infections, greater success is likely to be achieved using microbial strains that have been selected on a

Table 10.4.	Properties of *L. fermentum* RC-14 and *L. rhamnosus* GR-1 that make them suitable for use as probiotics for the prevention of urogenital tract infections

they are constituents of the indigenous microbiota of the urogenital tract of healthy females

they can adhere to epithelial cells

they can prevent the adhesion of exogenous microbes to epithelial cells

they produce a number of substances that are antagonistic to exogenous microbes

they appear to be neither pathogenic nor carcinogenic

they are both acid- and bile-tolerant

they are able to colonise the vagina after oral administration

more rational basis utilising our greatly increased knowledge of the composition of the microbial communities colonising the human GIT. Furthermore, a number of promising lines of research may open up other possibilities for the use of probiotics. For example, suitable microbial strains could be genetically engineered to synthesise novel immunogens. Secretion of the immunogen following colonisation of the GIT with the organism (or its repeated ingestion) would enable immunisation of the host to occur. A number of such organisms have been prepared, including *L. plantarum* (able to secrete the gp41E protein of human immunodeficiency virus – HIV) and a *Lactococcus lactis* strain (which secretes a fragment of tetanus toxin). *Lac. lactis* has also been engineered to secrete a number of cytokines, including interleukin (IL)-2, IL-6, and IL-10. IL-10 is an important anti-inflammatory cytokine which can suppress the release of the inflammatory cytokines responsible for inducing the expression of matrix metalloproteinases, which are responsible for the tissue damage accompanying inflammatory bowel disease. Administration of IL-10-secreting *Lac. lactis* has been shown to prevent and cure colitis in mice.

10.1.2 Probiotics and vaginal health

The vagina has a complex microbiota which is dominated by *Lactobacillus* spp. The organisms most frequently detected are *L. jensenii*, *L. iners*, *L. crispatus*, *L. gasseri*, *L. cellobiosus*, and *L. fermentum*, although at least seven other species are often found (Section 6.4.3.1). High levels of lactobacilli, particularly hydrogen-peroxide-producing strains, suppress the growth of potentially pathogenic members of the vaginal microbiota, including *Can. albicans* (responsible for vaginal candidiasis or vaginitis) and organisms responsible for vaginosis – *G. vaginalis*, *Prevotella* spp. (and/or other Gram-negative obligate anaerobes), *Peptostreptococcus* spp., *Mobiluncus* spp., *Myc. hominis*, and *U. urealyticum*. However, antibiotic therapy (and other factors) can disturb the microbiota, reducing the proportions of lactobacilli and resulting in vaginitis or vaginosis.

Although a number of different lactobacilli have been investigated for their ability to act as probiotics in the prevention of urogenital infections in women, two strains appear to be particularly promising in this respect – *L. fermentum* RC-14 and *L. rhamnosus* GR-1. These strains have a number of properties that render them suitable as probiotics for this purpose, and these are listed in Table 10.4. In a study involving thirty-two females administered a mixture of RC-14 and GR-1 orally once daily for 60 days, significant beneficial changes were found in the vaginal microbiota. Hence, in subjects receiving the probiotics, the number of lactobacilli increased, whereas the number of yeasts and coliforms decreased, compared with thirty-two controls who received a placebo. Twenty-four percent of the control group, but none of the subjects given the probiotics, developed vaginosis during the course of the trial. Furthermore, in those subjects who had a microbiota indicative of asymptomatic vaginosis at the start of the trial, 13% of the controls and 37% of those receiving probiotics had normal microbiotas by the end of the trial. A number of other studies using either of these probiotics applied directly to the vagina have also demonstrated their effectiveness at restoring the vaginal microbiota to one that is dominated by lactobacilli and their benefit in patients with vaginitis or vaginosis.

Other strains that may prove to be useful as probiotics for maintaining vaginal health include *L. brevis* CD2, *L. salivarius* FV2, and *L. gasseri* MB335 because these all are able to adhere to epithelial cells, can displace vaginal pathogens from epithelial cells, produce high levels of hydrogen peroxide, co-aggregate with pathogens, and inhibit the growth of *G. vaginalis*. In contrast, two other widely used probiotics – *L. rhamnosus* GG and *L. acidophilus* – are unable to colonise the vagina. A lactobacillus-dominated vaginal microbiota is important not only for the prevention of vaginitis and vaginosis, but also confers resistance to sexually transmitted diseases. Hence, it has been shown that lactobacilli, especially hydrogen-peroxide-producing strains, can kill HIV and inhibit the growth of *N. gonorrhoeae*. The beneficial effects of probiotics may, therefore, extend to providing protection against diseases due to these sexually transmitted pathogens.

10.1.3 Prevention of dental caries using probiotics

L. rhamnosus GG produces low-molecular mass, heat-stable compounds that inhibit the growth of cariogenic streptococci, and the organism is able to colonise the oral cavity. Studies have shown that consumption of milk containing this organism achieves a significant reduction in caries risk in children. Hence, in a randomised, double-blind, placebo-controlled study involving 594 children aged between 1 and 6 years, one group received milk containing *L. rhamnosus* GG and the other group received only milk for 5 days a week over a 7-month period. The children given milk containing the probiotic showed less dental caries and lower counts of *Strep. mutans* in their saliva than those receiving only milk.

Cheese consumption is known to have beneficial consequences for oral health in that it enhances remineralisation and reduces demineralisation of teeth. There are also reports that it can reduce the levels of *Strep. mutans* in saliva. In a study of the effects of the daily consumption of cheese with and without supplementation with *L. rhamnosus* GG and *L. rhamnosus* LC 705 for 3 weeks, reductions in the salivary count of *Strep. mutans* were achieved in approximately 20% of the individuals in both groups. However, the proportion of individuals showing decreased counts of *Strep. mutans* after the 3-week treatment period was significantly greater in the group of individuals who had taken the probiotic-containing cheese.

10.2 | Prebiotics

A prebiotic is defined as "a non-digestible food ingredient that beneficially affects the host by selectively stimulating the growth and/or activity of one or a limited number

Table 10.5.	The ideal attributes of a prebiotic

the compound must resist hydrolysis and absorption in the upper regions of the GIT

it must be fermented by only one or a limited number of potentially beneficial bacteria in the colon

it must induce an alteration in the composition of the colonic microbiota towards one that is "healthier"

it must induce effects that are beneficial to the host's health

Note: GIT = gastrointestinal tract.

of bacteria in the colon". Only those dietary constituents that reach the colon can, therefore, be potential prebiotics, and these include carbohydrates, some peptides and proteins, and some lipids. In order to be classified as a prebiotic, a compound must have the attributes listed in Table 10.5. The most likely compounds to reach the colon are non-digestible carbohydrates, such as fructo-oligosaccharides and the complex oligosaccharides present in milk. Fructo-oligosaccharides are polymers of β-D-fructose – those with between two and sixty residues (mean number = 12) are known as inulins, whereas those with fewer residues (between two and twenty, mean number = 4.8) are known as oligofructoses. Inulins are present in a wide variety of plants and have been the most extensively investigated for their prebiotic properties. Although these polymers are resistant to digestion by mammalian enzymes, they can be fermented by *Bifidobacterium* spp. and, to some extent, by other gut microbes, including lactobacilli. However, uniquely, bifidobacteria appear to selectively ferment inulins in preference to other carbohydrates, such as starch, pectins, and fructose. Consequently, inclusion of inulins in the diet could lead to increased proportions of bifidobacteria in faeces, and a number of *in vitro* and *in vivo* studies have shown that this does occur. In one such study, volunteers were fed a controlled diet supplemented with either sucrose or inulin for 15 days, and the viable counts of various bacterial genera in their faeces were determined. As can be seen in Figure 10.1, the counts of bifidobacteria increased to the greatest extent, while those of the other major groups of faecal bacteria were largely unaffected. Dietary constituents known to increase the proportions of bifidobacteria in the colonic microbiota are also known as "bifidogenic factors". One of the reasons proposed for the predominance of bifidobacteria in the GIT of breast-fed infants is that human breast milk contains a variety of bifidogenic factors. As well as inulin and fructo-oligosaccharides, other carbohydrates with prebiotic effects include galacto-oligosaccharides, lactulose, soybean oligosaccharides, and xylo-oligosaccharides.

Although prebiotics appear to be able to increase the proportion of bifidobacteria in the colonic microbiota, there are few studies of the consequences of this for human health. Administration of prebiotics increases calcium uptake in humans and other animals, and this is likely to be important in the prevention and treatment of osteoporosis. The increased calcium absorption may be a consequence of its increased solubility at the lower pH of the intestinal lumen resulting from the fermentation of the prebiotic. The SCFAs produced by metabolism of the prebiotic are also known to be associated with increased calcium absorption. There is some evidence from animal studies that prebiotics may be able to prevent cancer. This is attributed to the fact

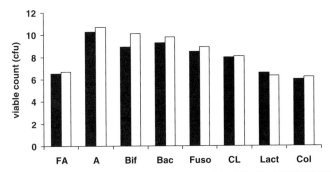

Figure 10.1 Effect of inulin (open bars) and sucrose (black bars) on the viable counts of key members of the faecal microbiota. FA = facultative anaerobes, A = obligate anaerobes, Bif = *Bifidobacterium* spp., Bac = *Bacteroides* spp., Fuso = *Fusobacterium* spp., CL = *Clostridium* spp., Lact = *Lactobacillus* spp., Col = coliforms.

that bifidobacteria have fewer of the enzymes (β-glucuronidase, azoreductase, nitroreductase, and nitrate reductase) that are able to convert pre-cancerous compounds into carcinogens. It has been suggested that an increase in the proportions of bifidobacteria or lactobacilli increases the colonisation resistance of the colonic microbiota due to nutrient competition and the production of antagonistic substances such as SCFAs and bacteriocins.

It has been suggested that a combination of a probiotic and a prebiotic may be of benefit to human health. Hence, the survival of a probiotic such as a *Bifidobacterium* species would be enhanced by the concomitant ingestion of inulin because this would provide a substrate which would selectively encourage the growth of the probiotic. Such a combination has been termed a "synbiotic".

10.3 | Inhibition of microbial adhesion

In order to maintain itself within its habitat, an organism must, in general, adhere to some substratum within that habitat. The exceptions to this generalisation are those microbes present in the lumen of the GIT – these are, of course, eventually expelled through the anus. One way of modifying the microbiota of a site, therefore, would be to prevent the adhesion of particular microbes to the substrata available at that site. Such an approach could be used to exclude potentially pathogenic members of the indigenous microbiota of the site (e.g., *N. meningitidis* in the nasopharynx, *Staph. aureus* in the anterior nares) or to prevent members of the microbiota of one site from colonising other sites where they induce pathology (e.g., *E. coli* in the urinary tract).

Adhesion involves a specific interaction between a molecule (adhesin) on the surface of the organism and its cognate receptor located on a host cell, the extracellular matrix, a host secretion (e.g., mucin), or a mineralised tissue (e.g., a tooth or bone). If the specific molecule (or epitope) involved in the ligand–receptor interaction which underpins the adhesion process is known, then it should be possible to design molecules which could be used to block their interaction and prevent adhesion (Figure 10.2). A number of approaches to this are possible: (1) the use of antibodies to the adhesin – this is an approach already used by the host because the immune response to

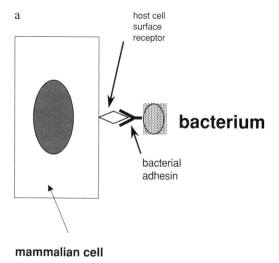

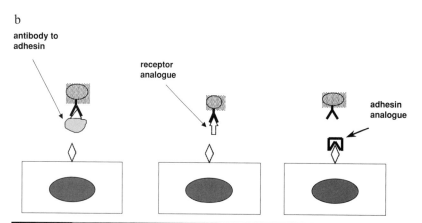

Figure 10.2 (a) Adhesion of a bacterium to a host cell is often mediated by a specific adhesin–receptor interaction. (b) Blocking either the adhesin or the receptor may result in inhibition of adhesion of the organism to the host cell or tissue, and there are a number of possible means of achieving this.

many infectious diseases involves the production of secretory IgA antibodies capable of preventing bacterial adhesion to mucosal surfaces; (2) the use of adhesins, or adhesin analogues, which would bind to the host receptor, thereby preventing access by the bacterial adhesin; and (3) the use of molecules which can bind to the bacterial adhesin; such molecules may be those which actually constitute the epitope of the receptor or may be analogues of these.

10.3.1 Inhibition of adhesion using antibodies

This is a form of passive immunisation in which the antibodies are directed specifically at the adhesin(s) of the target organism. This approach has been used with the greatest success to prevent colonisation of the human oral cavity by members of the oral microbiota associated with caries and periodontitis – *Strep. mutans* and *Por. gingivalis*, respectively. Dental caries is due primarily to *Strep. mutans* (Section 8.5.1), which adheres to salivary glycoproteins adsorbed on the tooth surface. Binding is mediated by a proteinaceous adhesin, streptococcal antigen I/II (SA I/II), which recognises a carbohydrate epitope on the salivary glycoproteins. A monoclonal antibody (mAb) raised against SA I/II has been used successfully to prevent colonisation of humans by *Strep. mutans*. *Strep. mutans* levels in subjects were reduced to undetectable levels by application of the antiseptic chlorhexidine, and then the mAb was applied directly to the teeth six times over a 3-week period. No oral recolonisation by *Strep. mutans* was found for at least 1 year in these subjects, whereas controls were recolonised within 2–3 months. The long-term protection observed in this study is surprising in view of the fact that the mAb persisted in the oral cavity for only 24 hours. However, it is likely that prevention of *Strep. mutans* colonisation enabled a non-cariogenic species to become established, which prevented recolonisation by *Strep. mutans* by competitive exclusion. Subsequent studies have shown that an IgA rather than an IgG version of the mAb persists for 3 days in the oral cavity, possibly because of its greater resistance to proteolysis, and was also able to prevent colonisation by *Strep. mutans*. Similar experiments carried out using a mAb against an adhesin of *Por. gingivalis* have demonstrated that prevention of recolonisation of the human oral cavity by this periodontopathogenic organism is also possible. Following mechanical removal of subgingival plaque (Section 8.5.2.3) and treatment with metronidazole to eliminate *Por. gingivalis* from the oral cavity, administration of the mAb to the subgingival region prevented recolonisation by the organism for up to 9 months.

Experiments with animals have demonstrated that this approach can be used to prevent vaginal colonisation of rats by *Can. albicans*. In this case, direct application to the vagina of mAbs against either a mannan or an aspartyl protease were able to reduce the degree of candidal colonisation when the rats were challenged with the yeast 30 minutes later. Reductions in yeast colonisation compared with control animals were maintained for up to 10 days.

10.3.2 Inhibition of adhesion using adhesins or adhesin analogues

Again, most work in this field has been carried out with *Strep. mutans*. Residues 1025–1044 of SA I/II constitute one of the main adhesin epitopes of the molecule, and a peptide (p1025) has been constructed that matches this region. Following elimination of oral *Strep. mutans* with chlorhexidine, the peptide was applied directly to the teeth of human volunteers (six times over 3 weeks) who also rinsed daily for 2 weeks with a mouthwash containing the peptide. None of the four subjects was recolonised after 88 days, and only one was recolonised after 120 days. In contrast, all of the control subjects (who received a non-inhibitory peptide corresponding to residues 1125–1144 of SA I/II, which did not block adhesion *in vitro*) had been recolonised after 21 days.

Several *in vitro* studies have shown that adhesins or their analogues block the binding of autochthonous microbes *in vitro*. For example, a recombinant peptide containing three repeats of a fibronectin-binding domain is able to inhibit binding of *Strep. pyogenes* to both fibronectin and epithelial cells. If such inhibition occurs *in vivo*, the peptide may be able to protect against pharyngeal colonisation by the organism.

10.3.3 Inhibition of adhesion using receptors or receptor analogues

A number of studies have demonstrated that receptor epitopes or their analogues are able to prevent adhesion of microbes to cells in tissue culture or in animal models,

Table 10.6. Examples of the use of receptors (or their analogues) to prevent bacterial adhesion

Organism	Receptor or analogue	Results obtained in laboratory or animal models
E. coli	methyl α-D-mannose	inhibited adhesion to urinary tract of mice
E. coli	non-immunoglobulin plasma glycoproteins	eliminated organism from infected calves
Strep. pneumoniae	sialylated oligosaccharides terminating with the disaccharide NeuAcα2-3(or 6)Galβ1	inhibited adhesion to respiratory epithelial cells
Strep. pneumoniae	lacto-N-neotetraose	decreased colonisation of the respiratory tract in rabbits
Strep. pneumoniae	N-acetylglucosamine	inhibited colonisation of mouse lungs
Hel. pylori	3'-sialyllactose	eliminated the organism from stomach of 2 of 6 monkeys
Hel. pylori	sulphated algal polysaccharides	inhibited adhesion to cells in vitro

and these are listed in Table 10.6. Interestingly, human milk is a particularly good source of complex oligosaccharides and glycoconjugates that can function as receptor analogues and thereby inhibit not only bacterial adhesion to epithelial cells, but also binding of some bacterial exotoxins to host cells. Unfortunately, only a few studies have been carried out in humans, and these have produced disappointing results. A sialylated form of lacto-N-neotetraose when administered by nasal spray to children was unable to protect them from nasopharyngeal colonisation by *Strep. pneumoniae* or *H. influenzae*.

Although the results of anti-adhesion approaches are generally encouraging, a number of problems may arise with their clinical use. Binding of microbes to host cells often triggers a signalling cascade, resulting in profound changes in the host cell. The binding of adhesin analogues to host-cell receptors may induce similar signaling processes, resulting in pathology. If this is found to occur, the peptide would have to be modified in some way to prevent such signal induction. Another problem is that many microbes are known to elaborate multiple adhesins, any one (or more) of which may enable binding of the organism to a host cell. Furthermore, the binding of receptor analogues to a microbial adhesin may induce the expression of alternative adhesins. Consequently, the blocking of only one adhesin–receptor interaction may not be sufficient to prevent microbial adhesion *in vivo* – it may be necessary to use a mixture of receptor analogues to achieve this.

10.3.4 Prevention of medical-device-associated infections

Anti-adhesion approaches have met with some success in preventing infections associated with the use of medical devices. Such infections are usually caused by members of the microbiota colonising the site of insertion of the device. The surfaces of devices, such as intravenous or urinary catheters, have been modified in a number of

Table 10.7.	Prevention of microbial adhesion to medical devices by modifying the surface properties of the device

Surface modification	Effect
hydrophilic coating of copolymer of poly(ethylene oxide) and poly(propylene oxide)	reduces adhesion of *Staph. epidermidis*
hydrophilic coating of poly(ethylene oxide)	reduces adhesion of *Staph. epidermidis* and *Staph. aureus*
hydrogel coating	reduces adhesion of *Staph. epidermidis* and *Staph. aureus*
heparin treatment	reduces adhesion of *Staph. aureus*
treatment with a biosurfactant from *L. fermentum*	reduces adhesion of *Staph. aureus*
treatment with a biosurfactant from *L. acidophilus*	reduces adhesion of *Staph. epidermidis, Ent. faecalis*, and *E. coli*
treatment with a surfactant from *Bac. subtilis*	reduces adhesion of *E. coli, Pr. mirabilis*, and *Sal. typhimurium*
treatment with Tween 80	reduces adhesion of *E. coli, Pr. mirabilis*, and *Sal. typhimurium*

ways in order to reduce microbial adhesion – these are summarised in Table 10.7. Biofilm formation on medical devices is known to involve quorum sensing systems, and inhibitors of such signaling systems may be able to prevent and/or disrupt biofilm formation. A group of natural halogenated furanones produced by the alga *Delisea pulchra*, as well as synthetic derivatives of these, have been shown to inhibit acyl-homoserine lactone-based quorum sensing systems and consequently biofilm formation by several organisms, including *Ps. aeruginosa, E. coli*, and *Bac. subtilis*. Such compounds offer another potential approach to preventing device-related and other biofilm-associated infections.

10.4 | Replacement therapy

Replacement therapy (also often referred to as "bacteriotherapy" or "bacterial interference") involves the use of a specific organism (an "effector" strain) to prevent the colonisation of a site by a potentially pathogenic species or to displace such a species from the site. The term "replacement therapy" is sometimes used interchangeably with "probiotics", with the latter term being more recent. While both approaches are similar in that they involve the use of live microbes for the prevention or treatment of an infectious disease, there are a number of differences between them. Firstly, probiotics are generally used as dietary supplements, whereas the effector strain used in replacement therapy is not ingested, but is applied directly to the site of the infection. Secondly, in replacement therapy, colonisation of the site by the effector strain is essential, whereas probiotics are able to exert a beneficial effect without permanently colonising the site. Thirdly, replacement therapy involves a dramatic and long-term change in the indigenous microbiota of a site, whereas this rarely happens with probiotics. Fourthly, unlike probiotics, replacement therapy is usually directed at displacing or preventing colonisation by a specific organism. Lastly, probiotics have been claimed to exert their beneficial effects by influencing the immune system in some way, whereas one of the

Table 10.8.	The ideal attributes of an effector organism for use in replacement therapy

active specifically against the target pathogen(s)
does not affect other members of the indigenous microbiota
member of the indigenous microbiota of the site of its proposed action
non-pathogenic for its host
susceptible to low-risk antibiotics, such as penicillin, so that it can later be
    eliminated if necessary
easy to grow and can be produced in a stable form for distribution
easy to identify among the indigenous microbiota of the host
does not cause systemic toxicity or immunological sensitisation in the host
capable of persisting in the host, thereby ensuring long-term protection

selection criteria for an effector strain in replacement therapy is that it has a minimal immunological impact.

Replacement therapy, therefore, involves colonisation of a site by a naturally occurring or laboratory-derived effector strain in order to prevent that site being colonised by a pathogenic species, to displace the pathogen from the site, or to prevent the growth of the pathogen at the site. These effects may be achieved by one or more of the following mechanisms: blocking attachment sites, competing for essential nutrients, and producing inhibitory or cidal compounds. The host, therefore, is protected for as long as the effector strain persists as a member of the indigenous microbiota – ideally, for the lifetime of the host. The ideal characteristics of an effector organism are listed in Table 10.8, and examples of the successful application of replacement therapy in the prevention or treatment of infections are listed in Table 10.9.

10.4.1 Prevention of infections due to *Staphylococcus aureus*

During the 1950s and 1960s, serious infections due to *Staph. aureus* among neonates reached epidemic proportions in a number of hospitals. Initially, these were caused by penicillin-sensitive strains of the organisms, but, as resistance to this antibiotic spread, penicillin-resistant strains began to predominate. Epidemiological surveillance at the New York Hospital in 1959 revealed that the presence of one strain of *Staph. aureus* at a site (e.g., the nasal mucosa or the umbilicus) could prevent colonisation by other strains. A study was then carried out in which a non-pathogenic strain of *Staph. aureus* (designated 502A – isolated from the nasal mucosa of a nurse at the hospital) was applied to the umbilicus and nasal mucosa of neonates – the hypothesis being that this would prevent colonisation by pathogenic strains of the organism. The strategy was effective at preventing infection with a pathogenic strain (phage type 80/81) and, therefore, this strategy was used to control staphylococcal infections in a number of hospital nurseries. Although the approach was very successful at preventing infections due to *Staph. aureus*, it was subsequently discovered that strain 502A was able to cause minor infections in a number of neonates, and one neonate died from meningitis due to this organism. The use of this approach was, therefore, abandoned.

The main sites of colonisation of *Staph. aureus* are the nares and antimicrobial agents, such as mupirocin, are frequently used to eradicate the organism from these sites. During surveillance studies of hospital staff, an inverse relationship had been observed between the presence of *Staph. aureus* and *Corynebacterium* spp. in the nares. Daily

| Table 10.9. | Examples of clinical trials that have demonstrated the effectiveness of replacement therapy in the prevention or treatment of an infectious disease |

Application	Effector organism
prevention of infections due to Staph. aureus	Staph. aureus 502A
eradication of nasal carriage of Staph. aureus	Corynebacterium sp. strain Co304
protection of neonates against oropharyngeal colonisation by Staph. aureus, enterobacteria, and Ps. aeruginosa	viridans streptococci
protection of children against recurrence of pharyngitis due to Strep. pyogenes	a mixture of Strep. sanguis and Strep. mitis
protection of children against recurrence of acute otitis media	a mixture of Strep. sanguis, Strep. oralis, and Strep. mitis
prevention of UTIs in catheterized patients	E. coli strain 83972
prevention of UTIs	L. casei ss. rhamnosus GR-1; a mixture of L. casei ss. rhamnosus GR-1 and L. fermentum B54
treatment of pseudo membranous colitis	faeces (as an enema); a mixture of Bacteroides spp., Clostridium spp., E. coli, Ent. faecalis, and Pep. productus; a non-toxigenic strain of Cl. difficile

Note: GIT = gastrointestinal tract.

inoculation of the nares with a *Corynebacterium* sp. (Co304) isolated from the anterior nares was found to eradicate *Staph. aureus* from the nares of 71% of hospital staff after approximately 15 days. There is uncertainty about the underlying mechanism, but it could involve displacement of *Staph. aureus* from adhesion sites because the corynebacterium was found to exhibit a greater binding affinity to nasal mucus than *Staph. aureus*. The *Corynebacterium* species used did not produce compounds inhibitory to *Staph. aureus*.

Staph. aureus is also a major cause of infections of burns and infections associated with medical devices. Studies in animals have shown that deliberate colonisation of burns by strains of *Staph. aureus* with low virulence potential can prevent subsequent colonisation by other strains when challenged with these organisms. Animal studies have also shown that *L. fermentum* can inhibit the adhesion of *Staph. aureus* to implant materials and also to significantly reduce surgical-implant-associated infections due to the organism.

10.4.2 Prevention of pharyngeal colonisation and infection

Studies of hospitalised neonates have shown that prior to the onset of a generalised infection, the oropharynx of the baby is often colonised by the infecting organism. In contrast, neonates with a normal oropharyngeal microbiota (i.e., one in which the predominant organisms are α-haemolytic streptococci) do not generally suffer from such infections. Implantation of an α-haemolytic streptococcus into the oropharynx of neonates with an abnormal oropharyngeal microbiota (i.e., one in which the predominant organism was not an α-haemolytic streptococcus and/or did not constitute >90% of the total viable count) has been found to protect them against subsequent infections by the colonising organism. The effector strain used was carefully selected by screening the oropharyngeal microbiota of healthy neonates for viridans streptococci able to inhibit the growth of organisms responsible for neonatal infections (e.g., *Staph. aureus*, enterobacteria, and *Pseudomonas* spp.). The chosen organism, therefore, had many of the desirable attributes of an effector strain – i.e., it was a member of the indigenous microbiota of the target site, was able to colonise that site, was effective at inhibiting the growth of the target organisms, and was considered to be harmless.

The pharynx is also the principal habitat of the important pathogen *Strep. pyogenes*, which is able to initiate a wide range of infections in children and adults (Section 4.5.7). Some members of the pharyngeal microbiota, particularly viridans streptococci, are able to produce compounds that inhibit or kill *Strep. pyogenes*, and are important in preventing colonisation and growth of the organism. It has been shown that children harbouring *Strep. pyogenes* have in their throat much lower proportions of viridans streptococci able to produce these bacteristatic or bactericidal compounds than children who resist colonisation. Pharyngitis due to *Strep. pyogenes* also occurs more frequently in children than in adults, and it has been found that twice as many adults than children harbour *Strep. pyogenes*-inhibitory α-haemolytic streptococci in their throat. These observations suggest that α-haemolytic streptococci may be suitable effector organisms for the prevention of infections with *Strep. pyogenes*, and a number of clinical trials have shown this to be the case. In a study involving patients with recurrent pharyngitis due to *Strep. pyogenes*, it was found that spraying the throat with *Strep. pyogenes*-inhibiting strains of α-haemolytic streptococci for 10 days after antibiotic therapy significantly reduced the risk of a recurrent infection. Only 2% of the fifty-one patients treated with the α-haemolytic streptococci experienced another infection due to *Strep. pyogenes*, whereas 23% of the sixty-one patients treated with a placebo were reinfected by the organism. The α-haemolytic streptococci used in the study were isolated from the throats of healthy individuals and consisted of three strains of *Strep. sanguis* and one strain of *Strep. mitis* – all of which exhibited strong inhibitory activity against *Strep. pyogenes*. In a larger study carried out in a similar manner, antibiotic therapy was followed by the use of a spray containing the same mixture of α-haemolytic streptococci as previously described. A total of 10^7–10^8 cfu was sprayed into the throat twice daily for 10 days. Twenty-four percent of the 189 individuals treated with the α-haemolytic streptococci became reinfected with *Strep. pyogenes*, while 42% of the 93 receiving a placebo were reinfected. These studies demonstrate the value of replacement therapy in preventing infections with *Strep. pyogenes*, an organism which can cause a wide spectrum of diseases, including ones with a high mortality.

10.4.3 Prevention of dental caries

The main aetiological agent of dental caries is *Strep. mutans*, and its pathogenicity is related to its ability to produce lactic acid (which degrades tooth enamel) and to survive at acid pHs. Ideally, a suitable effector strain for the prevention of dental caries should have the following characteristics: (1) a significantly reduced pathogenic potential (i.e., acid-producing ability); (2) be able to prevent colonisation of the tooth surface by *Strep. mutans*; (3) be able to displace indigenous strains of *Strep. mutans*, thereby enabling previously infected subjects to be treated effectively; and (4) be generally safe and not predispose the host to other diseases.

Although several attempts have been made to isolate suitable effector strains from the human oral cavity, the most successful approach to date appears to be that of Jeffrey Hillman at the University of Florida, who has produced a genetically modified strain of *Strep. mutans* with reduced acid-producing abilities. As a starting point, a strain (JH1140) producing high levels of a lantibiotic known as mutacin 1140 was used. Mutacin 1140 is able to kill virtually all of the *Strep. mutans* strains against which it has been tested and provides strain JH1140 with a significant colonisation advantage, enabling it to displace existing strains of *Strep. mutans* from the tooth surface. The gene encoding lactate dehydrogenase in this strain was replaced with the alcohol dehydrogenase (ADH) gene from *Zymomonas mobilis*. The resulting effector strain (BCS3-L1) had no detectable lactate dehydrogenase activity and had levels of ADH activity approximately 10-fold greater than that of the parent strain. Studies showed that the main metabolic end-products were the neutral compounds alcohol and acetoin – no lactic acid was produced. Growth of BCS3-L1 on a variety of sugars and polyols resulted in pH values between 0.4–1.2 pH units higher than those produced by the parent strain. Obviously, the genetic stability of the resulting strain is very important. The reacquisition of an acidogenic phenotype by spontaneous reversion is extremely unlikely because, essentially, the whole of the *ldh* open-reading frame was deleted and no acidogenic revertants have been observed. However, horizontal transmission of an *ldh* gene is possible by transduction or transformation. Transduction of *Strep. mutans* by bacteriophages has not yet been observed and is very unlikely. Although the organism is poorly transformable, the gene (*comE*) encoding the competence stimulating peptide has been deleted as an additional precaution. Studies in rodents have shown that (1) the strain is genetically stable and does not produce any deleterious side effects during prolonged colonisation; (2) it is significantly less cariogenic than wild-type *Strep. mutans* in gnotobiotic rats; and (3) the level of caries in conventional rats colonised with the organism was similar to that found in *Strep. mutans*-free animals, and caries levels in both groups were considerably lower than those observed in animals infected with wild-type *Strep. mutans*. Its strong colonisation properties suggest that a single application of the effector strain to human subjects should result in its permanent implantation and the displacement over time of indigenous, disease-causing *Strep. mutans* strains. Unfortunately, no studies in humans have yet been reported.

Another approach to caries prevention currently being explored is the construction of strains of oral viridans streptococci containing urease genes. Such strains have been produced and are able to generate ammonia from the urea normally present in human saliva. These could be implanted into the oral cavity and should be able to restore the pH of acidic dental plaques to neutrality, thereby reducing their cariogenic potential. By preventing the establishment of low pHs in dental plaque, they may also be able to alter the composition of the microbiota to produce one that is less cariogenic.

10.4.4 Prevention of otitis media

Acute otitis media (AOM) is a common infection of children (Section 4.5.4) and is caused by certain members of the indigenous nasopharyngeal microbiota – mainly

Strep. pneumoniae, non-typeable *H. influenzae*, and *Mor. catharralis*. Recurrence of the infection is common, and the risk of another disease episode within 1 month of the primary infection is approximately 35%. Many strains of α-haemolytic streptococci resident in the nasopharynx (mainly *Strep. mitis*, *Strep. oralis*, *Strep. sanguis*, *Strep. salivarius*, and *Strep. intermedius*) are known to produce substances that can either kill or inhibit the pathogens responsible for AOM and are thought to play a role in controlling their numbers in this ecosystem. Hence, not only do children with recurrent AOM have lower proportions of α-haemolytic streptococci than healthy children, but also the strains present have less inhibitory activity against AOM pathogens. Strains of α-haemolytic streptococci which produce substances inhibitory to AOM pathogens, therefore, should be suitable as effector organisms for replacement therapy of AOM. To test this hypothesis, a clinical trial has been carried out involving 108 children between the ages of 0.5 and 6 years. After a 10-day course of antibiotics, a suspension of α-haemolytic streptococci in skimmed milk was sprayed into the nostrils of half of the children twice a day for 10 days. This was then repeated after an interval of approximately 1 month. The spray contained a mixture of two strains of *Strep. sanguis*, two strains of *Strep. mitis*, and one strain of *Strep. oralis* in equal proportions. The organisms had been isolated from the children and selected for their ability to inhibit the growth of AOM pathogens. The remaining fifty-five children used a placebo spray consisting of skimmed milk. The proportions of children who suffered from a recurrence of AOM during the 3-month period after the start of treatment were 22% in the group treated with the α-haemolytic streptococci and 42% in the control group. The results of the study clearly indicated that recolonisation of the nasopharynx with inhibitory strains of α-haemolytic streptococci could prevent recurrence of AOM in susceptible children. Interestingly, in a very similar study which used the same mixture of α-haemolytic streptococci, but omitted the prior antibiotic therapy, no differences were observed between the test and control groups in terms of the number of episodes of AOM or of the prevalence of AOM pathogens in the nasopharynx. This implies that, while effector organisms can prevent colonisation by AOM pathogens, they are unable to displace these pathogens once they have become established in the nasopharynx. Therefore, in this ecosystem, replacement therapy would appear to be effective only at preventing AOM rather than treating the disease.

10.4.5 Prevention of urinary tract infections

It has long been known that the presence of bacteria in the urinary tract (i.e., asymptomatic bacteriuria) can provide protection against urinary tract infection (UTI). This observation suggested the possibility that implantation of an avirulent strain of *E. coli* in the urinary tract may be an effective form of prophylaxis against UTIs in those with a high risk of such infections. One candidate effector strain which has been investigated in a number of studies is a strain of *E. coli* isolated from an individual with asymptomatic bacteriuria. This strain, *E. coli* 83972, was isolated from a girl whose urinary tract had been colonised with the organism for 3 years. It was able to colonise the urinary tract of volunteers for up to 232 days, who experienced no fever or signs of any systemic infection.

Catheter-dependent spinal-cord-injured patients have a high risk of UTIs, with an annual incidence of 20% being reported. A group of such patients had their bladder inoculated with *E. coli* 83972, and persistent colonisation (longer than 1 month) with the organism was achieved in 13 of the 21 patients. The mean duration of colonisation was 12.3 months, and four patients remained colonised for 2–3 years. None of the patients had a UTI while they remained colonised by the implanted *E. coli*. In contrast, 5 of 7 patients who experienced spontaneous loss of colonisation had a UTI within 3.4 months. The same strain of *E. coli* has also been used in an attempt to prevent UTIs in patients with a neurogenic bladder (i.e., one that has been damaged because of some event originating in the nervous system) who suffered from recurrent UTIs. Bladder colonisation was achieved in 75% of these patients and persisted for a mean of 420 days. All of the patients colonised with *E. coli* remained free of UTIs as long as the organism persisted. With regard to the safety of the treatment, there have been no reports of deterioration in bladder or kidney function in individuals who have been continuously colonised with the organism for more than 3 years. In a series of *in vitro* studies, it has also been shown that inoculation of catheters with the organism prevents subsequent colonisation of the catheter by uropathogens such as *E. coli, Ent. faecalis, Providencia stuartii*, and *Can. albicans*.

The encouraging results of these preliminary studies suggest that this approach may be valuable in the prevention of UTIs in patient groups who have a high risk of acquiring such infections.

Because of its proximity to the anus, the vagina is often colonised by faecal bacteria, such as *E. coli*, and then acts as a reservoir for these organisms which can eventually reach and ascend the short urethra and initiate an infectious process in the bladder. In the 1970s, it was observed that women with few UTIs often had high proportions of lactobacilli in their vagina, whereas those with recurrent UTIs often had low counts of these organisms. Some, but not all, lactobacilli are able to exert a protective effect against colonisation of the vagina by uropathogens by virtue of their ability to (1) prevent the adhesion of uropathogens to epithelial cells, (2) displace adherent uropathogens from epithelial cells by means of a surfactant, and (3) inhibit the growth of, or kill, such organisms by producing hydrogen peroxide, fatty acids, and bacteriocins. A number of clinical trials have been carried out to assess the ability of *L. casei* ss. *rhamnosus* GR-1 to prevent UTIs because this organism has been shown to inhibit the growth of uropathogens and their adhesion to epithelial cells. Vaginal colonisation was achieved following administration of the organism directly into the vagina on a weekly basis, and three of the five patients did not experience any UTI for a 6-month period. However, two of the patients developed UTIs due to enterococci. A larger study involving 55 women given a vaginal suppository containing a mixture of GR-1 and *L. fermentum* B54 on a weekly basis for 1 year resulted in a decrease in the UTI rate from 6.0 in the year prior to the trial to 1.6 during the trial period. Lactobacilli may also be useful in preventing recurrence of UTIs once the primary infection has been treated. Hence, a group of women who had been administered antibiotics to cure their UTI were then treated intra-vaginally with a mixture of *L. casei* and *L. fermentum* B54. The rate of recurrence of UTI was lower in the lactobacillus-treated group than in a control group who had not received the lactobacilli.

Despite the limited number of trials in this area, it would appear that the application of certain species of lactobacilli directly to the vagina is of some benefit in preventing UTIs in females.

10.4.6 Treatment of infections due to *Clostridium difficile*

A number of different approaches using the administration of live bacteria to patients with pseudo membranous colitis have proved to be effective. The earliest forms of replacement therapy used faecal enemas and resulted in the cure of 16 of 18 patients. Success has also been achieved using rectal infusion of a mixture of ten faecal species – three *Bacteroides* spp., three *Clostridium* spp., two *E. coli* strains, and one strain each of *Ent. faecalis* and *Pep. productus*. Five patients were treated in this way, and all recovered. The *E. coli*, *Cl. bifermentans*, and *Pep. productus* in the mixture were all able to inhibit the growth of *Cl. difficile in vitro*. A non-toxigenic strain of *Cl. difficile*, administered orally after treatment with metronidazole and vancomycin, has also been used successfully to treat two patients with relapsing diarrhoea due to *Cl. difficile*.

10.5 | Localised modification of the host environment

Previous chapters have emphasised that the presence of a particular organism at a site depends upon the site having an environment favourable to the organism's growth and reproduction. One possible approach that could be used to prevent or control the growth of a particular organism at a site, therefore, would be to render the environment unsuitable for its growth. It is important, of course, that any environmental alterations do not harm the host tissues. The key environmental determinants affecting the growth of a microbe include nutrient availability, temperature, pH, atmospheric composition, and redox potential. Because many microbes colonising humans derive their nutrients from molecules present in host secretions (e.g., mucins, proteins, lipids), it would be impossible to deprive the microbes of essential nutrients without harming the host at the same time. Extremes of pH can certainly limit microbial growth, but human tissues have a narrow pH range for optimum functioning and would certainly also be damaged by any dramatic pH changes. However, some organisms have exacting atmospheric requirements (e.g., obligate aerobes, obligate anaerobes). While it would be difficult to deprive a body site of oxygen to prevent the growth of an obligate aerobe without damaging human tissues, the creation of an aerobic atmosphere at a site to prevent the growth of obligate anaerobes should not harm host tissues – providing that the oxygen levels used are not high enough to cause oxidative damage. A similar situation exists with respect to redox potential. The redox potential of respiring tissues is approximately +40 mV, but many obligate anaerobes require much lower redox potentials for growth. Raising the redox potential of a site, therefore, may provide a means of preventing the growth of anaerobes without damaging human tissues. Modifying the environment of a body site by creating aerobic conditions and/or a high redox potential should, theoretically, provide a means of controlling the growth of obligate anaerobes and/or microaerophiles.

Hyperbaric oxygen has long been used for the treatment of serious infections due to anaerobic bacteria (e.g., gangrene). As well as creating an environment unsuitable for the growth of the causative organism, a high oxygen concentration also increases the bactericidal effectiveness of polymorphonuclear leukocytes and stimulates fibroblast activity and angiogenesis, thereby speeding up the wound-healing process. The patient is exposed to 100% oxygen at 2–3 atmospheres pressure for 90 minutes on several occasions, and this results in a 20-fold increase in the oxygen content of the blood.

More than 1,200 cases of gas gangrene have been treated successfully using hyperbaric oxygen in conjunction with antibiotics and surgical debridement, and its use has reduced the mortality rate by more than 50%. Examples of other anaerobic infections successfully treated using hyperbaric oxygen include necrotising fasciitis, Fournier's gangrene of the scrotum, osteomyelitis due to *F. nucleatum*, and diabetic ulcers.

Oxygen and oxygenating compounds have also been investigated for their ability to control periodontitis, which is one of the most prevalent infections of humans and is due, primarily, to obligately anaerobic bacteria (Section 8.5.2.3). The environment within the periodontal pocket has a low oxygen content (approximately 13 mm Hg) and, as long ago as 1913, oxygen was applied to the gingival crevice under high pressure for therapeutic purposes. This continued to be popular in the treatment of periodontitis for the next two decades, and a number of studies reported promising results. More recently, the periodontal pockets of a group of periodontitis patients were oxygenated once a week for 8 weeks, and this resulted in significant reductions in the proportions of spirochaetes and motile rods in the pockets, and there was also evidence of a clinical improvement. As well as oxygen, a range of oxygenating compounds has also been used in the treatment of the disease with varying degrees of success. The compounds used include potassium permanganate, sodium perborate, zinc peroxide, hydrogen peroxide, monoxychlorosene, sodium peroxyborate, and urea peroxide. Unfortunately, many of these compounds – including sodium perborate, zinc peroxide, and hydrogen peroxide – are toxic to host tissues. Potassium permanganate causes discolouration of teeth, while monoxychlorosene appears to have little antimicrobial activity. A limited number of clinical studies have been carried out using sodium peroxyborate and urea peroxide and, in each case, some modest beneficial effects have been reported.

One of the problems with the use of oxygen and oxygenating compounds is likely to be that high oxygen concentrations are maintained in the periodontal pocket for only short periods of time. Such transient oxygenation may be insufficient to bring about the environmental changes needed to kill or prevent the growth of the resident anaerobes. It may be possible to overcome this problem by using agents which, while not generating oxygen, can raise the redox potential of the periodontal pocket, thus creating an environment inimicable to anaerobes. Because they are not oxygen-producing, they should be able to achieve a more long-lasting change in the periodontal environment. The redox potential of the healthy gingival crevice is approximately $+75$ mV, whereas that of the periodontal pocket can be as low as -300 mV (Figure 10.3). Although the maximum redox potential tolerated by most anaerobic periodontopathogens is not known, for *Treponema microdentium* this has been determined to be -185 mV. A series of *in vitro* studies have shown that methylene blue (with a redox potential of $+11$ mV) can raise the redox potential of the environment to a level sufficient to kill obligately anaerobic organisms, such as *Por. gingivalis* and *F. nucleatum*, which can cause periodontitis. Furthermore, incorporation of the compound into a slow-release device (which maintains high concentrations of the compound in the pocket for long periods of time) has been shown to shift the composition of the subgingival microbiota to one that is more compatible with gingival health and also to result in clinical improvements. Hence, reductions were observed in (1) the proportions of anaerobes and spirochaetes in periodontal pockets, (2) bleeding in the pockets, and (3) the depth of the

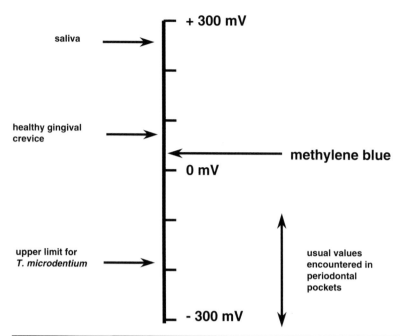

Figure 10.3 In infections such as periodontitis, which are caused mainly by obligate anaerobes, raising the redox potential of the disease lesion can inhibit or kill the organisms responsible, resulting in disease remission. The maximum redox potential tolerated by most periodontopathogenic species is not known, but for *Treponema microdentium*, it is −185 mV.

pockets. Such an ecological approach to disease treatment has great potential for the treatment of these widespread anaerobic infections that affect the majority of adults to some extent.

10.6 Further Reading

Books

Gibson, S.A.W. (ed.). (1994). *Human health: the contribution of microorganisms.* London: Springer-Verlag.

Tannock, G.W. (ed.). (2002). *Probiotics and prebiotics: where are we going?* Wymondham: Caister Academic Press.

Reviews and Papers

Ahola, A.J., Yli-Knuuttila, H., Suomalainen, T., Poussa, T., Ahlström, A., Meurman, J.H., and Korpela, R. (2002). Short-term consumption of probiotic-containing cheese and its effect on dental caries risk factors. *Archives of Oral Biology* **47**, 799–804.

Alvarez-Olmos, M.I. and Oberhelman, R.A. (2001). Probiotic agents and infectious diseases: a modern perspective on a traditional therapy. *Clinical Infectious Diseases* **32**, 1567–1576.

Bengmark, S. (2002). Gut microbial ecology in critical illness: is there a role for prebiotics, probiotics, and synbiotics? *Current Opinion in Critical Care* **8**, 145–151.

Blaut, M. (2002). Relationship of prebiotics and food to intestinal microflora. *European Journal of Nutrition* **41** (Suppl 1), I11–I16.

Bos, R., van der Mei, H.C., and Busscher, H.J. (1999). Physico-chemistry of initial microbial adhesive interactions – its mechanisms and methods for study. *FEMS Microbiology Reviews* **23**, 179–229.

Boyce, J.M. (1996). Treatment and control of colonisation in the prevention of nosocomial infections. *Infection Control and Hospital Epidemiology* **17**, 256–261.

Brook, I. (1999). Bacterial interference. *Critical Reviews in Microbiology* **25**, 155–172.

Brook, I. and Gober, A.E. (2000). *In vitro* bacterial interference in the nasopharynx of otitis media-prone and non-otitis media-prone children. *Archives of Otolaryngology – Head and Neck Surgery* **126**, 1011–1013.

Burne, R.A. and Marquis, R.E. (2000). Alkali production by oral bacteria and protection against dental caries. *FEMS Microbiology Letters* **193**, 1–6.

Collins, M.D. and Gibson, G.R. (1999). Probiotics, prebiotics, and synbiotics: approaches for modulating the microbial ecology of the gut. *American Journal of Clinical Nutrition* **69**, 1052S–1057S.

Comelli, E.M., Guggenheim, B., Stingele, F., and Neeser, J.-R. (2002). Selection of dairy bacterial strains as probiotics for oral health. *European Journal of Oral Sciences* **110**, 218–224.

Darouiche, R.O. and Hull, R.A. (2000). Bacterial interference for prevention of urinary tract infection: an overview. *Journal of Spinal Cord Medicine* **23**, 136–141.

de Vrese, M. and Schrezenmeir, J. (2002). Probiotics and non-intestinal infectious conditions. *British Journal of Nutrition* **88** (Suppl 1), S59–S66.

Dunne, C. (2001). Adaptation of bacteria to the intestinal niche: probiotics and gut disorder. *Inflammatory Bowel Diseases* **7**, 136–145.

Erickson, K.L. and Hubbard, N.E. (2000). Probiotic immunomodulation in health and disease. *Journal of Nutrition* **130**, 403S–409S, 2000.

Falck, G., Grahn-Hakansson, E., Holm, S.E., Roos, K., and Lagergren, L. (1999). Tolerance and efficacy of interfering alpha-streptococci in recurrence of streptococcal pharyngotonsillitis: a placebo-controlled study. *Acta Oto-Laryngologica* **119**, 944–948.

Famularo, G., Pieluigi, M., Coccia, R., Mastroiacovo, P., and De Simone, C. (2001). Microecology, bacterial vaginosis and probiotics: perspectives for bacteriotherapy. *Medical Hypotheses* **56**, 421–430.

Fooks, L.J. and Gibson, G.R. (2002). Probiotics as modulators of the gut flora. *British Journal of Nutrition* **88** (Suppl 1), S39–849.

Gan, B.S., Kim, J., Reid, G., Cadieux, P., and Howard, J.C. (2002). *Lactobacillus fermentum* RC-14 inhibits *Staphylococcus aureus* infection of surgical implants in rats. *Journal of Infectious Diseases* **185**, 1369–1372.

Gibson, G.R. and Roberfroid, M.B. (1995). Dietary modulation of the human colonic microbiota: introducing the concept of prebiotics. *Journal of Nutrition* **125**, 1401–1412.

Guzman-Murillo, M.A. and Ascencio, F. (2000). Anti-adhesive activity of sulphated exopolysaccharides of microalgae on attachment of red sore disease-associated bacteria and *Helicobacter pylori* to tissue culture cells. *Letters in Applied Microbiology* **30**, 473–478.

Hart, A.L., Stagg, A.J., Frame, M., Graffner, H., Glise, H., Falk, P., and Kamm, M.A. (2002). The role of the gut flora in health and disease, and its modification as therapy. *Alimentary Pharmacology and Therapeutics* **16**, 1383–1393.

Hillman, J.D. (2002). Genetically modified *Streptococcus mutans* for the prevention of dental caries. *Antonie Van Leeuwenhoek* **82**, 361–366.

Hillman, J.D., Brooks, T.A., Michalek, S.M., Harmon, C.C., Snoep, J.L., and Van Der Weijden, C.C. (2000). Construction and characterization of an effector strain of *Streptococcus mutans* for replacement therapy of dental caries. *Infection and Immunity* **68**, 543–549.

Holzapfel, W.H., Haberer, P., Snel, J., Schillinger, U., and Huis in't Veld, J.H.J. (1998). Overview of gut flora and probiotics. *International Journal of Food Microbiology* **41**, 85–101.

Isolauri, E., Kirjavainen, P.V., and Salminen, S. (2002). Probiotics: a role in the treatment of intestinal infection and inflammation? *Gut* **50** (Suppl 3), III54–III59.

Kailasapathy, K. and Chin, J. (2000). Survival and therapeutic potential of probiotic organisms with reference to *Lactobacillus acidophilus* and *Bifidobacterium* spp. *Immunology and Cell Biology* **78**, 80–88.

Kashket, S. and DePaola, D.P. (2002). Cheese consumption and the development and progression of dental caries. *Nutrition Reviews* **60**, 97–103.

Kaur, I.P., Chopra, K., and Saini, A. (2002). Probiotics: potential pharmaceutical applications. *European Journal of Pharmaceutical Sciences* **15**, 1–9.

Kelly, C.G. and Younson, J.S. (2000). Anti-adhesive strategies in the prevention of infectious disease at mucosal surfaces. *Expert Opinion on Investigational Drugs* **9**, 1711–1721.

Lina, G., Boutite, F., Tristan, A., Bes, M., Etienne, J., and Vandenesch, F. (2003). Bacterial competition for human nasal cavity colonization: role of staphylococcal agr alleles. *Applied and Environmental Microbiology* **69**, 18–23.

Macfarlane, G.T. and Cummings, J.H. (2002). Probiotics, infection, and immunity. *Current Opinion in Infectious Diseases* **15**, 501–506.

McCartney, A.L. (2002). Application of molecular biological methods for studying probiotics and the gut flora. *British Journal of Nutrition* **88** (Suppl 1), S29–S37.

Miller, J.L. and Krieger, J.N. (2002). Urinary tract infections, cranberry juice, underwear, and probiotics in the 21st century. *Urologic Clinics of North America* **29**, 695–699.

Ouwehand, A.C., Salminen, S., and Isolauri, E. (2002). Probiotics: an overview of beneficial effects. *Antonie van Leeuwenhoek* **82**: 279–289.

Pascual, A. (2002). Pathogenesis of catheter-related infections: lessons for new designs. *Clinical Microbiology and Infection* **8**, 256–264.

Reid, G. (2001). Probiotic agents to protect the urogenital tract against infection. *American Journal of Clinical Nutrition* **73** (2 Suppl), 437S–443S.

Reid, G. (2002). Probiotics for urogenital health. *Nutrition in Clinical Care* **5**, 3–8.

Reid, G. (2002). The potential role of probiotics in pediatric urology. *Journal of Urology* **168**, 1512–1517.

Reid, G., Howard, J., and Gan, B.S. (2001). Can bacterial interference prevent infection? *Trends in Microbiology* **9**, 424–428.

Reid, G. and Burton, J. (2002). Use of Lactobacillus to prevent infection by pathogenic bacteria. *Microbes and Infection* **4**, 319–324.

Reid, G., Charbonneau, D., Erb, J., Kochanowski, B., Beuerman, D., Poehner, R., and Bruce, A.W. (2003). Oral use of *Lactobacillus rhamnosus* GR-1 and *L. fermentum* RC-14 significantly alters vaginal flora: randomized, placebo-controlled trial in 64 healthy women. *FEMS Immunology and Medical Microbiology* **35**, 131–134.

Reid, G., Jass, J., Sebulsky, M.T., and McCormick, J.K. (2003). Manipulation of microbiota: potential uses of probiotics in clinical practice. *Clinical Microbiology Reviews* **16**, 658–672.

Riley, M.A. and Wertz, J.E. (2002). Bacteriocins: evolution, ecology, and application. *Annual Review of Microbiology* **56**, 117–137.

Roos, K., Hakansson, E.G., and Holm, S. (2001). Effect of recolonisation with "interfering" alpha streptococci on recurrences of acute and secretory otitis media in children: randomised placebo controlled trial. *British Medical Journal* **322**, 1–4.

Shanahan, F. (2003). Probiotics: a perspective on problems and pitfalls. *Scandinavian Journal of Gastroenterology Supplement* **237**, 34–36.

Sharon, N. and Ofek, I. (2000). Safe as mother's milk: carbohydrates as future anti-adhesion drugs for bacterial diseases. *Glycoconjugate Journal* **17**, 659–664.

Sharon, N. and Ofek, I. (2002). Fighting infectious diseases with inhibitors of microbial adhesion to host tissues. *Critical Reviews in Food Science and Nutrition* **42**, (3 Suppl), 267–272.

Sullivan, A. and Nord, C.E. (2002). Probiotics in human infections. *Journal of Antimicrobial Chemotherapy* **50**, 625–627.

Tagg, J.R. and Dierksen, K.P. (2003). Bacterial replacement therapy: adapting 'germ warfare' to infection prevention. *Trends in Biotechnology* **21**, 217–223.

Tannock, G.W. (1998). Studies of the intestinal microflora: a prerequisite for the development of probiotics. *International Dairy Journal* **8**, 527–533.

Tano, K., Grahn-Hakansson, E., and Holm, S.E., and Hellstrom, S. (2000). Inhibition of OM pathogens by alpha-hemolytic streptococci from healthy children, children with SOM, and children with rAOM. *International Journal of Pediatric Otorhinolaryngology* **56**, 185–190.

Tano, K., Grahn-Hakansson, E., Holm, S.E., and Hellstrom, S. (2002). A nasal spray with alpha-haemolytic streptococci as long-term prophylaxis against recurrent otitis media. *International Journal of Pediatric Otorhinolaryngology* **62**, 17–23.

Tano, K., Hakansson, E.G., Holm, S.E., and Hellstrom, S. (2002). Bacterial interference between pathogens in otitis media and alpha-haemolytic streptococci analysed in an *in vitro* model. *Acta Otolaryngologica* **122**, 78–85.

Tano, K., Olofsson, C., Grahn-Hakansson, E., and Holm, S.E. (1999). *In vitro* inhibition of *S. pneumoniae*, nontypable *H. influenzae*, and *M. catharralis* by alpha-hemolytic streptococci from healthy children. *International Journal of Pediatric Otorhinolaryngology* **47**, 49–56.

Teitelbaum, J.E. and Walker, W.A. (2002). Nutritional impact of pre- and probiotics as protective gastrointestinal organisms. *Annual Review of Nutrition* **22**, 107–138.

Trautner, B.W., Hull, R.A., and Darouiche, R.O. (2003). *Escherichia coli* 83972 inhibits catheter adherence by a broad spectrum of uropathogens. *Urology* **61**, 1059–1062.

Uehara Y., Nakama, H., Agematsu, K., Uchida, M., Kawakami, Y., Abdul Fattah, A.S.M., and Maruchi, N. (2000). Bacterial interference among nasal inhabitants: eradication of *Staphylococcus aureus* from nasal cavities by artificial implantation of *Corynebacterium* sp. *Journal of Hospital Infection* **44**, 127–133.

Vanmaele, R.P., Heerze, L.D., and Armstrong, G.D. (1999). Role of lactosyl glycan sequences in inhibiting enteropathogenic *Escherichia coli* attachment. *Infection and Immunity* **67**, 3302–3307.

Vaughan, E.E., de Vries, M.C., Zoetendal, E.G., Ben-Amor, K., Akkermans, A.D.L., and de Vos, W.M. (2002). The intestinal LABs. *Antonie van Leeuwenhoek* **82**, 341–352.

Index